Fields Institute Communications

The Fields Institute for Research in Mathematical Sciences

WIN—Women in Numbers

Research Directions in Number Theory

Alina-Carmen Cojocaru
Kristin Lauter
Rachel Pries
Renate Scheidler
Editors

American Mathematical Society
Providence, Rhode Island

The Fields Institute for Research in Mathematical Sciences
Toronto, Ontario

FIELDS

2010 *Mathematics Subject Classification.* Primary 11–06, 11–02, 14–06, 11Gxx, 11Mxx, 11Rxx, 11Yxx, 14Hxx, 14Kxx, 14Qxx.

Library of Congress Cataloging-in-Publication Data

Win—women in numbers : research directions in number theory / Alina-Carmen Cojocaru ... [et al.], editors.

p. cm. — (Fields Institute Communications, ISSN 1069-5265 ; 60)

Includes bibliographical references.

ISBN 978-0-8218-5226-2 (alk. paper)

1. Number theory. 2. Arithmetical algebraic geometry. 3. Women mathematicians. I. Cojocaru, Alina Carmen.

QA241.W628 2011

512.7—dc22

2010049877

∞ The paper used in this book is acid-free and falls within the guidelines
established to ensure permanence and durability.
This publication was prepared by the Fields Institute.
`http://www.fields.utoronto.ca`
Visit the AMS home page at `http://www.ams.org/`

10 9 8 7 6 5 4 3 2 1 16 15 14 13 12 11

Contents

Preface

This volume is a collection of papers on number theory which evolved out of the workshop *WIN—Women In Numbers*, held November 2–7, 2008, at the Banff International Research Station (BIRS) in Banff, Alberta, Canada. It includes articles showcasing outcomes from collaborative research initiated during the workshop as well as survey papers aimed at introducing graduate students and recent PhDs to important research topics in number theory.

The workshop and this volume are part of a broader WIN initiative, whose goals are to raise the profile of active female researchers in number theory and increase their participation in research activities in the field. Although the number of female number theorists has grown over the past 20 years, this group remains underrepresented at high profile conferences and among the tenured faculty at top research universities. This underrepresentation has profound negative consequences on the recruitment and training of future female mathematicians. Part of the purpose of the BIRS workshop and this volume was to address this issue.

The WIN workshop represented a unique effort to combine a top level technical research program with strong broad impact. The workshop brought together senior and junior female researchers in the field of number theory. The main goals were to highlight and increase the research activities of women in number theory, and to train female graduate students in number theory and related fields. Emphasis was placed on on-site collaboration on open research problems as well as student training. Students were introduced to areas of active research in collaborative group projects which connected them with senior female faculty and with a network of future potential collaborators.

The workshop included 41 female number theorists and was organized by the last three editors of this volume. The attendees were spread across all levels of seniority, ranging from graduate students to postdocs and senior researchers. Based on the participants' research interests and expertise, the workshop organizers formed eight research groups each consisting of 4-6 members; generally, this included 2 group leaders who planned the specific topics and 2-4 other members. The content spanned a wide range of topics in arithmetic geometry and algebraic, algorithmic, and analytic number theory. Many of the group projects initiated at the workshop led to new research results that are published in this volume and elsewhere. A follow-up BIRS workshop entitled *WIN-2—Women In Numbers 2* will take place November 6–11, 2011, at BIRS. Other future WIN events are in the planning stage at the time of this volume's publication.

Workshop project titles and lectures

The eight project groups each had one or two group leaders who developed the scientific content and gave introductory lectures for the projects. On the last two days of the conference, the other participants of each group gave presentations on the group's progress reports.

- *Computations on Hilbert modular surfaces*
  - Lecture 1: Helen Grundman

  - Lecture 2: Kristin Lauter
  - Project report: Jennifer Johnson-Leung, Adriana Salerno, Bianca Viray, Erica Wittenborn
- *L-functions and Frobenius distributions*
  - Lecture 1: Chantal David
  - Lecture 2: Alina-Carmen Cojocaru
  - Project report: Alina Bucur, Brooke Feigon, Matilde Lalín
- *Class groups of function fields*
  - Lecture 1: Renate Scheidler
  - Lecture 2: Yoonjin Lee
  - Project report: Lisa Berger, Jing Long Hoelscher, Jennifer Paulhus
- *Computation of pairings on hyperelliptic curves*
  - Lecture 1: Kirsten Eisentraeger
  - Lecture 2: Edlyn Teske
  - Project report: Jennifer Balakrishnan, Juliana Belding, Sarah Chisholm, Katherine Stange
- *Galois covers of curves in characteristic p*
  - Lecture 1: Katherine Stevenson
  - Lecture 2: Rachel Pries
  - Project report: Linda Gruendken, Laura Hall-Seelig, Bo-Hae Im, Ekin Ozman
- *Zeta functions of graphs*
  - Lecture 1: Audrey Terras
  - Lecture 2: Winnie Li
  - Project report: Shabnam Akhtari, Habiba Kadiri, Beth Malmskog, Michelle Manes
- *Modular forms*
  - Lecture 1: Stephanie Treneer
  - Lecture 2: Ling Long
  - Project report: Sharon Garthwaite, Holly Swisher
- *Galois representations*
  - Lecture: Mirela Ciperiani
  - Project report: Margaret Upton, Núria Vila

Contributions in this volume

In 2009, the editors invited WIN workshop participants to submit articles to this proceedings volume. Papers were also solicited from some female number theorists who had been unable to attend. The aim was to document new research which emerged from the BIRS workshop and to encourage the publication of survey papers that showcase active areas of research in number theory. Following a careful and thorough refereeing process by external experts, 16 submissions were accepted, largely in the areas of arithmetic geometry and algebraic number theory.

We grouped the contributions in this volume into five areas. Clusters of papers center around the four topics of moduli spaces and Shimura curves, curves and Jacobians over finite fields, Galois covers of function fields in positive characteristic, and zeta functions of graphs, with a fifth group of three individual articles on modular forms, Iwasawa theory, and Galois representations, respectively.

Moduli spaces and Shimura curves. Arithmetic intersection theory on moduli spaces and Shimura curves is a very active area of recent research in number theory. This volume includes four contributions to this field that focus predominantly on computational aspects of Shimura curves and modular curves and varieties. Grundman's paper *Hilbert modular variety computations* provides a survey of computational methods used in studying Hilbert modular varieties, with particular emphasis on determining numerical invariants used in their classification. Bayer's *Contributions to Shimura curves* presents new results on the computation of the automorphic functions of canonical models for Shimura curves as well as equations defining CM points on such curves. The six-author article entitled *Igusa class polynomials, embeddings of quartic CM fields, and arithmetic intersection theory* by Grundman et al. evolved from research conducted at the WIN workshop; it investigates the arithmetic intersection numbers conjectured by Bruinier and Yang and finds both numerical support for the conjecture and some anomalies in the general case. Finally, Mantovan's contribution *Cohomology of PEL-type Shimura varieties with non-trivial ℓ-adic coefficients* extends previous results on computing ℓ-adic cohomology of Shimura varieties from the case of constant ℓ-adic coefficients to a more general scenario.

Curves and Jacobians over finite fields. Curves over finite fields and their Jacobians have long been objects of intense number theoretic study. Much information can be gleaned from the zeta functions of these objects, and recent decades have seen the emergence of exciting cryptographic applications. All three articles in this area originate from WIN workshop group projects. Pairings on elliptic and hyperelliptic curves have seen a recent surge of research activity. The six-author paper *Pairings on hyperelliptic curves* by Balakrishnan et al. surveys constructions of pairing-friendly hyperelliptic curves and develops a unified framework for all hyperelliptic pairings proposed to date. The four-author contribution *Biased statistics for traces of cyclic p-fold covers over finite fields* by Bucur et al. provides statistics of the trace of Frobenius for a cyclic p-fold cover of the projective line, generalizing previous work by the authors that focused mainly on the case $p = 3$. The paper *The ℓ-rank structure of a global function field* by Berger and four co-authors analyzes in detail the behaviour of the ℓ-rank of the Jacobian associated to any curve over a finite field when the base field is gradually enlarged.

Galois covers of function fields in positive characteristic. There are many open problems related to Galois covers and fundamental groups of curves in positive characteristic. These arise, in large part, because of the phenomena of wild ramification. This area has close connections with the previous topic of curves and Jacobians. The volume contains three papers on this topic. Pries & Stevenson's *A survey of Galois theory of curves in characteristic p* provides a survey on Galois covers and fundamental groups of curves in positive characteristic. An open problem is to determine for a non-abelian quasi-p group G the smallest genus that can occur for a G-Galois cover of the affine line. Bouw's contribution *Covers of the affine line in positive characteristic with prescribed ramification* answers this problem for the case that $G = A_d$ is the alternating group on d letters, with $p+2 \leq d \leq 2p-1$, and the six authors of *Semi-direct Galois covers of the affine line* by Gruendken et al. provide the answer for the case $G = (\mathbb{Z}/\ell)^b \rtimes \mathbb{Z}/p$, with $b \in \mathbb{N}$ and ℓ a prime distinct from p. This last paper grew out of research conducted at the WIN workshop.

Zeta functions of graphs. Zeta functions of graphs represent a new area of number theory, with connections to combinatorial and analytic number theory. Three papers on this topic appear here. Terras' *Looking into a graph theory mirror of number theoretic zetas* surveys the Ihara zeta function for irregular graphs and its connections with number theoretic zeta functions. A second survey article by Li entitled *Zeta functions of group based graphs and complexes* focuses on graphs and complexes arising from quotients of the Bruhat-Tits buildings associated to $PGL_2(F)$ and $PGL_3(F)$ for a non-archimedean local field F with finite residue field. The contribution *Ramified covers of graphs and the Ihara zeta functions of certain ramified covers* by Malmkog and Manes has its basis in research conducted at the WIN conference. It revisits the zeta functions discussed by Terras, investigating the Ihara zeta functions of regular graphs as well as certain coverings of these graphs.

Other topics. Three other important areas of number theory are represented by contributions in this volume. Research on modular forms and Eisenstein series conducted at the WIN workshop spawned the four-author paper *Zeros of classical Eisenstein series and recent developments* by Garthwaite et al. In her expository article *On the μ-invariant in Iwasawa theory*, Sujatha reviews this important classical invariant attached to Iwasawa modules. Finally, the contribution entitled *Galois representations and the tame inverse Galois problem* by Arias-de-Reyna and Vila showcases progress made on a variant of the inverse Galois problem over the rational numbers through the analysis of the Galois representations arising from arithmetic-geometric objects.

Workshop website

`http://www.birs.ca/events/2008/5-day-workshops/08w5112`

Acknowledgments

We are grateful to the sponsoring organizations listed below for their support of the workshop and this volume. Their generous financial backing allowed us to provide full travel support to all the participants, many of whom would not have been able to attend otherwise. It was a pleasure to work with BIRS and the Fields Institute to organize the conference and prepare this volume.

- Banff International Research Station
- Fields Institute for Research in Mathematical Sciences
- Microsoft Research
- Pacific Institute for the Mathematical Sciences
- United States National Security Agency
- University of Calgary

We would also like to thank our many referees, whose effort helped the authors improve the papers for this volume.

WIN Editorial Committee November 2010

Alina-Carmen Cojocaru, University of Illinois at Chicago, USA
Kristin Lauter, Microsoft Research, USA
Rachel Pries, Colorado State University, USA
Renate Scheidler, University of Calgary, Canada

WIN—Women In Numbers, November 2008

Workshop Participants and Affiliations at the Time of the Workshop

Shabnam Akhtari, Queen's University, Canada
Jennifer Balakrishnan, Massachusetts Institute of Technology, USA
Juliana Belding, University of Maryland, USA
Lisa Berger, Stony Brook University, USA
Alina Bucur, Massachusetts Institute of Technology, USA
Sarah Chisholm, University of Calgary, Canada
Mirela Ciperiani, Columbia University, USA
Alina-Carmen Cojocaru, University of Illinois at Chicago, USA
Chantal David, Concordia University, Canada
Kirsten Eisentraeger, Pennsylvania State University, USA
Brooke Feigon, University of Toronto, Canada
Sharon Garthwaite, Bucknell University, USA
Linda Gruendken, University of Pennsylvania, USA
Helen Grundman, Bryn Mawr College, USA
Laura Hall-Seelig, University of Massachusetts, USA
Jing Long Hoelscher, University of Arizona, USA
Bo-Hae Im, Chung-Ang University, South Korea
Jennifer Johnson-Leung, University of Idaho, USA
Habiba Kadiri, University of Lethbridge, Canada
Matilde Lalín, University of Alberta, Canada
Kristin Lauter, Microsoft Research, USA
Yoonjin Lee, Ewha Woman's University, South Korea
Winnie Li, Pennsylvania State University, USA
Ling Long, Iowa State University, USA
Beth Malmskog, Colorado State University, USA
Michelle Manes, University of Hawaii, USA
Ekin Ozman, University of Wisconsin-Madison, USA
Jennifer Paulhus, Kansas State University, USA
Rachel Pries, Colorado State University, USA
Adriana Salerno, University of Texas, USA
Renate Scheidler, University of Calgary, Canada
Katherine Stange, Simon Fraser University/PIMS, Canada
Katherine Stevenson, California State University Northridge, USA
Holly Swisher, Oregon State University, USA
Audrey Terras, University of California San Diego, USA
Edlyn Teske, University of Waterloo, Canada
Stephanie Treneer, Western Washington University, USA
Margaret Upton, Texas A & M University, USA
Núria Vila, University of Barcelona, Spain
Bianca Viray, University of California Berkeley, USA
Erika Wittenborn, University of Colorado, USA

Part I – Moduli Spaces and Shimura Curves

Fields Institute Communications
Volume **60**, 2011

Hilbert Modular Variety Computations

Helen Grundman
Department of Mathematics
Bryn Mawr College
101 N. Merion Ave.
Bryn Mawr, PA 19010-2899 USA
grundman@brynmawr.edu

Abstract. Let k be a totally real number field of degree n over $\mathbf{Q}$, with ring of integers $\mathcal{O}_k$. The Hilbert modular variety of k (defined up to birational equivalence) is a desingularization of a natural toroidal compactification of $\mathrm{PSL}_2(\mathcal{O}_k)\backslash\mathcal{H}^n$. This paper is a survey of computational methods used in studying these varieties, with an emphasis on the determination of numerical invariants used in their classification.

1 Introduction

Let k be a totally real algebraic number field of degree n over $\mathbf{Q}$ with ring of integers $\mathcal{O}_k$. The Hilbert modular group of the field k is $\Gamma_k = \mathrm{PSL}_2(\mathcal{O}_k) = \mathrm{SL}_2(\mathcal{O}_k)/\{\pm I\}$. The group $\mathrm{SL}_2(\mathcal{O}_k)$ acts on $\mathfrak{H}^n$, the product of n copies of the complex upper half plane, by

$$Mz = \left(\frac{a^{(1)}z_1 + b^{(1)}}{c^{(1)}z_1 + d^{(1)}}, \frac{a^{(2)}z_2 + b^{(2)}}{c^{(2)}z_2 + d^{(2)}}, \ldots, \frac{a^{(n)}z_n + b^{(n)}}{c^{(n)}z_n + d^{(n)}}\right)$$

where $M = \begin{pmatrix} a & b \\ c & d \end{pmatrix} \in \mathrm{SL}_2(\mathcal{O}_k)$, $z = (z_1, z_2, \ldots, z_n) \in \mathfrak{H}^n$, and $x \mapsto x^{(i)}$ denotes the ith embedding of k into the real numbers. This gives a well-defined, effective action of Γ_k on $\mathfrak{H}^n$. The quotient space $\Gamma_k\backslash\mathfrak{H}^n$ is a normal algebraic variety which is a coarse moduli space of abelian varieties A of dimension n with $k \subseteq \mathrm{End}(A)\otimes\mathbf{Q}$.

The n embeddings of k into $\mathbf{R}$ induce an embedding of $\mathbf{P}^1(k)$ into $\mathbf{P}^1(\mathbf{R})^n$, and thus into $\mathbf{P}^1(\mathbf{C})^n$, allowing for a natural extension of the action of Γ_k to $\mathfrak{H}^n \cup \mathbf{P}^1(k)$, which we equip with the Satake topology. This yields the standard compactification, $X_k = \Gamma_k\backslash(\mathfrak{H}^n \cup \mathbf{P}^1(k))$ of $\Gamma_k\backslash\mathfrak{H}^n$. The points of $X_k - \Gamma_k\backslash\mathfrak{H}^n$, that is, the orbits of $\mathbf{P}^1(k)$ under Γ_k, are called the cusps of Γ_k.

For $n > 1$, X_k has singularities, some arising from the cusps and others arising from elliptic fixed points, points fixed by elements of finite order in Γ_k. By a general

---

2010 *Mathematics Subject Classification.* Primary 11F41, 14E08; Secondary 11Y40.

theorem of Hironaka, X_k can be desingularized to form a compact, complex, non-singular variety — the Hilbert modular variety of k. (See [5, 16, 27] for more detailed discussions of this construction.) Although, for $n > 1$, there are many different ways to carry out the desingularization, the birational equivalence class of X_k, and hence of its desingularization, is uniquely determined.

In the following sections, we present a survey of computational methods used in classifying Hilbert modular varieties. In Section 2, we state and prove a standard result on the number of cusps of Γ_k (needed for the results in Section 4). In Section 3, we introduce the arithmetic genus and describe its use and its computation in the cubic field case. The solution to the more demanding problem of computing the arithmetic genus in the quartic field case is detailed in Section 4. Finally, in Section 5, we introduce plurigenera and explain how computing bounds on these invariants has contributed significantly to the classification of Hilbert modular threefolds.

2 The number of cusps

When Blumenthal [2] published the first major work on Hilbert modular groups and Hilbert modular varieties in 1903, he mischaracterized the fundamental domains, basically assuming that the group had only one cusp. This error was corrected in 1940 by Maass [22] who showed that the number of cusps is equal to the class number of k. In modern terms, the proof is actually quite straight-forward and provides a one-to-one correspondence needed in Section 4.2.

Theorem 2.1 *The number of cusps of Γ_k is equal to the class number of k.*

Proof Let $\mathcal{C}_k$ be the class group of k and for each ideal $\mathfrak{a}$, let $[\mathfrak{a}]$ denote the ideal of class of $\mathfrak{a}$. Define $\varphi : X_k - \Gamma_k\backslash\mathfrak{H}^n \to \mathcal{C}_k$ by

$$\varphi(\alpha : \beta) = [(\alpha, \beta)],$$

where $(\alpha : \beta) \in \mathbf{P}^1(k)$ and, without loss of generality, $\alpha,\ \beta \in \mathcal{O}_k$. The map is well-defined, since, given $\sigma = (\alpha : \beta)$ and $\tau = \frac{a\sigma+b}{c\sigma+d}$ with $\begin{pmatrix} a & b \\ c & d \end{pmatrix} \in \Gamma_k$, we have $\varphi(\tau) = [(a\alpha + b\beta, c\alpha + d\beta)] = [(\alpha, \beta)]$. Further, φ maps onto $\mathcal{C}_k$, since every ideal of $\mathcal{O}_k$ can be generated by at most two elements of $\mathcal{O}_k$.

To see the φ is one-to-one, let $\sigma = (\alpha : \beta)$ and $\tau = (\gamma : \delta)$ with $[(\alpha, \beta)] = [(\gamma, \delta)]$. Then $\mu(\alpha, \beta) = \nu(\gamma, \delta) = \mathfrak{a}$ for some $\mu,\ \nu \in \mathcal{O}_k$. It follows that there exist $\hat{\alpha},\ \hat{\beta},\ \hat{\gamma},\ \hat{\delta} \in \mathfrak{a}^{-1}$ such that $\mu\alpha\hat{\alpha} + \mu\beta\hat{\beta} = 1$ and $\nu\gamma\hat{\gamma} + \nu\delta\hat{\delta} = 1$. Now,

$$\begin{pmatrix} r\alpha & \hat{\beta} \\ r\beta & \hat{\alpha} \end{pmatrix} \text{ and } \begin{pmatrix} r\delta & \hat{\gamma} \\ r\gamma & \hat{\delta} \end{pmatrix}$$

map $(1 : 0) \in \mathbf{P}^1(k)$ to σ and τ, respectively. Further,

$$\begin{pmatrix} r\alpha & \hat{\beta} \\ r\beta & \hat{\alpha} \end{pmatrix} \begin{pmatrix} r\delta & \hat{\gamma} \\ r\gamma & \hat{\delta} \end{pmatrix}^{-1} \in \Gamma_k.$$

So σ and τ are in the same Γ_k orbit, as desired.

Thus φ is a bijection and therefore the number of cusps is equal to the class number of k. □

Given a cusp $\sigma = (\alpha : \beta) \in \mathbf{P}^1(k)$, with α, $\beta \in \mathcal{O}_k$, let $\mathfrak{a} = (\alpha, \beta)$ and let $\hat{\alpha}$, $\hat{\beta} \in \mathfrak{a}^{-1}$ such that $\alpha\hat{\alpha} + \beta\hat{\beta} = 1$. Then the matrix

$$M = \begin{pmatrix} \alpha & \hat{\beta} \\ \beta & \hat{\alpha} \end{pmatrix}$$

maps $(1:0) = \infty$ to σ. We can, thus, view σ as the cusp at infinity for the group $M^{-1}SL(2, \mathcal{O}_k)M/\{\pm I\} =$

$$\left\{ \begin{pmatrix} a & b \\ c & d \end{pmatrix} \in SL(2,k) \,\middle|\, a, d \in \mathcal{O}_k, b \in \mathfrak{a}^{-2}, c \in \mathfrak{a}^2 \right\},$$

with isotropy subgroup

$$\left\{ \begin{pmatrix} \varepsilon^2 & \rho \\ 0 & 1 \end{pmatrix} \,\middle|\, \varepsilon \in \mathcal{E}_k^2, \rho \in \mathfrak{a}^{-2} \right\}.$$

We say that the cusp σ is of type $(\mathfrak{a}^{-2}, \mathcal{E}_k^2)$. When a field has class number one, the one cusp of $\Gamma_k \backslash \mathfrak{H}^n$ is already the cusp at infinity and so it has type $(\mathcal{O}_k, \mathcal{E}_k^2)$.

3 The arithmetic genus

A great deal of the work on classifying Hilbert modular varieties has involved computing birational invariants, particular the arithmetic genus (in the sense of Hirzebruch [17]). Given a variety V of complex dimension n, the arithmetic genus of V is defined by

$$\chi(V) = \sum_{i=0}^{n} (-1)^i g_i,$$

where g_i is the number of complex, linearly independent holomorphic differential forms on V of degree i. Defined in this way, the arithmetic genus is a birational invariant and the arithmetic genus of a rational variety (one that is birationally equivalent to complex projective space) is equal to one.

Hirzebruch, van de Ven, and Zagier [18, 19] used the arithmetic genus and other birational invariants in classifying Hilbert modular surfaces. They showed that exactly ten quadratic fields have Hilbert modular varieties with arithmetic genus equal to one and that, in fact, those ten surfaces are all rational.

Considering Hilbert modular varieties of higher dimensions, Tsuyumine [26] showed that those of dimension greater than six are all of general type. But for dimensions between two and seven, comparatively little is known.

Restricting to the Galois cubic case, Weisser [29] used the following formula, derived from the work of Hirzebruch [16] and Vignéras [28], to compute the arithmetic genus of Hilbert modular threefolds.

Theorem 3.1 *For a totally real algebraic number field k of degree n over $\mathbf{Q}$ containing a unit of norm -1, the arithmetic genus of the Hilbert modular variety of the field k is given by*

$$\chi(k) = 2^{-n} \left(2\zeta_k(-1) + \sum_{r \geq 2} \frac{r-1}{r} a_r \right), \tag{3.1}$$

where ζ_k is the Dedekind zeta function of k and a_r is the number of equivalence classes of fixed points of Γ_k in $\mathfrak{H}^n$ with isotropy subgroups of order r.

Note that $\zeta_k(-1) \in \mathbf{Q}$ and, as will be illustrated below, the sum in (3.1) is necessarily finite. Also, Theorem 3.1 clearly applies to any number field k of odd degree, since $-1 \in k$ is a unit of norm -1.

Weisser used this theorem to show that there are exactly four Galois cubic fields with Hilbert modular varieties of arithmetic genus one. At this point, it is still unknown whether or not any of these four varieties is in fact rational. Using the same methods, Keqin [20] showed that there is exactly one Galois quintic field with Hilbert modular variety of arithmetic genus one.

For non-Galois cubics, the author [7, 8, 12] and the author with Lippincott [13] extended these methods to prove that exactly 33 cubic fields have Hilbert modular varieties of arithmetic genus one.

For fields of degree four, the existence of a unit of norm -1 is not guaranteed and thus Theorem 3.1 may not apply. In the absence of such a unit, the analysis is more involved, as we now describe in detail.

4 Hilbert modular fourfolds

In this section we discuss the computation of the arithmetic genus of Hilbert modular varieties of quartic number fields and the proof of the following theorem:

Theorem 4.1 *Exactly three Hilbert modular fourfolds have arithmetic genus one.*

As described above, the derivations of the analogues of Theorem 4.1 for fields of odd degree are similar, but made easier by the use of Theorem 3.1. In its place, we use a more general formula for the arithmetic genus of the Hilbert modular variety of a field k of degree n over $\mathbf{Q}$:

$$\chi(k) = (-1)^n \mathrm{vol}(\Gamma_k \backslash \mathfrak{H}^n) + \sum_{\eta} E(\Gamma_k, \eta) + \sum_{\kappa} L(\Gamma_k, \kappa), \tag{4.1}$$

where the first summation is over a complete set of representatives of the Γ_k-equivalence classes of elliptic fixed points of Γ_k and the second is over the set of cusps [6]. Restricting, for simplicity, to the case of $n = 4$, the quantities $E(\Gamma_k, \eta)$ and $L(\Gamma_k, \kappa)$ are defined as follows.

Given an elliptic fixed point $z \in \mathfrak{H}^4$, let $M \in \mathrm{SL}_2(\mathcal{O}_k)$ be a representative of a nonidentity element $\overline{M} \in \Gamma_k$ with $Mz = z$. Conjugating $\overline{M}$ by the unique fractional linear transformation mapping z to 0, the complex conjugate of z to ∞, and ∞ to 1, the action of $\overline{M}$ transforms to a product of rotations on a 4-fold product of unit disks. The action is that of multiplication by the rotation factor of $\overline{M}$, $\zeta = (\zeta_1, \zeta_2, \zeta_3, \zeta_4)$ where each ζ_j is an rth root of unity, for $r \geq 2$ equal to the order of the fixed group of z under Γ_k. Then the definition of $E(\Gamma_k, [z])$ is given by

$$E(\Gamma_k, [z]) = \frac{1}{|\Gamma_z|} \sum_{\substack{\overline{M} \in \Gamma_z \\ \overline{M} \neq \overline{I}}} \prod_{i=1}^{4} \left(\frac{1}{1 - \zeta_i} \right), \tag{4.2}$$

where $\zeta = (\zeta_1, \zeta_2, \zeta_3, \zeta_4)$ is the rotation factor of $\overline{M}$.

To define $L(\Gamma_k, \kappa)$, let $\mathcal{E}_k$ be the group of units of $\mathcal{O}_k$ and again let $\mathcal{C}_k$ be the ideal class group. As shown in the proof of Theorem 2.1, the set of cusps of $\Gamma_k \backslash \mathfrak{H}^4$ is in one-to-one correspondence with $\mathcal{C}_k$. Letting $[\mathfrak{a}]$ be the ideal class corresponding to the cusp κ, we define $L(\Gamma_k, \kappa)$ by

$$L(\Gamma_k, \kappa) = L(\Gamma_k, [\mathfrak{a}]) = \frac{1}{16\pi^4}\sqrt{d(\mathfrak{a}^{-2})} \sum_{a \in \mathfrak{a}^{-2}/\mathcal{E}_k^2} \frac{\operatorname{sign}(N(a))}{|N(a)|},$$

with $d(\mathfrak{a}^{-2})$ the discriminant of $\mathfrak{a}^{-2}$.

The first term of equation (4.1), the volume of a fundamental domain of Γ_k acting on $\mathfrak{H}^4$, is well known to equal $\frac{1}{8}\zeta_k(-1)$. We now discuss the computation of each of the other two terms. (See [14, 15] for further details.)

4.1 Computing $\sum E(\Gamma_k, [z])$. Let z be an elliptic fixed point of Γ_k. Prestel's work [23] enables the determination of the possible rotation factors of elements of the fixed group Γ_z. The key is a one-to-one correspondence between Γ_k-equivalence classes of elliptic fixed points and Γ_k-conjugacy classes of orders of a certain form.

More concretely, let r be the order of the fixed point z, in other words, the order of the fixed group Γ_z, which is easily shown to be finite and cyclic. Let $\overline{M}$ be a generator of Γ_z. Let $M_2(k)$ be the ring of 2×2 matrices over k and let $k' = k[M]$ in $M_2(k)$. Then the Γ_k-equivalence class of z corresponds to the Γ_k-conjugacy class of the order $k' \cap M_2(\mathcal{O}_k)$.

This immediately gives us the possible values of r, as follows. Letting s be the trace of M, we have $k' = k[M] \cong k\left(\sqrt{s^2-4}\right)$, which is necessarily a totally imaginary quadratic extension of k. Since M is multiplicatively of finite order in k', M must be a root of unity. So $k[M] \cong k(e^{\frac{2\pi i}{m}})$, where, as shown in [29], $m = r$ or $2r$ and $s = e^{\frac{2\pi i}{m}} + e^{-\frac{2\pi i}{m}}$. Since k' is of degree eight over $\mathbf{Q}$, $m \in \{3, 4, 5, 6, 8, 10, 12, 15, 16, 20, 24, 30\}$. This yields the possible values of r.

Lemma 4.2 *Let r be the order of a fixed point of Γ_k for some totally real quartic field k. Then $r \in \{2, 3, 4, 5, 6, 8, 10, 12, 15\}$.*

Let δ be the relative discriminant of k' over k. Following Prestel, let $\mathfrak{a}_0$ be the ideal of $\mathcal{O}_k$ such that $\mathfrak{a}_0^2\delta = (s^2-4)\mathcal{O}_k$. Prestel showed that $\mathfrak{a}_0$ is characterized as being the smallest ideal in $\mathcal{O}_k$ such that $\mathfrak{a}_0^2|(s^2-4)\mathcal{O}_k$ and $\mathfrak{a}_0^2|(a_0^2 - a_0 s + 1)\mathcal{O}_k$ for some $a_0 \in \mathcal{O}_k$. Further, he demonstrated that any order $\mathcal{O}$ of k' containing an element M with trace s and determinant 1 is of the form

$$\mathcal{O} = \mathcal{O}_k \oplus \mathfrak{a}_0^{-1}\mathfrak{a}(M - a_0) \tag{4.3}$$

for some ideal $\mathfrak{a}|\mathfrak{a}_0$ with $\mathfrak{a}$ and a_0 as above.

Two orders of this form, $\mathcal{O}_1$ and $\mathcal{O}_2$, are isomorphic if and only if there exists an invertible matrix $N \in M_2(k)$ with $\mathcal{O}_1 = N\mathcal{O}_2N^{-1}$. Each isomorphism class of orders corresponds to a different ideal $\mathfrak{a}$ in equation (4.3). Note that Γ_k-conjugate orders are necessarily isomorphic and that all conjugacy classes of orders in the same isomorphism class correspond to fixed points with fixed groups of the same order, denoted $r = w(\mathfrak{a})$.

We can now use Prestel's work to determine the number of conjugacy classes in a given equivalence class. Let $\mathcal{E}_k$ and $\mathcal{E}_{k'}$ be the groups of units and h_k and $h_{k'}$ the class numbers of $\mathcal{O}_k$ and $\mathcal{O}_{k'}$ respectively. The following theorem can be derived from Theorems 3 and 4 of [23].

Theorem 4.3 *Let $\mathcal{O} = \mathcal{O}_k \oplus \mathfrak{a}_0^{-1}\mathfrak{a}(M - a_0)$, as in equation (4.3). The number of conjugacy classes of orders isomorphic to $\mathcal{O}$ is*

$$\ell(\mathcal{O}) = \frac{[\mathcal{E}_k : \mathcal{E}_k^2]h_{k'}}{2[\mathcal{E}_{k'} : \mathcal{E}_k]h_k} N_{k/\mathbf{Q}}(\mathfrak{a})w(\mathfrak{a}) \prod_{\mathfrak{p}|\mathfrak{a}} \left(1 - \left(\frac{k'}{\mathfrak{p}}\right) N_{k/\mathbf{Q}}(\mathfrak{p}^{-1})\right),$$

where $\left(\frac{k'}{\mathfrak{p}}\right)$ is the Artin symbol.

Let w be the number of roots of unity in k' and let R_k and $R_{k'}$ be the regulators of k and k', respectively. It is easy to show that $[\mathcal{E}_k : \mathcal{E}_k^2] = 16$ and that $[\mathcal{E}_{k'} : \mathcal{E}_k] = 4wR_k/R_{k'}$. Thus, for any particular order $\mathcal{O}$, Theorem 4.3 reduces the value of $\ell(\mathcal{O})$ to computable terms.

Prestel [23] proved that in any given isomorphism class of these orders, the number of conjugacy classes having a given rotation factor is the same for all rotation factors that occur in that isomorphism class. His work also allows us to determine the complete set of rotation factors occurring in each isomorphism class.

Fix $\mathcal{O} = \mathcal{O}_k \oplus \mathfrak{a}_0^{-1}\mathfrak{a}(M - a_0)$. Set

$$\alpha_j = \left(\frac{(s^{(j)})^2}{2} - 1\right) - i\ \mathrm{sign}(s^{(j)})\sqrt{1 - \left(\frac{(s^{(j)})^2}{2} - 1\right)^2}.$$

For a non-zero ideal $\mathfrak{c} \subseteq \mathcal{O}_k$, let $\left(\frac{k'}{\mathfrak{c}}\right)$ be the Artin symbol extended multiplicatively to all nonzero integral ideals. Prestel proved that if $\delta \neq \mathcal{O}_k$, then the set of rotation factors that occur in the isomorphism class of $\mathcal{O}$ is

$$\{(\alpha_1, \alpha_2^{\epsilon_2}, \alpha_3^{\epsilon_3}, \alpha_4^{\epsilon_4}) | \epsilon_i = \pm 1\}$$

and if $\delta = \mathcal{O}_k$, the set is instead

$$\left\{(\alpha_1, \alpha_2^{\epsilon_2}, \alpha_3^{\epsilon_3}, \alpha_4^{\epsilon_4}) \middle| \epsilon_i = \pm 1 \text{ and } \epsilon_2\epsilon_3\epsilon_4 = \left(\frac{k'}{\mathfrak{a}_0\mathfrak{a}^{-1}}\right)\right\}.$$

When $r = 2$, there is only one possible rotation factor for $\overline{M}$, $(-1, -1, -1, -1)$. For other values of r, there are either four or eight rotation factors that occur, depending on the value of the relative discriminant δ.

Using this and equation (4.2), we can compute the average value of $E(\Gamma_k, [z])$ over the equivalence classes, $[z]$, corresponding to Γ_k-conjugacy classes of orders isomorphic to $\mathcal{O}$. The results are give in Table 1. Notice that each average value is completely determined by the order, r, of a fixed point (which determines the possible values of s) and the set of rotation factors that occur.

Table 1 Average Values of $E(\Gamma_k, [z])$

	$r:2$	3	4	5	6	8	10	12	15
$\delta \neq \mathcal{O}_k$	$\frac{1}{32}$	$\frac{1}{24}$	$\frac{3}{64}$	$\frac{1}{20}$	$\frac{5}{96}$	$\frac{7}{128}$	$\frac{9}{160}$	$\frac{11}{192}$	$\frac{7}{120}$
$\delta = \mathcal{O}_k, \left(\frac{k'}{\mathfrak{a}_0\mathfrak{a}^{-1}}\right) = +1$	$\frac{1}{32}$	$\frac{5}{108}$	$\frac{5}{64}$	$\frac{3}{50}$	$\frac{209}{864}$	$\frac{13}{128}$	$\frac{149}{800}$	$\frac{317}{1728}$	$\frac{59}{625}$
$\delta = \mathcal{O}_k, \left(\frac{k'}{\mathfrak{a}_0\mathfrak{a}^{-1}}\right) = -1$	$\frac{1}{32}$	$\frac{1}{27}$	$\frac{1}{64}$	$\frac{1}{25}$	$-\frac{119}{864}$	$\frac{1}{128}$	$-\frac{59}{800}$	$-\frac{119}{1728}$	$\frac{79}{2700}$

For each order, r, the corresponding field k' and the ideal $\mathfrak{a}_0$ can be found. The divisors $\mathfrak{a}$ of $\mathfrak{a}_0$ determine the isomorphism classes of orders of the form in equation (4.3). It is then reasonably straightforward to establish which of these isomorphism classes correspond to fixed points of order r.

For each isomorphism class, the number of conjugacy classes in the isomorphism class can be determined using Theorem 4.3, and Table 1 provides the average value of $E(\Gamma_k, \eta)$ over the conjugacy classes in that class. The value of the contribution to the sum from points of order r is then computed by taking the product and summing over all of the isomorphism classes corresponding to fixed points of order r.

Significantly, further analysis of the possible isomorphism classes of orders that can occur when $r = 6$, 10, or 12 leads directly to the fact that the three negative values in Table 1 never occur. Hence the average value of $E(\Gamma_k, [z])$ for any isomorphism class is always positive. Therefore the sum of $E(\Gamma_k, [z])$ over all equivalence classes of fixed points must be non-negative, thus proving the following lemma, needed for the proof of Theorem 4.8.

Corollary 4.4 *$\sum_\eta E(\Gamma_k, \eta) \geq 0$, where η ranges over the Γ_k-equivalence classes of fixed points of Γ_k.*

4.2 Computing $\sum_\kappa L(\Gamma_k, \kappa)$. Recall that each cusp κ corresponds to a unique ideal class $[\mathfrak{a}]$ under the one-to-one correspondence given in the proof of Theorem 2.1. By definition, $L(\Gamma_k, \kappa) = L(\Gamma_k, [\mathfrak{a}])$ and so

$$\sum_\kappa L(\Gamma_k, \kappa) = \sum_{[\mathfrak{a}] \in \mathcal{C}_k} L(\Gamma_k, [\mathfrak{a}]) = \sum_{[\mathfrak{a}] \in \mathcal{C}_k} \frac{1}{16\pi^4} \sqrt{d(\mathfrak{a}^{-2})} \sum_{a \in \mathfrak{a}^{-2}/\mathcal{E}_k^2} \frac{\operatorname{sign}(N(a))}{|N(a)|}.$$

Lemma 4.5 *If $\mathcal{O}_k$ contains a unit of norm -1 or an integral ideal $\mathfrak{b}$ for which $\mathfrak{b}^2 = (\beta)$ is a principal ideal with $N(\beta) < 0$, then $\sum_\kappa L(\Gamma_k, \kappa) = 0$.*

Proof First suppose that $\mathcal{O}_k$ contains a unit ε of norm -1. Then for each ideal class $[\mathfrak{a}] \in \mathcal{C}_k$, $a \in \mathfrak{a}^{-2}$ if and only if $\varepsilon a \in \mathfrak{a}^{-2}$. So $L(\Gamma_k, [\mathfrak{a}]) = -L(\Gamma_k, [\mathfrak{a}])$, implying that $L(\Gamma_k, [\mathfrak{a}]) = 0$ and thus $\sum_\kappa L(\Gamma_k, \kappa) = 0$.

Now suppose that $\mathcal{O}_k$ does not contain a unit of norm -1. Then there exists a character on the group of non-zero fractional ideals of k, $\psi : \mathcal{I}_k \to \mathbf{C}^*$, satisfying $\psi((a)) = \operatorname{sign}(N(a))$ for each nonzero $a \in \mathcal{O}_k$. This yields,

$$L(\Gamma_k, [\mathfrak{a}]) = \frac{\sqrt{d_k}}{\pi^4} \overline{\psi}(\mathfrak{a}^2) \lim_{s \to 1} \sum_{\substack{\mathfrak{c} \subseteq \mathcal{O}_k \\ \mathfrak{c} \in [\mathfrak{a}^2]}} \frac{\psi(\mathfrak{c})}{N(\mathfrak{c})^s}.$$

(See [6] for details of this derivation.)

Suppose further that there is some integral ideal $\mathfrak{b}$ for which $\mathfrak{b}^2 = (\beta)$ is a principal ideal with $N(\beta) < 0$. Then for each $\mathfrak{a} \subseteq \mathcal{O}_k$, $[(\mathfrak{a}\mathfrak{b})^2] = [\mathfrak{a}^2]$ while $\overline{\psi}((\mathfrak{a}\mathfrak{b})^2) = -\overline{\psi}(\mathfrak{a}^2)$. Hence $L(\Gamma_k, [\mathfrak{a}\mathfrak{b}]) = -L(\Gamma_k, [\mathfrak{a}])$ and therefore $\sum_{[\mathfrak{a}] \in \mathcal{C}_k} L(\Gamma_k, [\mathfrak{a}]) = 0$. □

As before, for k' an extension field of k, let $h_k, h_{k'}, R_k, R_{k'}, d_k, d_{k'}$ denote the class numbers, regulators, and absolute values of the discriminants of $\mathcal{O}_k$ and $\mathcal{O}_{k'}$ respectively and let $w_k, w_{k'}$ denote the numbers of roots of unity in these rings. Theorem 4.6 extends Lemma 4.5, providing a computable formula for $\sum_\kappa L(\Gamma_k, \kappa)$ in all cases.

Theorem 4.6 *If $\mathcal{O}_k$ contains a unit of norm -1 or an integral ideal $\mathfrak{b}$ for which $\mathfrak{b}^2 = (\beta)$ is a principal ideal with $N(\beta) < 0$, then $\sum_\kappa L(\Gamma_k, \kappa) = 0$. Otherwise,*

$$\sum_\kappa L(\Gamma_k, \kappa) = \frac{2}{h_k R_k} \sum_{k'} \frac{h_{k'} R_{k'}}{w_{k'}} \tag{4.4}$$

where k' ranges through the set of unramified complex quadratic extensions of k.

Proof Assume that $\mathcal{O}_k$ contains neither a unit of norm -1 nor an integral ideal whose square is generated by an element of norm less than zero. It follows, then that the character ψ in the proof of Lemma 4.5 can be chosen to be real. Taking ψ real, gives us $\overline{\psi}(\mathfrak{a}^2) = 1$ and so, letting $\mathfrak{c}$ vary over integral ideals in $\mathcal{O}_k$,

$$\begin{aligned} \sum_{[\mathfrak{a}] \in \mathcal{C}_k} L(\Gamma_k, [\mathfrak{a}]) &= \frac{\sqrt{d_k}}{\pi^4} \lim_{s \to 1} \sum_{[\mathfrak{a}] \in \mathcal{C}_k} \sum_{\mathfrak{c} \in [\mathfrak{a}^2]} \frac{\psi(\mathfrak{c})}{N(\mathfrak{c})^s} \\ &= \frac{\sqrt{d_k}}{\pi^4} \lim_{s \to 1} \sum_{[\mathfrak{c}] \in \mathcal{C}_k^2} \frac{\psi(\mathfrak{c})}{N(\mathfrak{c})^s} |\mathcal{C}_k / \mathcal{C}_k^2|. \end{aligned}$$

Next, let Φ be the set of characters $\phi : \mathcal{I}_k \to \pm 1$ induced by a complete set of characters on the group $\mathcal{C}_k / \mathcal{C}_k^2$. We get

$$\begin{aligned} \sum_{[\mathfrak{a}] \in \mathcal{C}_k} L(\Gamma_k, [\mathfrak{a}]) &= \frac{\sqrt{d_k}}{\pi^4} \lim_{s \to 1} \sum_{\mathfrak{c}} \frac{\psi(\mathfrak{c})}{N(\mathfrak{c})^s} \sum_{\phi \in \Phi} \phi(\mathfrak{c}) \\ &= \frac{\sqrt{d_k}}{\pi^4} \sum_{\phi \in \Phi} \lim_{s \to 1} \sum_{\mathfrak{c}} \frac{\psi\phi(\mathfrak{c})}{N(\mathfrak{c})^s} \\ &= \frac{\sqrt{d_k}}{\pi^4} \sum_{\phi \in \Phi} L(1, \psi\phi). \end{aligned}$$

Now, since each ϕ is trivial on principal ideals, each $\psi\phi$ is a quadratic character on $\mathcal{I}_k$ satisfying $\psi\phi((a)) = \text{sign}(N(a))$ for all nonzero $a \in \mathcal{O}_k$. Hence, each $\psi\phi$ corresponds, via class field theory, to a quadratic extension k' of k, unramified at each finite place and ramified at each infinite place. In fact, as ϕ ranges through Φ, the k' range through all such field extensions of k, that is, unramified complex quadratic extensions of k.

Finally, given $\phi \in \Phi$, let k' be the complex quadratic extension of k corresponding to $\psi\phi$. Then

$$L(1, \psi\phi) = L(1, k'/k) = \frac{\pi^4 h_{k'} R_{k'} w_k \sqrt{d_k}}{h_k R_k w_{k'} \sqrt{d_{k'}}}. \tag{4.5}$$

Noting that $d_{k'} = d_k^2$ and $w_k = 2$ completes the proof. □

Corollary 4.7 $\sum_\kappa L(\Gamma_k, \kappa) \geq 0$, *where κ ranges over the cusps of $\Gamma_k \backslash \mathfrak{H}^4$.*

4.3 Implementation and Bounds. The above results allowed for the implementation of a computer algorithm for calculating the arithmetic genus of a Hilbert modular fourfold. In doing so, we used PARI [1] to compute the values of $\zeta_k(-1)$ needed for computing the first term of equation (4.1) and wrote a KASH [4] program to perform the remainder of the calculations. As input, we used data from the University of Bordeaux's number fields web site [3].

In order to prove Theorem 4.1, it remains to prove the existence of an upper bound on the size of the discriminant of a Hilbert modular fourfold of arithmetic genus one.

Theorem 4.8 *Let k be a totally real quartic number field with discriminant $d_k > 11348$. Then the Hilbert modular variety of k has arithmetic genus greater than one.*

Proof Let k be a totally real quartic number field. By Corollaries 4.4 and 4.7, the final two terms of equation (4.1) are nonnegative and thus the arithmetic genus of the Hilbert modular variety of k is bounded below by $\mathrm{vol}(\Gamma_k\backslash\mathfrak{H}^n)$ or equivalently, $\frac{1}{8}\zeta_k(-1)$. As shown in [11], since $\zeta_k(-1) = \frac{1}{16}\pi^{-8}d_k^{\frac{3}{2}}\zeta_k(2)$ and $\zeta_k(2) > \zeta(8) = \frac{\pi^8}{9450}$, $\zeta_k(-1) > \frac{1}{151200}d_k^{\frac{3}{2}}$. Also, if $d_k > 11348$, then, in fact, we have $d_k \geq 11661$ [3]. Therefore, $\chi(k) > \frac{1}{8}\frac{1}{151200}11661^{3/2} > 1$, as desired. □

5 Bounding the plurigenera

In general, for a variety to be rational, it is necessary but not sufficient that it have arithmetic genus equal to one. As noted earlier, for Hilbert modular surfaces, arithmetic genus one is actually equivalent to being rational. The goal of this section is to show how bounding the plurigenera of certain Hilbert modular threefolds has been used to prove that the two conditions are not always equivalent for Hilbert modular varieties. For more details, see [9, 10].

Theorem 5.1 *There exist Hilbert modular varieties of arithmetic genus one that are not rational. In particular, the Hilbert modular threefolds of the cubic fields of discriminants 697, 761, and 985 have arithmetic genus one, but are not rational.*

We begin with the definition of plurigenera. Let V be a nonsingular variety and let $\mathcal{K}$ be the canonical sheaf of holomorphic differential forms on V. For $q \geq 1$, the qth plurigenus of V is defined by

$$\rho_q = \dim_{\mathbf{C}} \Gamma\left(V, \mathcal{K}^{\otimes q}\right).$$

The plurigenera are birational invariants of V and if V is a rational variety, then, for all q, $\rho_q = 0$. For a totally real field k, let $\rho_q(k)$ denote the qth plurigenus of the Hilbert modular variety of k.

The plurigenera $\rho_q(k)$ are closely related to the dimensions of spaces of Hilbert modular forms over k. Let $n > 1$ be the degree of k over $\mathbf{Q}$. A holomorphic function $f : \mathfrak{H}^n \to \mathbf{C}$ is a Hilbert modular form of weight $2q$ over k, if for all $\gamma \in \Gamma_k$ and $z \in \mathfrak{H}^n$,

$$f(\gamma z) = \prod_{i=1}^{n}\left(c^{(i)}z_i + d^{(i)}\right)^{2q} f(z),$$

where c and d are the entries in the bottom row of a representative matrix of γ. (For $n > 1$, such an f is automatically holomorphic at the cusps.) Let $M_q(k)$ denote the complex dimension of the space of Hilbert modular forms of weight $2q$ over k.

For cubic fields, we can determine the values of $M_q(k)$ using formulas due to Thomas and Vasquez [25]. The following is an immediate consequence of [25, (2.1), (2.11), and Theorem 2.16].

Theorem 5.2 *For k a totally real cubic field with $d_k > 81$ and $q \geq 0$,*

$$M_q(k) = h_k - \frac{(2q-1)^3}{4}\zeta_k(-1) + \frac{a_2}{16}\phi_q + \frac{a_3}{12}\psi_q, \tag{5.1}$$

where ϕ_q and ψ_q are defined by

$$\sum_{q\geq 0} \phi_q t^q = \frac{1}{1+t} \quad \text{and} \quad \sum_{q\geq 0} \psi_q t^q = \frac{1-t}{1-t^3}.$$

For cubic fields k with Hilbert modular varieties of arithmetic genus one (which are the ones that interest us here), the values of $\zeta_k(-1)$, a_2, and a_3 are given in [13, Table 1], and so the values of $M_q(k)$ are easily computed.

Now, a Hilbert modular form, f, of weight $2q$ defines a holomorphic q-canonical form on $X_k = \Gamma_k \backslash (\mathfrak{H}^n \cup \mathbf{P}^1(k))$ minus its singularities. The qth plurigenus, however, measures the space of forms on the entire (compact, desingularized) variety. To get a lower bound on $\rho_q(k)$, we want to measure the extendability of the Hilbert modular forms across the resolved singularities. For this, we define the defect of a singularity.

Let X be a normal complex space with singularity $x \in X$ and q a positive integer. The qth defect at x is defined by

$$\delta_x(q) = \dim_{\mathbf{C}} \Gamma(\tilde{U} - S, \mathcal{K}^q)/\Gamma(\tilde{U}, \mathcal{K}^q),$$

where $S = g^{-1}(x)$ for a resolution $g : \tilde{X} \to X$ of x, U is a small Stein neighborhood of x, $\tilde{U} = g^{-1}(U)$, and $\mathcal{K}$ is the sheaf of holomorphic differential forms on $\tilde{X}$. For a totally real field k, let $d_q(k) = \sum \delta_x(q)$, summing over all singularities of X_k.

It follows from the work of Knöller [21] (see also [24]) that

$$M_q(k) - d_q(k) \leq \rho_q(k) \leq M_q(k). \tag{5.2}$$

The left-hand inequality will allow us to prove that in certain cases $\rho_q(k) > 0$, implying that the Hilbert modular variety of k is not rational.

The computation of $d_q(k)$ is greatly simplified by a result of Knöller's giving that all singularities of X_k arising from elliptic fixed points have defects equal to zero [21]. Thus when the class number of k is one, $d_q(k) = \delta_x(q)$, with x the unique cusp.

In [24], Thomas defines the defect series for a singularity x to be the formal power series:

$$\Delta(x, \lambda) = 1 + \sum_{q=1}^{\infty} \delta_x(q)\lambda^q \quad \text{in} \quad \mathbf{Z}[[\lambda]].$$

He computed the defect series for an infinite family of singularities and used the results with equation (5.2) to determine bounds on the plurigenera of certain Hilbert modular varieties of Galois cubic fields of class number one. These methods were extended by the author to additional families of cusps, ones determined by non-Galois cubic fields [9, 10].

As an example, let k be the cubic field with $d_k = 697$. Then $h_k = 1$ and the Hilbert modular variety of k has arithmetic genus one [7]. Using equation (5.1), since $a_2 = 16$, $a_3 = 8$ and $\zeta_k(-1) = -\frac{8}{3}$, we have $M_2(k) = 20$. In [9], the author worked with a parametrized family of cusps which included one of type $(\mathcal{O}_k, \mathcal{E}_k^+)$, where $\mathcal{E}_k^+$ is the group of totally positive units in $\mathcal{O}_k$ and, providing an explicit resolution for the entire family of cusps, deduced a parametrized formula for the defect series of the cusps. Restricting to this one cusp, this reduces to the following theorem.

Theorem 5.3 *Let k be the cubic field with $d_k = 697$. The defect series for the cusp $(\mathcal{O}_k, \mathcal{E}_k^+)$ is given by*

$$\Delta((\mathcal{O}_k, \mathcal{E}_k^+), \lambda) = \frac{1 - 3\lambda + 11\lambda^2 + 8\lambda^3 - \lambda^4 + 3\lambda^5 - 12\lambda^6 - 7\lambda^7}{(1-\lambda)^4(1-\lambda^4)}. \tag{5.3}$$

Combining equation (5.3) with the equation

$$\Delta\left((\mathcal{O}_k, \mathcal{E}_n^2), \lambda\right) = 2 \cdot \Delta\left((\mathcal{O}_k, \mathcal{E}_n^+), \lambda\right) - \frac{1}{1-\lambda},$$

(see [9]), it is straightforward to compute the defect series for the cusp $(\mathcal{O}_k, \mathcal{E}_k^2)$ and from that, the value of the qth defect for any $q > 0$. In particular, we find that $d_2(k) = 17$. Applying equation (5.2), yields

$$\rho_2(k) \geq M_2(k) - d_2(k) = 20 - 17 = 3 > 0.$$

Thus, the Hilbert modular threefold of k is not rational. Similar computations (found in [10]) completes the proof of Theorem 5.1.

References

[1] C. Batut, K. Belabas, D. Benardi, H. Cohen, and M. Olivier, *User's guide to PARI-GP*, 1998, ⟨ftp://megrez.math.u-bordeaux.fr/pub/pari⟩.

[2] O. Blumenthal, *Über Modulfunktionen von mehereren Veränderlichen*, Math. Ann. **56** (1903), 509–548.

[3] J. Buchmann, F. Diaz y Diaz, D. Ford, P. Létard, M. Olivier, M. Pohst, and A. Schwarz, Tables of number fields of low degree: ⟨ftp://megrez.math.u-bordeaux.fr/pub/numberfields/⟩.

[4] M. Daberkow, C. Fieker, J. Klüners, M. Pohst, K. Roegner, and K. Wildanger, *Kant v4*, Symbolic Comput. **24** (1997), 267–283, ⟨http://www.math.TU-Berlin.DE/~kant/kash.html⟩.

[5] F. Ehlers, *Eine Klasse komplexer Mannigfaltigkeiten und die Auflösung einiger isolierter Singularitäten*, Math. Ann. **218** (1975), 127–156.

[6] E. Freitag, *Hilbert modular forms*, Springer-Verlag, Berlin, 1980.

[7] H. G. Grundman, *The arithmetic genus of Hilbert modular varieties over non-Galois cubic fields*, J. Number Theory **37** (1991), 343–365.

[8] ———, *On the classification of Hilbert modular threefolds*, Manuscripta Math. **72** (1991), 297–305.

[9] ———, *Defects of cusp singularities and the classification of Hilbert modular threefolds*, Math. Ann. **292** (1992), 1–12.

[10] ———, *Defect series and nonrational Hilbert modular threefolds*, Math. Ann. **300** (1994), 77–88.

[11] ———, *Hilbert modular varieties over Galois quartic fields*, J. Number Theory **63** (1997), 47–58.

[12] ———, *Nonrational Hilbert modular threefolds*, J. Number Theory **83** (2000), 50–58.

[13] H. G. Grundman and L. E. Lippincott, *Hilbert modular threefolds of arithmetic genus one*, J. Number Theory **95** (2002), 72–76.

[14] ———, *Hilbert modular fourfolds of arithmetic genus one*, High Primes and Misdemeanours: lectures in honour of the 60th birthday of Hugh Cowie Williams, Fields Institute Communications Series, 41, Amer. Math. Soc., 2004, pp. 217–226.

[15] ———, *Computing the arithmetic genus of Hilbert modular fourfolds*, Math. Comp. **75** (2006), 1553–1560.

[16] F. Hirzebruch, *Hilbert modular surfaces*, Enseign. Math. (2) **19** (1973), 183–281.

[17] ———, *Topological methods in algebraic geometry*, 3rd ed., Springer-Verlag, Berlin, 1978.

[18] F. Hirzebruch and A. Van de Ven, *Hilbert modular surfaces and the classification of algebraic surfaces*, Invent. Math. **23** (1974), 1–29.

[19] F. Hirzebruch and D. Zagier, *Classification of Hilbert modular surfaces*, Complex Analysis and Algebraic Geometry, University Press and Iwanami Shoten, Cambridge, 1977, pp. 43–77.

[20] F. Keqin, *On arithmetic genus of Hilbert modular varieties on cyclic number fields*, Sci. Sinica Ser. A **27** (1984), 576–584.

[21] F. W. Knöller, *Beispiele dreidimensionaler Hilbertsher Modulmannigfaltigkeiten von allgemeinem Typ*, Manuscripta Math. **37** (1982), 135–161.

[22] H. Maass, *Über Gruppen von hyperabelschen Transformationen*, Sitzungsber. Heidelb. Akad Wiss (1940), 3–26.

[23] A. Prestel, *Die elliptischen Fixpunkte der Hilbertschen Modulgruppen*, Math. Ann. **177** (1968), 181–209.

[24] E. Thomas, *Defects of cusp singularities*, Math. Ann. **264** (1983), 397–411.

[25] E. Thomas and A. T. Vasquez, *Rings of Hilbert modular forms*, Compos. Math. **48** (1983), 139–165.

[26] S. Tsuyumine, *On the Kodaira dimensions of Hilbert modular varieties*, Invent. Math. **80** (1985), 269–281.

[27] G. van de Geer, *Hilbert modular surfaces*, Springer-Verlag, Berlin, 1988.

[28] M.-F. Vignéras, *Invariants numériques des groupes de Hilbert*, Math. Ann. **224** (1976), 189–215.

[29] D. Weisser, *The arithmetic genus of the Hilbert modular variety and the elliptic fixed points of the Hilbert modular group*, Math. Ann. **257** (1981), 9–22.

Fields Institute Communications
Volume **60**, 2011

Contributions to Shimura Curves

Pilar Bayer
Departament d'Àlgebra i Geometria
Facultat de Matemàtiques
Universitat de Barcelona
08007 Barcelona, Spain
bayer@ub.edu

Abstract. Shimura curves are a generalization of the classical modular curves, and they have been studied since the time of Klein, Poincaré and Fricke. Their arithmetical significance was recognized by Shimura in his outstanding paper [35].

As is well-known, modular curves are coarse moduli spaces for elliptic curves, together with some additional level structures. Shimura curves obtained from non-split quaternion algebras over totally real number fields are coarse moduli spaces for abelian surfaces with quaternionic multiplication, also called fake elliptic curves, endowed with some additional level structures.

A convenient generalization of the classical complex multiplication theory led Shimura to his theory of the canonical models. These are models defined over specific number fields with a distinguished family of algebraic points provided by the fake elliptic curves with complex multiplication. As in the classical setting, CM points play a key role in the construction of class fields and their behavior under the Galois action is governed by a reciprocity law.

The computational study of Shimura curves has lagged behind that of modular curves, partially owing to the lack of cusps. However, beginning with the work of Ihara [18], Hashimoto-Murabayashi [17] and Elkies [12], among others, this area has recently seen renewed interest and progress. Our paper provides an outline of joint work with Alsina, Guàrdia and Travesa devoted to making Shimura curves more accessible to computations.

Introduction

Let H be a quaternion algebra defined over a totally real number field F. Suppose that H is unramified at exactly one real place of F and fix an embedding

2010 *Mathematics Subject Classification.* Primary 11G18, 14G35, 30F35, 33E30; Secondary 30F10, 32N15.

Partially supported by MCYT MTM2009-07024.

of H into $\mathrm{M}(2,\mathbb{R})$. Given an order $\mathcal{O} \subseteq H$, the group $\Gamma(H,\mathcal{O})$ of the units of norm equal to one is a Fuchsian group of the first kind: it acts discontinuously on the upper half-plane $\mathcal{H}$ and the quotient $\Gamma\backslash\mathcal{H}$ is a Riemann surface.

A Fuchsian group $\Gamma \subseteq \mathrm{SL}(2,\mathbb{R})$ is called arithmetic when it is commensurable with some $\Gamma(H,\mathcal{O})$. The first examples of arithmetic Fuchsian groups are provided by the congruence subgrups of the modular group $\mathrm{SL}(2,\mathbb{Z})$. In this case, the defining quaternion algebra is just the algebra $M(2,\mathbb{Q})$, of the rational 2×2 matrices, and the order is $M(2,\mathbb{Z})$. As is well known, these groups Γ uniformize the modular curves which, alternatively, can be described as coarse moduli spaces of elliptic curves with level structure, depending on Γ.

Shimura curves are Riemann surfaces uniformized by arithmetic Fuchsian groups. In [35], Shimura described these curves as moduli spaces of principally polarized abelian surfaces with quaternionic multiplication and level structure. One of Shimura's main contributions to the topic is his theory of the canonical models [36]. As in the modular case, the canonical models of Shimura curves are uniquely defined (up to isomorphisms) by automorphic functions that generate class fields when evaluated at the complex multiplication (CM) points of the curve.

A far-reaching generalization of Shimura curves is provided by Shimura varieties associated to congruence subgroups of algebraic reductive groups. Shimura attacked the problem of finding canonical models for these objects in a series of papers published between 1964 and 1970. An axiomatic treatment of the existence of canonical models in this setting was given by Deligne in [10] and [11]. The interested reader can find an introduction to the topic in Milne's home-page [26].

Langlands has made Shimura varieties a central part of his program as a test for the conjecture that all motivic L-functions should be automorphic. In this sense, the first spectacular application of Shimura varieties has already occurred in dimension 1, namely in the proof of Shimura-Taniyama-Weil modularity conjecture by Wiles *et al.*

From an algorithmic point of view, the fact that elliptic curves are more amenable to explicit computation than abelian surfaces has implied that there is a fundamental difference between the treatment of modular curves and Shimura (non-modular) curves. The uniformization of Shimura curves and the computation of their automorphic functions have been long-standing problems. Some of the first papers that we find in this direction are those of Ihara [18], Kurihara [23], Michon [25], Jordan [21], Hashimoto-Murabayashi [17] and Elkies [12].

In general, algorithms used in the modular case do not transfer directly to the quaternionic case. The main difficulties are due to the fact that while modular curves contain cusps, Shimura curves attached to non-split quaternion algebras lack them. In consequence, automorphic functions admit q-developments (with $q = e^{2\pi i z}$) only in the modular case.

In this paper, we provide an outline of joint results obtained with Montserrat Alsina, Jordi Guàrdia and Artur Travesa (cf. [1], [5], [6] and [7]) concerning the computation of explicit developments for the automorphic functions of the canonical models and that of equations defining CM points on Shimura curves. To simplify the exposition, we formulate our results in the simplest non-modular case, given by the rational quaternion algebra of discriminant $D = 6$.

Recent contributions concerning explicit computations in Shimura curves can also be found in the following papers: [2], [3], [13], [14], [19], [20], [22], [27], [38], and [39].

ACKNOWLEDGEMENTS. I am grateful to the editors of the volume and to the anonymous referee for their time and effort in providing constructive comments on my submitted original.

1 Basic concepts on rational quaternion algebras

For simplicity, we only consider quaternion algebras defined over the rationals. The reader can find more general cases and the basic theory on quaternion algebras in [1] or in [40].

We consider a quaternion algebra $H = H(a, b)$ over the rational number field $\mathbb{Q}$ with basis $\{1, i, j, k\}$ and product defined by

$$i^2 = a, \quad j^2 = b, \quad ij = -ji = k,$$

where $a, b \in \mathbb{Q}^*$. The algebra H has a conjugation: $\omega \mapsto \overline{\omega} = x - yi - zj - tk$, which is a linear anti-involution. A quaternion $\omega = x + yi + zj + tk$ is called pure if $\overline{\omega} = -\omega$. The trace and the norm are the maps from H to $\mathbb{Q}$ defined by $\mathrm{tr}(\omega) = \omega + \overline{\omega}$, $\mathrm{n}(\omega) = \omega\overline{\omega}$.

We shall suppose that H is indefinite, i. e. $H \otimes \mathbb{R} \simeq M(2, \mathbb{R})$, and that $a > 0$. Throughout we fix the embedding

$$\begin{array}{rccc} \Phi : & H & \hookrightarrow & M(2, \mathbb{R}) \\ & x + yi + xj + tk & \mapsto & \begin{bmatrix} x + y\sqrt{a} & z + t\sqrt{a} \\ b(z - t\sqrt{a}) & x - y\sqrt{a} \end{bmatrix}. \end{array}$$

A place p is said to ramify in H if $H \otimes \mathbb{Q}_p$ is a division algebra over the field $\mathbb{Q}_p$ of the p-adic numbers, when p is finite, or over the field $\mathbb{R}$, when $p = \infty$. In general, the discriminant D of a quaternion algebra is defined as the product of all the places where the algebra ramifies. Since we have supposed that H is indefinite, H is unramified at ∞ and, by class field theory, D equals the product of an even number of different rational primes. Since, up to isomorphism, H is determined by D, we shall also use the notation $H = H_D$.

Examples 1.1 a) The algebra $H(1, -1)$ is isomorphic to the rational matrix algebra $M(2, \mathbb{Q})$ and its discriminant is $D = 1$. This is the only case in which H is not a division algebra and has zero divisors.

b) The quaternion algebra $H_A(p) = H(p, -1)$, where $p \equiv 3 \pmod 4$ is a prime number, has discriminant $D = 2p$.

c) The quaternion algebra $H_B(p, q) = H(p, q)$, where p, q are prime numbers, $q \equiv 1 \pmod 4$ and the Legendre symbol $\left(\dfrac{p}{q}\right)$ is equal to -1, has discriminant $D = pq$. $\square$

An order of a quaternion algebra is a subring $\mathcal{O} \subseteq H$ which is a free $\mathbb{Z}$-module of rank 4. Eichler orders are obtained as the intersection of two maximal orders. They are classified by their level, which is a positive integer N prime to D.

In the case of rational indefinite quaternion algebras, Eichler orders of a given level are determined up to conjugation in H. We shall denote by $\mathcal{O}(D, N)$ a representative of the conjugacy class of Eichler orders of level N in H_D. The orders $\mathcal{O}(D) = \mathcal{O}(D, 1)$ are maximal.

Examples 1.2 The following examples correspond to some Eichler orders of square-free level N for the quaternion algebras considered above (cf. Alsina-Bayer

[1, Sect. 1.2]). (We remark that these are only some Eichler orders, not all orders for the algebras H_A or H_B arise in this way.)

a) If $H = M(2,\mathbb{Q})$, take

$$\mathcal{O}(1,N) = \left\{ \begin{bmatrix} a & b \\ cN & d \end{bmatrix} : a,b,c,d \in \mathbb{Z} \right\}.$$

b) If $H = H_A(p)$, take

$$\mathcal{O}(2p,N) = \mathbb{Z}\left[1, i, Nj, \frac{1+i+j+k}{2}\right], \quad N | \frac{p-1}{2}.$$

c) If $H = H_B(p,q)$, take

$$\mathcal{O}(pq,N) = \mathbb{Z}\left[1, Ni, \frac{1+j}{2}, \frac{i+k}{2}\right], \quad N | \frac{q-1}{4}, \quad \gcd(N,p) = 1.$$

For an Eichler order $\mathcal{O}(D,N)$, we denote by $\mathcal{O}(D,N)_1^*$ the group of quaternion units whose norm is equal to 1 and define

$$\Gamma(D,N) = \Phi(\mathcal{O}(D,N)_1^*) \subseteq \mathrm{SL}(2,\mathbb{R}),$$

obtaining in this way an arithmetic Fuchsian group.

Examples 1.3 a) The Fuchsian group $\Gamma(1,N)$ obtained from the order $\mathcal{O}_0(1,N)$ is equal to the congruence group $\Gamma_0(N)$.

b) The Fuchsian group $\Gamma(2p,1)$ obtained from the order $\mathcal{O}(2p,1)$ is given by

$$\left\{ \gamma = \frac{1}{2}\begin{bmatrix} \alpha & \beta \\ -\beta' & \alpha' \end{bmatrix} : \alpha \equiv \beta \equiv \alpha\sqrt{p} \pmod{2}, \det\gamma = 1 \right\},$$

where $\alpha, \beta \in \mathbb{Z}[\sqrt{p}]$.

c) The Fuchsian group $\Gamma(pq,1)$ obtained from the order $\mathcal{O}(pq,1)$ is given by

$$\left\{ \gamma = \frac{1}{2}\begin{bmatrix} \alpha & \beta \\ q\beta' & \alpha' \end{bmatrix} : \alpha \equiv \beta \pmod{2},\ 2 | \frac{\alpha - \alpha' - \beta + \beta'}{2\sqrt{p}}, \det\gamma = 1 \right\},$$

where $\alpha, \beta \in \mathbb{Z}[\sqrt{p}]$.

In all the above examples, $(-)'$ denotes the Galois conjugate of an element in the quadratic field.

2 The Shimura curves X_D

Let $\mathcal{O}(D) = \mathcal{O}(D,1)$ be a maximal order of H_D. The group $\Gamma_D = \Gamma(D,1)$ acts on the upper half-plane $\mathcal{H}$ as a group of orientation-preserving isometries; the action factorizes through its image $\overline{\Gamma}_D \subseteq \mathrm{PSL}(2,\mathbb{R})$ and the quotient $\overline{\Gamma}_D \setminus \mathcal{H}$ is a Riemann surface, which is compact when $D > 1$.

A Shimura's canonical model (X_D, j_D) consists of an algebraic curve X_D defined over $\mathbb{Q}$ and a birregular isomorphism

$$j_D : \overline{\Gamma}_D \backslash \mathcal{H} \longrightarrow X_D(\mathbb{C}).$$

Up to birregular isomorphisms, the model is characterized by the fact that for any imaginary quadratic field F contained in H_D, of conductor 1 with respect to $\mathcal{O}(D)$, and with τ as its fixed point, the value $j_D(\tau)$ is algebraic and the field

$$F(j_D(\tau))$$

is the Hilbert class field of F.

In order to provide a moduli interpretation for the Shimura curves, it is necessary to make the notion of a PEL-type precise. According to Shimura [35, Sect. 4], a PEL-type

$$\Omega = (H,\ \Phi,\ *;T,\ \mathcal{O};\ V)$$

consists of an embedding of $\mathbb{Q}$-algebras $\Phi : H \hookrightarrow M(2,\mathbb{R})$, a positive anti-involution $*$ of H; a maximal order $\mathcal{O} \subseteq H$, a nondegenerate alternating form T on H, integer valued over $\mathcal{O}$; and a level structure $V = (v_1,\dots,v_s)$ given by finitely many ordered quaternion classes $v_i \in \mathcal{O}\otimes\mathbb{Q}/\mathcal{O} = H/\mathcal{O}$.

From the Skolem-Noether theorem, it follows that there exists a quaternion $\mu \in H_D$ such that $\alpha^* = \mu^{-1}\overline{\alpha}\mu$ for all $\alpha \in H_D$. The element μ turns out to be a pure quaternion for which $\mu^2 < 0$, because $*$ is positive. Since the field $\mathbb{Q}(\sqrt{-D})$ always splits H_D, it can be embedded in H_D and μ may be chosen so that $\mu \in \mathcal{O}(D)$ and $\mu^2 = -D$. Then the lattice $\mathcal{O}(D)$ becomes stable under $*$.

Definition 2.1 A fake elliptic curve $(A,\iota,\mathcal{L},W)$ over $\mathbb{C}$ of PEL-type Ω consists of

(i) An abelian surface $A/\mathbb{C}$.
(ii) An injective ring homomorphism $\iota : \mathcal{O} \hookrightarrow \mathrm{End}(A)$ such that $H_1(A,\mathbb{Z})$ regarded as a left $\mathcal{O}$-module is isomorphic to $\mathcal{O}$.
(iii) A principal polarization $\mathcal{L}$ on A such that the associated Rosati anti-involution, $\phi_{\mathcal{L}} : \mathrm{End}^0(A) \to \mathrm{End}^0(A)$, restricts to the anti-involution $*$ on $\iota(\mathcal{O})$.
(iv) A level structure $W \subset H_1(A,\mathbb{Q})$ given by the image of V under the isomorphism $H \simeq H_1(A,\mathbb{Q})$ obtained from (ii).

An isomorphism of a fake elliptic curve $(A,\iota,\mathcal{L},W)$ into $(A',\iota',\mathcal{L}',W')$ is an isomorphism of the abelian surfaces compatible with the whole quadruple. We denote by $[A,\iota,\mathcal{L},W]$ the corresponding isomorphism class. When $W = 0$, we drop W of the notation and simply write $[A,\iota,\mathcal{L}]$.

The complex points of X_D are in one to one correspondence with the isomorphism classes $[A,\iota,\mathcal{L}]$ of fake elliptic curves of PEL-type given by

$$(H_D,\ \Phi,\ *;T,\ \mathcal{O}(D);0).$$

The level structures W will be used in section 5 to obtain Weierstrass models for CM fake elliptic curves.

Now, let us consider an imaginary quadratic field $F = \mathbb{Q}(\sqrt{d}), d < 0$, and let $R = R(d,m) \subseteq F$ be a quadratic order of conductor m. Suppose that there is an embedding $\varphi : F \hookrightarrow H$ or, equivalently, that the field F splits H (cf. Shimura [35, Lemma 2.3]). A fake elliptic curve $(A,\iota,\mathcal{L})$ is said to have complex multiplication (CM) by R if the embedding φ restricts to an optimal embedding $\varphi : R \hookrightarrow \mathcal{O}$ (in the sense that $R = \varphi^{-1}(\varphi(F)\cap\mathcal{O})$), and

$$\mathrm{End}(A,\iota,\mathcal{L}) \simeq R,$$

where $\mathrm{End}(A,\iota,\mathcal{L})$ denotes the endomorphisms ring of the triple. If this is the case, then $\mathrm{End}^0(A,\iota,\mathcal{L}) \simeq F$, $\mathrm{End}^0(A) \simeq H\otimes_{\mathbb{Q}} F \simeq M(2,F)$, and the abelian variety A is isogenous to the square of an elliptic curve with complex multiplication. Then the class $[A,\iota,\mathcal{L}]$ defines a CM point on the Shimura curve X_D.

CM points on Shimura curves give rise to class fields. More precisely, the embedding φ determines a point $\tau \in \mathcal{H}$ which is fixed by all transformations in $\Phi(\varphi(F))$, and $j_D\circ\pi(\tau) \in X_D(F^{ab})$, where $\pi : \mathcal{H} \to \Gamma_D\backslash\mathcal{H}$ denotes the canonical

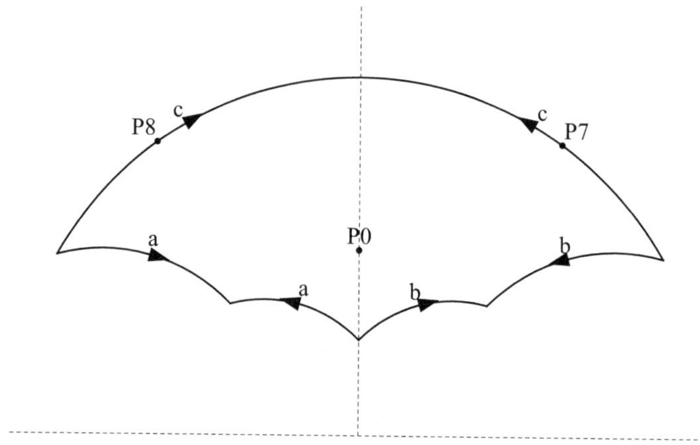

$$P_0 = \frac{i(\sqrt{6}-\sqrt{2})}{2}, \quad P_7 = \frac{\sqrt{3}+i\sqrt{6}}{3} \sim P_8 = \frac{-\sqrt{3}+i\sqrt{6}}{3}$$

Figure 1 SCM points in X_6 $(D = 6, N = 1)$.

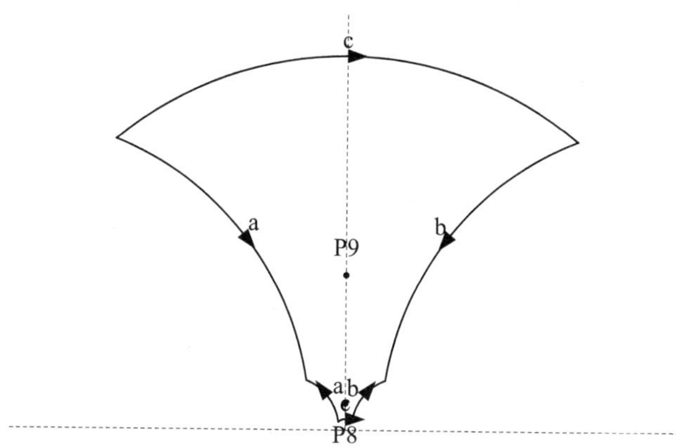

$$P_8 = \frac{i(3\sqrt{5}-2\sqrt{10})}{5}, \quad P_9 = \frac{i\sqrt{5}}{5}$$

Figure 2 SCM points in X_{10} $(D = 10, N = 1)$.

projection and F^{ab} stands for the maximal abelian extension of F. As we shall see in section 5, the PEL-type allows to define an isomorphism class of fake elliptic curves attached to τ, which will be denoted by $[A_\tau, \iota_\tau, \mathcal{L}_\tau]$. The behaviour of the CM points under the Frobenius automorphisms is governed by Shimura's reciprocity law (cf. [35, Main Theorem II(3.5)]).

A CM point $\tau \in \mathcal{H}$ is called special (SCM) if it is the fixed point of some fractional linear transformation $\gamma \in \Phi \circ \varphi(R(d,m))$ of determinant equal to D. The binary normic form $n_F(X,Y) = X^2 - dY^2$ of the field $F = \mathbb{Q}(\sqrt{d})$ yields a binary quadratic form $n_R(X,Y)$ when restricted to an order $R(d,m)$. The SCM points

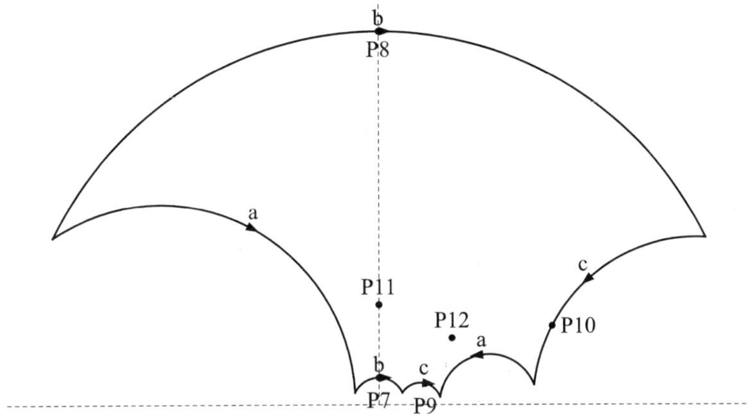

$$P_7 = \frac{i(2\sqrt{5} - \sqrt{15})}{5} \sim P_8, \quad P_9 = \frac{15 - 5\sqrt{3} + i(3\sqrt{5} - \sqrt{15})}{30} \sim P_{10},$$
$$P_{11} = \frac{i\sqrt{5}}{5}, \quad P_{12} = \frac{5 + 2i\sqrt{5}}{15}$$

Figure 3 SCM points in X_{15} $(D = 15, N = 1)$.

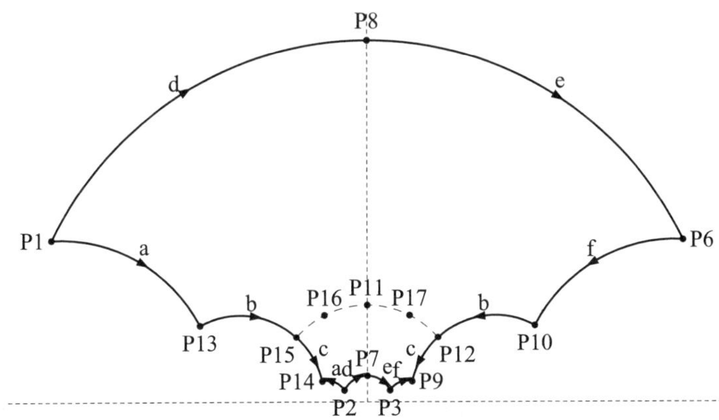

$$P_7 = \frac{i(2\sqrt{5} - \sqrt{15})}{5} \sim P_8, \quad P_9 = \frac{15 - 5\sqrt{3} + i(3\sqrt{5} - \sqrt{15})}{30} \sim P_{10},$$
$$P_{11} = \frac{i\sqrt{5}}{5}, \quad P_{12} = \frac{5 + 2i\sqrt{5}}{15}$$

Figure 4 SCM points in X_{15} $(D = 15, N = 1)$.

are obtained from quadratic orders R for which D has an integral representation by $n_R(X, Y)$. Moreover, their orbits under Γ_D are finite in number.

The pictures of the fundamental domains for Γ_6, Γ_{10} and Γ_{15} and the SCM points in figures 1, 2 and 3 are taken from Alsina-Bayer [1, Ch. 6]. Of course, the shape of the fundamental domain depends on the set of generators chosen for the Fuchsian group Γ. Figure 4 shows a second fundamental domain for Γ_{15} computed recently by Travesa from the one given in 3.

3 Canonical models

Our next goal is to explain the procedure followed in Bayer-Travesa [6] to obtain the automorphic functions parametrizing the canonical models of the Shimura curve X_6 and its quotients by involutions. The functions have expansions with algebraic integer coefficients around the elliptic points and the SCM points, which have been explicitly computed. The main tool in our approach is the use of Fuchs theory together with an ordinary differential operator obtained by a suitable modification of the Schwarzian derivative, introduced by Schwarz in 1873 (cf. [34]).

Other constructions of automorphic functions for $D = 6$ and $D = 10$, based on the use of Hilbert modular forms, have been obtained by Baba-Granath in [3].

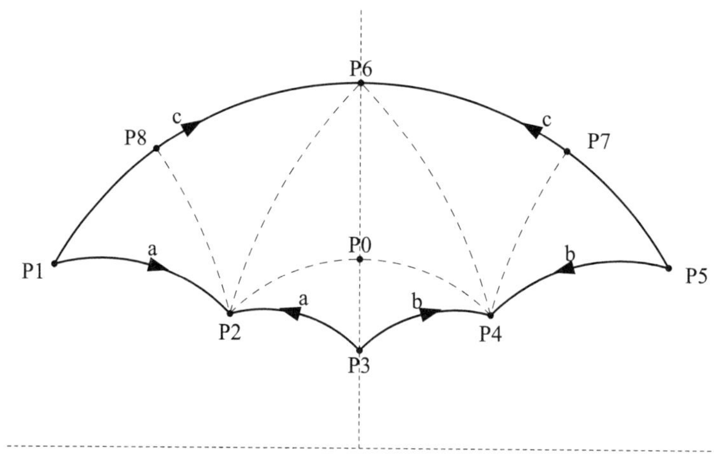

Figure 5 Fundamental domain for Γ_6.

Let $N(\mathcal{O}_6)$ be the normalizer of $\mathcal{O}_6$ in H_6. The elements of $N(\mathcal{O}_6)$ of positive norm define a subgroup whose image in $\mathrm{GL}(2,\mathbb{R})$ will be denoted by Γ_6^+. The group Γ_6 is contained in Γ_6^+ as a normal subgroup and the quotient $\overline{\Gamma}_6^+/\overline{\Gamma}_6$ is isomorphic to $(\mathbb{Z}/2\mathbb{Z})^2$. The classes of this quotient are represented by elements $w_r \in \mathcal{O}_6$ of norm r dividing 6. We can take $w_2 = 1 + j$, $w_3 = (-3 - i - 3j + k)/2$ and $w_6 = w_2 w_3 = -3j + k$ as representatives. They define involutions ω_r of the curve X_6.

Together with the curve X_6, we consider its quotients $X_6^{(r)} = X_6/\langle \omega_r \rangle$ and $X_6^+ = X_6/\langle \omega_r : r|6 \rangle$. All these curves fit in a diagram of Galois covers of degree two:

$$\begin{array}{ccc} & X_6 & \\ \swarrow & \downarrow & \searrow \\ X_6^{(2)} & X_6^{(3)} & X_6^{(6)} \\ \searrow & \downarrow & \swarrow \\ & X_6^{+}. & \end{array}$$

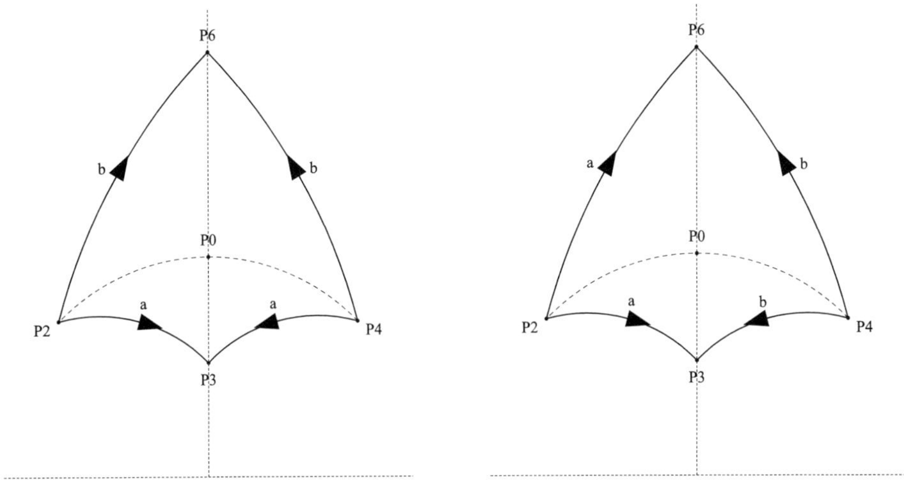

Figure 6 Fundamental domains for $\Gamma_6^{(2)}$ and $\Gamma_6^{(3)}$.

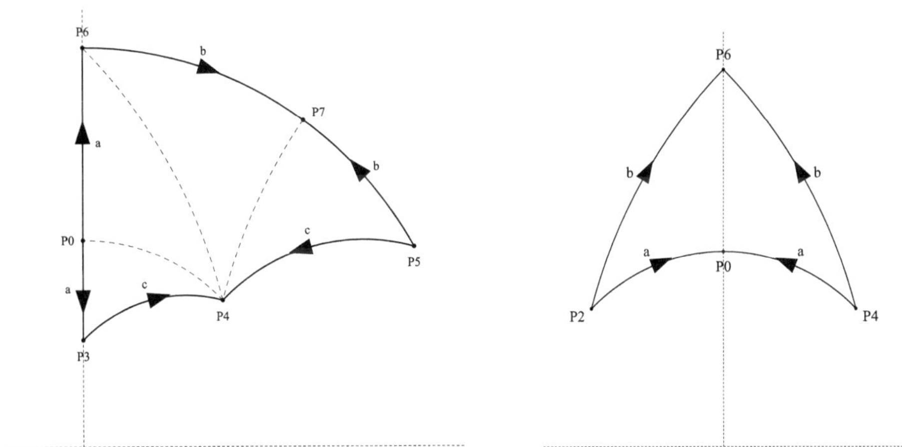

Figure 7 Fundamental domains for $\Gamma_6^{(6)}$ and Γ_6^{+}.

From the genus formula (cf. [37, Ch. 2]), it follows that the curves we are dealing with are all of genus zero (cf. [1, Ch. 2]). Thus there exist automorphic functions t_6, $t_6^{(2)}$, $t_6^{(3)}$, $t_6^{(6)}$, t_6^{+} on $\mathcal{H}$, which uniformize the curves X_6, $X_6^{(2)}$, $X_6^{(3)}$, $X_6^{(6)}$, X_6^{+}, respectively, and uniquely defined by imposing their values at three points. We choose these functions in accordance with table 1. The values a and b will be determined in Proposition 3.1.

Table 1 Values of the uniformizing functions at the vertices.

t	P_0	P_2	P_3	P_4	P_6	P_7
t_6	$*$	a	0	1	∞	$*$
$t_6^{(2)}$	$*$	$*$	0	1	∞	$*$
$t_6^{(3)}$	$*$	0	$*$	1	∞	$*$
$t_6^{(6)}$	0	$*$	$*$	1	∞	b
t_6^{+}	0	$*$	$*$	1	∞	$*$

Proposition 3.1 *The uniformizing functions defined in table 1 fulfil the following algebraic relations:*

(1) $4t_6^+t_6^{(2)} = (t_6^{(2)}+1)^2$. (5) $t_6^+ + t_6^{(6)}(t_6^{(6)}-2) = 0$.
(2) $t_6^+ = (2t_6^{(3)}-1)^2$. (6) $2t_6t_6^{(6)} = i(t_6-i)^2$.
(3) $t_6^2 = t_6^{(2)}$. (7) $4t_6^2t_6^+ = (t_6^2+1)^2$.
(4) $4t_6t_6^{(3)} = (t_6+1)^2$.

Moreover,

(11) $t_6^{(2)}(P_0) = -1$. (12) $t_6^{(3)}(P_0) = \frac{1}{2}$. (13) $b = t_6^{(6)}(P_7) = 2$.
(14) $a = t_6(P_2) = -1$. (15) $t_6(P_0) = i$. (16) $t_6(P_7) = -i$.

For a non-constant smooth function $f(z)$, let $f'(z)$ be its derivative. The automorphic derivative of f is defined as

$$D(f,z) = \frac{2f'(z)f'''(z) - 3f''(z)^2}{f'(z)^4}.$$

The automorphic derivative $D(f,z)$ of a Γ-automorphic function $f(z)$ is again a Γ-automorphic function; that is to say, the following equality holds: $D(f,\gamma(z)) = D(f,z)$, for every $\gamma \in \Gamma$, $z \in \mathcal{H}$.

When $X(\Gamma)$ is a curve of genus 0 and t is a generator of the field of Γ automorphic functions, there exists a rational function $R(t) \in \mathbb{C}(t)$ such that

$$D(t,z) + R(t) = 0.$$

Suppose that the fundamental domain admits a fundamentla half domain (see [6] for the precise definition of this concept) that the values of the function t at the vertices of the fundamental half domain belong to $\mathbf{P}^1(\mathbb{R})$ and that $\alpha_i\pi$ are the internal angles at the vertices of the fundamental half domain. Then

$$R(t) = \sum \frac{1-\alpha_i^2}{(t-a_i)^2} + \sum \frac{B_i}{t-a_i},$$

where the B_i are constants and the summation extends over all the vertices of the fundamental half domain where the function t takes finite values a_i. Moreover, if the values of t at all the vertices are finite, then

(a) $\sum B_i = 0$,
(b) $\sum a_iB_i + \sum(1-\alpha_i^2) = 0$,
(c) $\sum a_i^2B_i + \sum a_i(1-\alpha_i^2) = 0$.

But if ∞ is the value of t at a vertex with internal angle $\alpha\pi$, then

(a) $\sum B_i = 0$,
(b) $\sum a_iB_i + \sum(1-\alpha_i^2) - (1-\alpha^2) = 0$.

Table 2 Automorphic derivatives of the functions.

Curve	function t	vertices angles	$-D(t,z)$
X_6	t_6	$[P_2, P_3, P_4, P_6]$ $[\pi/3, \pi/2, \pi/3, \pi/2]$	$\dfrac{27t^4+74t^2+27}{36t^2(t^2-1)^2}$
$X_6^{(2)}$	$t_6^{(2)}$	$[P_3, P_4, P_6]$ $[\pi/4, \pi/3, \pi/4]$	$\dfrac{135t^2-142t+135}{144t^2(t-1)^2}$
$X_6^{(3)}$	$t_6^{(3)}$	$[P_2, P_4, P_6]$ $[\pi/6, \pi/6, \pi/2]$	$\dfrac{27t^2-27t+35}{36t^2(t-1)^2}$
$X_6^{(6)}$	$t_6^{(6)}$	$[P_0, P_4, P_7, P_6]$ $[\pi/2, \pi/3, \pi/2, \pi/2]$	$\dfrac{27t^4-108t^3+211t^2-206t+108}{36t^2(t^2-3t+2)^2}$
X_6^+	t_6^+	$[P_0, P_4, P_6]$ $[\pi/2, \pi/6, \pi/4]$	$\dfrac{135t^2-103t+108}{144t^2(t-1)^2}$

Proposition 3.2 *The functions defined by table 1 satisfy the differential equations $D(t,z)+R(t)=0$, where the rational functions $R(t)$ are listed in table 2.*

In general, the above relations between the constants B_i, the angles α_i, and the values a_i do not suffice to determine all the constants. In our case, the relations between our functions fully determine the constants B_i for those associated to $t_6^{(2)}$, $t_6^{(3)}$, and t_6^+. To determine the constants that remain, we formally integrate the differential equation computing some coefficients of the expansion in a neighbourhood of the point P_6 and compare the solutions using the equations for the covers that we have obtained in 3.1.

Theorem 3.3 *The functions $j_6^+, j_6^{(2)}, j_6^{(3)}, j_6^{(6)} : \mathcal{H} \to \mathbb{P}^1(\mathbb{C})$ given by*

$$j_6^+ = t_6^+, \qquad j_6^{(2)} = u_2 = i\frac{1+t_6^{(2)}}{1-t_6^{(2)}},$$

$$j_6^{(3)} = u_3 = i\sqrt{3}\,(1-2t_6^{(3)}), \quad j_6^{(6)} = u_6 = i\sqrt{3}\,(1-t_6^{(6)}),$$

and the function $j_6 : \mathcal{H} \to \mathbb{P}^2(\mathbb{C})$ given in homogeneous coordinates by

$$j_6 = (u_3 : u_6 : 1) = (u_2 : 1 : 1/u_6) = (1 : 1/u_2 : 1/u_3)$$

define canonical models over $\mathbb{Q}$ of the curves X_6^+, $X_6^{(2)}$, $X_6^{(3)}$, $X_6^{(6)}$, X_6, respectively. For the last one, we have

$$\mathbb{Q}(X_6) = \mathbb{Q}(j_6) = \mathbb{Q}(u_3, u_6) = \mathbb{Q}(u_2, u_6) = \mathbb{Q}(u_2, u_3).$$

The equations

$$u_3^2 + u_6^2 + 3 = 0, \quad u_2^2 + 1 + 3u_6^{-2} = 0, \quad 1 + u_2^{-2} + 3u_3^{-2} = 0$$

provide a set of affine rational charts for the projective curve $X_6/\mathbb{Q}$. Moreover,

$$j_6(P_0) = (0 : -i\sqrt{3}, 1), \quad j_6(P_7) = (0 : i\sqrt{3}, 1). \ \square$$

For the proofs of propositions 3.1, 3.2 and theorem 3.3 we refer the reader to Bayer-Travesa [6].

4 Local expansions

We shall now explain how to compute explicit expansions for the uniformizing functions determined in section 3. We shall show the results only for t_6.

Definition 4.1 Let $\overline{\Gamma}_P \subseteq \mathrm{PSL}(2,\mathbb{R})$ be the isotropy group under Γ at a point $P \in \mathcal{H}$ and let e_P be its order. A local parameter at a point P for the $\overline{\Gamma}_P$-action is any function

$$q(z) = \left(k_P \frac{z-P}{z-\overline{P}}\right)^{e_P},$$

where $k_P \in \mathbb{C}$ is a constant. The local parameter q is said to be adapted to a function $t = \sum_{n=m}^{\infty} a_{ne} q^n$ when, moreover, $a_{re} = 1$ if $t - a_0$ has a zero of order re at $z = P$, or $a_{-re} = 1$ it t has a pole of order re at $z = P$.

Case $t(P) \neq \infty$. We consider expansions of the shape

$$t(z) = \sum_{n=0}^{\infty} b_n q^n = \sum_{n=0}^{\infty} b'_n \frac{q(z)^n}{(en)!}, \quad b'_1 = e!$$

We define the factor $\mathfrak{n}_\nu = \nu e!$, where $\nu = t(P)$, and normalize the generating function t by $t(P, q_P; z) = \mathfrak{n}_\nu^{-1} t(z)$ so that

$$t(P, q_P; z) = \sum_{n=0}^{\infty} c_n \frac{q_P(z)^n}{(en)!}, \quad c_1 = 1, \quad q_P(z) = \frac{1}{\nu_P}\left(k_P \frac{z-P}{z-\overline{P}}\right)^{e_P},$$

where the values of e_P, k_P and ν_P are listed in table 3.

Case $t(P) = \infty$. We consider expansions of the shape

$$t(z) = \sum_{n=-1}^{\infty} b'_n \frac{q(z)^n}{(2e(n+2))!}, \quad b'_{-1} = (2e)!$$

Then we define the factor $\mathfrak{n}_\infty = \nu(2e)!$ and normalize the generating function t by $t(P, q_P; z) = \mathfrak{n}_\infty^{-1} t(z)$ so that

$$t(P, q_P; z) = \sum_{n=-1}^{\infty} c_n \frac{q_P(z)^n}{(2e(n+2))!}, \quad c_{-1} = 1, \quad q_P(z) = \frac{1}{\nu_P}\left(k_P \frac{z-P}{z-\overline{P}}\right)^{e_P},$$

where the values of e_P, k_P and ν_P are listed in table 3.

As an example, we provide the starting coefficients for some series expansions associated to the complex uniformizing function t_6 of the curve X_6.

Table 3 Adapted local constants for the function t_6.

t	P	e_P	$t(P)$	ν_P	k_P
t_6	P_0	1	i	$2^2\cdot 3$	$i\frac{\sqrt{2}+\sqrt{3}}{2}\,\frac{\Gamma(\frac{7}{24})\Gamma(\frac{11}{24})}{\Gamma(\frac{19}{24})\Gamma(\frac{23}{24})}$
t_6	P_2	3	-1	3^{-1}	$\frac{1+(2+\sqrt{3})i}{6\sqrt[3]{2}}\,\frac{\Gamma(\frac{1}{3})^2\Gamma(\frac{7}{12})}{\Gamma(\frac{2}{3})^2\Gamma(\frac{11}{12})}$
t_6	P_3	2	0	$3\cdot 2^{-2}$	$\frac{(1+\sqrt{3})(1+i)}{8}\,\frac{\Gamma(\frac{1}{4})\Gamma(\frac{5}{12})}{\Gamma(\frac{3}{4})\Gamma(\frac{11}{12})}$
t_6	P_4	3	1	3^{-1}	$\frac{2+\sqrt{3}-i}{6\sqrt[3]{2}}\,\frac{\Gamma(\frac{1}{3})^2\Gamma(\frac{7}{12})}{\Gamma(\frac{2}{3})^2\Gamma(\frac{11}{12})}$
t_6	P_6	2	∞	2^4	$\frac{\sqrt{3}(1-i)}{4\sqrt{2}}\,\frac{\Gamma(\frac{1}{3})\Gamma(\frac{2}{3})\Gamma(\frac{1}{4})}{\Gamma(\frac{3}{4})\Gamma(\frac{7}{12})\Gamma(\frac{11}{12})}$

Coefficients c_n $(1 \le n \le 10)$ of $t_6(P_3, q_{P_3}; z)$:

$$
\begin{aligned}
1 &= 1\\
0 &= 0\\
-48 &= -2^4\cdot 3\\
0 &= 0\\
27504 &= 2^4\cdot 3^2\cdot 191\\
0 &= 0\\
-64498392 &= -2^3\cdot 3^2\cdot 7\cdot 127973\\
0 &= 0\\
436272183216 &= 2^4\cdot 3^4\cdot 23\cdot 229\cdot 63913\\
0 &= 0
\end{aligned}
$$

•

Coefficients c_n $(0 \le n \le 10)$ of $t_6(P_4, q_{P_4}; z)$:

$$
\begin{aligned}
1/2 &= 2^{-1}\\
1 &= 1\\
20 &= 2^2\cdot 5\\
1356 &= 2^2\cdot 3\cdot 113\\
227040 &= 2^5\cdot 3\cdot 5\cdot 11\cdot 43\\
74611380 &= 2^2\cdot 3\cdot 5\cdot 1243523\\
42574294080 &= 2^6\cdot 3^2\cdot 5\cdot 17\cdot 19\cdot 45767\\
38683567274400 &= 2^5\cdot 3^2\cdot 5^2\cdot 5372717677\\
52554612744944640 &= 2^{10}\cdot 3^2\cdot 5\cdot 11\cdot 23\cdot 4507937111\\
101782604056899960000 &= 2^6\cdot 3^4\cdot 5^4\cdot 139\cdot 226002762361\\
270629344957362042528000 &= 2^8\cdot 3^4\cdot 5^3\cdot 29\cdot 16126171\cdot 223259851
\end{aligned}
$$

•

Coefficients c_n $(0 \leq n \leq 10)$ of $t_6(P_0, q_{P_0}; z)$:

$$\begin{aligned}
i/12 &= -i\cdot(1+i)^{-4}\cdot 3^{-1}\\
1 &= 1\\
-12i &= i\cdot(1+i)^4\cdot 3\\
-226 &= -2\cdot 113\\
5664i &= (1+i)^{10}\cdot 3\cdot 59\\
160728 &= 2^3\cdot 3\cdot 37\cdot 181\\
-5467296i &= -(1+i)^{10}\cdot 3\cdot 56951\\
-211472208 &= -2^4\cdot 3^5\cdot 109\cdot 499\\
9193300992i &= -i\cdot(1+i)^{20}\cdot 3^2\cdot 571\cdot 1747\\
445513958784 &= 2^7\cdot 3^3\cdot 128910289\\
-23734590202368i &= -(1+i)^{18}\cdot 3^4\cdot 15919\cdot 35951
\end{aligned}$$

•

Coefficients c_n $(-1 \leq n \leq 10)$ of $t_6(P_6, q_{P_6}; z)$:

$$\begin{array}{r}
1\\
0\\
18480\\
0\\
12803590800\\
0\\
-817993722627081000\\
0\\
-156078929845326558019950000\\
0\\
12285995340772011067924117938034500\\
0
\end{array}$$

and their factorizations:

$$\begin{array}{l}
1\\
0\\
2^4\cdot 3\cdot 5\cdot 7\cdot 11\\
0\\
2^4\cdot 3^2\cdot 5^2\cdot 7\cdot 11^2\cdot 13\cdot 17\cdot 19\\
0\\
-2^3\cdot 3^3\cdot 5^3\cdot 7^3\cdot 11\cdot 13\cdot 17\cdot 19\cdot 23\cdot 29\cdot 47\cdot 61\\
0\\
-2^4\cdot 3^4\cdot 5^5\cdot 7^3\cdot 11^2\cdot 13\cdot 17\cdot 19^2\cdot 23\cdot 29\cdot 31\cdot 41\cdot 13729\\
0\\
2^3\cdot 3^5\cdot 5^4\cdot 7^4\cdot 11^4\cdot 13^2\cdot 17\cdot 19\cdot 23\cdot 29\cdot 31\cdot 37\cdot 41\cdot 43\cdot 661\cdot 59107\\
0
\end{array}$$

•

Let K be any subfield of $\mathbb{C}$. Given $a, b \in \mathbb{C}^*$, we write $a \sim_K b$ when their quotient is an algebraic number over K.

In [7, Sect. 6], it is proved that the constants k_P that normalize the local parameters at CM-points agree, up to the product by algebraic elements, with some specific constants π_d which only depend on the discriminant d of the field of complex multiplication. (When $D = 1$ and P is the cusp at infinity, this constant is the number π in $q(z) = e^{2\pi i z}$.)

Given a discriminant $d < 0$, we define

$$\pi_d = \prod_{m=1}^{|d|} \left\{ \Gamma\left(\frac{m}{|d|}\right)^{\left(\frac{d}{m}\right)} \right\}^{\frac{w}{2h}}.$$

Theorem 4.2 (Bayer-Travesa, [7, Thms. 6.3 and 6.6]) *The local constants k_P adapted to the function t_6 at the elliptic points and at the SCM points of X_6 are trascendental. Moreover,*

(i) $k(P_3, X_6, t_6) \sim \dfrac{\Gamma(\frac{1}{4})\Gamma(\frac{5}{12})}{\Gamma(\frac{3}{4})\Gamma(\frac{11}{12})} \sim \pi_{-4} = \dfrac{\Gamma(\frac{1}{4})^2}{\Gamma(\frac{3}{4})^2}.$

(ii) $k(P_4, X_6, t_6) \sim \dfrac{\Gamma(\frac{1}{3})^2\Gamma(\frac{7}{12})}{\Gamma(\frac{2}{3})^2\Gamma(\frac{11}{12})} \sim \pi_{-3} = \dfrac{\Gamma(\frac{1}{3})^3}{\Gamma(\frac{2}{3})^3}.$

(iii) $k(P_0, X_6, t_6) \sim \dfrac{\Gamma(\frac{7}{24})\Gamma(\frac{11}{24})}{\Gamma(\frac{19}{24})\Gamma(\frac{23}{24})} \sim \pi_{-6} = \left(\dfrac{\Gamma(\frac{1}{24})\Gamma(\frac{5}{24})\Gamma(\frac{7}{24})\Gamma(\frac{11}{24})}{\Gamma(\frac{13}{24})\Gamma(\frac{17}{24})\Gamma(\frac{19}{24})\Gamma(\frac{23}{24})}\right)^{1/2}.$

Proof The proof of this result is based on the use of the LLL-algorithm for recognizing the numbers (cf. [9, Ch. 2]), transcendence results due to G. Chudnovsky [8], and Chowla-Selberg's formula [33]. □

5 Fake elliptic curves

In this section we explain the method followed in Bayer-Guàrdia [5] in order to obtain equations for genus 2 curves whose Jacobian are fake elliptic curves with complex multiplication. As we shall see, all the entries defining a PEL-type are necessary to obtain models with algebraic coefficients.

Let $\mathcal{H}_2$ denote the Siegel upper half-space whose elements are the 2×2 symmetric matrices with positive definite imaginary part

$$\mathcal{H}_2 = \{Z \in M(2, \mathbb{C}) : Z = {}^tZ, \ \mathrm{Im}Z > 0\}.$$

We consider the Riemann theta function

$$\vartheta(z; Z) = \sum_{n \in \mathbb{Z}^2} \exp(\pi i\, {}^tnZn + 2\pi i\, {}^tnz) \quad (z \in \mathbb{C}^2, Z \in \mathcal{H}_2).$$

Given a pair (r, s) of column vectors, $r, s \in \{0, 1/2\}^2$, let $c = {}^t[r, s]$. The theta function with half-integral characteristic c is defined by

$$\vartheta[c](z; Z) = \exp\left(\pi i\, {}^trZr + 2\pi i\, {}^tr(z + s)\right) \vartheta(z + Zr + s; Z).$$

The function $\vartheta[c](z; Z)$ is even or odd with respect to z according to the parity of c:

$$\vartheta[c](z; Z) \text{ is } \left\{ \begin{array}{c} \text{even} \\ \text{odd} \end{array} \right. \iff 4\, {}^tr \cdot s \equiv \left\{ \begin{array}{c} 0 \\ 1 \end{array} \right. \pmod 2.$$

Now suppose that we are given a point $Z \in \mathcal{H}_2$ and consider the torus $J_Z = \mathbb{C}^2/\langle 1_2, Z\rangle$ attached to it. Then, the half-integral characteristics are in bijection with the 2-torsion points in J_Z by means of

$$c = {}^t[r, s] \longleftrightarrow w = Zr + s \in J_Z[2].$$

Accordingly, we define the parity of a 2-torsion point as the parity of its characteristic. Given a pair (c_1, c_2) of different odd characteristics and the corresponding pair (w_1, w_2) of odd 2-torsion points in J_Z, we define

$$P(c_1, c_2) = P(w_1, w_2) = \begin{bmatrix} \dfrac{\partial\vartheta[c_1]}{\partial z_1}(0; Z) & \dfrac{\partial\vartheta[c_1]}{\partial z_2}(0; Z) \\ \dfrac{\partial\vartheta[c_2]}{\partial z_1}(0; Z) & \dfrac{\partial\vartheta[c_2]}{\partial z_2}(0; Z) \end{bmatrix}$$

and

$$p(c_1, c_2) = p(w_1, w_2) = \det P(w_1, w_2).$$

If $[1_2|Z]$ is the normalized period matrix of a hyperelliptic curve C/k, the following theorem, whose proof is based on theorems due to Jacobi, Weil and Shimura, tells us how to compute an equation for this curve with coefficients in an algebraic closure $\overline{k}$.

Theorem 5.1 (Guàrdia, [15], [16]) *Let C/k be a genus 2 curve and let $Z \in \mathcal{H}_2$ define a normalized period matrix for C. Fix a pair of points (w_1, w_2), $w_i \in J_Z[2]^{\mathrm{odd}}$, $w_1 \neq w_2$.*

(i) *For any $w \in J_Z[2]^{\mathrm{odd}}$, the point in $\mathbb{P}^1(\mathbb{C})$ defined by*

$$\begin{bmatrix} \dfrac{\partial\vartheta}{\partial z_1}(w; Z) & \dfrac{\partial\vartheta}{\partial z_2}(w; Z) \end{bmatrix} P(w_1, w_2)^{-1} \begin{bmatrix} X_1 \\ X_2 \end{bmatrix} = 0$$

has projective coordinates $(p(w_1, w) : p(w_2, w))$.

(ii) *Let $J_Z[2]^{\mathrm{odd}} = \{w_i\}_{1 \le i \le 6}$. The ratios*

$$\ell_{12j} = \frac{p(w_2, w_j)}{p(w_1, w_j)}, \quad 1 \le j \le 6,$$

are equal to ∞ for $j = 1$ and are algebraic over k, otherwise.

(iii) *The curve C admits the hyperelliptic model*

$$Y^2 = X(X - \ell_{123})(X - \ell_{124})(X - \ell_{125})(X - \ell_{126}),$$

which is defined over a finite extension of k.

Note that the above theorem yields thirty models for the isomorphism class of the curve C, which correspond to the thirty possible choices of ordered pairs (w_1, w_2).

We now consider the Shimura curves X_D and their SCM points $\tau \in \mathcal{H}$. We will obtain equations for the genus 2 curves whose Jacobians are the CM fake elliptic curves attached to them, once the PEL-type is fixed.

The SCM fake elliptic curve $[A_\tau, \iota_\tau, \mathcal{L}_\tau, W_\tau]$ will be obtained as follows:

$$\Lambda_\tau = \Phi(\mathcal{O}_D) \begin{bmatrix} \tau \\ 1 \end{bmatrix}, \quad \mathbb{C}^2/\Lambda_\tau \simeq A_\tau(\mathbb{C}).$$

We define $T(\alpha, \beta) = \mathrm{Tr}(\mu\alpha\overline{\beta})$, where $\mu \in \mathcal{O}_D$, $\mu^2 = -D$ and $\alpha, \beta \in H$. We choose in $\Phi(\mathcal{O}_D)$ a symplectic basis with respect to the bilinear form $E(\Phi(\alpha), \Phi(\beta)) = \delta^{-1}T(\alpha, \beta)$, where $\delta = (\mathrm{Pf}(T))^{1/2}$ denotes the positive square root of the Pfaffian of T. The principal polarization $\mathcal{L}_\tau$ will be given by the Riemann form

$$\begin{array}{rccc} E_\tau : & \Lambda_\tau \times \Lambda_\tau & \longrightarrow & \mathbb{C} \\ & (\Phi(\alpha)\tau, \Phi(\beta)\tau) & \longrightarrow & E(\Phi(\alpha), \Phi(\beta)). \end{array}$$

The level structure $V = (v_1, v_2)$ will be associated with an ordered pair (v_1, v_2), $v_i \in H/\mathcal{O}$, such that $w_i = \Phi(v_i)\tau \in \mathbb{C}^2/\Lambda_\tau[2]^{\text{odd}}$; then $W = (w_1, w_2)$.

The next example is computed for $D = 6$ and illustrates theorem 5.1. In it, we have fixed one of the thirty possible level structures.

Examples 5.2 Let $H(3, -1) = \langle 1, i, j, k \rangle,\ D = 6$.
We consider the PEL-type given by the following data:

$$\Phi(i) = \begin{bmatrix} \sqrt{3} & 0 \\ 0 & -\sqrt{3} \end{bmatrix},\ \Phi(j) = \begin{bmatrix} 0 & 1 \\ -1 & 0 \end{bmatrix},$$

$$\mu = i + 3j, \quad \delta = 6,$$

$$\mathcal{O}_6 = \mathbb{Z}\left[1, i, j, \frac{1+i+j+k}{2}\right],$$

$$V = ((\tfrac{1}{4}, \tfrac{1}{4}, -\tfrac{3}{4}, \tfrac{1}{4}), (\tfrac{1}{4}, \tfrac{1}{4}, -\tfrac{1}{4}, \tfrac{1}{4})),$$

$$\Phi(\mathcal{O}_6) = \mathbb{Z}\left[1_2, \begin{bmatrix} \frac{-1+\sqrt{3}}{2} & \frac{-1+\sqrt{3}}{2} \\ \frac{1+\sqrt{3}}{2} & \frac{-1-\sqrt{3}}{2} \end{bmatrix}, \begin{bmatrix} 0 & 1 \\ -1 & 0 \end{bmatrix}, \begin{bmatrix} 1-\sqrt{3} & 0 \\ 0 & 1+\sqrt{3} \end{bmatrix}\right].$$

We have two SCM points, which have CM by the maximal order $R(-6, 1)$:

$$\tau_1 = \frac{i(\sqrt{6} - \sqrt{2})}{2}, \quad \tau_2 = \frac{\sqrt{3} + i\sqrt{6}}{3}.$$

They define the points in the Siegel upper half-space

$$Z_1 = \begin{bmatrix} \frac{-1+3i\sqrt{6}}{5} & \frac{-2+i\sqrt{6}}{5} \\ \frac{-2+i\sqrt{6}}{5} & \frac{6+2i\sqrt{6}}{5} \end{bmatrix},\ Z_2 = \begin{bmatrix} \frac{-1+2i\sqrt{6}}{5} & \frac{-2-i\sqrt{6}}{5} \\ \frac{-2-i\sqrt{6}}{5} & \frac{6+3i\sqrt{6}}{5} \end{bmatrix},$$

and yield the genus 2 curves

$$Y^2 = X^5 - 2i\sqrt{2}X^4 - \frac{11}{3}X^3 + 2i\sqrt{2}X^2 + X,$$

$$Y^2 = X^5 + 2\sqrt[4]{-5 + 2\sqrt{6}}X^4 - \frac{10}{3}i\sqrt{2}X^3 + 2i\sqrt[4]{-5 - 2\sqrt{6}}X^2 + X.$$

The cases $D = 10$ and $D = 15$ can be found in [5].

6 Suggestions for further research

The results presented in the paper are only a small part of the computations that could be carried out in the context of Shimura curves. In general, these should deal with curves $X(D, N)$, for any given discriminant and level of reasonable size. Nevertheless, the task in this direction is not easy and the difficulties increase considerably when the genus becomes higher, mainly due to the presence of accessory parameters in the differential equations attached to the fundamental domains.

For a given curve $X(D, N)$, and in the complex setting, the computations should aim to provide: algebraic equations and uniformizing automorphic functions for the canonical models, as well as automorphic eigenforms of low weight. The procedures should compute the first coefficients of the functions, the first eigenvalues of the Hecke operators, and the first coefficients of the eigenforms in the neighbourhood of the CM points. The information so obtained could then be used to improve results in explicit class field theory. Under a theoretical point of view, it might not be uninteresting to know some diophantine interpretation, if any, of the coefficients of the eigenforms.

In connection with the Jacobians $J(D, N)$ of the above Shimura curves, a lot of things deserve to be better understood, specially some aspects oriented to make more explicit and comprehensible the isogenies appearing in the Jacquet-Langlands-Shimizu correspondence.

Finally, I would like to mention the interest of making modularity (or automorphy) more accessible and effective. Namely, once we know that some Galois representations or other objects are modular (or automorphic), we should be able of numerically visualizing the forms or the functions giving rise to the phenomenon, in the spirit of the very first examples that were computed by Serre and Mestre.

References

[1] Alsina, M.; Bayer, P.: *Quaternion orders, quadratic forms and Shimura curves.* CRM Monograph Series 22. AMS, 2004.

[2] Baba, S.; Granath, H.: Primes of superspecial reduction for QM abelian surfaces. *Bull. Lond. Math. Soc.* **40** (2008), no. 2, 311–318.

[3] Baba, S.; Granath, H.: Genus 2 curves with quaternionic multiplication. *Canad. J. Math.* **60** (2008), no. 4, 734–757.

[4] Bayer, P.: Uniformization of certain Shimura curves. In *Differential Galois Theory*, T. Crespo and Z. Hajto (eds.). Banach Center Publications 58 (2002), 13–26.

[5] Bayer, P.; Guàrdia, J.: On equations defining fake elliptic curves. *J. Théor. Nombres Bordeaux* 17 (2005), no. 1, 57–67.

[6] Bayer, P.; Travesa, A.: Uniformizing functions for certain Shimura curves, in the case $D = 6$. *Acta Arith.* 126 (2007), no. 4, 315–339.

[7] Bayer, P.; Travesa, A.: On local constants associated to arithmetical automorphic functions. *Pure Appl. Math. Q.* 4 (2008), no. 4, part 1, 1107–1132.

[8] Chudnovsky, G. V.: Contributions to the theory of transcendental numbers. *Mathematical Surveys and Monographs*, 19. American Mathematical Society, Providence, RI, 1984. xi+450 pp.

[9] Cohen, H.: *A course in computational algebraic number theory.* Graduate Texts in Mathematics, 138. Springer-Verlag, Berlin, 1993. xii+534 pp. ISBN: 3-540-55640-0.

[10] Deligne, P.: Travaux de Shimura. *Séminaire Bourbaki*, 23ème année (1970/71), Exp. No. 389, pp. 123–165. Lecture Notes in Math., Vol. 244, Springer, Berlin, 1971.

[11] Deligne, P.: Variétés de Shimura: interprétation modulaire, et techniques de construction de modèles canoniques. Automorphic forms, representations and L-functions (Proc. Sympos. Pure Math., Oregon State Univ., Corvallis, Ore., 1977), Part 2, pp. 247–289, Proc. Sympos. Pure Math., XXXIII, Amer. Math. Soc., Providence, R.I., 1979.

[12] Elkies, N. D.: Shimura curves computations. Lecture Notes in Computer Sciences 1423, Springer 1998, 1–49.

[13] Elkies, N. D.: Shimura curves for level-3 subgroups of the $(2, 3, 7)$ triangle group, and some other examples. *Algorithmic number theory,* 302–316. Lecture Notes in Comput. Sci., 4076. Springer, 2006.

[14] Elkies, N. D.: Shimura curve computations via $K3$ surfaces of Néron-Severi rank at least 19. *Algorithmic number theory,* 196–211. Lecture Notes in Comput. Sci., 5011. Springer, 2008.

[15] Guàrdia, J.: Jacobian nullwerte and algebraic equations. *Journal of Algebra* 253 (2002), 112-132.

[16] Guàrdia, J.: Jacobi Thetanullwerte, periods of elliptic curves and minimal equations. *Math. Res. Lett.* 11 (2004), no. 1, 115–123.

[17] Hashimoto, K.; Murabayashi, N.: Shimura curves as intersections of Humbert surfaces and defining equations of QM-curves of genus two, *Tôhoku Math. J.* 47 (1995), 271–296.

[18] Ihara, Y.: Schwarzian equations. *J. Fac. Sci. Univ. Tokyo Sect. IA Math.* 21 (1974), 97–118.

[19] Jetchev, D.; Lauter, K.; Stein, W.: Explicit Heegner points: Kolyvagin's conjecture and non-trivial elements in the Shafarevich-Tate group. *J. Number Theory* 129(2009), no. 2, 284–302.

[20] Johansson, S.: On fundamental domains of arithmetic Fuchsian groups. *Math. Comp.* 69 (2000), no. 229, 339–349.

[21] Jordan, B. W.: *On the Diophantine arithmetic of Shimura curves.* Harvard PhD. Thesis, 1981.

[22] Kohel, D.; Verrill, H.: Fundamental domains for Shimura curves. Les XXIIèmes Journées Arithmetiques (Lille, 2001). *J. Théor. Nombres Bordeaux* 15 (2003), no. 1, 205–222.

[23] Kurihara, A.: On some examples of equations definig Shimura curves and the Mumford uniformization. *J. Fac. Sci. Univ. Tokyo* 25 (1979), 277–301.

[24] Michon, J-F.: Courbes de Shimura hyperelliptiques. *Bull. Soc. Math. France* 109 (1981), no. 2, 217–225.

[25] Michon, J-F.: Courbes de Shimura. *Seminar on number theory*, Paris 1982–83 (Paris, 1982/1983), 185–197. Progr. Math., 51, Birkhuser Boston, Boston, MA, 1984.

[26] Milne, J. S.: Books, Course Notes, Expository Notes, etc. in www.jmilne.org/math/.

[27] Molina, S.: Specialization of Heegner points and applications. Thesis. Universitat Politècnica de Catalunya, 2010.

[28] Nehari, Z.: *Conformal mapping.* Dover Publications, 1952.

[29] Rotger, V.: *Abelian varieties with quaternionic multiplication and their moduli.* Thesis. Universitat de Barcelona, 2002.

[30] Rotger, V.: Quaternions, polarizations and class numbers. *J. reine u. angew. Math.* 2003, 177–197.

[31] Rotger, V.: Modular Shimura varieties and forgetful maps. *Trans. Amer. Math. Soc.* 356, 1535–1550.

[32] Sansone, G.; Gerretsen, J.: *Lectures on the Theory of Functions of a Complex Variable. II Geometric Theory.* Wolters–Noordhoff Publishing Groningen, 1969.

[33] Selberg, A.; Chowla, S.: On Epstein's zeta-function. *J. Reine Angew. Math.* 227 (1967), 86–110.

[34] Schwarz, H. A.: Ueber diejenigen Fälle, in welchen die Gaussische hypergeometrische Reihe eine *algebraische* Function ihres vierten Elementes darstellt. *J. für reine u. angew. Math.* 75 (1873), 292–335.

[35] Shimura, G.: Construction of class fields and zeta functions of algebraic curves. *Annals of Math.* 85 (1967), 58-159. Collected papers, Vol. II, 1967-1977, 67b. Springer, 2002. xiv+831 pp.

[36] Shimura, G.: On canonical models of arithmetic quotients of bounded symmetric domains. *Annals of Math.* 91 (1970), 144-222. Collected papers, Vol. II, 1967-1977, 70a. Springer, 2002. xiv+831 pp.

[37] Shimura, G.: *Introduction to the Arithmetic Theory of Automorphic Functions.* Iwanami Shoten and Princeton University Press, 1971.

[38] Sijsling, J. R.: *Equations for arithmetic pointed tori.* Thesis. Universiteit Utrecht, 2010.

[39] Voight, J.: Shimura curves of genus at most two. *Math. Comp.* 78 (2009), no. 266, 1155–1172.

[40] Vignéras, Marie-France: *Arithmétique des algbres de quaternions.* Lecture Notes in Mathematics, 800. Springer, Berlin, 1980. vii+169 pp. ISBN: 3-540-09983-2.

[41] Yoshida, M.: *Fuchsian Differential Equations.* Max-Planck-Institut für Mathematik. Vieweg & Sohn, 1987.

Fields Institute Communications
Volume **60**, 2011

Igusa Class Polynomials, Embeddings of Quartic CM Fields, and Arithmetic Intersection Theory

Helen Grundman
Department of Mathematics, Bryn Mawr College
101 North Merion Avenue, Bryn Mawr, PA 19010-2899 USA
grundman@brynmawr.edu

Jennifer Johnson-Leung
Department of Mathematics, University of Idaho
303 Brink Hall, Moscow, ID 83844 USA
jenfns@uidaho.edu

Kristin Lauter
Microsoft Research
One Microsoft Way, Redmond, WA 98052 USA
klauter@microsoft.com

Adriana Salerno
Department of Mathematics, Bates College
3 Andrews Rd., Lewiston, ME 04240-6028 USA
asalerno@bates.edu

Bianca Viray
Mathematics Department, Box 1917
Brown University
Providence, RI 02912 USA
bviray@math.brown.edu

Erika Wittenborn
Department of Mathematics, University of Colorado at Boulder
Boulder, CO 80309-0395 USA
erika.frugoni@colorado.edu

2010 *Mathematics Subject Classification.* Primary 11G15; Secondary 14G40, 11G20, 14K15, 14K22.

The fifth author was partially support by NSF Grant DMS-0841321, Microsoft Research, and a Ford Foundation Dissertation Fellowship.

Abstract. Bruinier and Yang conjectured a formula for an intersection number on the arithmetic Hilbert modular surface, $\mathcal{CM}(K).T_m$, where $\mathcal{CM}(K)$ is the zero-cycle of points corresponding to abelian surfaces with CM by a primitive quartic CM field K, and T_m is the Hirzebruch-Zagier divisors parameterizing products of elliptic curves with an m-isogeny between them. In this paper, we examine fields not covered by Yang's proof of the conjecture. We give numerical evidence to support the conjecture and point to some interesting anomalies. We compare the conjecture to both the denominators of Igusa class polynomials and the number of solutions to the embedding problem stated by Goren and Lauter.

1 Introduction

Hilbert class polynomials are polynomials whose zeros are j-invariants of elliptic curves with complex multiplication (CM). Similarly, Igusa class polynomials are polynomials whose zeros are Igusa invariants of genus 2 curves whose Jacobians have CM by a quartic CM field K. Igusa class polynomials can be hard to compute, mostly because recovering the coefficients from approximations requires a bound on the denominators. A precise description of the denominators would greatly improve the estimated running time of current algorithms to compute Igusa class polynomials [Spa94, vW99, Wen03, EL, GHK$^+$06].

The first attempt to understand the primes appearing in the factorization of the denominators was given in [Lau], where it was conjectured that the primes are bounded by the absolute value of the discriminant d_K of the field K, and that the primes divide $d_K - x^2$, for some integer x. The divisibility condition actually follows from work of Goren [Gor97] along with the observation that primes in the denominators are primes of superspecial reduction for the abelian surfaces with CM by K. In [GL07], Goren and Lauter proved a bound on the primes which appear in the factorization of the denominators. In recent work [GL], they have given a bound on the denominators (i.e. bounding the powers of the primes which appear); however, this bound is not sharp.

Igusa invariants can be expressed as rational functions in Eisenstein series and cusp forms. Using this description it can be shown that the primes dividing the denominators are (up to cancellation) the primes p such that there exists a genus 2 curve C with CM by K and a prime ideal $\mathfrak{p}|p$ with

$$\mathrm{Jac}(C) \cong E \times E' \pmod{\mathfrak{p}}.$$

Goren and Lauter obtained the bound on such primes by showing that these primes are exactly the primes for which there is a solution to the *embedding problem*: the problem of giving an embedding $\mathcal{O}_K \hookrightarrow \mathrm{End}(E \times E') \subset M_2(\mathbb{B}_{p,\infty})$, where E, E' are supersingular elliptic curves over $\overline{\mathbb{F}}_p$ and $\mathbb{B}_{p,\infty}$ is the quaternion algebra over $\mathbb{Q}$ ramified at only p and ∞. Such decomposable Jacobians can be viewed as points on the Siegel moduli space where the arithmetic intersection of the CM(K)-cycle with the divisor of the cusp form χ_{10} is nonzero at p.

This arithmetic intersection number on the Siegel moduli space can be related to a sum of intersection numbers of Hirzebruch-Zagier divisors and CM(K)-cycles on the Hilbert modular surface associated to the real quadratic subfield of K. In [BY06], Bruinier and Yang conjectured a formula for the intersection number

$\mathcal{CM}(K).\mathcal{T}_m$, where $\mathcal{CM}(K)$ parameterizes abelian surfaces with complex multiplication by a primitive quartic CM field K, and $\mathcal{T}_m$ is a Hirzebruch-Zagier divisor parameterizing products of elliptic curves with an m-isogeny between them when m is split in the real quadratic subfield of K. The results of Bruinier and Yang can be viewed as a generalization of the celebrated formula of Gross and Zagier [GZ85]. Bruinier and Yang conjectured their formula for these arithmetic intersection numbers under the assumption that the real quadratic subfield of K has prime discriminant. Their formula relates the intersection numbers to the number of ideals of a certain norm in the reflex field of K. In [Yana, Yanb], Yang gives a detailed treatment of the embedding problem and uses it, along with other techniques, to prove the conjectured intersection formula under certain conditions on the CM field K.

For a given CM field K, the conjectural formula of Bruinier and Yang is straightforward to compute. Thus, if the Bruinier-Yang formula held for all K, the formula would give a way to compute the denominators precisely (up to cancellation).

In this paper, we compare the number of solutions to the embedding problem, the denominators in the Igusa class polynomials, and the conjectural formula of Bruinier and Yang for the 13 CM fields found in [vW99]. Only 6 of these fields are covered by Yang's proof of the conjecture. We give numerical evidence to support the conjecture and point to some interesting anomalies and some possible explanations for these discrepancies. Such a project was suggested in the Introduction and Section 9 of [BY06], and we give here the first numerical evidence comparing the conjecture with denominators of Igusa class polynomials and counting solutions to the embedding problem. This comparison required developing an algorithm to count the number of solutions to the embedding problem for a given primitive quartic CM field K and prime number p. The algorithm is given in Section 6.1 and the `Magma` code for computing the values is given in the Appendix.

The paper is organized as follows. Section 2 contains background on abelian surfaces with complex multiplication by a primitive quartic CM field K. Section 3 contains a description of Igusa invariants and Igusa class polynomials. It also gives a brief introduction to the algorithms for computing Igusa class polynomials and their connection with generating genus 2 hyperelliptic curves for cryptography. Section 4 introduces arithmetic intersection theory on the arithmetic Hilbert moduli space and explains how it relates to the denominators of Igusa class polynomials. In this section, we also describe the Bruinier-Yang conjecture and the cases in which it is known. Section 5 explains how the denominators of Igusa class polynomials relate to the embedding problem. Section 6 gives an algorithm to count the number of solutions to the embedding problem. Section 7 presents our numerical data and analysis of the findings.

This paper is concerned with computing these intersection numbers from three different points of view: denominators of Igusa class polynomials, arithmetic intersection numbers via the conjectured Bruinier-Yang formula, and embeddings of a CM number field into a matrix algebra over a quaternion algebra. Thus the paper necessarily requires the introduction of a number of objects from different points of view, and the background is spread over many different references and papers. Therefore, for the readers' convenience, we have tried to include here as much background as possible. Most sections are self-contained, but Section 4 leaves some of the background material to the references.

2 Abelian surfaces with CM by $\mathcal{O}_K$

A *CM field* is a totally imaginary quadratic extension of a totally real number field of finite degree. For a CM field K, let K^+ denote the totally real index 2 subfield of K. If $\{\phi_1, \ldots, \phi_{2n}\}$ are the $2n$ complex embeddings of K, then a *CM-type* Φ for K is a choice of n distinct complex embeddings, no two of which are complex conjugates of one another. When $K^+ = \mathbb{Q}$, CM fields are imaginary quadratic fields.

An *abelian variety with CM by K* is a pair (A, ι) where A is an abelian variety and ι is an isomorphism of K into $\mathrm{End}^0(A) := \mathrm{End}(A) \otimes \mathbb{Q}$. We say (A, ι) has type (K, Φ) if A has CM by K and the analytic representation of $\mathrm{End}^0(A)$ is equivalent to the direct sum of the n isomorphisms $\phi_i \in \Phi$. One can define a CM-type to be *primitive* if all of the abelian varieties over $\mathbb{C}$ of that type are *simple*, i.e. have no abelian subvarieties other than $\{0\}$ and itself. A CM-type (K, Φ) is primitive if and only if $K = K^+(\xi)$, where ξ is an element of K such that ξ^2 is totally negative and $\mathrm{Im}(\xi^{\phi_i}) > 0$ for $i = 1, \ldots, n$ [Shi98, p. 61]. For quartic CM fields, the only case where non-primitive CM-types arise is the case when K is biquadratic [Shi98, p. 64]. We will restrict ourselves to the primitive (i.e. non-biquadratic) case in this article.

Let (K, Φ) denote a quartic CM field K with CM-type Φ. Define the reflex field of (K, Φ) to be

$$K_\Phi = \mathbb{Q}\left(\sum_{\phi \in \Phi} x^\phi \mid x \in K\right).$$

If K is Galois cyclic, then we have $K = K_\Phi$ ([Shi98, §8.4]).

Shimura gives a construction of a principally polarized complex abelian variety of a given CM type in [Shi98, §6.1, Theorem 2]: to any ideal $I \subseteq \mathcal{O}_K$, associate the abelian variety $A_I = \mathbb{C}^2/\Phi(I)$ of dimension 2. It has endomorphism ring $\mathcal{O}_K$ and we say it has *CM by $\mathcal{O}_K$*. All abelian varieties with CM by $\mathcal{O}_K$ are obtained via this construction. A *principal polarization* of A_I is an isomorphism between A_I and its *dual* $\widehat{A_I} = \mathbb{C}^2/\Phi(\overline{I}^{-1}\mathcal{D}_K^{-1})$, where

$$\mathcal{D}_K^{-1} = \{x \in K \mid \mathrm{Tr}_{K/\mathbb{Q}}(x\mathcal{O}_K) \subseteq \mathbb{Z}\}$$

is the inverse different. If π is a totally imaginary element of K which satisfies $\Phi(\pi) \in (i\mathbb{R}_{>0})^2$ and $\pi I\overline{I} = \mathcal{D}_K^{-1}$, then the map $A_I \to \widehat{A}_I$ given by

$$(z_1, z_2) \mapsto (\phi_1(\pi)z_1, \phi_2(\pi)z_2)$$

is an isomorphism ([Shi98, p. 102–104]).

Let K be a fixed primitive quartic CM-field and fix a CM-type $\Phi : K \to \mathbb{C}^2$. Let $A/\mathbb{C}$ be a principally polarized abelian surface of CM-type (K, Φ) with endomorphism ring isomorphic to $\mathcal{O}_K$. We define a group $\mathfrak{C}(K) = \{(\mathfrak{a}, \alpha)\}/\sim$ where $\mathfrak{a}$ is a fractional ideal of $\mathcal{O}_K$ such that $\mathfrak{a}\bar{\mathfrak{a}} = (\alpha)$, with $\alpha \in K^+$ totally positive. Two pairs $(\mathfrak{a}, \alpha)$ and $(\mathfrak{b}, \beta)$ are equivalent if and only if there exists a unit $u \in K^*$ with $\mathfrak{b} = u\mathfrak{a}$ and $\beta = u\bar{u}\alpha$. The multiplication is defined componentwise, and $(\mathcal{O}_K, 1)$ is the identity element of $\mathfrak{C}(K)$.

The group $\mathfrak{C}(K)$ naturally acts on the finite set $S(K, \Phi)$ of isomorphism classes of principally polarized abelian surfaces that have CM by $\mathcal{O}_K$ of a given type Φ. Every principally polarized abelian variety with CM by K has an analytic representation A_I determined by an ideal I and a Φ-positive element $\pi \in K$ giving the principal polarization. We now put

$$(\mathfrak{a}, \alpha) \cdot (I, \pi) = (\mathfrak{a}I, \alpha\pi)$$

for $(\mathfrak{a}, \alpha) \in \mathfrak{C}(K)$. By [Shi98, §14.6], for a given Φ the action of $\mathfrak{C}(K)$ on $S(K, \Phi)$ is transitive and free. In particular, we have $|\mathfrak{C}(K)| = |S(K, \Phi)|$, and the sets are often identified.

Abelian surfaces A of type (K, Φ) defined by an ideal $\mathfrak{a}$ have potentially zero, one, or two principal polarizations. There are zero principal polarizations only if there is no totally positive element $\alpha \in \mathcal{O}_{K^+}$ with $\mathfrak{a}\bar{\mathfrak{a}} = (\alpha)$. Otherwise there are $[U : U_1]$ principal polarizations on A, where U is the group of all totally positive units of K^+, and U_1 is the subgroup of norms of units from K.

For quartic primitive CM fields K, there are two different possible CM types up to complex conjugation. Let ρ be complex conjugation, and let ϕ be a non-trivial embedding of K into $\mathbb{C}$ other than complex conjugation. A priori, the 4 possible CM types are: $\Phi = (1, \phi)$, $\overline{\Phi} = (\rho, \rho\phi)$, $\Phi' = (1, \overline{\phi})$, $\overline{\Phi}' = (\rho, \rho\overline{\phi})$. We denote by K_Φ (resp. $K_{\Phi'}$, etc.) the set of principally polarized abelian varieties of CM type Φ (resp. Φ', etc.).

The description above identifies isomorphism classes of principally polarized abelian varieties of a given type Φ with the set $\{(\mathfrak{a}, \alpha)\}/\sim$, where $\mathfrak{a}$ is an ideal of $\mathcal{O}_K$ such that $\mathfrak{a}\bar{\mathfrak{a}} = (\alpha)$, α a totally positive element of $\mathcal{O}_{K_0}$. The associated abelian variety is $A = \mathbb{C}^2/\Phi(\mathfrak{a})$ and the embedding of $\mathcal{O}_K$ into $\mathrm{End}(A)$ sends $\beta \in \mathcal{O}_K$ to the diagonal matrix with $(\beta^{\phi_1}, \beta^{\phi_2})$ on the diagonal acting by multiplication on $\mathbb{C}^2$.

We can show that $K_\Phi = K_{\overline{\Phi}}$ by sending $(\mathfrak{a}, \alpha)$ to $(\bar{\mathfrak{a}}, \alpha)$. For the same reason, $K_{\Phi'} = K_{\overline{\Phi}'}$. If K is cyclic, then ϕ^2 is complex conjugation and sending $(\mathfrak{a}, \alpha)$ to $(\mathfrak{a}^\phi, \alpha^\phi)$ shows that $K_\Phi = K_{\Phi'}$.

3 Genus 2 curves with CM

In cryptography, it is useful to be able to construct genus 2 curves over a finite field $\mathbb{F}_q$ with a given number of points on its Jacobian, since they can be used to implement discrete-log cryptosystems. Many algorithms to generate such curves have been developed using complex multiplication, all of which require computing certain invariants first studied by Igusa [Igu60]. The closely related Igusa class polynomials have rational coefficients. The known algorithms for computing Igusa class polynomials all require a good bound on the primes appearing in the denominators of these invariants. One of the main goals of this project is to better understand exactly which primes appear in those denominators. In this section, we will review the definitions of the Igusa invariants and explain several algorithms for computing Igusa class polynomials.

3.1 Generating genus 2 curves for cryptography. To construct a genus 2 curve C over a finite field $\mathbb{F}_q$ with a given number of points N on its Jacobian, one method is to compute a related quartic CM field K such that a curve C with $\#J(C)(\mathbb{F}_q) = N$ has CM by K.

For an ordinary genus 2 curve C over a prime field $\mathbb{F}_q$, let $N_1 = \#C(\mathbb{F}_q)$ and $N_2 = \#C(\mathbb{F}_{q^2})$. Then

$$\#J(C)(\mathbb{F}_q) = (N_1^2 + N_2)/2 - q. \tag{3.1}$$

To find a curve C over $\mathbb{F}_q$ such that $\#J(C) = N$, first find N_1 and N_2 in the Hasse-Weil intervals for $\mathbb{F}_q$ and $\mathbb{F}_{q^2}$ satisfying relation (3.1). Next, set

$$s_1 := q + 1 - N_1$$

and

$$s_2 := \frac{1}{2}\left(s_1^2 + N_2 - 1 - q^2\right).$$

Then the quartic polynomial $f(t) = t^4 - s_1t^3 + s_2t^2 - qs_1t + q^2$ is the Weil polynomial of a genus 2 curve as long as the exceptional cases listed in [HNR09, Theorem 1.2] are avoided. Note that if s_2 is prime to q, then the Jacobian is ordinary [How95, p.2366]. Under those conditions and assuming that $f(t)$ is an irreducible polynomial in $\mathbb{Q}[t]$, the Jacobian of the curve has endomorphism ring equal to an order in the quartic CM field $K = \mathbb{Q}[t]/(f(t))$. Also, if K can be written in the form $K = \mathbb{Q}(i\sqrt{a + b\sqrt{d}})$, with $a, b, d \in \mathbb{Z}$ and d and (a, b) square-free, then K is a primitive CM field if and only if $a^2 - b^2d$ is not a square [KW89, p.135].

To find a curve with CM by K, one method is to compute its Igusa invariants by evaluating certain Siegel modular functions at CM points associated to K in the Siegel moduli space. Igusa invariants will be explained in the next section.

3.2 Igusa invariants and Igusa class polynomials. Recall that the j-invariant of an elliptic curve can be calculated in two ways. One can compute it as a value of the modular j-function on a lattice defining the elliptic curve as a complex torus over $\mathbb{C}$, or directly from the coefficients of the equation defining the elliptic curve. Similarly, the three Igusa invariants of a genus 2 curve can be calculated in two different ways.

Let $y^2 = f(x)$ be a hyperelliptic curve, where $f(x)$ is a sextic with roots $\alpha_1, \ldots, \alpha_6$ and leading coefficient a_6. Igusa defined invariants A, B, C, D as certain symmetric functions of the roots [Igu60]. For example, $D = a_6^{10}\prod_{i<j}(\alpha_i - \alpha_j)^2$ is the discriminant. The ring of rational functions of the coarse moduli space for hyperelliptic curves of genus 2 is generated by the absolute Igusa invariants, which can be defined as:

$$i_1 := \frac{A^5}{D}, \quad i_2 := \frac{A^3B}{D}, \quad i_3 := \frac{A^2C}{D}. \tag{3.2}$$

This choice of generators is not unique. These invariants are a generalization of the j-invariant for elliptic curves in the sense that, if i_1 is non-zero and the characteristic is not 2 or 3, these invariants agree for two smooth genus 2 curves if and only if the two curves are isomorphic over an algebraically closed field. Note that if $f(x)$ defines a smooth genus 2 curve, then i_1, i_2, i_3 are well defined.

The invariants can also be defined in terms of modular functions on the Siegel moduli space. In [Igu62, p. 195], Igusa defined normalized Siegel modular cusp forms χ_{10} and χ_{12} of weights 10 and 12 in terms of the Siegel-Eisenstein series E_w for $w = 4, 6, 10, 12$. There he used the term *normalized* to mean that the leading coefficient of the Fourier expansion is 1. For our purposes, we will want to work with modular forms that have integral Fourier coefficients which are relatively prime. By working with $4\chi_{10}$ and $12\chi_{12}$, we obtain modular forms normalized in this second sense.

In [GL07, Section 5.2], it was noted that the absolute Igusa invariants defined above, i_1, i_2, i_3, can also be expressed in terms of these Eisenstein series and cusp forms as follows:

$$i_1 = 2 \cdot 3^5 \frac{\chi_{12}^5}{\chi_{10}^6}, \quad i_2 = 2^{-3} \cdot 3^3 \frac{E_4 \chi_{12}^3}{\chi_{10}^4}, \quad i_3 = 2^{-5} \cdot 3 \left(\frac{E_6 \chi_{12}^2}{\chi_{10}^3} + 2^2 \cdot 3 \frac{E_4 \chi_{12}^3}{\chi_{10}^4} \right),$$

(where the parentheses in the expression for i_3 were mistakenly left out in [GL07]). It was also remarked there that it follows from the formulas given by Igusa in [Igu67, p. 848] that these i_1, i_2, i_3 coincide with the invariants defined in terms of theta functions and used for computation by [vW99, Wen03, Lau]. However, it is worth noting that in order to work with invariants which are quotients of normalized modular forms in the second sense, one should actually work with $2^{-3}i_1$, $2i_2$, $2^3 i_3$ as Spallek did [Spa94].

Igusa class polynomials are the genus 2 analogue of Hilbert class polynomials. Just as we need three Igusa invariants to define an isomorphism class instead of one j-invariant, we have a triple of Igusa class polynomials for a CM field K instead of one Hilbert class polynomial. This triple of polynomials is defined as follows.

$$H_\ell(X) := \prod_\tau (X - i_\ell(\tau)), \quad \ell = 1, 2, 3, \tag{3.3}$$

where the i_ℓ are the absolute Igusa invariants defined above, and the product is taken over all $\tau \in \mathrm{Sp}(4, \mathbb{Z}) \backslash \mathfrak{h}_2$ such that the associated principally polarized abelian variety has CM by $\mathcal{O}_K$. Igusa class polynomials have rational coefficients [Spa94, Satz 5.8] (as opposed to integral coefficients as in the case of Hilbert class polynomials). In the next section, we describe how this plays a key role in the difficulty of computing these polynomials.

3.3 Algorithms for computing Igusa class polynomials. The complex analytic method for computing genus 2 curves with a given number of points on the Jacobian is analogous to the Atkin-Morain algorithm for elliptic curves [AM93]. Recall that the Atkin-Morain algorithm computes the Hilbert class polynomial of an imaginary quadratic field K by evaluating the modular j-invariants of all elliptic curves with CM by K to high precision. The complex analytic method [Spa94, vW99, Wen03] for generating genus 2 curves takes as input a primitive quartic CM field and evaluates the Igusa invariants to high precision to form the class polynomials with coefficients in $\mathbb{Q}$. Computing the Igusa invariants to high precision is difficult to do efficiently. To give an algorithm requires a bound on the denominators of the class polynomials, and no such bound was known at the time of those papers.

Given a triple of roots of these polynomials, Mestre gave an algorithm to recover the equation of a genus 2 curve from its invariants [Mes91]. So to find a genus 2 curve, first find roots of the polynomials modulo a prime number p and then use Mestre's algorithm to construct a genus 2 curve over $\mathbb{F}_p$ whose Jacobian has CM by K.

In [EL], Lauter and Eisenträger present a *CRT algorithm*, an alternative to the complex analytic algorithm to compute the Igusa class polynomials. The CRT algorithm takes as input a quartic CM field K *and* a constant c such that all denominators divide c. The algorithm first computes the Igusa class polynomials modulo small primes ℓ by finding all curves with the right zeta function and the right endomorphism ring modulo ℓ. It then forms the three polynomials whose roots are the invariants of those curves. Finally, it uses the Chinese Remainder Theorem

(CRT) and the constant c to obtain the Igusa class polynomials with coefficients in $\mathbb{Q}$.

There is another alternative approach to computing Igusa class polynomials using p-adic arithmetic ([GHK$^+$06]). The method is similar to the complex analytic algorithm, except the values are calculated to a high p-adic precision, instead of a complex precision. This algorithm also requires a bound on the denominators.

Notice that all three algorithms require a bound on the denominators, and their running time is directly affected by the sharpness of the bound. In the complex analytic and p-adic algorithms, it is required to bound the amount of precision needed in the approximation of the j_i-values. In the CRT algorithm, a common multiple of the denominators is needed. Thus, an exact formula for the denominators would be very useful.

4 Arithmetic intersection theory and denominators of Igusa class polynomials

In this section we describe how the denominators of Igusa class polynomials are related to arithmetic intersection numbers on the Siegel moduli space of principally polarized abelian surfaces and explain how this relates to intersection numbers on the Hilbert modular surface and the conjectural formula of Bruinier and Yang for the intersection number $\mathcal{CM}(K).\mathcal{T}_m$.

It is a classical fact that the zero locus of χ_{10} on the coarse moduli space of abelian surfaces consists of exactly those abelian surfaces that decompose as a product of elliptic curves with the product polarization. The arithmetic analogue of this statement was proved in [GL07, Cor 5.1.2], that if a prime $\mathfrak{p}$ divides the denominator of $(f/(4\chi_{10})^k)(\tau)$, for τ a CM point corresponding to a smooth curve C and f a Siegel modular form of weight $10k$ with integral Fourier coefficients with gcd 1, then C has bad reduction modulo $\mathfrak{p}$.

The absolute Igusa invariants are defined above as rational functions of the Siegel modular forms χ_{10}, χ_{12}, ψ_4, and ψ_6, with only powers of χ_{10} in the denominators. Although this definition of the invariants is given in terms of genus 2 curves over $\mathbb{C}$, the Igusa invariants can also be defined for genus 2 curves over finite fields (as in formula 3.2 above). Furthermore, the Igusa invariants evaluated at CM points are algebraic numbers and can be reduced modulo suitable primes. Avoiding bad primes in the denominators, we have [EL, Thm 2]

$$i_\ell(C \bmod \mathfrak{p}) = i_\ell(C) \bmod \mathfrak{p}.$$

Thus the valuation of the denominator of the constant term of H_1 at a prime p is the sum of the order of the poles of $2 \cdot 3^5 \frac{\chi_{12}^5}{\chi_{10}^6}$ at the CM-points over $\overline{\mathbb{F}}_p$. We have an analogous statement for the Igusa invariants H_2 and H_3. Since χ_{12}, ψ_4, and ψ_6 have no poles, the orders of the poles of i_ℓ are bounded above by suitable multiples of the order of zeros of χ_{10}. Experimentally, we observe that usually χ_{12}, ψ_4, and ψ_6 do not share zeros of high multiplicity with χ_{10}, so the order of the poles should be close to the order of zeros of a suitable power of χ_{10}. Computing the order of zeros of χ_{10} is equivalent to computing the arithmetic intersection number $\mathrm{div}(\chi_{10}).\mathcal{CM}(K)$ (explained below) of the divisor of χ_{10} with the cycle of CM points associated to K.

To compute the arithmetic intersection number $\mathrm{div}(\chi_{10}).\mathcal{CM}(K)$, first note that, up to a power of 2, $\chi_{10} = \Psi_{1,\mathrm{Siegel}}^2$ ([BY06, Remark 9.3]), where $\Psi_{1,\mathrm{Siegel}}$ is the normalized Siegel-Borcherds product of discriminant 1 whose divisor is G_1, the Humbert surface of discriminant 1. Thus at all primes except possibly 2, we need to compute $2G_1.\mathcal{CM}(K)$. The Humbert surface G_1 can be pulled back to a sum of Hirzebruch-Zagier divisors T_m on the Hilbert moduli space, and so the quantity to be computed can be expressed as a sum of intersection numbers of the form $\mathcal{CM}(K).\mathcal{T}_m$ on the Hilbert moduli space.

We now explain all of this in more detail. The presentation in the rest of this section follows closely the treatment in [Yana, Yanb, BBGK07]. For further background, good references are the books of [vdG88, Gor02, KRY06]. In Section 4.1, we describe the Hilbert moduli space and the moduli interpretation of Hirzebruch-Zagier divisors and CM-cycles. In Section 4.2, we explain arithmetic intersection theory and the meaning of $\mathcal{CM}(K).\mathcal{T}_m$. In Section 4.3, we explain the relationship between the Humbert surface and Hirzebruch-Zagier divisors. In Section 4.4, we describe the Bruinier-Yang conjectural formula for these intersection numbers.

4.1 The Hilbert moduli space and Hirzebruch-Zagier divisors. Fix F, a real quadratic extension of $\mathbb{Q}$ with different $\mathcal{D}_F$. The Hilbert moduli space parametrizes principally polarized abelian surfaces A with *real multiplication by* $\mathcal{O}_F$, i.e. with a fixed embedding $\iota\colon \mathcal{O}_F \hookrightarrow \mathrm{End}(A)$. More precisely, the arithmetic Hilbert moduli space $\mathcal{M}$ is the moduli stack that associates a scheme S over $\mathbb{Z}$ to the set of triples (A, ι, λ) where

1. A is an abelian surface over S,
2. $\iota\colon \mathcal{O}_F \hookrightarrow \mathrm{End}_S(A)$ gives the structure of real multiplication, and
3. $\lambda\colon \mathcal{D}_F^{-1} \to \mathrm{Hom}_{\mathcal{O}_F}(A, A^\vee)^{\mathrm{Sym}}$ is a homomorphism of $\mathcal{O}_F$-modules that takes totally positive elements to $\mathcal{O}_F$-linear symmetric polarizations and satisfies the Deligne-Pappas condition that $\mathcal{D}_F^{-1} \otimes A \to A^\vee$ is an isomorphism of abelian schemes.

The $\mathbb{C}$-points of $\mathcal{M}$ can be identified with the usual Hilbert moduli space $M = \mathrm{SL}_2(\mathcal{O}_F)\backslash\mathfrak{H}^2$ described in the article of Grundman in this volume [Gru].

Now suppose that m is a prime number that splits in F. Let $Y_0(m) = \Gamma_0(m)\backslash\mathfrak{H}$ be the modular curve parameterizing pairs of elliptic curves with an m-isogeny. Let $\mathcal{Y}_0(m)$ be the moduli stack over $\mathbb{Z}[1/m]$ of pairs of elliptic curves with an m-isogeny. Bruinier, Burgos Gil, and Kühn [BBGK07, Prop 5.11, 5.13] construct a map from $\mathcal{Y}_0(m)$ into $\mathcal{M}$. They prove that the image of the minimal compactification of $\mathcal{Y}_0(m)$ is the Zariski closure of the Hirzebruch-Zagier divisor, $\mathcal{T}_m$.

Let K be a primitive quartic CM number field with $K^+ = F$. Let $\mathcal{CM}(K)$ be the moduli stack over $\mathbb{Z}$ that represents the moduli problem assigning a base scheme S to the set of triples (A, ι, λ), where $\iota\colon \mathcal{O}_K \to \mathrm{End}_S(A)$ gives the $\mathcal{O}_K$-action on A such that the Rosati involution defined by λ induces complex conjugation on $\mathcal{O}_K$, and $(A, \iota|_{\mathcal{O}_F}, \lambda) \in \mathcal{M}(S)$. Bruinier and Yang also refer to the direct image of $\mathcal{CM}(K)$ in $\mathcal{M}$ by $\mathcal{CM}(K)$.

4.2 Arithmetic Intersection Theory. Following [KRY06, Chapter 2] and [Yana, Yanb], we define the arithmetic intersection number of two cycles on a Deligne-Mumford stack of dimension n that is proper and flat over $\mathbb{Z}$. Let $\mathcal{Z}_1$ and $\mathcal{Z}_2$ be two cycles that intersect properly, i.e. their codimensions are r_1 and r_2, with $r_1 + r_2 = n$, and $\mathcal{Z}_1 \cap \mathcal{Z}_2 = \mathcal{Z}_1 \times_{\mathcal{M}} \mathcal{Z}_2$ is a Deligne-Mumford stack of dimension

0. Then the arithmetic intersection number (written logarithmically) is

$$\begin{aligned}\mathcal{Z}_1.\mathcal{Z}_2 &= \sum_p \sum_{x \in \mathcal{Z}_1 \cap \mathcal{Z}_2(\overline{\mathbb{F}}_p)} \frac{1}{\#\operatorname{Aut}(x)} \log \#\widetilde{\mathcal{O}}_{\mathcal{Z}_1 \cap \mathcal{Z}_2, x} \\ &= \sum_p \sum_{x \in \mathcal{Z}_1 \cap \mathcal{Z}_2(\overline{\mathbb{F}}_p)} \frac{1}{\#\operatorname{Aut}(x)} i_p(\mathcal{Z}_1, \mathcal{Z}_2, x) \log p,\end{aligned}$$

where $\widetilde{\mathcal{O}}_{\mathcal{Z}_1 \cap \mathcal{Z}_2, x}$ is the local ring of $\mathcal{Z}_1 \cap \mathcal{Z}_2$ at x, and $i_p(\mathcal{Z}_1, \mathcal{Z}_2, x) = \text{Length } \tilde{\mathcal{O}}_{\mathcal{Z}_1 \cap \mathcal{Z}_2, x}$ is the local intersection at x.

We are interested in the arithmetic intersection of the cycles $\mathcal{CM}(K)$ and $\mathcal{T}(m)$ on the moduli stack $\mathcal{M}$. Using the moduli interpretation of the cycles $\mathcal{CM}(K)$ and $\mathcal{T}_m$, it is clear that a geometric point x in $\mathcal{CM}(K) \cap \mathcal{T}_m = \mathcal{CM}(K) \times_{\mathcal{M}} \mathcal{T}_m$ is indexed by a pair $(\phi\colon E \to E', \iota)$, where $\phi \in \mathcal{Y}_0(m)(\overline{\mathbb{F}_p})$ and $\iota\colon \mathcal{O}_K \to \operatorname{End}_{\mathcal{O}_F}(B)$ is an embedding such that the Rosati involution associated to λ induces complex conjugation on the image of K and where (B, λ) is the polarized abelian surface in $\mathcal{T}_m$ which is the image of ϕ. Denote by $i_p(\phi, \iota) = i_p(\mathcal{CM}(K), \mathcal{T}(m), x)$ the local intersection number at a point $x = (\phi, \iota)$. It follows from deformation theory that $i_p(\phi, \iota) = \text{Length}\,(\mathbb{W}[[t, t']]/I)$, where $\mathbb{W}$ is the Witt ring of $\overline{\mathbb{F}}_p$ and I is the minimal ideal such that ϕ can be lifted to an isogeny modulo I and the embedding ι can be lifted to an embedding modulo I [Yanb, Eq. (3.7)].

4.3 The relationship between the Hilbert and Siegel moduli spaces. Let M be the Hilbert modular surface associated to the real quadratic field of discriminant D, $F = \mathbb{Q}(\sqrt{D})$. Recall that forgetting real multiplication gives a morphism ϕ_D from the Hilbert moduli space M to the Siegel moduli space $X = \operatorname{Sp}(4, \mathbb{Z})\backslash\mathfrak{h}_2$. The image $G_D \subset X$ of M is known as the Humbert surface of discriminant D. For every G_D, there is a normalized integral Siegel Borcherds product $\Psi_{D,Siegel}$ whose divisor is equal to G_D (up to multiplication by ± 1). Let $\mathcal{G}_D$ denote the flat closure of G_D in the Siegel moduli stack representing principally polarized abelian surfaces.

The pullback to the Hilbert modular surface M of a Humbert surface $G_{D'}$ under the morphism ϕ_D is a linear combination of Hirzebruch-Zagier divisors [vdG88, p.215]:

$$\phi_D^*(G_{D'}) = \sum_{\substack{x \in \mathbb{Z}_{\geq 0} \\ x < \sqrt{DD'} \\ x^2 \equiv DD'(4)}} T_{(DD'-x^2)/4}, \tag{4.1}$$

for D' a positive discriminant, and the same is true of the flat closures of these cycles in $\mathcal{M}$. By Theorem 2.8 in [BY06] we also know that the pullback to M of the Borcherds product $\Psi_{D',\text{Siegel}}$ is a Hilbert-Borcherds product with divisor (4.1). In our case, the denominators of the Igusa functions are powers of χ_{10}, and $\chi_{10} = \Psi_{1,\text{Siegel}}^2$ up to a power of 2 and its divisor is $2G_1$. So we can now express the quantity that we want to compute, $2\mathcal{G}_1.\mathcal{CM}(K)$ as a sum

$$2 \sum_{\substack{x > 0 \\ \frac{D-x^2}{4} \in \mathbb{Z}_{\geq 0}}} \mathcal{CM}(K).\mathcal{T}_{(D-x^2)/4}.$$

4.4 Bruinier-Yang Conjecture. Let $K = F(\sqrt{\Delta})$ be a non-biquadratic quartic CM field K, let Φ be a CM type, and let $(\widetilde{K}, \widetilde{\Phi})$ be a fixed reflex field and type. Bruinier and Yang assume that $F = \mathbb{Q}(\sqrt{D})$ is a real quadratic field with D a prime congruent to 1 modulo 4. The real quadratic subfield of $\widetilde{K}$ is $\widetilde{F} = \mathbb{Q}(\sqrt{\Delta\Delta'})$, where Δ' is the real conjugate of Δ. Let $\widetilde{D}$ be the norm of the relative discriminant of K/F and let $d_{\widetilde{K}/\widetilde{F}}$ be the relative discriminant of $\widetilde{K}/\widetilde{F}$. Bruinier and Yang's conjectural formula for the arithmetic intersection number on the Hilbert moduli space is:

$$\mathcal{CM}(K).\mathcal{T}_m = \frac{1}{2}\sum_p b_m(p)\log p,$$

where

$$b_m(p)\log p = \sum_{\mathfrak{p}|p} \sum_{t=\frac{n+m\sqrt{\widetilde{D}}}{2D}\in d^{-1}_{\widetilde{K}/\widetilde{F}}, |n|<m\sqrt{\widetilde{D}}} B_t(\mathfrak{p}).$$

The ideals $\mathfrak{p}|p$ are ideals of $\mathcal{O}_{\widetilde{F}}$ and

$$B_t(\mathfrak{p}) = \begin{cases} 0 & \text{if } \mathfrak{p} \text{ is split in } \widetilde{K} \\ (\operatorname{ord}_{\mathfrak{p}} t+1)\rho(td_{\widetilde{K}/\widetilde{F}}\mathfrak{p}^{-1})\log|\mathfrak{p}| & \text{if } \mathfrak{p} \text{ is not split in } \widetilde{K} \end{cases}$$

where $\rho(\mathfrak{a}) = \#\{\mathfrak{A} \subset \mathcal{O}_{\widetilde{K}} \mid N_{\widetilde{K}/\widetilde{F}}\mathfrak{A} = \mathfrak{a}\}$.

Yang has proved the conjecture [Yana, Yanb] under the assumptions that $D \equiv 1 \mod 4$ is prime, $\Delta\Delta' \equiv 1 \mod 4$ is prime, and $\mathcal{O}_K$ is a free $\mathcal{O}_F$-module generated by 1 and $\frac{1}{2}\left(w+\sqrt{\Delta}\right)$, for some $w \in \mathcal{O}_{\widetilde{F}}$. Note that Yang sets $\widetilde{D} = \Delta\Delta'$, which is equal to the norm of the relative discriminant of K/F under his assumptions. They are not necessarily equal in general, in particular, they differ for many of the fields listed in [vW99]. In our computations, we use the convention from [BY06], that $\widetilde{D} = \mathrm{N}_{F/\mathbb{Q}}(\mathrm{Disc}_{K/F})$ and we find that with this convention, the predicted arithmetic intersection number agrees with the computed values for the number of embeddings and the denominators of Igusa class polynomials in many cases.

5 Quaternion algebras and embeddings of CM fields

In this section, we explain the connection given in [GL07] between denominators of Igusa class polynomials and certain embeddings of the ring of integers of K into the matrix algebra $\mathrm{M}_2(\mathbb{B}_{p,\infty})$, where p is a prime number and $\mathbb{B}_{p,\infty}$ is the unique (up to isomorphism) quaternion algebra over $\mathbb{Q}$ which is ramified at only p and ∞.

Let C be a smooth projective genus 2 curve such that $\mathrm{Jac}(C)$ has CM by $\mathcal{O}_K$. There is a number field L such that all of the endomorphisms of $\mathrm{Jac}(C)$ are defined over L and thus $\mathrm{Jac}(C)$ has everywhere good reduction. It is observed in [GL07, §4.1] that if K is a primitive CM field and C has bad reduction at a rational prime p, then for a prime $\mathfrak{p}|p$ of L, the reduction of $\mathrm{Jac}(C)$ modulo $\mathfrak{p}$ is isomorphic over $\mathcal{O}_L/\mathfrak{p}$ to a product of supersingular elliptic curves $E_1 \times E_2$ with the product polarization. Moreover, we have that E_1 is isogenous to E_2 as otherwise

$$\mathrm{End}(E_1 \times E_2) = \mathrm{End}(E_1) \times \mathrm{End}(E_2)$$

contradicting $K \hookrightarrow \mathrm{End}^0(\mathrm{Jac}(C))$ with K primitive. Goren and Lauter prove that in fact the existence of a smooth projective genus 2 curve C that has CM by $\mathcal{O}_K$

and has bad reduction at $\mathfrak{p}$ is equivalent to the existence of an embedding

$$\iota : \mathcal{O}_K \hookrightarrow \mathrm{End}(E_1 \times E_2)$$

such that the Rosati involution coming from the product polarization induces complex conjugation on $\mathcal{O}_K$ [GL07, Lemma 4.1.1 and Thm. 4.2.1].

5.1 Relating $\mathrm{End}(E_1 \times E_2)$ to a subring of $M_2(\mathbb{B}_{p,\infty})$. We view an element $f \in \mathrm{End}(E_1 \times E_2)$ as a matrix $\begin{pmatrix} f_{1,1} & f_{1,2} \\ f_{2,1} & f_{2,2} \end{pmatrix}$ where $f_{i,j} \in \mathrm{Hom}(E_j, E_i)$. Then the composition of endomorphisms $f \circ g$ corresponds to multiplication of the matrices and the Rosati involution $f \mapsto f^\vee$ corresponds to

$$\begin{pmatrix} f_{1,1} & f_{1,2} \\ f_{2,1} & f_{2,2} \end{pmatrix} \mapsto \begin{pmatrix} f_{1,1}^\vee & f_{2,1}^\vee \\ f_{1,2}^\vee & f_{2,2}^\vee \end{pmatrix}$$

Fix a supersingular elliptic curve $E/\overline{\mathbb{F}}_p$ and an isomorphism $\psi\colon \mathrm{End}(E) \to \mathcal{O}$ where $\mathcal{O}$ is a maximal order in $\mathbb{B}_{p,\infty}$. Note that by using ψ we can view elements of $\mathrm{End}(E) \otimes \mathbb{Q}$ as elements of $\mathbb{B}_{p,\infty}$. Given an element $\phi \in \mathrm{Hom}(E, E')$, we obtain an embedding $\mathrm{Hom}(E', E) \to \mathrm{End}(E)$ by mapping $f \mapsto f \circ \phi$. Thus we can view $\mathrm{Hom}(E', E)$ as an ideal of $\mathrm{End}(E)$ or, by using the isomorphism ψ, as a left ideal I of $\mathcal{O}$. Note that if we replace ϕ by a different endomorphism ϕ', this corresponds to right multiplying I by $\psi(\phi^{-1}\phi') \in \mathbb{B}_{p,\infty}$ (where $\phi^{-1}\phi' = (\phi^\vee\phi') / \deg(\phi)$ can be viewed as an element of $\mathrm{End}(E) \otimes \mathbb{Q}$). Thus a different choice of ϕ results in a different ideal, but one in the same *ideal class*. In fact, the map

$$\begin{array}{rcl} \left\{\text{supersingular curves over } \overline{\mathbb{F}}_p\right\} & \to & \{\text{left ideal classes of } \mathcal{O}\} \\ E' & \mapsto & \psi(\mathrm{Hom}(E', E)\phi) \text{ for some } \phi \in \mathrm{Hom}(E, E') \end{array}$$

is a bijection. Given a morphism ϕ and the ideal $I = \psi(\mathrm{Hom}(E', E)\phi)$, we can also identify the group $\mathrm{Hom}(E, E')$ with the ideal I^{-1} by the map

$$\mathrm{Hom}(E, E') \to \mathbb{B}_{p,\infty}, \quad g \mapsto \psi\left(\phi^{-1}g\right).$$

We would like to view $\mathrm{End}(E')$ as a subring of $\mathbb{B}_{p,\infty}$ in a compatible way. Let ϕ denote a fixed element of $\mathrm{Hom}(E, E')$. Then we obtain a map

$$\mathrm{End}(E') \to \mathbb{B}_{p,\infty}, \quad g \mapsto \psi\left(\phi^\vee g\phi\right) / \left(\deg(\phi)\right).$$

Since this map is an embedding, the image $\mathcal{O}'$ is isomorphic to $\mathrm{End}(E')$ and therefore is a maximal order in $\mathbb{B}_{p,\infty}$. One can check that $\mathcal{O}'$ is the right order of $I := \psi(\mathrm{Hom}(E', E)\phi)$. Thus, given a fixed $\phi \in \mathrm{Hom}(E, E')$ and a fixed isomorphism $\psi\colon \mathrm{End}(E) \to \mathcal{O} \subseteq \mathbb{B}_{p,\infty}$, we can identify $\mathrm{End}(E \times E')$ with the subring of $M_2(\mathbb{B}_{p,\infty})$,

$$\begin{pmatrix} \mathcal{O} & I \\ I^{-1} & \mathcal{O}' \end{pmatrix},$$

where I and $\mathcal{O}'$ are defined as above. Under this identification the Rosati involution can be described as

$$\begin{pmatrix} x & y \\ z & w \end{pmatrix}^\vee = \begin{pmatrix} x^\vee & \mathrm{N}(I)z^\vee \\ y^\vee / \mathrm{N}(I) & w^\vee \end{pmatrix},$$

where $N(I) := \gcd\{\gamma\gamma^\vee \mid \forall\gamma \in I\}$ is the norm of I. From this point forward, we will refer interchangeably to elements of $\mathrm{End}(E \times E')$ and elements of $\mathcal{R}_{(E,E',\phi)} := \begin{pmatrix} \mathcal{O} & I \\ I^{-1} & \mathcal{O}' \end{pmatrix}$.

5.2 Isomorphisms of principally polarized CM abelian varieties. Two principally polarized abelian varieties with complex multiplication (A, ι_1, λ_1), (A, ι_2, λ_2) are isomorphic either if ι_1 and ι_2 differ by an automorphism of $\mathcal{O}_K$ or if there exists $\sigma \in \mathrm{Aut}(A)$ such that $\sigma^{-1}\iota_1\sigma = \iota_2$ and $\lambda_1 = \sigma^\vee \lambda_2 \sigma$. We would like to describe the set of $\sigma \in \mathrm{Aut}(E \times E')$ such that conjugation by σ preserves the polarization. Preserving the polarization is equivalent to

$$\left(\sigma^{-1}\psi\sigma\right)^\vee = \sigma^{-1}\psi^\vee\sigma \quad \text{for all } \psi \in \mathrm{End}(E \times E').$$

Thus $\sigma\sigma^\vee = \pm 1$. Let $\begin{pmatrix} p & q \\ r & s \end{pmatrix} \in \mathcal{R}_{(E,E',\phi)}$ be the element associated to σ. The previous condition implies that $q = r = 0$ and both p and s are isomorphisms, or $p = s = 0$ and q and r are isomorphisms. Note that the latter possibility only occurs if E and E' are isomorphic. Let $\mathcal{U}_{(E,E',\phi)}$ denote the set of elements in $\mathcal{R}_{(E,E',\phi)}$ of this form, i.e. satisfying the condition that $\begin{pmatrix} p & q \\ r & s \end{pmatrix}\begin{pmatrix} p & q \\ r & s \end{pmatrix}^\vee = \pm 1$.

6 Counting the number of embeddings

Let $\mathcal{O}_K$ denote the ring of integers of K. If the real quadratic subfield F of K has class number 1, then there exists an $\eta \in \mathcal{O}_K$ such that $\mathcal{O}_K = \mathcal{O}_F[\eta]$. Let D denote the discriminant of F and let $\omega := \frac{1}{2}\left(D + \sqrt{D}\right)$, a generator of $\mathcal{O}_F$. Denote $\mathrm{Tr}(\eta) = \alpha_0 + \alpha_1\omega$ and $\mathrm{N}(\eta) = \beta_0 + \beta_1\omega$.

Proposition 6.1 *Let E, E' be supersingular curves over $\overline{\mathbb{F}}_p$. Fix an isomorphism $\psi\colon \mathrm{End}(E) \to \mathcal{O}$ where $\mathcal{O}$ is a maximal order and an element $\phi \in \mathrm{Hom}(E, E')$. Determining an embedding*

$$\iota\colon \mathcal{O}_K \hookrightarrow \mathrm{End}(E \times E')$$

is equivalent to giving two elements $\Lambda_1, \Lambda_2 \in \mathcal{R}_{(E,E',\phi)}$ such that

$$\begin{aligned} \Lambda_1\Lambda_2 &= \Lambda_2\Lambda_1 \\ \Lambda_1^2 - D\Lambda_1 + \frac{D^2 - D}{4} &= 0 \\ \Lambda_2 + \Lambda_2^\vee &= \alpha_0 + \alpha_1\Lambda_1 \\ \Lambda_2\Lambda_2^\vee &= \beta_0 + \beta_1\Lambda_1. \end{aligned} \tag{*}$$

Proof Let ι be an embedding. Then $\Lambda_1 = \iota(\omega)$ and $\Lambda_2 = \iota(\eta)$ satisfy the conditions above. In the other direction, it is clear that the map that sends $\omega \mapsto \Lambda_1$ and $\eta \mapsto \Lambda_2$ gives a well-defined homomorphism from $\mathcal{O}_K$ to $\mathcal{R}_{(E,E',\phi)}$. Therefore it remains to show that the map is injective. Assume that the map is not injective and let J denote the kernel. If J is nontrivial then $\mathcal{O}_K/J$ embeds into $\mathcal{R}_{(E,E',\phi)}$. However, since $\mathcal{O}_K/J$ is a finite abelian group and $\mathbb{B}_{p,\infty}$ is a torsion-free $\mathbb{Z}$-module, this is impossible. □

We note that the idea behind this lemma is similar to the ideas in [GL07] which used the embedding problem to obtain a bound on the primes dividing the denominator. Goren and Lauter studied pairs of matrices M_1, M_2 in $\mathcal{R}_{(E,E',\phi)}$ where M_1 satisfies the minimal polynomial of $\sqrt{D}$ and M_2 satisfies the minimal polynomial of $\sqrt{\Delta}$ where $K = F(\sqrt{\Delta})$.

In light of Proposition 6.1 and Section 5.2, the number of pairs of $(E \times E', \iota)$, counted up to isomorphism, is equal to

$$\sum_{\substack{\text{unordered pairs}\\ E,E'}} \# \frac{\left\{(\Lambda_1, \Lambda_2) \in \mathcal{R}^2_{(E,E',\phi)} \text{ satisfying } *\right\}}{(\Lambda_1, \Lambda_2) \sim (U\Lambda_1 U^{-1}, U\Lambda_2 U^{-1})}$$

where $\phi \in \operatorname{Hom}(E, E')$ and $U \in \mathcal{U}_{(E,E',\phi)}$.

We use this formula to count the number of embeddings. Our algorithm is sketched below and our implementation in `Magma` can be found in Appendix A. We make no claims on the running time of the algorithm or the implementation; the algorithm was only implemented to obtain the numerical evidence in Table 1.

6.1 Algorithm for counting solutions to the embedding problem.

1. (Initialize variables) Input $p, D, \alpha_0, \alpha_1, \beta_0, \beta_1, \widetilde{D}$.
2. (Create representation of $\mathbb{B}_{p,\infty}$) Let $\mathbb{B}_{p,\infty} = (a, -p)$ where a is the maximal negative integer such that $\mathbb{B}_{p,\infty}$ is ramified only at p and ∞.
3. (Create representations of $\mathcal{R}_{(E,E',\phi)}$ for each unordered pair of elliptic curves (E, E')) We represent each $\mathcal{R}_{(E,E',\phi)}$ by a triple (B_1, B_2, B_3) where B_1 is a basis for $\mathcal{O}$, B_2 is a basis for I and B_3 is a basis for $\mathcal{O}'$. For each isomorphism class of supersingular elliptic curves, let $\mathcal{O}$ be a maximal order isomorphic to the endomorphism ring. Then for each left ideal class of $\mathcal{O}$ let I be an integral representative and let $\mathcal{O}'$ be the right order. Take bases to give representatives of $\mathcal{R}_{(E,E',\phi)}$ for each unordered pair (E, E').
   (a) (Find possibilities for Λ_1) The condition on Λ_1 implies that it is of the form
   $$\begin{pmatrix} s_{1,1} & s_{1,2} \\ s_{1,2}^{\vee}/N(I) & D - s_{1,1} \end{pmatrix},$$
   where $s_{1,1}$ is an integer in $[(D - \sqrt{D})/2, (D + \sqrt{D})/2]$ and $s_{1,2}$ is an element in I with norm equal to
   $$-s_{1,1}^2 + Ds_{1,1} - (D^2 - D)/4.$$
   There are only finitely many possibilities for each of these, so we iterate over all of them.
      (i) (Find possibilities for Λ_2) Denote the entries of Λ_2 by $t_{i,j}$. The condition on $\Lambda_2 + \Lambda_2^{\vee}$ implies that $t_{2,1}$ is determined by $t_{1,2}$ and Λ_1 and that $t_{1,1}$ and $t_{2,2}$ have fixed trace. The condition on $\Lambda_2\Lambda_2^{\vee}$ implies that $t_{1,1}, t_{1,2}, t_{2,2}$ have bounded norm. This gives only finitely many possibilities for $t_{1,1}, t_{1,2}, t_{2,2}$ and we iterate through them all, checking whether the pair (Λ_1, Λ_2) satisfies all of the conditions in $(*)$. If $(*)$ is satisfied, then keep (Λ_1, Λ_2) in the list, otherwise discard the pair.
      (ii) (Identify elements that differ by conjugation or automorphisms of $\mathcal{O}_K$) There are only finitely many elements in $\mathcal{U}_{(E,E',\phi)}$. For a fixed pair (Λ_1, Λ_2), iterate through all elements U in $\mathcal{U}_{(E,E',\phi)}$ and all $\sigma \in \operatorname{Aut}(\mathcal{O}_K)$ and remove all pairs $(\sigma(U\Lambda_1U^{-1}), \sigma(U\Lambda_2U^{-1}))$.
4. (Count the embeddings) Count the number of pairs (Λ_1, Λ_2).

7 Numerical Data

In Table 1 we examine the 13 fields studied by van Wamelen [vW99]. These 13 fields are the only fields K such that there exists a genus 2 curve C *over* $\mathbb{Q}$ with CM by K, and they are all quartic Galois cyclic extensions of $\mathbb{Q}$. For each field K, we list

1. the factorization of the rational sixth root of the denominator of $\frac{\chi_{12}^5}{\chi_{10}^6}(\tau)$, where τ is a CM-point of K,
2. the conjectural value of $2\phi^* G_1.\mathcal{CM}(K)$ as predicted by the Bruinier-Yang formula (this was computed via a `Magma` program), and
3. the number of solutions to the embedding problem, computed using Algorithm 6.1 and the `Magma` code listed in Appendix A.

The three values listed above should agree if the number of solutions to the embedding problem is counted *with multiplicity*. Unfortunately, our representation of embeddings does not count multiplicity. However, Yang's proof suggests that the value $(\operatorname{ord}_{\mathfrak{p}} t + 1)/2$ is the multiplicity of an embedding, under certain conditions.

We write the Bruinier-Yang values as iterated exponents, where the outer exponent denotes $(\operatorname{ord}_{\mathfrak{p}} t + 1)/2$. If there is no outer exponent, then this value is 1. This is the case in most examples, so it appears that most of the embeddings have no multiplicity. We investigate the cases where the Bruinier-Yang conjecture does not agree with the numerical data for the other columns in Section 7.1.

Remark 7.1 Since all of the fields in [vW99] are Galois, there is only one CM-type up to isomorphism. Therefore we multiply the Bruinier-Yang formula by 1/2 since the CM-cycle in [BY06] is $\mathrm{CM}(K, \Phi)+\mathrm{CM}(K, \Phi')$ where Φ is not the complex conjugate of Φ'.

Remark 7.2 In the data we present in Table 1, we set $\widetilde{D} = \mathrm{N}_{F/\mathbb{Q}}(\mathrm{Disc}_{K/F})$ as in the original conjecture [BY06]. In his proofs, Yang sets $\widetilde{D} = \Delta\Delta'$ which is equal to the norm of the relative discriminant of K/F under his assumptions, but which is not equal in general.

7.1 Analysis of data. The first remark to make is that, in every example in the table, the powers of the primes in the denominators of the Igusa invariants match the number of solutions to the embedding problem, except for the power of 2 in four examples and the power of 3 in the last row. The power of 3 in the last row appears to be accounted for by multiplicity. The power of 2 is different in the four rows 2, 4, 9, and 11, and we explain why that is accounted for by cancellation in more detail in Section 7.2.

The rest of this section is devoted to explaining the discrepancies when comparing the Bruinier-Yang conjecture with the denominators of the Igusa class polynomials. Note that the six rows which are marked with a $(*)$ are cases where the Bruinier-Yang conjecture is already proved by Yang, since in those cases both D and $\widetilde{D}$ are primes congruent to 1 (mod 4). The two rows with a $(**)$ in the table are not covered by the conjecture, since in those cases the real quadratic subfield has discriminant which is divisible by a power of 2. Next we remark that in all the cases which are covered by the Bruinier-Yang conjecture, that is, where $D \equiv 1 \pmod 4$ is prime, the numerical data confirms the conjecture and shows that the 2 quantities computed for each field match, except for the power of 2 in the examples

Table 1: Numerical data: precise descriptions of the values computed can be found in Section 7.
$*$: Conjecture has been proved in this case since $\widetilde{D} \equiv 1 \pmod 4$ is a prime
$**$: Conjecture (as stated) does not include this case; [BY06] assumed that the real quadratic subfield has prime discriminant

$\widetilde{D}$	CM field	denominators of Igusa invariants	Bruinier-Yang formula (written multiplicatively)	Number of embeddings (written multiplicatively)
5	$\mathbb{Q}(\zeta_5)^*$	1	1	1
2^5	$\mathbb{Q}\left(\sqrt{-2+\sqrt{2}}\right)^{**}$	1	$(2^4)^{-1/2}(2^8)^{-3/2}$	2
13	$\mathbb{Q}\left(\sqrt{-13+2\sqrt{13}}\right)^*$	1	1	1
$2^6 5$	$\mathbb{Q}\left(\sqrt{-5+\sqrt{5}}\right)$	$1,\ 11^2$	$(2^2)^{3/2}11^2$	$2^2 \cdot 11^2$
$5 \cdot 13^2$	$\mathbb{Q}\left(\sqrt{-65+26\sqrt{5}}\right)$	$11^2,\ 31^2 41^2$	$11^2 31^2 41^2$	$11^2 31^2 41^2$
29	$\mathbb{Q}\left(\sqrt{-29+2\sqrt{29}}\right)^*$	5^2	5^2	5^2
$5 \cdot 17^2$	$\mathbb{Q}\left(\sqrt{-85+34\sqrt{5}}\right)$	$71^2,\ 11^2 41^2 61^2$	$11^2 41^2 61^2 71^2$	$11^2 41^2 61^2 71^2$
37	$\mathbb{Q}\left(\sqrt{-37+6\sqrt{37}}\right)^*$	$3^2 11^2$	$3^2 11^2$	$3^2 11^2$
$2^5 5^2$	$\mathbb{Q}\left(\sqrt{-10+5\sqrt{2}}\right)^{**}$	$7^2 23^2,\ 7^2 17^2 23^2$	$(2^{28})^{-3/2}(2^2)^{-1}(2^{12})^{-1/2}$ $\times(5^{32})^{1/2}5^{14}7^6$ $\times 17^4 23^4 31^2 41^2 79^2$	$2^2 7^4 17^2 23^4$
$5^2 13$	$\mathbb{Q}\left(\sqrt{-65+10\sqrt{13}}\right)$	$3^2,\ 3^2 53^2$	$3^4 53^2$	$3^4 53^2$
$2^6 13$	$\mathbb{Q}\left(\sqrt{-13+3\sqrt{13}}\right)$	$3^2 23^2,\ 3^2 23^2 131^2$	$(2^2)^{\frac{3}{2}} 3^4 23^2 131^2$	$2^2 3^4 23^4 131^2$
53	$\mathbb{Q}\left(\sqrt{-53+2\sqrt{53}}\right)^*$	$17^2 29^2$	$17^2 29^2$	$17^2 29^2$
61	$\mathbb{Q}\left(\sqrt{-61+6\sqrt{61}}\right)^*$	$3^4 5^2 41^2$	$\left(3^2\right)^2 5^2 41^2$	$3^2 5^2 41^2$

already mentioned above and the power of 23 in row 11. We cannot explain the discrepancy in the power of 23 in row 11.

Other than the power of 23, the discrepancies in the table occur in the four fields where $\widetilde{D}$ is highly divisible by 2. In those cases, negative and/or fractional powers of 2 appear in the Bruinier-Yang formula, plus other stray primes and powers in row nine. To explain that, observe that in [Yana, Yanb], Yang shows that the multiplicity is equal to

$$\frac{1}{2}\left(\operatorname{ord}_p \frac{m^2\widetilde{D}-n^2}{4D}+1\right)$$

and proves under his assumptions that this value is equal to

$$\frac{1}{2}\left(\operatorname{ord}_{\mathfrak{p}} \frac{n+m\sqrt{\widetilde{D}}}{2D}+1\right).$$

In particular, $\operatorname{ord}_p \frac{m^2\widetilde{D}-n^2}{4D}$ is odd. However, these statements no longer hold for those 4 examples in the table, and this leads to the fractional exponents.

When D is also even in addition to $\widetilde{D}$ being even, Bruinier and Yang did not conjecture a formula, so the different values do not give a counterexample to the conjecture. However, in the aim of obtaining a formula for all CM-fields, we suggest a possible explanation for the differences and a possible correction. In Yang's papers, he proves that one should consider all pairs m, n such that $\frac{m^2\widetilde{D}-n^2}{4D}$ is a positive integer divisible by p. This proof relies on the fact that $D \equiv 1 \pmod 4$. Note that if one imposes the additional condition that $8m+n \equiv 0 \pmod{16}$, then one obtains intersection numbers 1 and $7^4 17^2 23^4$ for the fields $\mathbb{Q}\left(\sqrt{-2+\sqrt{2}}\right)$ and $\mathbb{Q}\left(\sqrt{-10+5\sqrt{2}}\right)$ respectively. Away from 2, these numbers agree with both the number of embeddings and the denominators. This suggests that when the real quadratic subfield has discriminant congruent to 0 modulo 4, additional congruence conditions on the pairs m, n are needed to correct the intersection formula.

7.2 Powers of 2. In four examples in the table, rows 2, 4, 9, and 11, the power of 2 in the denominator does not match the number of embeddings at the prime 2. The reason for that is cancellation of powers of 2 in the numerator and denominator of Igusa invariants. It was noted in [GL07, §6.2] that the phenomenon of cancellation could occur, especially for the prime 2. It was explained there that this is due to the fact that there are no smooth superspecial genus 2 curves in characteristic 2, despite the fact that superspecial primes are prevalent. For these four fields, the prime 2 is ramified, and so it follows from a generalization of the work of Goren [Gor97] given in [GL, Section 3] that 2 is a superspecial prime, and thus these curves have bad reduction at 2. So indeed we do find solutions to the embedding problem at 2 in those cases, and the only reason that 2 does not appear in the denominator is because of cancellation with powers of 2 in the numerator.

Acknowledgements

This work was started as our group project at the WIN workshop "Women In Numbers", held at the Banff International Research Station in November 2008. We thank BIRS for hosting us and for the excellent working conditions there. We also thank the Fields Institute, PIMS, NSA, Microsoft Research, and the University of Calgary for their generous sponsorship of the conference. We thank Christophe Ritzenthaler for many detailed comments to improve the paper and the third author thanks Eyal Goren and Tonghai Yang for many discussions.

Appendix A Magma code for counting the number of embeddings

```
//Given a list of 4 rational numbers, q1, q2, q3, q4
//return the minimal positive integer N such that
//Nq1, Nq2 Nq3, Nq4 are integers
function Denom(L)

    denominators := [];
    for i := 1 to 4 do //hardcoded the length
        Append(~denominators, Denominator(L[i]));
    end for;
    return LCM(denominators);

end function;

/*********************************************************************/
//Check whether the element x is in the integral span of elements
//in the basis B
function InIdeal(x, B)
    A := Matrix([Coordinates(B[1]), Coordinates(B[2]),
                 Coordinates(B[3]), Coordinates(B[4])]);
    v := Vector(Coordinates(x));
    s := Solution(A,v);

    retval := true;
    for n := 1 to 4 do
        if not IsIntegral(s[n]) then
            retval := false;
        end if;
    end for;
    return retval;

end function;

/*********************************************************************/
//Return the Rosati involution of M.
//Notice this depends on the Norm of the ideal I
function Rosati(M,NI)

    return Matrix([ [ Conjugate(M[1][1]), Conjugate(M[2][1])*NI],
                    [ Conjugate(M[1][2])/NI, Conjugate(M[2][2])]]);
end function;

/*********************************************************************/
//Find all elements in the integral span of the basis B with
//norm = Normx and trace = Tracex
function FixedNormAndTrace(Normx, Tracex, B)

    xvals := [];
```

```
    Q := Parent(B[1]);
    isq := -Norm(Q.1);
    jsq := -Norm(Q.2);
    A := Transpose( Matrix([ Coordinates(B[1]), Coordinates(B[2]),
                       Coordinates(B[3]), Coordinates(B[4])]));

    denom1 := Denom(A[2]);
    denom2 := Denom(A[3]);
    denom3 := Denom(A[4]);
    x0 := Tracex/2;
    temp3 := Normx - x0^2;

    if temp3 ge 0 then
        for num3 := Ceiling(-Sqrt(Rationals()!(temp3*denom3^2/(isq*jsq))))
            to Floor(Sqrt(temp3*denom3^2/(isq*jsq))) do

            temp2 := temp3 - (num3/denom3)^2*isq*jsq;
            for num2 := Ceiling(-Sqrt(temp2*denom2^2/(-jsq)))
                to Floor(Sqrt(temp2*denom2^2/(-jsq))) do

                temp1 := temp2 - (num2/denom2)^2*(-jsq);
                for num1 := Ceiling(-Sqrt(temp1*denom1^2/(-isq)))
                    to Floor(Sqrt(temp1*denom1^2/(-isq))) do

                    x := x0 + num1/denom1*Q.1 +
                         num2/denom2*Q.2 + num3/denom3*Q.3;
                    if (Norm(x) eq Normx) and (Trace(x) eq Tracex)
                        and InIdeal(x, B) then
                        Append(~xvals, x);
                    end if;
                end for;
            end for;
        end for;
    end if;

    return xvals;
end function;


/**********************************************************************/
function RemoveConjElts(solns, E, alpha0, alpha1, beta0, beta1, D)

    n := 1;
    while n le #solns do

        // Loop over automorphisms of E
        for r in FixedNormAndTrace(1,  0, E[1]) cat
                 FixedNormAndTrace(1,  1, E[1]) cat
                 FixedNormAndTrace(1, -1, E[1]) cat
                 FixedNormAndTrace(1,  2, E[1]) cat
                 FixedNormAndTrace(1, -2, E[1]) do

            // Loop over automorphisms of E'
```

```
for s in FixedNormAndTrace(1,  0, E[3]) cat
     FixedNormAndTrace(1,  1, E[3]) cat
     FixedNormAndTrace(1, -1, E[3]) cat
     FixedNormAndTrace(1,  2, E[3]) cat
     FixedNormAndTrace(1, -2, E[3]) do

    //Calculating NormI so that we can compute Rosati involution
    I := LeftIdeal(QuaternionOrder(E[1]), E[2]);
    NormI := Norm(I);

    U := Matrix([[r,0],[0,s]]);
    Uinv := Matrix([[Conjugate(r), 0],[0, Conjugate(s)]]);

    //Remove embedding which is conjugate to given embedding by U
    // First check that U does not fix embedding
    if not (solns[n][1]*U eq U*solns[n][1] and
            solns[n][2]*U eq U*solns[n][2]) then

        Exclude(~solns, [U*solns[n][1]*Uinv,
            U*solns[n][2]*Uinv]);

    end if;

    //Remove embedding which is conjugate to the complex conjugate
    //of given embedding by U
    // First check that U does not fix complex conjugate of
    // embedding
    if not (solns[n][1]*U eq U*solns[n][1] and
            solns[n][2]*U eq U*Rosati(solns[n][2], NormI)) then

        size := #solns;
        Exclude(~solns, [U*solns[n][1]*Uinv,
                         U*Rosati(solns[n][2], NormI)*Uinv]);
    end if;

    //if E and E' are isomorphic then remove embedding which are
    // conjugate and have the order of E, E' switched
    if E[1] eq E[2] then

        U := Matrix([[0, r], [s, 0]]);
        Uinv := Matrix([[0, Conjugate(s)], [Conjugate(r), 0]]);

        if  U*Uinv ne Matrix([[1,0],[0,1]]) or
            Uinv*U ne Matrix([[1,0],[0,1]]) then
            print "Error - Uinv incorrect!!", r, s;
        end if;

        if not (solns[n][1]*U eq U*solns[n][1] and
                solns[n][2]*U eq U*solns[n][2]) then

            size := #solns;
            Exclude(~solns, [U*solns[n][1]*Uinv,
                             U*solns[n][2]*Uinv]);
```

```
                end if;
                if not (solns[n][1]*U eq U*solns[n][1] and
                        solns[n][2]*U eq U*Rosati(solns[n][2], NormI)) then
                    size := #solns;
                    Exclude(~solns, [U*solns[n][1]*Uinv,
                                    U*Rosati(solns[n][2], NormI)*Uinv]);
                end if;
            end if;
        end for;
    end for;

    n := n + 1;

end while;

return solns;

end function;

/****************************************************************/
// find Lambda1, Lambda2 satsifying embedding conditions and such that
// entries have fixed norm
function FindSolns(E, alpha0, alpha1, beta0, beta1, D,
                  Norms12, Normt12, Normt22, Tracet22,
                  Normt11, Tracet11, s11);

    solns := [];


    R1 := QuaternionOrder(E[1]);
    R2 := QuaternionOrder(E[3]);
    I := LeftIdeal(R1, E[2]);

    // Create list of possible values in I with Norm = Norms12
    s12traces := [-Floor(Sqrt(4*Norms12))..Floor(Sqrt(4*Norms12))];
    s12vals := [];
    for T in s12traces do
        s12vals := s12vals cat FixedNormAndTrace(Norms12, T, E[2]);
    end for;

    for s12 in s12vals do

        // loop over values in R1 with Norm = Normt11 and Trace = Tracet11
        for t11 in FixedNormAndTrace(Normt11, Tracet11, E[1]) do

            // loop over values in R2 with Norm = Normt22 and Trace = Tracet22
            for t22 in FixedNormAndTrace(Normt22, Tracet22, E[3]) do

                // Create list of possible values in I with Norm = Normt12
                t12traces := [-Floor(Sqrt(4*Normt12))..Floor(Sqrt(4*Normt12))];
                t12vals := [];
                for T in t12traces do
```

```
                t12vals := t12vals cat FixedNormAndTrace(Normt12, T, E[2]);
            end for;

            for t12 in t12vals do

                t21 := 1/Norm(I)*(alpha1*Conjugate(s12) - Conjugate(t12));

                //define Lambda1 = omega and Lambda2 = eta
                omega := Matrix([[s11, s12],
                                 [Conjugate(s12)/Norm(I), D - s11]]);
                eta := Matrix([[t11, t12], [t21, t22]]);

                //check omega and eta live in correct subring
                // and satisfy all of the embedding conditions
                if InIdeal(s12, E[2]) and InIdeal(t12, E[2]) and
                    (omega*eta eq eta*omega) and
                    eta + Rosati(eta, Norm(I)) eq alpha0 + alpha1*omega and
                    eta*Rosati(eta, Norm(I)) eq beta0 + beta1*omega then

                    Append(~solns, [omega, eta]); //add to list of solutions
                end if;
            end for;
        end for;
    end for;
  end for;
  return solns;

end function;



/****************************************************************/
// Lists all possible pairs of matrices Lambda1, Lambda2 in R_{E, E'}
// such that
// Lambda1*Lambda2 = Lambda2*Lambda1
// Lambda1^2 - D*Lambda1 + 1/4*(D^2 - D) = 0
// Lambda2 + Lambda2^{\vee} = alpha0 + alpha1*Lambda1
// Lambda2*Lambda2^{\vee} = beta0 + beta1*Lambda1

function ListAllEmb(alpha0, alpha1, beta0, beta1, D, p)

    // Define constants associated to CM field
    F := QuadraticField(D);
    D := Discriminant(F);

    trw := D;  //Trace(omega)
    nw := Integers()!(1/4*(D^2 - D));  //Norm(omega)

    Dtilde := 1/16*D^4*alpha1^4 + 1/2*D^3*alpha0*alpha1^3 - 1/8*D^3*alpha1^4
        - D^3*alpha1^2*beta1 + 3/2*D^2*alpha0^2*alpha1^2
        - 1/2*D^2*alpha0*alpha1^3 - 4*D^2*alpha0*alpha1*beta1
        + 1/16*D^2*alpha1^4 - 2*D^2*alpha1^2*beta0 + D^2*alpha1^2*beta1
        + 4*D^2*beta1^2 + 2*D*alpha0^3*alpha1 - 1/2*D*alpha0^2*alpha1^2
```

```
    - 4*D*alpha0^2*beta1 - 8*D*alpha0*alpha1*beta0
    + 4*D*alpha0*alpha1*beta1 - 2*D*alpha1^2*beta0 + 16*D*beta0*beta1
    - 4*D*beta1^2 + alpha0^4 - 8*alpha0^2*beta0 + 16*beta0^2;

// Give a presentation of B_{p, infinity}
if p ne 2 then
    jsq := -p;
    isq := -1;
    while (IsSquare(FiniteField(p)!isq)) or
        (#RamifiedPrimes(QuaternionAlgebra<Rationals() | isq, jsq>)
        gt 1) do
        isq := isq - 1;
    end while;
else
    jsq := -1;
    isq := -1;
end if;

QQ := Rationals();
Q<i, j, k> := QuaternionAlgebra<QQ | isq, jsq>;
ksq := -isq*jsq;


// Orders = { O : O maximal order isomorphic to End(E) for E supersingular
// curve over Fpbar}
// Note: if E and E' are not isomorphic, but have isomorphic endomorphism
// rings, Orders contains two different maximal orders, one is identified
// with End(E), the other with End(E')
OO := MaximalOrder(Q);
LI := LeftIdealClasses(OO);
Orders := [];
for I in LI do
    Append(~Orders, RightOrder(I));
end for;

// Ends = list of triples, one triple for each (ordered) pair (E, E')
// Let O be a maximal order which is isomorphic to End(E), I a left ideal
// of O which can be identified with Hom(E', E), and O' the right order of I
// which is identified with End(E').  Then the triple associated to (E, E')
// is (Basis(O), Basis(I), Basis(O')).  The correspondence between
// (O, I, O') and (E, E') is explained more in the paper.
Ends := [];
for O in Orders do
    for I in LeftIdealClasses(O) do
        Append(~Ends, [Basis(O), Basis(I), Basis(RightOrder(I))]);
    end for;
end for;

// initialize list of solutions
solns := [];
number := 0;
//look for Lambda1, Lambda2 in R_{(E, E')} satisfying above conditions
// this is equivalent to searching for s11, t11 in O, s12, t12 in I,
```

```
// s22, t22 in O'.  Then
// Lambda1 =
// [s11 s12]
// [s12^{\vee} s22]
// Lambda2 =
// [t11 t12]
// [t12^{\vee} t22]
for E in Ends do

    NormI := Norm(LeftIdeal(QuaternionOrder(E[1]), E[2]));

    //possible values for s11.  s22 = D - s11
    s11vals := [Ceiling( (D - Sqrt(D))/2 )..Floor( (D + Sqrt(D))/2)];

    Esolns := [];

    for s11 in s11vals do

        a := s11; //remnants of old notation
        delta := -a^2 + a*trw - nw;
        m := delta;
        Tracet11 := alpha0 + alpha1*s11;  //conditions from Lambda + Lambda*
        Tracet22 := alpha0 + alpha1*(D - s11);

        for n := Ceiling(-Sqrt(Max(m^2*Dtilde - 4*D, 0))) to
            Floor(Sqrt(Max(m^2*Dtilde - 4*D, 0))) do

            //parts of old notation
            Normx := 1/(2*D)*(n - (1/4*a^2*D^2*alpha1^2
            - 1/4*a*D^3*alpha1^2 + 1/16*D^4*alpha1^2 - 1/4*a^2*D*alpha1^2
            + 1/4*a*D^2*alpha1^2 - 1/8*D^3*alpha1^2
            + a^2*D*alpha1*alpha0 - a*D^2*alpha1*alpha0
            + 1/4*D^3*alpha1*alpha0 + 1/16*D^2*alpha1^2
            - 1/4*D^2*alpha1*alpha0 + a^2*alpha0^2 - a*D*alpha0^2
            + 1/4*D^2*alpha0^2 - 2*a^2*D*beta1 + 2*a*D^2*beta1
            - 1/2*D^3*beta1 - 1/4*D*alpha0^2 - 2*a*D*beta1
            + 1/2*D^2*beta1 - 4*a^2*beta0 + 4*a*D*beta0
            - D^2*beta0 - D*beta0));

            Normu := delta*(beta0 + beta1*a) - delta*Normx;

            Tracexuc := beta1*delta - (D - 2*a)/delta*Normu;

            Normv := delta^2*Normx + delta*(D - 2*a)*Tracexuc
                        + (D - 2*a)^2*Normu;

            //conditions coming from fixed n
            Normt11 := Normx;
            Normt12 := Normu*NormI/delta;
            Normt22 := Normv/delta^2;
            Norms12 := NormI*delta;

            //check n can give a solution which is integral
```

```
          if  IsIntegral((m^2*Dtilde - n^2)/(4*D*p)) and
              IsIntegral(Normt11) and IsIntegral(Normt12) and
              IsIntegral(Normt22) then

              Esolns := Esolns cat FindSolns(E,
                        alpha0, alpha1, beta0, beta1, D,
                        Norms12, Normt12, Normt22, Tracet22,
                        Normt11, Tracet11, s11);
                        // finds all solutions with
                        // entries of Lambda2 having
                        // fixed norm and fixed
                        // trace (for some entries)
          end if;
       end for;
    end for;

    // remove elements which are conjugate
    Esolns := RemoveConjElts(Esolns, E, alpha0, alpha1, beta0, beta1, D);

    if E[1] eq E[2] then
       number := number + #Esolns;
    else
       number := number + #Esolns/2;
    end if;
    Append(~solns, Esolns);
 end for;

 return solns, Ends, Integers()!number;

end function;
```

References

[AM93] A. O. L. Atkin and F. Morain, *Elliptic curves and primality proving*, Math. Comp. **61** (1993), no. 203, 29–68. MR **1199989 (93m:**11136)

[BBGK07] Jan H. Bruinier, José I. Burgos Gil, and Ulf Kühn, *Borcherds products and arithmetic intersection theory on Hilbert modular surfaces*, Duke Math. J. **139** (2007), no. 1, 1–88. MR **2322676 (2008h:**11059)

[BY06] Jan Hendrik Bruinier and Tonghai Yang, *CM-values of Hilbert modular functions*, Invent. Math. **163** (2006), no. 2, 229–288. MR **2207018 (2008b:**11053)

[EL] Kirsten Eisenträger and Kristin Lauter, *A CRT algorithm for constructing genus 2 curves over finite fields.* SMF Séminaires et Congrès, Proceedings of Arithmetic, Geometry, and Coding Theory, AGCT-10, Marseille (2005), **21**, 161–176, arXiv:0405305.

[GHK+06] P. Gaudry, T. Houtmann, D. Kohel, C. Ritzenthaler, and A. Weng, *The 2-adic CM method for genus 2 curves with application to cryptography*, Advances in cryptology—ASIACRYPT 2006, Lecture Notes in Comput. Sci., vol. 4284, Springer, Berlin, 2006, pp. 114–129. MR **2444631 (2009j:**94110)

[vdG88] Gerard van der Geer, *Hilbert modular surfaces*, Ergebnisse der Mathematik und ihrer Grenzgebiete (3) [Results in Mathematics and Related Areas (3)], vol. 16, Springer-Verlag, Berlin, 1988. MR **930101 (89c:**11073)

[Gor02] Eyal Z. Goren, *Lectures on Hilbert modular varieties and modular forms*, CRM Monograph Series, vol. 14, American Mathematical Society, Providence, RI, 2002. With the assistance of Marc-Hubert Nicole. MR **1863355 (2003c:**11038)

[Gor97] ———, *On certain reduction problems concerning abelian surfaces*, Manuscripta Math. **94** (1997), no. 1, 33–43. MR **1468933 (98m:**14048)

[GL07] Eyal Z. Goren and Kristin E. Lauter, *Class invariants for quartic CM fields*, Ann. Inst. Fourier (Grenoble) **57** (2007), no. 2, 457–480 (English, with English and French summaries). MR **2310947 (2008i:**11075)

[GL] ———, *Genus 2 curves with complex multiplication*, Preprint 2010, pp 51, `arXiv:1003.4759v1`.

[GZ85] Benedict H. Gross and Don B. Zagier, *On singular moduli*, J. Reine Angew. Math. **355** (1985), 191–220. MR **772491 (86j:**11041)

[Gru] Helen Grundman, *Hilbert Modular Variety Computations*. This volume.

[HNR09] Everett W. Howe, Enric Nart, and Christophe Ritzenthaler, *Jacobians in isogeny classes of abelian surfaces over finite fields*, Ann. Inst. Fourier **59** (2009), 239–289.

[How95] Everett W. Howe, *Principally polarized ordinary abelian varieties over finite fields*, Trans. Amer. Math. Soc. **347** (1995), no. 7, 2361–2401. MR **1297531 (96i:**11065)

[Igu60] Jun-ichi Igusa, *Arithmetic Variety of Moduli for Genus Two*, Ann. Math. **72** (1960), 612–649.

[Igu62] ———, *On Siegel modular forms of genus two.*, Amer. J. Math. **84** (1962), 175–200.

[Igu67] ———, *Modular forms and projective invariants*, Amer. J. Math. **89** (1967), 817–855. MR 0229643 (37 #5217)

[KW89] Luise-Charlotte Kappe and Bette Warren, *An elementary test for the Galois group of a quartic polynomial*, Amer. Math. Monthly **96** (1989), no. 2, 133–137. MR **992075 (90i:**12006)

[KRY06] Stephen S. Kudla, Michael Rapoport, and Tonghai Yang, *Modular forms and special cycles on Shimura curves*, Annals of Mathematics Studies, vol. 161, Princeton University Press, Princeton, NJ, 2006. MR **2220359 (2007i:**11084)

[Lau] Kristin Lauter, *Primes in the denominators of Igusa Class Polynomials*. Preprint 2003 `arXiv:0301240`.

[Mes91] Jean-François Mestre, *Construction de courbes de genre 2 à partir de leurs modules*, Effective methods in algebraic geometry (Castiglioncello, 1990), Progr. Math., vol. 94, Birkhäuser Boston, Boston, MA, 1991, pp. 313–334 (French). MR **1106431 (92g:**14022)

[Shi98] Goro Shimura, *Abelian varieties with complex multiplication and modular functions*, Princeton Mathematical Series, vol. 46, Princeton University Press, Princeton, NJ, 1998. MR **1492449 (99e:**11076)

[Spa94] Anne-Monika Spallek, *Kurven vom Geschlecht 2 und ihre Anwendung in Public-Key-Kryptosystemen*, 1994. Universität Gesamthochschule Essen, Ph. D. Thesis.

[vW99] Paul van Wamelen, *Examples of genus two CM curves defined over the rationals*, Math. Comp. **68** (1999), no. 225, 307–320. MR **1609658 (99c:**11079)

[Wen03] Annegret Weng, *Constructing hyperelliptic curves of genus 2 suitable for cryptography*, Math. Comp. **72** (2003), no. 241, 435–458 (electronic). MR **1933830 (2003i:**14029)

[Yana] Tonghai Yang, *An arithmetic intersection formula on Hilbert modular surfaces*. 2007, Preprint, to appear in *Amer. J. Math.*

[Yanb] ———, *Arithmetic intersection on a Hilbert modular surface and the Faltings height*. 2007, Preprint.

Fields Institute Communications
Volume **60**, 2011

l-Adic Étale Cohomology of PEL Type Shimura Varieties with Non-Trivial Coefficients

Elena Mantovan
Mathematics 253-37
Caltech
Pasadena, CA 91125 USA
mantovan@caltech.edu

Abstract. Given a Shimura datum (G, h) of PEL type, let p be an odd prime at which G is unramified. In [13], we established a formula computing the l-adic cohomology of the associated Shimura varieties (regarded as a representation of the adelic points of G and of the local Weil group at p) in terms of that of their local models at p (the associated Rapoport-Zink spaces) and of the corresponding Igusa varieties. In this paper we extend those results (which are for cohomology with constant l-adic coefficients) to the general case of coefficients in a lisse étale sheaf attached to a finite dimensional l-adic representation of the group G.

1 Introduction

To any reductive group G defined over a number field, one can associate a projective system of differentiable manifolds endowed with an action of $G(\mathbb{A}^\infty)$, the finite adelic points of G. Furthermore, to each algebraic representation ρ of G on a finite dimensional complex vector space, one can attach a local system $\mathfrak{L}_\rho$ on the corresponding manifolds and consider their complex cohomology spaces with coefficients in $\mathfrak{L}_\rho$ as representations of $G(\mathbb{A}^\infty)$. These complex cohomology spaces can then be interpreted in terms of automorphic forms on G. Moreover, for any cuspidal automorphic representation Π of $G(\mathbb{A})$, satisfying certain "algebraic" conditions at infinity, there is a suitable choice of a representation ρ such that Π^∞ will appear in the associated cohomology spaces. For a certain class of G's, which was isolated by Shimura, these differential manifolds have a natural algebraic structure defined over a number field E. In most cases, the algebraic structure is obtained by interpreting the manifolds as classifying spaces over $\mathbb{C}$ for abelian varieties with additional structures. E.g., in the PEL case, the Shimura varieties arise as moduli spaces of abelian varieties endowed with polarizations, prescribed endomorphisms and level structures. Furthermore, to each algebraic representation ξ of G on a

2010 *Mathematics Subject Classification.* Primary 11G18, 11F70; Secondary 14G22, 14L05.
Partially supported by the National Science Foundation under Grant DMS-0701310.

finite dimensional l-adic vector space, one can attach a lisse étale l-adic sheaf $\mathcal{L}_\xi$ on the corresponding varieties and their l-adic étale cohomology groups with coefficients in $\mathcal{L}_\xi$ are naturally representations of $G(\mathbb{A}^\infty) \times \Gamma_E$, for Γ_E the absolute Galois group of E. Again, the finite component of a cuspidal automorphic representation Π of $G(\mathbb{A})$, satisfying certain algebraic conditions at infinity, will appear in some l-adic étale cohomology group for a suitable choice of ξ. A conjecture of Langlands predicts that the Galois representation associated to an "algebraic" cuspidal automorphic representation Π of $G(\mathbb{A})$ under the (conjectural) global Langlands correspondences can be detected by looking at the Π^∞-isotypic component Σ_Π of these cohomology spaces. In view of the compatibility between global and local Langlands correspondences, one also expects that, for any prime number p, the restriction of Σ_Π to W_{E_v}, the local Weil group at a prime $v|p$, depends only on the p-adic component of Π. This latter prediction is the main motivation for our work, which focus on the study of the l-adic étale cohomology groups of Shimura varieties of PEL type as representations of $G(\mathbb{A}^\infty) \times W_{E_v}$, for v a prime of E dividing a fixed prime number p which we assume to be unramified for the group G. More precisely, in this paper we compute the l-adic étale cohomology groups of the Shimura varieties, with coefficients in $\mathcal{L}_\xi$ (for ξ an irreducible finite dimensional l-adic representation of G), in terms of the l-adic étale cohomology groups of their local models, constructed by Rapoport and Zink, and those of the associated Igusa varieties. The latter are classifying spaces for abelian varieties in positive characteristic, whose defining moduli problems are closely related to those of the Shimura varieties. In the resulting formula, both the actions of the p-adic points of G and of the local Weil group are trivial on the l-adic étale cohomology groups of the Igusa varieties but not on those of the Rapoport-Zink spaces. Moreover, only the l-adic étale cohomology groups of the Igusa varieties have coefficients in a local system depending on ξ (namely, the pullback of $\mathcal{L}_\xi$) while those of the Rapoport-Zink spaces have trivial l-adic coefficients. These two facts make the formula heuristically compatible with Langlands' conjecture, in the sense explained above, and also with a conjecture of Kottwitz which predicts that the local Langlands correspondences at p (for the groups G's in the class considered in this papers) are realized by the l-adic étale cohomology of the Rapoport-Zink spaces, with trivial coefficients. We remark that a special case of this formula, namely the case when ξ is the trivial representation, was previously established in [13]. The goal of this paper is to generalized those results to any ξ. In view of the general philosophy, the case of l-adic étale cohomology with constant coefficients is too restrictive for most applications.

1.1 Outline. Our strategy closely follows that of [13], where the final formula (in the case of trivial l-adic coefficients) is deduced by results about the local geometry of the Shimura varieties at the chosen prime $v|p$. In particular, the formula in [13] can be regarded as an application of the appropriate Hochschild-Serre spectral sequence and Künneth formula, and reflects a description of (a stratification of) the Shimura varieties at v as quotients by the action of a group of the products of their local models and Igusa varieties. Similarly, the formula with non-constant coefficients proved in this paper (and more precisely, its special case for ξ the standard representation) reflects analogous statements about the local geometry at v of the universal abelian scheme over the Shimura varieties (and its multi-self-products). In particular, we prove that the universal abelian schemes over the Shimura varieties at v and those over the Igusa varieties become isogenous once pull-backed over the products of the Rapoport-Zink local models and Igusa varieties. We remark that while it would be possible to reproduce in this new context all of the constructions of [13], we chose not to. Instead, we preferred to deduce the new results from those in [13]. In Section 2, we recall results and constructions from the theory of integral models of Shimura varieties of PEL type, following [10] for the cases of good reduction and [13] for the general cases. In Section 3, we state the main theorem of this paper and outline a strategy for its proof. In Section 4, we show that the general case of this theorem can be deduced from a special case, namely the case of ξ the standard representation. Finally, in Section 5, we prove the latter case. Throughout the paper, our notations follows those in [7] and [13].

2 Shimura varieties of PEL type

2.1 Integral models at unramified primes. In this section, we recall the definition of Shimura varieties of PEL type as moduli spaces for abelian varieties with additional structures. We focus in particular on the cases of good reduction at a chosen prime p. Our exposition follows [10] (Section 5, pp. 389–392) and we refer to it for details. A warning about notations: in certain cases the moduli spaces here described are not the canonical models of the corresponding Shimura varieties but disjoint unions of finitely many copies of them (see [10], Sec. 8, pp. 398–400). In the following, we will ignore this issue and refer to these moduli spaces as Shimura varieties.

2.1.1 We fix data of the following type:

- B a finite dimensional simple algebra over $\mathbb{Q}$;
- $*$ a positive involution on B over $\mathbb{Q}$;
- V a nonzero finitely generated left B-module;
- $\langle\,,\rangle$ a non degenerate $\mathbb{Q}$-valued $*$-hermitian alternating pairing on V.

To the above data, we associate an algebraic group G over $\mathbb{Q}$ as follows. We define G as the group of the B-linear automorphisms of V which preserve the pairing $\langle\,,\rangle$ up to scalar multiple, i.e. for any $\mathbb{Q}$-algebra R

$$G(R) = \{(c,g) \in R^\times \times \mathrm{Aut}_{B\otimes_{\mathbb{Q}}R}(V \otimes_{\mathbb{Q}} R) | \langle gv, gw\rangle = c\langle v,w\rangle \text{ for all } v,w \in V\otimes_{\mathbb{Q}} R\}.$$

In this paper we are occupied with the class of Shimura varieties associated with an algebraic group G of this form. These varieties are called *of PEL type*. We remark that any group G of this form is equipped with a canonical homomorphism $\nu : G \to \mathbb{G}_m$, which to a B-linear automorphism of V preserving the pairing $\langle\,,\rangle$ up to scalar multiple associates the latter, i.e. for any $\mathbb{Q}$-algebra R and any element $(c,g) \in G(R)$ we define $\nu(R)(c,g) = c$. Let (G,h) be a Shimura datum associated with some data $(B,*,V,\langle\,,\rangle)$ as above. As usual, h denotes a conjugacy class of cocharacters $\mu_h : \mathbb{G}_m \to G_{\mathbb{C}}$ satisfying the appropriate positivity condition. Then, the Shimura field $E = E(G,h) \subset \mathbb{C}$, associated with the pair (G,h), is the field of definition of the isomorphism class of the complex representation V_1 of B, where V_1 is the subspace of $V_{\mathbb{C}}$ on which $\mu_h(z)$ acts as z. The Shimura varieties attached to the datum (G,h) have a natural structure of algebraic varieties defined over E. More precisely, let $\mathbb{A}^\infty$ denote the ring of the finite adeles of $\mathbb{Q}$, and K an open compact subgroup of $G(\mathbb{A}^\infty)$. For K sufficiently small, the associated Shimura variety Sh_K is a smooth quasi-projective scheme defined over E, and has a moduli interpretation as a classifying space for abelian varieties. As K varies, the varieties Sh_K form a projective system, naturally endowed with an action of $G(\mathbb{A}^\infty)$. We refer to [10] for the definition of PEL type moduli problems for abelian varieties over the reflex field E. Here, we focus on the cases of good reduction at a chosen prime p, and introduce the corresponding moduli problems over $\mathcal{O}_{E,(p)}$, the localization at p of the ring of integers of E.

2.1.2 Let p be an odd prime. We assume that the PEL data $(B, *, V, \langle\, ,\rangle)$ satisfy the following three conditions:

1. there exists a $\mathbb{Z}_{(p)}$-order $\mathcal{O}_B$ in B which is preserved by $*$ and whose p-adic completion is a maximal order in $B_{\mathbb{Q}_p}$;
2. there exists a lattice Λ in $V_{\mathbb{Q}_p}$ which is self-dual for $\langle\, ,\rangle$ and is preserved by $\mathcal{O}_B$;
3. *(unramified hypothesis)* $B_{\mathbb{Q}_p}$ is a product of matrix algebras over unramified extensions of $\mathbb{Q}_p$.

Under the above conditions, the group G has an hyperspecial maximal compact subgroup $K_{p,0}$ of $G(\mathbb{Q}_p)$, namely the stabilizer of Λ. In the following, we write $G(\mathbb{Z}_p) = K_{p,0}$, $G(\mathbb{Q}_p)^\dagger = \langle G(\mathbb{Z}_p), p\rangle$ for the subgroup of $G(\mathbb{Q}_p)$ generated by $G(\mathbb{Z}_p)$ and multiplication by p on V, and $G(\mathbb{A}^\infty)^\dagger = G(\mathbb{A}^{\infty,p}) \times G(\mathbb{Q}_p)^\dagger$. We also write $\mathcal{O}_{B_{\mathbb{Q}_p}}$ for the p-adic completion of $\mathcal{O}_B$. Finally, we choose a basis $\alpha_1, \dots \alpha_t$ for the free $\mathbb{Z}_{(p)}$-module $\mathcal{O}_B$, and let $X_1, \dots X_t$ be indeterminates. Then the polynomial $\det(X_1\alpha_1 + \cdots + X_t\alpha_t; V_1)$ is homogeneous of degree $\dim_{\mathbb{C}}(V_1)$, with coefficients in $\mathcal{O}_{E,(p)}$. We restrict our attention to open compact subgroups K of $G(\mathbb{A})$ of the form $K = K^pK_{p,0}$, where K^p is an open compact subgroup of $G(\mathbb{A}^{\infty,p})$. We remark that if K^p is sufficiently small, then so is K. We assume K^p is sufficiently small. Under the above conditions, the Shimura varieties $\mathrm{Sh}_{K^pK_{p,0}}$ admit integral models S_{K^p} over $\mathcal{O}_{E,(p)}$ satisfying the following properties. For all K^p, the schemes S_{K^p} are smooth and quasi-projective over $\mathcal{O}_{E,(p)}$, and as K^p varies they form a projective system endowed with an action of $G(\mathbb{A}^{\infty,p})$. Moreover, there exist canonical isomorphisms $\mathrm{Sh}_{K^pK_{p,0}} \cong S_{K^p} \times_{\mathrm{Spec}(\mathcal{O}_{E,(p)})} \mathrm{Spec}(E)$ which form a compatible system as the level K^p varies, and are equivariant for the action of $G(\mathbb{A}^{\infty,p})$. The schemes S_{K^p} have a moduli interpretation as classifying spaces for abelian varieties. We recall their definition. To any open compact subgroup $K^p \subset G(\mathbb{A}^{\infty,p})$, we associate a set-valued contravariant functor $\mathcal{F}_{K^p}$ on the category of locally Noetherian $\mathcal{O}_{E,(p)}$-schemes. It suffices to define $\mathcal{F}_{K^p}(S)$ for S connected, in which case we first define $\mathcal{F}_{K^p}(S) = \mathcal{F}_{K^p}(S, s)$ for a choice of a geometric point $s \in S$, and then observe that such set is independent on the choice of $s \in S$. For any pair (S, s), we define $\mathcal{F}_{K^p}(S, s)$ as the set of equivalence classes of quadruples $(A, \lambda, i, \bar{\eta}^p)$ where:

- A is an abelian scheme over S;
- $\lambda : A \to A^\vee$ is a prime-to-p polarization;
- $i : \mathcal{O}_B \hookrightarrow \mathrm{End}(A)\otimes_{\mathbb{Z}}\mathbb{Z}_{(p)}$ is a morphism of $\mathbb{Z}_{(p)}$-algebras such that $\lambda \circ i(b^*) = i(b)^\vee \circ \lambda$, and $\det(X_1\alpha_1 + \cdots + X_t\alpha_t; \mathrm{Lie}(A)) = \det(X_1\alpha_1 + \cdots + X_t\alpha_t; V_1)$ (the latter is an equality of homogeneous polynomial with coefficients in the global sections of $\mathcal{O}_S$, known as the *determinant condition*);
- $\bar{\eta}^p$ is a $\pi_1(S, s)$-invariant K^p-orbit of isomorphisms of $B \otimes_{\mathbb{Q}} \mathbb{A}^{\infty,p}$-modules $\eta^p : V \otimes_{\mathbb{Q}} \mathbb{A}^{\infty,p} \to V^pA_s$ which takes the pairing $\langle\, ,\rangle$ on $V \otimes_{\mathbb{Q}} \mathbb{A}^{\infty,p}$ to a $(\mathbb{A}^{\infty,p})^\times$-scalar multiple of the λ-Weil pairing. (Here, V^pA_s denotes the Tate space of A_s away from p).

Two quadruples $(A, \lambda, i, \bar{\eta}^p)$ and $(A', \lambda', i', (\bar{\eta}^p)')$ are equivalent if there exists a prime-to-p isogeny $\beta : A \to A'$ which takes λ to a $\mathbb{Z}_{(p)}^\times$-multiple of λ', i to i' and $\bar{\eta}$

to $(\bar{\eta}^p)'$. For K^p sufficiently small, the functor $\mathcal{F}_{K^p}$ on the category of locally Noetherian $\mathcal{O}_{E,(p)}$-schemes is represented by a smooth quasi-projective $\mathcal{O}_{E,(p)}$-scheme. For each K^p sufficiently small, the integral model S_{K^p} is defined as the $\mathcal{O}_{E,(p)}$-scheme representing $\mathcal{F}_{K^p}$ ([10], Sec. 5, p. 391).

2.1.3 We now focus on the problem of defining integral models for Sh_K, in the case when the level K is not maximal at p. We consider exclusively the case of K of the form K^pK_p where K^p is any sufficiently small open compact subgroup of $G(\mathbb{A}^{\infty,p})$ and K_p is an open compact subgroup of $G(\mathbb{Q}_p)$, $K_p \subset K_{p,0}$, of the form

$$K_p = K_{p,m} = \{g \in K_{p,0} | g_\Lambda \equiv 1 \mod p^m\Lambda\},$$

for some integer $m \geq 0$. As m varies, the subgroups $K_{p,m}$ form a fundamental system of open neighborhoods of the identity in $G(\mathbb{Q}_p)$, and similarly, as both m and K^p vary, the subgroups $K = K^pK_{p,m}$ describe a fundamental system of open compact subgroups of $G(\mathbb{A}^\infty)$. Thus, it is without loss of generality that through the rest of this paper we assume the open compact subgroups K of $G(\mathbb{A}^\infty)$ to be of this form. For any such K, $K = K^pK_p$, we define the corresponding integral model of Sh_K as a S_{K^p}-scheme, and denote it by S_K. The key ingredient is Katz and Mazur's notion of a full set of sections for finite flat schemes ([9], Section 1.8.2, p.33). Details and proofs of the construction can be found in [13] (Section 6, pp. 597–601). For any $m \geq 1$, and $K = K^pK_{p,m}$, we define S_K as the S_{K^p}-scheme representing the following functor $\mathcal{S}_m$. Let A denote the universal abelian scheme over S_{K^p}. We write $A[p^m]$ for its p^m-torsion subgroup, $A[p^\infty]$ for its p-divisible part, and μ_{p^m} for the group scheme of the p^m-th roots of unity. We recall that the additional structures on A induce corresponding structures on the Barsotti-Tate group $A[p^\infty]$, namely (in our case) a quasi-polarization (which we denote by ℓ) and a compatible action of $\mathcal{O}_{B_{\mathbb{Q}_p}}$. To any S_{K^p}-scheme T, the functor $\mathcal{S}_m$ associates the set of pairs (α_0, α) where $\alpha_0 : p^{-m}\mathbb{Z}/\mathbb{Z} \to \mu_{p^m}(T)$ and $\alpha : p^{-m}\Lambda/\Lambda \to A[p^m](T)$ are two group homomorphisms satisfying the conditions:

1. $\{\alpha_0(x) | x \in p^{-m}\mathbb{Z}/\mathbb{Z}\}$ is a full set of sections of $\mu_{p^m,T}/T$;
2. $\{\alpha(x) | x \in p^{-m}\Lambda/\Lambda\}$ is a full set of sections of $A[p^m]_T/T$;
3. α is $\mathcal{O}_{B_{\mathbb{Q}_p}}$-equivariant;
4. the following diagram commutes

$$\begin{array}{ccc} p^{-m}\Lambda/\Lambda \times p^{-m}\Lambda/\Lambda & \xrightarrow{\langle,\rangle} & p^{-m}\mathbb{Z}/\mathbb{Z} \\ \downarrow {\scriptstyle \alpha\times\alpha} & & \downarrow {\scriptstyle \alpha_0} \\ A[p^m](T) \times A[p^m](T) & \xrightarrow[(,)_\ell]{} & \mu_{p^m}(T) \end{array}$$

(here, with abuse of notations $\langle\, ,\rangle$ denotes the pairing on $p^{-m}\Lambda/\Lambda$ induced by that on Λ, and $(\, ,)_\ell$ is the ℓ-Weil pairing on $A[p^m](T)$).

For any $m \geq 1$, the functor $\mathcal{S}_m$ is representable. The representing scheme S_K is finite over S_{K^p}, and naturally endowed with an action of $G(\mathbb{Z}_p)$, trivial on $K_{p,m}$. For $m = 0$, we write $S_K = S_{K^p}$. As K varies, the schemes S_K form a projective system naturally endowed with an action of $G(\mathbb{A}^\infty)^\dagger = G(\mathbb{A}^{\infty,p}) \times G(\mathbb{Q}_p)^\dagger$. Moreover, for each K, there is a canonical isomorphism $\mathrm{Sh}_K \cong S_K \times_{\mathrm{Spec}(\mathcal{O}_{E,(p)})} \mathrm{Spec}(E)$, such that, as K varies, they form compatible system which is $G(\mathbb{A}^\infty)^\dagger$-equivariant with respect to the natural action of $G(\mathbb{A}^\infty)$ on the Shimura varieties. We point out that in general the action of the whole $G(\mathbb{A}^\infty)$ on the Shimura varieties does not extend to the above integral models. On the other hand, for a slightly more general class of integral models, the associated correspondences do ([13], Proposition 17, p. 600).

2.2 Cohomology and vanishing cycle sheaves. Let l be a prime number, $l \neq p$. Any continuous l-adic representation of $G(\mathbb{A}^\infty)$ gives rise to lisse étale l-adic sheaves $\mathcal{F}_K$ over Sh_K, for all sufficiently small open compact subgroups K of $G(\mathbb{A}^\infty)$ ([10], Section 6, pp. 392-393; or [7], Section III.2, pp. 94–105). We apply this construction to ξ an algebraic finite dimensional l-adic representation of G.

2.2.1 Let ξ be an algebraic representation of G on a finite dimensional $\mathbb{Q}_l^{\mathrm{ac}}$-vector space W_ξ, and denote the corresponding sheaves on the Shimura varieties by $\mathcal{L}_\xi = \mathcal{L}_{\xi,K}$. As K varies, the sheaves $\mathcal{L}_{\xi,K}$ form a compatible system, naturally endowed with an action of $G(\mathbb{A}^\infty)$. More precisely, for each $\gamma \in G(\mathbb{A}^\infty)$ and level K, let $K_\gamma = K \cap \gamma K \gamma^{-1}$, $f : \mathrm{Sh}_{K_\gamma} \to \mathrm{Sh}_K$ denote the natural projection between Shimura varieties corresponding to the inclusion $K_\gamma \subset K$, and $\gamma : \mathrm{Sh}_{K_\gamma} \to \mathrm{Sh}_{\gamma^{-1}K_\gamma\gamma} \to \mathrm{Sh}_K$ the composition of the action of γ with the natural projection between Shimura varieties. Then there is a canonical isomorphism between the pullbacks $\gamma^*\mathcal{L}_{\xi,K}$ and $f^*\mathcal{L}_{\xi,K}$ over Sh_{K_γ} ([10], Section 6, pp. 392-393). It follows that, for each $i \geq 0$, the $\mathbb{Q}_l^{\mathrm{ac}}$-spaces

$$H_c^i(\mathrm{Sh}, \mathcal{L}_\xi) = \varinjlim{}_K\, H_c^i(\mathrm{Sh}_K \times_E E^{\mathrm{ac}}, \mathcal{L}_{\xi,K})$$

have a canonical structure of $G(\mathbb{A}^\infty) \times \mathrm{Gal}(E^{\mathrm{ac}}/E)$-modules. In fact, they are admissible/continuous $G(\mathbb{A}^\infty) \times \mathrm{Gal}(E^{\mathrm{ac}}/E)$-modules. We write

$$H_c(\mathrm{Sh}, \mathcal{L}_\xi) = \sum_i (-1)^i H_c^i(\mathrm{Sh}, \mathcal{L}_\xi) \in \mathrm{Groth}(G(\mathbb{A}^\infty) \times \mathrm{Gal}(E^{\mathrm{ac}}/E)).$$

In this paper we focus on their restriction to $G(\mathbb{A}^\infty) \times W_{E_v}$, for W_{E_v} the Weil group of the completion of E at a prime v dividing p.

2.2.2 Let v be prime of E dividing p. We write E_v for the completion of E at v, $\mathcal{O}_{E_v}$ for its ring of integers, and k for it residue field, $\#k = q = p^f$. We choose an algebraic closure k^{ac} of k, and denote by E_0 the fraction field of $W(k^{\mathrm{ac}})$, the ring of Witt vector of k^{ac}, and by σ the q-th power of Frobenius on E_0. The unramified hypothesis in 2.1.2 implies that v is unramified, i.e. $E_v \subset E_0$. We fix an algebraic closure E_v^{ac} of E_v, containing E_0, and an embedding $\bar{v} : E^{\mathrm{ac}} \to E_v^{\mathrm{ac}}$, extending the identity of E. These data allow us to identify the Weil group of E_v with the decomposition group of $\bar{v}$ in $\mathrm{Gal}(E^{\mathrm{ac}}/E)$, i.e. $W_{E_v} \subset \mathrm{Gal}(E^{\mathrm{ac}}/E)$. For each $i \geq 0$, we regard the cohomology spaces $H_c^i(\mathrm{Sh}, \mathcal{L}_\xi)$ as $G(\mathbb{A}^\infty) \times W_{E_v}$-modules via restriction. They are admissible/continuous $G(\mathbb{A}^\infty) \times W_{E_v}$-modules which can be computed as

$$H_c^i(\mathrm{Sh}, \mathcal{L}_\xi) = \varinjlim{}_K\, H_c^i(\mathrm{Sh}_K \times_E E_v^{\mathrm{ac}}, \mathcal{L}_{\xi,K}).$$

In the following, for each level K, we replace the E-scheme Sh_K by the E_v-scheme $\mathrm{Sh}_K \times_{\mathrm{Spec}(E)} \mathrm{Spec}(E_v)$, which with abuse of notations we still denote by Sh_K.

2.2.3 Let K be an open compact subgroup of $G(\mathbb{A}^\infty)$ as in Section 2.1.3, and consider the $\mathcal{O}_{E_v}$-scheme $S_K \times_{\mathrm{Spec}(\mathcal{O}_{E,(p)})} \mathrm{Spec}(\mathcal{O}_{E_v})$, which again with abuse of notations we still denote by S_K. We write $\overline{S}_K$ for its reduction in positive characteristic p, i.e. $\overline{S}_K = S_K \times_{\mathrm{Spec}(\mathcal{O}_{E_v})} \mathrm{Spec}(k)$, and $R\Psi_\eta^S = R\Psi_\eta^{S_K}$ for its vanishing cycle functor. For all integers $p, q \geq 0$, we define

$$\varinjlim{}_K H_c^p(\overline{S}_K \times_k k^{\mathrm{ac}}, R^q\Psi_\eta^S(\mathcal{L}_\xi)) = H_c^p(\overline{S}, R^q\Psi_\eta^S(\mathcal{L}_\xi)).$$

It follows from the above constructions that these l-adic spaces have a natural structure of $G(\mathbb{A}^\infty)^\dagger \times W_{E_v}$-modules. In fact, they are admissible/continuous $G(\mathbb{A}^\infty)^\dagger \times W_{E_v}$-modules. We define

$$H_c(\overline{S}, R\Psi_\eta^S(\mathcal{L}_\xi)) = \sum_{p,q}(-1)^{p+q} H_c^p(\overline{S}, R^q\Psi_\eta^S(\mathcal{L}_\xi)).$$

The main goal of this paper is to study these virtual $G(\mathbb{A}^\infty)^\dagger \times W_{E_v}$-modules.

2.2.4 Let K be of the form $K = K^p K_{p,0}$. Then, the above vanishing cycle sheaves are easy to compute. More precisely, it follows from the construction of the integral models $S_K = S_{K^p}$ that in this case the lisse étale sheaf $\mathcal{L}_\xi$ over Sh_K extends canonically to a lisse étale sheaf over S_{K^p} (which we also denote by $\mathcal{L}_\xi = \mathcal{L}_{\xi,K^p}$). Furthermore, given that the schemes S_{K^p} are smooth, the associated vanishing cycle sheaves $R^q\Psi_\eta^S(\mathcal{L}_\xi)$ over $\overline{S}_{K^p} \times_{\mathrm{Spec}(k)} \mathrm{Spec}(k^{\mathrm{ac}})$ vanish for all $q \geq 1$, and for $q = 0$ they are equal to the restriction of sheaf $\mathcal{L}_\xi/S_{K^p}$ to the geometric special fiber $\overline{S}_{K^p} \times_{\mathrm{Spec}(k)} \mathrm{Spec}(k^{\mathrm{ac}})$. In the following, we study the sheaves $R^q\Psi_\eta^{S_K}(\mathcal{L}_\xi)$, when the level K is of the form $K = K^p K_{p,m}$, for an integer $m \geq 1$.

2.2.5 We remark that in the case of proper Shimura varieties, the theory of vanishing cycle sheaves ([4]) enable us to express the cohomology of the Shimura varieties in terms of the above cohomology spaces. More precisely, we deduce an equality of virtual admissible/continuous $G(\mathbb{A}^\infty)^\dagger \times W_{E_v}$-modules

$$H(\mathrm{Sh}, \mathcal{L}_\xi) = H(\overline{S}, R\Psi_\eta^S(\mathcal{L}_\xi)),$$

where the action of $G(\mathbb{A}^\infty)^\dagger \times W_{E_v}$ on the left hand side is obtained by restriction from the action of $G(\mathbb{A}^\infty) \times W_{E_v}$. Furthermore, in [13] (Section 8, p. 607), we showed that, under the assumption of properness, the action of $G(\mathbb{A}^\infty)^\dagger \times W_{E_v}$ on the right hand side canonically extends to the whole group $G(\mathbb{A}^\infty) \times W_{E_v}$, and that the above equality holds as an equality of $G(\mathbb{A}^\infty) \times W_{E_v}$-modules. To be precise, in loc. cit. we prove these results for $\xi = \mathbb{Q}_l^{\mathrm{ac}}$ but the same arguments apply to any ξ. These results rely on a construction of integral correspondences which model the action of the whole $G(\mathbb{A}^\infty)$ on the Shimura varieties, as mentioned at the end of Section 2.1.3.

2.3 Newton stratification. We study the modules $H_c(\overline{S}, R\Psi_\eta^S(\mathcal{L}_\xi))$ following the strategy carried out in [13], in the case of $\xi = \mathbb{Q}_l^{\mathrm{ac}}$ the trivial representation. In this section, we recall the aspects of the general theory of local models for Shimura varieties of PEL type (following the work of Rapoport and Zink in [18]) and of the constructions in [13] which are relevant to extending the results to the case of ξ any finite dimensional l-adic representation. We refer to [13] (Section 5, pp. 589–596) for details.

2.3.1 To the global Shimura datum (G, h), we associate the local datum $(G_{\mathbb{Q}_p}, \mu_{\mathbb{Q}_p^{\mathrm{ac}}})$ where $G_{\mathbb{Q}_p} = G \otimes_{\mathbb{Q}} \mathbb{Q}_p$ and $\mu_{\mathbb{Q}_p^{\mathrm{ac}}}$ denotes the conjugacy class of cocharacters of $G_{\mathbb{Q}_p}$ determined by h and our choice of an embedding $\bar{v} : E^{\mathrm{ac}} \to E_v^{\mathrm{ac}}$, i.e. the class of cocharacters $\bar{v} \circ \mu_h$, for μ_h in h (here, $\mathbb{Q}_p^{\mathrm{ac}} = E_v^{\mathrm{ac}}$). In [11], to a local datum $(G_{\mathbb{Q}_p}, \mu_{\mathbb{Q}_p^{\mathrm{ac}}})$ Kottwitz associates a partially ordered finite set $B(G_{\mathbb{Q}_p}, \mu_{\mathbb{Q}_p^{\mathrm{ac}}})$. For $G_{\mathbb{Q}_p}$ of PEL type (which includes all cases coming from a global Shimura datum of PEL type), the set $B(G_{\mathbb{Q}_p}, \mu_{\mathbb{Q}_p^{\mathrm{ac}}})$ is canonically identified with the set of isogeny classes of Barsotti-Tate groups over k^{ac} with $\mu_{\mathbb{Q}_p^{\mathrm{ac}}}$-compatible $G_{\mathbb{Q}_p}$-structures. In our cases, these include (and conjecturally are exactly) the Barsotti-Tate groups arising as the p-divisible parts of the abelian varieties over k^{ac} classified by the

Shimura varieties associated with the corresponding global Shimura datum. Equivalently, $B(G_{\mathbb{Q}_p}, \mu_{\mathbb{Q}_p^{\mathrm{ac}}})$ is identified with a subset of the set of Newton polygons with end-points prescribed by $\mu_{\mathbb{Q}_p^{\mathrm{ac}}}$. Under the latter identification, the partial order on $B(G_{\mathbb{Q}_p}, \mu_{\mathbb{Q}_p^{\mathrm{ac}}})$ corresponds to the partial order "lying on or below" among convex polygons ([17], Section 1, pp. 155–164).

2.3.2 Let K be a sufficiently small open compact subgroup of $G(\mathbb{A}^\infty)$, and $\overline{S}_K$ the reduction modulo p of the corresponding Shimura variety. For each geometric point x of $\overline{S}_K$, we write A_x for the corresponding abelian varieties, and b_x for the isogeny class of the Barsotti-Tate group with additional structure $A_x[p^\infty]$. To each $b \in B(G_{\mathbb{Q}_p}, \mu_{\mathbb{Q}_p^{\mathrm{ac}}})$, we associate the set

$$\overline{S}_K(b) = \{x \in \overline{S}_K | b_x = b\}.$$

The set $\overline{S}_K(b)$ is a locally closed subspace of $\overline{S}_K$, and has a natural structure of locally closed subscheme defined over k. Furthermore, as $b \in B(G_{\mathbb{Q}_p}, \mu_{\mathbb{Q}_p^{\mathrm{ac}}})$ varies, the subschemes $\overline{S}_K(b)$ form a stratification of $\overline{S}_K$ which is known as the *Newton stratification.* (The definition of the Newton stratification is due to Rapoport and Richartz in [17]. Their work generalizes to the context of Shimura varieties a classical result of Grothendieck, in [6]). We recall that it is not known in general whether the strata $\overline{S}_K(b)$ are non-empty for all $b \in B(G_{\mathbb{Q}_p}, \mu_{\mathbb{Q}_p^{\mathrm{ac}}})$. In the following, we only consider elements $b \in B(G_{\mathbb{Q}_p}, \mu_{\mathbb{Q}_p^{\mathrm{ac}}})$ which are associated with non-empty strata. We remark that this condition is independent of the level K. It is an easy observation that, for each $b \in B(G_{\mathbb{Q}_p}, \mu_{\mathbb{Q}_p^{\mathrm{ac}}})$, as the level K varies, the strata $\overline{S}_K(b)$ form a projective system stable under the action of $G(\mathbb{A}^\infty)^\dagger$. It follows that, for all $p, q \geq 0$, the $\mathbb{Q}_l^{\mathrm{ac}}$-spaces

$$H_c^p(\overline{S}(b), R^q\Psi_\eta^S(\mathcal{L}_\xi)) = \varinjlim{}_K H_c^p(\overline{S}_K(b) \times_k k^{\mathrm{ac}}, R^q\Psi_\eta^S(\mathcal{L}_\xi)_{|\overline{S}_K(b)})$$

have a natural structure of admissible/continuous $G(\mathbb{A}^\infty)^\dagger \times W_{E_v}$-modules. For each $b \in B(G_{\mathbb{Q}_p}, \mu_{\mathbb{Q}_p^{\mathrm{ac}}})$, we define

$$H_c(\overline{S}(b), R\Psi_\eta^S(\mathcal{L}_\xi)) = \sum_{p,q}(-1)^{p+q} H_c^p(\overline{S}(b), R^q\Psi_\eta^S(\mathcal{L}_\xi)).$$

Then, there is an equality of virtual admissible/continuous $G(\mathbb{A}^\infty)^\dagger \times W_{E_v}$-modules

$$H_c(\overline{S}, R\Psi_\eta^S(\mathcal{L}_\xi)) = \sum_{b \in B(G_{\mathbb{Q}_p}, \mu_{\mathbb{Q}_p^{\mathrm{ac}}})} H_c(\overline{S}(b), R\Psi_\eta^S(\mathcal{L}_\xi)).$$

Furthermore, it follows from the remark in Section 2.2.5 that in the case of proper Shimura varieties the above equality holds as an equality of $G(\mathbb{A}^\infty) \times W_{E_v}$-modules.

2.4 Local models. Given the latter equality, we shift our focus to studying the modules $H_c(\overline{S}(b), R\Psi_\eta^S(\mathcal{L}_\xi))$, for $b \in B(G_{\mathbb{Q}_p}, \mu_{\mathbb{Q}_p^{\mathrm{ac}}})$. Our strategy is to establish a connection between these spaces and the cohomology of the corresponding local models. In the following, we recall the main aspects of the theory of local models for Shimura varieties of PEL type. We refer to [18] for a complete discussion of this theory, and to [12] (Section 7.2, pp. 298–302, and Section 8.2, pp. 317–327) for details and proof of the results in Sections 2.4.2 and 2.4.3.

2.4.1 Let $b \in B(G_{\mathbb{Q}_p}, \mu_{\mathbb{Q}_p^{ac}})$. In [18], Rapoport and Zink associate to the data $(G_{\mathbb{Q}_p}, \mu_{\mathbb{Q}_p^{ac}}, b)$ a projective system of rigid analytic spaces $\mathcal{M}_{b,K_p}$ over E_0, indexed by the open compact subgroups $K_p \subset K_{p,0}$ of $G(\mathbb{Q}_p)$, and naturally endowed with an action of $G(\mathbb{Q}_p)$ and a σ-semi-linear automorphism. For $K_p = K_{p,0}$, the rigid analytic space $\mathcal{M}_{b,K_{p,0}}$ is defined as the generic fiber of the formal $W(k^{ac})$-scheme representing the following set-valued functor M_b on the category of formal $W(k^{ac})$-schemes. We first define $M_b = M_{b,\mathbb{X}}$ as depending on the choice of a Barsotti-Tate group with additional structure $\mathbb{X}/k^{ac}$ in the isogeny class determined by b. We later observe that the resulting functor is up to isomorphism independent of such a choice. We remark that it suffices to define $M_{b,\mathbb{X}}(S)$ for any $W(k^{ac})$-scheme S where p is locally nilpotent. Let $M_{b,\mathbb{X}}(S)$ be the set of isomorphism classes of pairs (H, β) where H is a Barsotti-Tate group with additional structures defined over S, and $\beta : \mathbb{X}_{\bar{S}} \to H_{\bar{S}}$ is a quasi-isogeny defined over the subscheme $\bar{S} = Z(p)$ of S. (Here, $Z(p)$ denotes the subscheme defined by the equation $p = 0$, regarded as a k^{ac}-scheme.) The functor $M_{b,\mathbb{X}}$ is represented by a formal $W(k^{ac})$-scheme $\mathcal{M}_{b,\mathbb{X}}$, which is formally smooth and formally locally of finite type. The formal scheme $\mathcal{M}_{b,\mathbb{X}}$ is naturally endowed with an action of the group of quasi-self-isogenies of $\mathbb{X}$, by right translations. We denote the group of quasi-self-isogenies of $\mathbb{X}$ by $J_b(\mathbb{Q}_p)$, since it may be canonically identified with the $\mathbb{Q}_p$-points of an algebraic group $J_b/\mathbb{Q}_p$, depending only on $b \in B(G_{\mathbb{Q}_p}, \mu_{\mathbb{Q}_p^{ac}})$. Furthermore, the scheme $\mathcal{M}_{b,\mathbb{X}}$ is also endowed with a σ-semi-linear automorphism, which is defined by left translation by the inverse of the q-th power of the Frobenius of $\mathbb{X}$, and which commutes with the previous action of $J_b(\mathbb{Q}_p)$. In the following, we refer to this automorphism as an (unramified) action of W_{E_v}. Finally, for each pair of non-negative integers $n, d \geq 0$, we define the truncated Rapoport-Zink space $\mathcal{M}_{b,\mathbb{X}}^{n,d}$ to be the formal subscheme of $\mathcal{M}_{b,\mathbb{X}}$ classifying pairs (H, β) such that the quasi-isogenies $p^n\beta$ and $p^{d-n}\beta^{-1}$ are isogenies. For all n, d, the natural inclusions $\mathcal{M}_{b,\mathbb{X}}^{n,d} \to \mathcal{M}_{b,\mathbb{X}}$ are closed embeddings and, as n, d vary, they define a direct system. Let $\mathcal{M}_{b,K_{p,0}}$ denote the generic fiber of the formal $W(k^{ac})$-scheme $\mathcal{M}_{b,\mathbb{X}}$, regarded as a rigid analytic E_0-space. For each $K_p \subset K_{p,0}$, we define $\mathcal{M}_{b,K_p}$ as the space over $\mathcal{M}_{b,K_{p,0}}$ parameterizing K_p-level structures on the universal Barsotti-Tate group $H/\mathcal{M}_{b,K_{p,0}}$ (here, by a K_p-level structure on H we mean the K_p-orbit of a trivialization of the Tate module of H). For each $K_p \subset K_{p,0}$, the rigid analytic space $\mathcal{M}_{b,K_p}$ is a finite étale Galois cover of $\mathcal{M}_{b,K_{p,0}}$. In particular, it is a smooth rigid analytic space over E_0, of dimension $D_b = \dim \mathcal{M}_{b,K_{p,0}}$.

2.4.2 As K_p varies, the spaces $\mathcal{M}_{b,K_p}$ form a projective system naturally endowed with an action of $G(\mathbb{Q}_p) \times J_b(\mathbb{Q}_p) \times W_{E_v}$. For each $i \geq 0$, we define

$$H_c^i(\mathcal{M}_b, \mathbb{Q}_l^{ac}) = \varinjlim{}_{K_p} H_c^i(\mathcal{M}_{b,K_p} \times_{E_0} E_v^{ac}, \mathbb{Q}_l^{ac}).$$

Then, they are smooth/continuous $G(\mathbb{Q}_p) \times J_b(\mathbb{Q}_p) \times W_{E_v}$-modules, but it is not known whether they are also admissible/continuous. On the other hand, to each admissible representation ρ of $J_b(\mathbb{Q}_p)$, we may associated the $G(\mathbb{Q}_p) \times W_{E_v}$-modules $\mathrm{Ext}^p_{J_b(\mathbb{Q}_p)}(H_c^q(\mathcal{M}_b, \mathbb{Q}_l^{ac}), \rho)$. These modules are admissible/continuous and vanish for almost all $p, q \geq 0$. For each $b \in B(G_{\mathbb{Q}_p}, \mu_{\mathbb{Q}_p^{ac}})$, we define a functor

$$\mathcal{E}_b : \mathrm{Groth}(J_b(\mathbb{Q}_p)) \to \mathrm{Groth}(G(\mathbb{Q}_p) \times W_{E_v})$$

which to an admissible virtual representation ρ of $J(\mathbb{Q}_p)$ associates

$$\mathcal{E}_b(\rho) = \sum_{p,q} (-1)^{p+q} \mathrm{Ext}^p_{J_b(\mathbb{Q}_p)}(H^q_c(\mathcal{M}_b, \mathbb{Q}_l^{\mathrm{ac}}), \rho)(-D_b).$$

In the following we also denote by $\mathcal{E}_b$ the functor from $\mathrm{Groth}(J_b(\mathbb{Q}_p) \times G(\mathbb{A}^{\infty,p}))$ to $\mathrm{Groth}(G(\mathbb{A}^\infty) \times W_{E_v})$ which is obtain by extending the above functor by the identity on $\mathrm{Groth}(G(\mathbb{A}^{\infty,p}))$.

2.4.3 It follows from Berkovich's theory of vanishing cycles of formal schemes ([3], Corollary 2.5, p. 373), together with the fact that all the irreducible components of the reduced scheme associated with $\mathcal{M}_{b,\mathbb{X}}$ are projective k^{ac}-schemes ([18], Proposition 2.32, p. 63), that the above functor can also be computed in terms of the cohomology groups of the reduced closed fiber of $\mathcal{M}_{b,\mathbb{X}}$ with coefficients in the appropriate vanishing cycle sheaves. Indeed, let us restrict our attention to levels K_p of the form $K_p = K_{p,m}$, $m \geq 0$. Then, similarly to the case of the integral models of Shimura varieties of PEL type, we can apply Katz and Mazur's notion of full set of sections for finite flat group schemes to construct integral models for the Rapoport-Zink spaces. More precisely, for each $K_p = K_{p,m}$, $m \geq 0$, there exists a formal $W(k^{\mathrm{ac}})$-scheme $\mathcal{M}_{b,\mathbb{X},K_p}$ satisfying the following properties. It is finite over $\mathcal{M}_{b,\mathbb{X}} = \mathcal{M}_{b,\mathbb{X},K_{p,0}}$, and its generic fiber is canonically isomorphic to $\mathcal{M}_{b,K_p}$. Furthermore, as K_p varies, the formal schemes $\mathcal{M}_{b,\mathbb{X},K_p}$ form a projective system endowed with an action of $J_b(\mathbb{Q}_p) \times G(\mathbb{Q}_p)^\dagger \times W_{E_v}$, which is compatible under the above isomorphisms with the action of $J_b(\mathbb{Q}_p) \times G(\mathbb{Q}_p) \times W_{E_v}$ on the rigid analytic Rapoport-Zink spaces $\mathcal{M}_{b,K_p}$. For each $K_p = K_{p,m}$, we write $\overline{\mathcal{M}}_{b,K^p} = \overline{\mathcal{M}}_{b,\mathbb{X},K_p}$ for the reduced scheme associated with $\mathcal{M}_{b,\mathbb{X},K^p}$, regarded as a k^{ac}-scheme, and denote by $R\Psi_\eta^{\mathcal{M}} = R\Psi_\eta^{\mathcal{M}_{K_p}}$ its vanishing cycle functor. Then, for each $p, q \geq 0$, we define the $\mathbb{Q}_l^{\mathrm{ac}}$-space

$$H^p_c(\overline{\mathcal{M}}_b, R^q\Psi_\eta^{\mathcal{M}}(\mathbb{Q}_l^{\mathrm{ac}})) = \varinjlim{}_{K_p} H^p_c(\overline{\mathcal{M}}_{b,K_p}, R^q\Psi_\eta^{\mathcal{M}_{K_p}}(\mathbb{Q}_l^{\mathrm{ac}})).$$

For all $p, q \geq 0$, these are smooth/continuous $G(\mathbb{Q}_p) \times J_b(\mathbb{Q}_p) \times W_{E_v}$-modules. Moreover, for each admissible representation ρ of $J_b(\mathbb{Q}_p)$, the $G(\mathbb{Q}_p)^\dagger \times W_{E_v}$-modules $\mathrm{Tor}^i_{J_b(\mathbb{Q}_p)}(H^p_c(\overline{\mathcal{M}}_b, R^q\Psi_\eta^{\mathcal{M}}(\mathbb{Q}_l^{\mathrm{ac}})), \rho)$ are admissible/continuous and vanish for almost all $i, p, q \geq 0$. Similarly to the case of integral models for Shimura varieties, by considering a class of integral models for the Rapoport-Zink spaces larger than that introduced here, it is possible to canonically extend the smooth/continuous (resp. admissible /continuous) action of $G(\mathbb{Q}_p)^\dagger \times W_{E_v}$ on the above $\mathbb{Q}_l^{\mathrm{ac}}$-spaces to a smooth/continuous (resp. admissible/continuous) action of $G(\mathbb{Q}_p) \times W_{E_v}$. Furthermore, for each ρ, there is an equality of virtual $G(\mathbb{Q}_p) \times W_{E_v}$-modules

$$\mathcal{E}_b(\rho) = \sum_{i,p,q} (-1)^{i+p+q} \mathrm{Tor}^i_{J_b(\mathbb{Q}_p)}(H^p_c(\overline{\mathcal{M}}_b, R^q\Psi_\eta^{\mathcal{M}}(\mathbb{Q}_l^{\mathrm{ac}})), \rho).$$

2.5 Igusa varieties. In order to establish a connection between the cohomology of the Shimura varieties and that of their local models, we introduce a new kind of moduli spaces for abelian varieties in positive characteristic, the Igusa varieties. Igusa curves were first introduced by Igusa in the theory of modular curves ([8], or [9], chapters 12 and 13, pp. 344–450), and later generalized in the context of some simple Shimura varieties by Harris and Taylor ([7]). Their more general definition in the context of Shimura varieties of PEL type relies on work of Oort on foliations in moduli spaces of abelian varieties in positive characteristic ([15]), and on results of Zink ([19]), and of Oort and Zink ([16]) on the existence of the slope filtration for p-divisible groups. We refer to [13] (Section 4, pp. 584–589) for details.

2.5.1 Let $b \in B(G_{\mathbb{Q}_p}, \mu_{\mathbb{Q}_p^{\mathrm{ac}}})$, and fix a level $K^p \subset G(\mathbb{A}^{\infty,p})$. We assume $\overline{S}_{K^p}(b)$ is not empty, and choose x a closed geometric point of $\overline{S}_{K^p}(b)$, defined over k^{ac}. We write $\mathbb{X} = A_x[p^\infty]$. The Barsotti-Tate group $\mathbb{X}/k^{\mathrm{ac}}$ admits a unique slope decomposition, $\mathbb{X} = \oplus_1^t \mathbb{X}^i$, into isoclinic Barsotti-Tate subgroups $\mathbb{X}^i$, each endowed with the appropriate inherited additional structures (i.e. an action of $\mathcal{O}_{B_{\mathbb{Q}_p}}$, and isomorphisms $\ell^i : \mathbb{X}^i \cong (\mathbb{X}^{t+1-i})^\vee$ induced by the quasi-polarization ℓ of $\mathbb{X}$). As before, we write $J_b(\mathbb{Q}_p)$ for the group of quasi-self-isogenies of $\mathbb{X}$. We recall that the associated algebraic group $J_b/\mathbb{Q}_p$ is an inner form of the Levi subgroup of $G_{\mathbb{Q}_p}$ associated with the slope decomposition of the isocrystals of $\mathbb{X}$.

2.5.2 We say that a Barsotti-Tate group with additional structures Y defined over field $k' \supset k^{\mathrm{ac}}$ is geometrically isomorphic to $\mathbb{X}$ (we write $Y \cong_g \mathbb{X}$) if there exists a field extension $k'' \supset k'$ over which the two become isomorphic. We define $C = C_x = C_{\mathbb{X}}$, the *central leaf* through x (or equivalently, associated to $\mathbb{X}$), as

$$C_x = \{y \in \overline{S}_{K^p}(b) | A_y[p^\infty] \cong_g \mathbb{X}\}.$$

Then, C_x is a closed subspace of $\overline{S}_{K^p}(b)$ defined over k^{ac}, which is a smooth scheme when endowed with the reduced subscheme structure. We regard C as a smooth closed k^{ac}-subscheme of $\overline{S}_{K^p}(b)$. For each K^p, we also write $C = C_{K^p} \subset \overline{S}_{K^p}(b)$. Then, as K^p varies, the schemes C_{K^p} naturally form a projective system endowed with an action of $G(\mathbb{A}^{\infty,p}) \subset G(\mathbb{A}^\infty)^\dagger$.

2.5.3 Let $C = C_{\mathbb{X},K^p}$. Without loss of generality, we assume $\mathbb{X}$ is completely slope divisible. We write H/C for the p-divisible part of the restriction of the universal abelian scheme $A/\overline{S}_{K^p}(b)$ to C. Then, it follows from the work of Zink ([19]), and of Oort and Zink ([16]), that the Barsotti-Tate group H/C is also completely slope divisible. More precisely, H admits a canonical slope filtration $0 = H_0 \subset H_1 \subset \cdots H_t = H$, into Barsotti-Tate subgroups with slope divisible Barsotti-Tate subquotients $H^i = H_i/H_{i-1}$. Each H^i/C is a Barsotti-Tate group endowed with appropriate additional structures inherited by those of H. Moreover, it follows from the definition of C that, for each point $y \in C$, the Barsotti-Tate groups with additional structures H^i_y are geometrically isomorphic to $\mathbb{X}^i$, for all i. We write $H^{sp} = \oplus_1^t H^i$, also regarded as a Barsotti-Tate group with additional structure (its additional structures controlled by the Levi subgroup of $G_{\mathbb{Q}_p}$ which is an inner form of J_b). We observe that, for each $y \in C$, the Barsotti-Tate group H^{sp}_y is geometrically isomorphic (via a canonical isomorphism) to H_y. (This follows from the fact that the slope filtration of a Barsotti-Tate group canonically splits over an algebraically closed field.)

2.5.4 For each $m \geq 1$, we define the *Igusa variety of level* m as the space $\mathrm{Ig}_m = \mathrm{Ig}_{\mathbb{X},K^p,m}$ over $C_{\mathbb{X},K^p}$ parameterizing isomorphisms of truncated Barsotti-Tate groups with additional structures $j_m : \mathbb{X}[p^m]_C \simeq H^{sp}[p^m]$. (Here, by an isomorphism of truncated Barsotti-Tate group with additional structures, we mean an isomorphism of the underlying finite flat group schemes, which commutes with the additional structures and which extends étale locally to any truncation of higher depth $m' \geq m$.) It follows from the definition that, for each $m \geq 1$, the Igusa variety Ig_m is finite étale and Galois over C. Thus, in particular, it is a smooth scheme over k^{ac}. Ig_m is naturally endowed with an action of the group of automorphism of $\mathbb{X}$. Further more, as K^p and m vary, the schemes $\mathrm{Ig}_{K^p,m}$ naturally form a projective system endowed with commuting actions of $G(\mathbb{A}^{\infty,p}) \subset G(\mathbb{A}^\infty)^\dagger$ and the group of automorphism of $\mathbb{X}$. If we regard the group of automorphism of $\mathbb{X}$ as an open compact subgroup of $J_b(\mathbb{Q}_p)$, then its action on the projective system of Igusa varieties naturally extends to a larger subgroup which we denote by $J_b(\mathbb{Q}_p)^\dagger$ ([13], Section 4, Lemma 5, pp. 586–587).

2.5.5 Let ξ be a finite dimensional l-adic representation of G and $\mathcal{L}_\xi$ the associated lisse étale sheaf over the models S_{K^p} as in Section 2.2.4. For all K^p, m as above, we write $\mathcal{L}_{\xi\,|\mathrm{Ig}_{\mathbb{X},K^p,m}}$ for the pullback of the sheaf $\mathcal{L}_\xi/S_{K^p}$ under the natural morphism $\mathrm{Ig}_{\mathbb{X},K^p,m} \to C_{\mathbb{X},K^p} \subset \bar{S}_{K^p} \subset S_{K^p}$. For each $i \geq 0$, we define

$$H^i_c(\mathrm{Ig}_{\mathbb{X}}, \mathcal{L}_\xi) = \varinjlim{}_{K^p,m} H^i_c(\mathrm{Ig}_{\mathbb{X},K^p,m}, \mathcal{L}_{\xi\,|\mathrm{Ig}_{\mathbb{X},K^p,m}}).$$

These $\mathbb{Q}_l^{\mathrm{ac}}$-vector spaces have a natural structure of $G(\mathbb{A}^{\infty,p}) \times J_b(\mathbb{Q}_p)^\dagger$-modules, which canonically extends to a structure of $G(\mathbb{A}^{\infty,p}) \times J_b(\mathbb{Q}_p)$-modules. Further more, as $G(\mathbb{A}^{\infty,p}) \times J_b(\mathbb{Q}_p)$-modules they are admissible. We define

$$H_c(\mathrm{Ig}_{\mathbb{X}}, \mathcal{L}_\xi) = \sum_i (-1)^i H^i_c(\mathrm{Ig}_{\mathbb{X}}, \mathcal{L}_\xi) \in \mathrm{Groth}(G(\mathbb{A}^{\infty,p}) \times J_b(\mathbb{Q}_p)).$$

We remark that the above modules depend only on the isogeny class of $\mathbb{X}$. In the following, given $b \in B(G_{\mathbb{Q}_p}, \mu_{\mathbb{Q}_p^{\mathrm{ac}}})$, we sometimes write $H_c(\mathrm{Ig}_b, \mathcal{L}_\xi) = H_c(\mathrm{Ig}_{\mathbb{X}}, \mathcal{L}_\xi)$, for any choice of a Barsotti-Tate group $\mathbb{X}$ in the isogeny class determined by b.

3 Statement of the main results

We are finally read to state the main theorem of this paper, as well as the intermediate results necessary to adapt to this more general context the strategy of [12] and [13].

Theorem 3.1 *Maintaining the above notations. Let ξ be a representation of G on a finite dimensional $\mathbb{Q}_l^{\mathrm{ac}}$-vector space. For each $b \in B(G_{\mathbb{Q}_p}, \mu_{\mathbb{Q}_p^{ac}})$, there is an equality of virtual $G(\mathbb{A}^\infty)^\dagger \times W_{E_v}$-modules*

$$H_c(\overline{S}(b), R\Psi^S_\eta(\mathcal{L}_\xi)) = \mathcal{E}_b(H_c(Ig_b, \mathcal{L}_\xi)),$$

where $\mathcal{E}_b$ is the functor associated with the local models of the Shimura varieties attached to b. In the case of proper Shimura varieties, the above equalities can be extended to the equalities of $G(\mathbb{A}^\infty) \times W_{E_v}$-modules, for all $b \in B(G_{\mathbb{Q}_p}, \mu_{\mathbb{Q}_p^{ac}})$. In particular, we deduce an equality of $G(\mathbb{A}^\infty) \times W_{E_v}$-modules

$$H(Sh, \mathcal{L}_\xi) = \sum_{b \in B(G_{\mathbb{Q}_p}, \mu_{\mathbb{Q}_p^{ac}})} \mathcal{E}_b(H(Ig_b, \mathcal{L}_\xi)).$$

For $\xi = \mathbb{Q}_l^{\mathrm{ac}}$ the trivial representation, the above theorem was proven in [13] (Theorem 22, p. 607). While the above formula was previously only established for cohomology with trivial coefficients, many of the constructions and intermediate results in [12] and [13] were carried out in larger generality. In the following, we recall those constructions and point out the results about the local geometry of the Shimura varieties and their vanishing cycle sheaves which underlie the formula.

3.1 Strategy. We fix K^p a sufficiently small open compact subgroup of $G(\mathbb{A}^{\infty,p})$. Let $b \in B(G_{\mathbb{Q}_p}, \mu_{\mathbb{Q}_p^{\mathrm{ac}}})$ and assume $\overline{S}(b) = \overline{S}_{K^p}(b)$ is not empty. We choose $\mathbb{X}$ a Barsotti-Tate group with additional structures associated with a closed geometric point of $\overline{S}(b)$, and write $\overline{\mathcal{M}} = \overline{\mathcal{M}}_{b,\mathbb{X}}$ (resp. $\overline{\mathcal{M}}^{n,d} = \overline{\mathcal{M}}^{n,d}_{b,\mathbb{X}}$, $n, d \geq 0$) for the reduced closed fiber of the local model $\mathcal{M}_{b,\mathbb{X}}$ (resp. $\mathcal{M}^{n,d}_{b,\mathbb{X}}$). For each $m \geq 0$, we also write $\mathrm{Ig}_m = \mathrm{Ig}_{\mathbb{X},K^p,m}$ for the Igusa variety of level m, associated with our choice of $b, \mathbb{X}$, and K^p. For all $m, n, d \geq 0$, we define

$$q_1 : \mathrm{Ig}_m \times \overline{\mathcal{M}}^{n,d} \to \mathrm{Ig}_0 \text{ and } q_2 : \mathrm{Ig}_m \times \overline{\mathcal{M}}^{n,d} \to \overline{\mathcal{M}},$$

respectively as the first projection of the product composed with the cover map $\mathrm{Ig}_m \to \mathrm{Ig}_0$, and the second projection of the product composed with the natural inclusion $\overline{\mathcal{M}}^{nd} \hookrightarrow \overline{\mathcal{M}}$. Finally, for any level K of the form $K = K^p K_p$, with $K_p = K_{p,m} \subset K_{p,0}$, we write $K_0 = K^p K_{p,0}$, and denote by $f_{K_p} : \mathrm{Sh}_K \to \mathrm{Sh}_{K_0}$ and $g_{K_p} : \mathcal{M}_{b,K_p} \to \mathcal{M}_{b,K_{p,0}}$ the cover maps between Shimura varieties and Rapoport-Zink spaces corresponding respectively to the inclusions $K \subset K_0$ and $K_p \subset K_{p,0}$.

3.1.1 In [13] (Proposition 9, p.592) we construct a system of finite k^{ac}-morphisms

$$\pi_{(N,m,n,d)} : \mathrm{Ig}_m \times \overline{\mathcal{M}}^{n,d} \to \overline{S}(b)_{k^{\mathrm{ac}}} = \overline{S}(b) \times_k k^{\mathrm{ac}},$$

indexed by quadruples of non-negative integers (N, m, n, d) satisfying certain conditions (namely, we require that m, N are sufficiently large with respect to d). As m, n, d vary, the morphisms $\pi_{(N,m,n,d)}$ form a system compatible under the natural projections among Igusa varieties and inclusion of truncated Rapoport-Zink spaces, and as N varies, they are compatible up to powers of the relative Frobenius F of $\overline{S}(b)/k$. Moreover, as a system, they are $J_b(\mathbb{Q}_p)^\dagger$-equivariant, for the natural action of $J_b(\mathbb{Q}_p)^\dagger \subset J_b(\mathbb{Q}_p)$ on the system of products $\mathrm{Ig}_m \times \overline{\mathcal{M}}^{n,d}$, and the trivial action on $\overline{S}(b)$. In the following, with abuse of notations we sometime write $\pi_N = \pi_{(N,m,n,d)}$. (We postpone recalling their definition to Section 5, proof of proposition 5.2.) The special case $\xi = \mathbb{Q}_l^{\mathrm{ac}}$ of Theorem 3.1 is proved in [13] by combining together the following results.

Theorem 1 *[[12], Theorem 5.11, p. 287] Let $\mathcal{L}$ be an abelian torsion étale sheaf over $\overline{S}(b)_{k^{\mathrm{ac}}}$, with torsion orders prime to p and endowed with an action of W_{E_v}. There exists a W_{E_v}-equivariant spectral sequence*

$$E_2^{p,q} = H_q(J_b(\mathbb{Q}_p), \varinjlim{}_{m,n,d} H_c(Ig_m \times \overline{\mathcal{M}}^{n,d}, \pi_N^*\mathcal{L})) \Rightarrow H_c^{p+q}(\overline{S}(b) \times_k k^{ac}, \mathcal{L}).$$

Theorem 2 *[[12], Theorem 5.13, p. 281] Let $r \geq 1$, and $\mathcal{L}/\overline{S}(b)_{k^{\mathrm{ac}}}$, $\mathcal{G}/\overline{\mathcal{M}}$ be two étale sheaves of $\mathbb{Z}/l^r\mathbb{Z}$-modules, which are endowed with an action of W_{E_v}. We assume that there exists a system of W_{E_v}-equivariant isomorphisms, indexed by the quadruples (N, m, n, d),*

$$\pi_N^*\mathcal{L} \cong q_2^*\mathcal{G},$$

which are compatible under the morphisms induced by the projections among the Igusa varieties, the inclusion of truncated Rapoport-Zink spaces and the action of $J_b(\mathbb{Q}_p)^\dagger$. Then, there exists a W_{E_v}-equivariant spectral sequence

$$E^{p,q} = \oplus_{t+s=q} Tor^p_{J_b(\mathbb{Q}_p)}(H_c^s(\overline{\mathcal{M}}, \mathcal{G}_2), H_c^t(Ig, \mathbb{Z}/l^r\mathbb{Z})) \Rightarrow H_c^{p+q}(\overline{S}(b) \times_k k^{ac}, \mathcal{L}).$$

Theorem 3 *[[13], Proposition 21, p. 605] Let $K = K^pK_p$ be a level as in Section 2.1.3, $b \in B(G_{\mathbb{Q}_p}, \mu_{\mathbb{Q}_p^{ac}})$. For each $r \geq 1$, there exists a system of W_{E_v}-quasi-isomorphism of complexes, indexed by the quadruples (N, m, n, d),*

$$\pi_N^*(R\Psi_\eta^S Rf_{K_p*}(\mathbb{Z}/l^r\mathbb{Z})_{|\overline{S}(b)_{k^{ac}}}) \cong q_2^*(R\Psi_\eta^{\mathcal{M}} Rg_{K_p*}(\mathbb{Z}/l^r\mathbb{Z})),$$

which are compatible under the morphisms induced by the projections among the Igusa varieties, the inclusion of truncated Rapoport-Zink spaces and the action of $J_b(\mathbb{Q}_p)^\dagger$. Furthermore, they are also compatible with the action of $G(\mathbb{Q}_p)^\dagger$, as the level K varies.

3.1.2 We state the appropriate analogues of Theorems 2 and 3 in the context of this paper.

Theorem 3.2 *Let $r \geq 1$, and $\mathcal{L}/\overline{S}(b)_{k^{ac}}$, $\mathcal{G}_1/Ig_0$ and $\mathcal{G}_2/\overline{\mathcal{M}}$ be some étale sheaves of $\mathbb{Z}/l^r\mathbb{Z}$-modules, which are endowed with an action of W_{E_v}. We assume that there exists a system of W_{E_v}-equivariant isomorphisms, indexed by the quadruples (N, m, n, d),*

$$\pi_N^*\mathcal{L} \cong q_1^*\mathcal{G}_1 \otimes q_2^*\mathcal{G}_2,$$

which are compatible under the morphisms induced by the projections among the Igusa varieties, the inclusion of truncated Rapoport-Zink spaces and the action of $J_b(\mathbb{Q}_p)^\dagger$. Then, there exists a W_{E_v}-equivariant spectral sequence

$$E^{p,q} = \oplus_{t+s=q} Tor^p_{J_b(\mathbb{Q}_p)}(H_c^s(\overline{\mathcal{M}}, \mathcal{G}_2), H_c^t(Ig, \mathcal{G}_1)) \Rightarrow H_c^{p+q}(\overline{S}(b) \times_k k^{\mathrm{ac}}, \mathcal{L}).$$

Theorem 3.3 *Let ξ be a representation of G on a finite dimensional $\mathbb{Q}_l^{ac}$-vector space, $b \in B(G_{\mathbb{Q}_p}, \mu_{\mathbb{Q}_p^{ac}})$, and $K = K^pK_p$ a level as in Section 2.1.3. There exists a system of W_{E_v}-quasi-isomorphism of complexes, indexed by the quadruples (N, m, n, d),*

$$\pi_N^*(R\Psi_\eta^S Rf_{K_p*}(\mathcal{L}_\xi)_{|\overline{S}(b)_{k^{ac}}}) \cong q_1^*(\mathcal{L}_{\xi\,|Ig_0}) \otimes q_2^*(R\Psi_\eta^{\mathcal{M}} Rg_{K_p*}(\mathbb{Q}_l^{ac})),$$

which are compatible under the morphisms induced by the projections among the Igusa varieties, the inclusion of truncated Rapoport-Zink spaces and the action of $J_b(\mathbb{Q}_p)^\dagger$. Furthermore, they are also compatible with the action of $G(\mathbb{Q}_p)^\dagger$, as the level K varies.

Proof of Theorem 3.1 In view of the results discussed in Section 2, Theorem 3.1 is an immediate consequence of Theorems 3.2 and 3.3. In particular, in the case of proper Shimura varieties, following the discussions in Sections 2.2.5 and 2.4.3, we can deduce the second statement from Theorems 3.2 and 3.3, as long as we include in Theorem 3.3 the case of the vanishing cycle sheaves associated with a larger class of integral models. □

Proof of Theorem 3.2 The techniques used in [12] to prove Theorem 2 apply directly to this more general case. (Indeed, Theorem 2 is a special case of Theorem 3.2, namely the case of $\mathcal{G}_1 = \mathbb{Q}_l^{\mathrm{ac}}$.) More precisely, after applying the Künneth formula for étale cohomology to the products of Igusa varieties and Rapoport-Zink spaces, the statement follows from Theorem 1. □

The next two sections are dedicated to proving Theorem 3.3.

4 Reduction to a special case

In the previous section, we stated our results for ξ a finite dimensional l-adic representation of G. It is an easy observation that in fact it is sufficient to consider the cases when ξ is irreducible, since the general case would then follow from those. In this section, we explain how the special case of ξ equal to the exterior algebra of sums of copies of the dual of the representation V of G implies all the cases of ξ irreducible and thus the general statement. Further more, we explain how these cases are related to the study of the local geometry of the multi-self products of the universal abelian scheme over the Shimura varieties.

4.1 Preliminaries. Let V_0 denote the dual of the representation $V \otimes \mathbb{Q}_l^{\mathrm{ac}}$ of G. We write $\mathbb{Q}_l^{\mathrm{ac}}(1)$ for the l-adic representation of G defined by the character ν. Thus, we have $V_0 = V \otimes \mathbb{Q}_l^{\mathrm{ac}}(-1)$. For any integer $s \geq 1$, let $\wedge^\bullet V_0^s$ be the exterior algebra of the sum of s copies of V_0. (In the following, we also allow $s = 0$ in which case $\wedge^\bullet V_0^0$ denotes the trivial representation $\mathbb{Q}_l^{\mathrm{ac}}$.) These l-adic representations of G play a special role in the theory, since the associated lisse étale sheaves have a natural geometric interpretation, as described below.

4.1.1 Let K be a sufficiently small open compact subgroup of $G(\mathbb{A}^\infty)$. We denote by $\alpha_K : A_K \to \mathrm{Sh}_K$ the universal abelian scheme over Sh_K, and by $\alpha_K^s : A_K^s \to \mathrm{Sh}_K$ its s-fold self-product over Sh_K, for all $s \geq 1$. (In the following, we also allow $s = 0$ in which case by A_K^0 we simply mean Sh_K. We also sometime omit the index K from the notations.) We define $\mathcal{L} = \mathcal{L}_K = R^1\alpha_{K*}\mathbb{Q}_l^{\mathrm{ac}}$. Then, $\mathcal{L}$ is a lisse étale sheaf over Sh_K, and more precisely it is the lisse étale sheaf associated with the representation V_0. More generally, for any integers $s, t \geq 0$, the sheaf $\mathcal{L}_{s,t} = \mathcal{L}_{s,t,K} = R^t\alpha_{K*}^s\mathbb{Q}_l^{\mathrm{ac}}$ over Sh_K is the lisse étale sheaf associated with the representation $\wedge^t V_0^s$.

4.1.2 For any integer $s \geq 0$, let E_s be the $\mathbb{Q}_l^{\mathrm{ac}}$-algebra $\mathrm{End}(\wedge^\bullet V_0^s)$. We recall that an element $u \in E_s$ is said to be homogeneous of degree t if its image is contained in $\wedge^t V_0^s \subset \wedge^\bullet V_0^s$. Weyl's theorem on complete reducibility implies that every finite dimensional representation of G over $\mathbb{Q}_l^{\mathrm{ac}}$ is completely reducible. Furthermore, in our context, every irreducible representation of G appears in an exterior algebra of a sum of copies of V_0. ([5], Section 15.3, Proposition 15.15 for type A_{n-1}; Section 17.3, Theorems 17.11, 17.21 for type C_n; Section 19.5, Theorems 19.19, 19.22 for type D_n. In loc. cit., these results are stated for isometry groups but can be adapted to similitude groups.) More precisely, for any irreducible representation ξ of G on a finite dimensional $\mathbb{Q}_l^{\mathrm{ac}}$-vector space W_ξ, there exist two integers $s = s_\xi \geq 0$, $t = t_\xi \geq 0$ and an idempotent $\epsilon = \epsilon_\xi \in E_s$ such that ϵ is homogeneous of degree t, and

$$W_\xi = \epsilon(\wedge^\bullet V_0^s).$$

For each ξ, the idempotent ϵ_ξ lies in the subalgebra of E_s, $s = s_\xi$, generated by $\mathrm{End}(V_0^s)$ and certain endomorphisms of $\wedge^\bullet V_0^s$ associated with the bilinear form on V, regarded as an element in $\wedge^2 V_0^?(1)$.

4.1.3 We deduce that, for each irreducible finite dimensional representation ξ of G, there exists a natural isomorphism of lisse étale sheaves over Sh_K

$$\mathcal{L}_\xi \simeq \epsilon_\xi(R^{t_\xi}\alpha_*^{s_\xi}\mathbb{Q}_l^{\mathrm{ac}}),$$

and moreover $\epsilon_\xi(R^t\alpha_*^{s_\xi}\mathbb{Q}_l^{\mathrm{ac}}) = 0$ for all $t \neq t_\xi$. Furthermore, for each irreducible representation ξ, the action of the associated idempotent $\epsilon_\xi \in E_s$ on the sheaves $R^t\alpha_*^s\mathbb{Q}_l^{\mathrm{ac}}$, for $t \geq 0$, $s = s_\xi$, has a natural geometric interpretation. It can be defined using the endomorphism structure, polarizations, permutations and multiplication by scalars on the s-fold self-product of the universal abelian scheme. This observation justifies what follows in Sections 4.2 and 4.3 where these idempotents are used in both the generic and special fibers.

Proposition 4.1 *Maintaining the above notations. For each $i \geq 0$, there exists a canonical Galois equivariant isomorphism of $\mathbb{Q}_l^{ac}$-spaces*

$$H_c^i(Sh_K \times_E E^{ac}, \mathcal{L}_\xi) \simeq \epsilon_\xi H_c^{i+t_\xi}(A_K^{s_\xi} \times_E E^{ac}, \mathbb{Q}_l^{ac}),$$

and an equality of virtual admissible/continuous $G(\mathbb{A}^\infty) \times Gal(E^{ac}/E)$-modules

$$H_c(Sh, \mathcal{L}_\xi) = (-1)^{t_\xi}\epsilon_\xi H_c(A^{s_\xi}, \mathbb{Q}_l),$$

where $H_c(A^{s_\xi}, \mathbb{Q}_l) = \sum_i (-1)^i \varinjlim_K H_c^i(A_K^{s_\xi} \times_E E^{ac}, \mathbb{Q}_l(t_\xi))$.

Proof The proposition is an immediate consequence of the aforementioned properties of the idempotents ϵ_ξ and the properness of $\alpha_K : A_K \to \mathrm{Sh}_K$. □

4.2 Cohomology groups with coefficients in $\mathcal{L}_\xi$. In the following, we proceed to adapt the previous observations to the context of the modules associated with the cohomology groups of the reduction of the Shimura varieties, of their Newton strata and of the Igusa varieties.

4.2.1 Let K be an open compact subgroup of $G(\mathbb{A}^\infty)$ as in Section 2.1.3. For each $s \geq 0$, we write A_K^s for the s-fold self product of the universal abelian scheme over S_K, regarded as a scheme over $\mathcal{O}_{E_v}$. We denote by $\overline{A}_K^s$ its reduction in positive characteristic p, i.e. $\overline{A}_K^s = A_K^s \times_{\mathrm{Spec}(\mathcal{O}_{E_v})} \mathrm{Spec}(k)$, and by $R\Psi_\eta^A = R\Psi_\eta^{A_K^s}$ its vanishing cycle functor. For all $p, q \geq 0$, the spaces

$$H_c^p(\overline{A}^s \times_k k^{\mathrm{ac}}, R^q\Psi_\eta^A(\mathbb{Q}_l^{\mathrm{ac}})) = \varinjlim_K H_c^p(\overline{A}_K^s \times_k k^{\mathrm{ac}}, R^q\Psi_\eta^A(\mathbb{Q}_l^{\mathrm{ac}}))$$

are admissible/continuous $G(\mathbb{A}^\infty)^\dagger \times W_{E_v}$-modules, and we define

$$H_c(\overline{A}^s \times_k k^{\mathrm{ac}}, R\Psi_\eta^A(\mathbb{Q}_l^{\mathrm{ac}})) = \sum_{p,q}(-1)^{p+q} H_c^p(\overline{A}^s \times_k k^{\mathrm{ac}}, R^q\Psi_\eta^A(\mathbb{Q}_l^{\mathrm{ac}})).$$

Proposition 4.2 *Let ξ be an irreducible finite dimensional l-adic representation of G. For each level K, there is an equality of virtual continuous W_{E_v}-modules*

$$H_c(\overline{S}_K \times_k k^{ac}, R\Psi_\eta^S(\mathcal{L}_\xi)) = (-1)^{t_\xi}\epsilon_\xi H_c(\overline{A}_K^{s_\xi} \times_k k^{\mathrm{ac}}, R\Psi_\eta^A(\mathbb{Q}_l^{ac})).$$

Further more, there is an equality of virtual admissible/continuous $G(\mathbb{A}^\infty)^\dagger \times W_{E_v}$-modules

$$H_c(\overline{S} \times_k k^{ac}, R\Psi_\eta^S(\mathcal{L}_\xi)) = (-1)^{t_\xi}\epsilon_\xi H_c(\overline{A}^{s_\xi} \times_k k^{\mathrm{ac}}, R\Psi_\eta^A(\mathbb{Q}_l^{ac})).$$

Proof In order to deduce the latter equalities from the above constructions, it suffices to recall that the vanishing cycle functor and the derived pushforward by a proper morphism commute ([4], Section 2.1.7, p. 20). We deduce that, for all integers $s \geq 0$, there is a canonical quasi-isomorphism of derived functors

$$R\Psi_\eta^S \circ R(\alpha_\eta^s)_* \simeq R(\bar{\alpha}^s)_* \circ R\Psi_\eta^{A^s},$$

where α_η^s and $\bar{\alpha}^s$ denote respectively the restrictions of $\alpha^s : A^s \to S$ to the generic and special fibers of $\mathcal{O}_{E_v}$ (here, $\alpha^0 = \mathrm{id}_S$). □

4.2.2 Let $b \in B(G_{\mathbb{Q}_p}, \mu_{\mathbb{Q}_p^{ac}})$. For all levels K and integers $s \geq 0$, we write $\overline{A}_K^s(b)$ for the restriction of $\overline{A}_K^s/\overline{S}_K$ to the Newton stratum $\overline{S}_K(b)$, and with definition analogous to those above, introduce the W_{E_v}-module $H_c(\overline{A}_K^s(b) \times_k k^{ac}, R\Psi_\eta^A(\mathbb{Q}_l^{ac}))$ and the admissible/continuous $G(\mathbb{A}^\infty)^\dagger \times W_{E_v}$-module $H_c(\overline{A}^s(b) \times_k k^{ac}, R\Psi_\eta^A(\mathbb{Q}_l^{ac}))$. They are endowed with a natural action of the idempotents ϵ_ξ, for all ξ with $s = s_\xi$.

Proposition 4.3 *Let $b \in B(G_{\mathbb{Q}_p}, \mu_{\mathbb{Q}_p^{ac}})$, and ξ an irreducible finite dimensional l-adic representation of G. For each level K, there is an equality of virtual continuous W_{E_v}-modules*

$$H_c(\overline{S}_K(b) \times_k k^{ac}, R\Psi_\eta^S(\mathcal{L}_\xi)) = (-1)^{t_\xi} \epsilon_\xi H_c(\overline{A}_K^{s_\xi}(b) \times_k k^{ac}, R\Psi_\eta^A(\mathbb{Q}_l^{ac})).$$

Further more, there is an equality of virtual admissible/continuous $G(\mathbb{A}^\infty)^\dagger \times W_{E_v}$-modules

$$H_c(\overline{S}(b) \times_k k^{ac}, R\Psi_\eta^S(\mathcal{L}_\xi)) = (-1)^{t_\xi} \epsilon_\xi H_c(\overline{A}^{s_\xi}(b) \times_k k^{ac}, R\Psi_\eta^A(\mathbb{Q}_l^{ac})).$$

Proof This is an immediate consequence of the definition of the idempotents ϵ_ξ, proposition 4.2 and the Proper Base Change Theorem. □

4.2.3 Let K be of the form $K = K^p K_{p,0}$. We assume that the Newton stratum $\overline{S}_K^p(b)$ is not empty, and let $\mathbb{X}$ be a Barsotti-Tate group with additional structures associated with a closed geometric point of $\overline{S}_{K^p}(b)$. For each $m \geq 1$, we write $\mathrm{Ig}_{\mathbb{X},K^p,m}$ for the associated Igusa variety of level m over $\overline{S}_{K^p}(b)$, and $A^s_{b,K^p} = A^s_{\mathbb{X},K^p,m}$ for the pullback of the s-fold self-product of universal abelian scheme $\overline{A}_{K^p}(b)/\overline{S}_{K^p}(b)$ over $\mathrm{Ig}_{K^p,m}$, for all $s \geq 0$. For all $i \geq 0$, the spaces

$$H_c^i(A_b^s, \mathbb{Q}_l^{ac}) = \varinjlim_{K^p,m} H_c^i(A^s_{\mathbb{X},K^p,m}, \mathbb{Q}_l^{ac})$$

are admissible/continuous $G(\mathbb{A}^{\infty,p}) \times J_b(\mathbb{Q}_p)$-modules, and we define

$$H_c(A_b^s, \mathbb{Q}_l^{ac}) = \sum_i (-1)^i H_c^i(A_b^s, \mathbb{Q}_l^{ac}).$$

Proposition 4.4 *Let $b \in B(G_{\mathbb{Q}_p}, \mu_{\mathbb{Q}_p^{ac}})$ be an element associated with a non-empty Newton stratum, and ξ an irreducible finite dimensional l-adic representation of G. There is an equality of virtual admissible/continuous $G(\mathbb{A}^{\infty,p}) \times J_b(\mathbb{Q}_p)$-modules*

$$H_c(Ig_b, \mathcal{L}_\xi) = (-1)^{t_\xi} \epsilon_\xi H_c(A_b^{s_\xi}, \mathbb{Q}_l^{ac}).$$

Proof This is also a consequence of the definition of the idempotents ϵ_ξ, proposition 4.2 and of the Proper Base Change Theorem. □

4.3 Cohomology of the universal abelian scheme. In view of the observation of Section 4.2, we now restate Theorems 3.1 and 3.3 in the special case of the representations $\wedge^\bullet V_0^s$, $s \geq 0$, as results about the cohomology of the multi-self-products of the universal abelian scheme over the Shimura varieties.

Theorem 4.5 *Let $s \geq 0$. For each $b \in B(G_{\mathbb{Q}_p}, \mu_{\mathbb{Q}_p^{ac}})$ (associated with a non-empty Newton stratum), there is an equality of virtual $G(\mathbb{A}^\infty)^\dagger \times W_{E_v} \times S_s$-modules*

$$H_c(\overline{A}^s(b) \times_k k^{ac}, R\Psi_\eta^A(\mathbb{Q}_l)) = \mathcal{E}_b(H_c(A_b^s, \mathbb{Q}_l)),$$

where $\mathcal{E}_b$ is the functor associated with the local models of the Shimura varieties attached to b. In the case of proper Shimura varieties, the above equalities can be *extended to the equalities of* $G(\mathbb{A}^\infty)\times W_{E_v}\times S_s$*-modules, for all* $b\in B(G_{\mathbb{Q}_p},\mu_{\mathbb{Q}_p^{ac}})$*. In particular, we deduce an equality of* $G(\mathbb{A}^\infty)\times W_{E_v}\times S_s$*-modules*

$$H_c(A^s,\mathbb{Q}_l)=\sum_{b\in B(G_{\mathbb{Q}_p},\mu_{\mathbb{Q}_p^{ac}})}\mathcal{E}_b(H_c(A_b^s,\mathbb{Q}_l)).$$

Theorem 4.6 *Let* $s\geq 0$*,* $b\in B(G_{\mathbb{Q}_p},\mu_{\mathbb{Q}_p^{ac}})$ *(associated with a non-empty Newton stratum), and* $K=K^pK_p$ *a level of the form* $K_p=K_{p,m}$*, for* $m\geq 0$*. There exists a system of* $W_{E_v}\times S_s$*-quasi-isomorphism of complexes, indexed by the quadruples* (N,m,n,d)*,*

$$\begin{aligned}&\pi_N^*(R\Psi_\eta^S Rf_{K_p\,*}(R\alpha^s_{K\,*}(\mathbb{Q}_l^{ac}))_{|\overline{S}(b)_{\bar{k}^{ac}}})\\ &\cong\; q_1^*(R\alpha^s_{K^p\,*}(\mathbb{Q}_l^{ac})_{|Ig_0})\otimes q_2^*(R\Psi_\eta^{\mathcal{M}} Rg_{K_p\,*}(\mathbb{Q}_l^{ac})),\end{aligned}$$

which are compatible under the morphisms induced by the projections among the Igusa varieties, the inclusion of truncated Rapoport-Zink spaces and the action of $J_b(\mathbb{Q}_p)^\dagger$*. Furthermore, they are also compatible with the action of* $G(\mathbb{A}^\infty)^\dagger$*, as the level* K *varies.*

Proof of Theorem 4.5 It follows from the discussion in Section 3.1 that this statement is implied by Theorem 4.6 together with Theorem 3.2. Again, in the case of proper varieties, the second equality follows from a version of Theorem 4.6 which includes the case of the vanishing cycle sheaves associated with a larger class of integral models. □

Proof of Theorem 3.3 It is an immediate consequence of the properties of the idempotents ϵ_ξ that Theorem 4.6 implies Theorem 3.3. □

4.3.1 We point that, in view of Propositions 4.1, 4.2, 4.3 and 4.4, the latter argument also proves that Theorem 3.1 follows directly from Theorem 4.5. Indeed, it suffices to observe that the action of the idempotent ϵ_ξ on the modules $H_c(A_b^s,\mathbb{Q}_l)$, for all s,ξ, commutes with the functor $\mathcal{E}_b$, and this follows from the fact that the actions of $G(\mathbb{Q}_l)$ and $J_b(\mathbb{Q}_p)$ commute.

5 Proof of Theorem 4.6

5.1 Preliminaries. Let $s\geq 0$, $K^p\subset G(\mathbb{A}^{\infty,p})$, $b\in B(G_{\mathbb{Q}_p},\mu_{\mathbb{Q}_p^{ac}})$ and $\mathbb{X}$ a Barsotti-Tate group associated with b be all as above. We write

$$\bar{\alpha}=\bar{\alpha}_{K^p}:\overline{A}^s_{K^p}(b)\to\overline{S}_{K^p}(b)$$

for the multi-self-product of the universal abelian scheme over the Newton stratum associated with b, and

$$\beta=\beta_{K^p}:A^s_{b,K^p}\to \mathrm{Ig}_{b,K^p,0}=\mathrm{Ig}_{\mathbb{X},K^p,0}$$

for the pullback of $\overline{A}^s_{K^p}(b)\to\overline{S}_{K^p}(b)$ over the corresponding Igusa variety of level 0, under the natural inclusion $\mathrm{Ig}_{b,K^p,0}\subset\overline{S}_{K^p}(b)$. (For simplicity, we are here omitting the upper index s in the notation of the structure morphisms, differently than in Section 4.)

5.1.1 Let m, n, d, N be non-negative integers as in Section 3.1.1. We write $\mathrm{Ig}_{b,K^p,m} = \mathrm{Ig}_{\mathbb{X},K^p,m}$, and $\overline{\mathcal{M}}_b^{n,d}$ for the reduced closed fiber of the truncated Rapoport-Zink space $\mathcal{M}_{b,\mathbb{X}}^{n,d}$. We introduce two distinct abelian schemes defined over the product $\mathrm{Ig}_{b,K^p,m} \times \overline{\mathcal{M}}_b^{n,d}$. The first one is the pullback of the abelian scheme $A_{b,K^p}^s \to \mathrm{Ig}_{b,K^p,0}$ under the projection $q_1 : \mathrm{Ig}_{b,K^p,m} \times \overline{\mathcal{M}}_{b,K_{p,0}}^{n,d} \to \mathrm{Ig}_{b,K^p,0}$, we denote it by

$$\gamma = \gamma_{K^p} = \beta \times 1 : \mathcal{A}_{b,K^p}^s = q_1^* A_{b,K^p}^s = A_{b,K^p}^s \times \overline{\mathcal{M}}_b^{n,d} \to \mathrm{Ig}_{b,K^p,m} \times \overline{\mathcal{M}}_b^{n,d}.$$

The second one is the pullback of $\overline{A}_{K^p}^s(b) \to \overline{S}_{K^p}(b)$ under the map $\pi_N = \pi_{(N,m,n,d)}$ introduced in Section 3.1.1, we write

$$\delta = \delta_{K^p,N} : \mathcal{B}_{b,K^p,N}^s = \pi_N^* \overline{A}_{K^p}^s(b) \to \mathrm{Ig}_{b,K^p,m} \times \overline{\mathcal{M}}_b^{n,d}.$$

In the following, with abuse of notation, we also denote by π_N the corresponding morphism $\mathcal{B}_{b,K^p,N}^s \to \overline{A}_{K^p}^s(b)$. We remark that, given the moduli interpretation of the scheme $\overline{S}_{K^p}(b)$, the datum of the abelian scheme $\mathcal{B}_{b,K^p,N}$ over $\mathrm{Ig}_{b,K^p,m} \times \overline{\mathcal{M}}_b^{n,d}$ is equivalent to the datum of the corresponding morphism π_N. We will recall the explicit definition of the schemes $\mathcal{B}_{b,K^p,N}$ in the proof of proposition 5.2.

5.1.2 To extend the previous definitions to the cases of general levels K, of the form $K = K^p K_p$ as in Section 2.1.3, we consider the morphisms $1 \times g_{K_p}$: $\mathrm{Ig}_{b,K^p,m} \times \overline{\mathcal{M}}_{b,K_p}^{n,d} \to \mathrm{Ig}_{b,K^p,m} \times \overline{\mathcal{M}}_b^{n,d}$, where $g_{K_p} : \overline{\mathcal{M}}_{b,K_p}^{n,d} \to \overline{\mathcal{M}}_b^{n,d}$ denote the natural projections associated with the change of level. (For each n, d, we denote by $\overline{\mathcal{M}}_{b,K_p}^{n,d}$ the pullback over $\overline{\mathcal{M}}_b^{n,d}$ of the cover $\overline{\mathcal{M}}_{b,K_p} \to \overline{\mathcal{M}}_b$.) For each level K, we define

$$\gamma_K : \mathcal{A}_{b,K}^s = A_{b,K^p}^s \times \overline{\mathcal{M}}_{b,K_p}^{n,d} \to \mathrm{Ig}_{b,K^p,m} \times \overline{\mathcal{M}}_{b,K_p}^{n,d}$$

and

$$\delta_K = \delta_{K,N} : \mathcal{B}_{b,K}^s \to \mathrm{Ig}_{b,K^p,m} \times \overline{\mathcal{M}}_{b,K_p}^{n,d}$$

respectively as the pullbacks of γ_{K^p} and $\delta_{K^p,N}$ under the map $1 \times g_{K_p}$. With abuse of notations, we denote the natural projections $\mathcal{A}_{b,K}^s \to \mathcal{A}_{b,K^p}^s$ and $\mathcal{B}_{b,K}^s \to \mathcal{B}_{b,K^p}^s$ also by g_{K_p}. We recall that, for all level K, $\bar{\alpha}_K : \overline{A}_K^s(b) \to \overline{S}_K(b)$ denotes the pullback of $\bar{\alpha} = \bar{\alpha}_{K^p}$ under the map $f_{K_p} : \overline{S}_K(b) \to \overline{S}_{K^p}(b)$. In the following, we also write f_{K_p} for the associated projection $\overline{A}_K^s(b) \to \overline{A}_{K^p}^s(b)$.

5.1.3 We observe that, for all levels K, the schemes $\overline{A}_K^s(b)$ arise as the special fiber of integral abelian schemes, namely (the restriction of) the corresponding multi-self-products of the universal abelian schemes over the integral models of the Shimura varieties. The same does not hold for the abelian schemes $\mathcal{A}_{b,K}^s$ and $\mathcal{B}_{b,K}^s$, which are defined only in positive characteristic. In particular, while over $\overline{A}_K^s(b)$ it is only natural to consider the vanishing cycle sheaves associated with the corresponding integral abelian schemes in characteristic zero, the analogous construction is not available over $\mathcal{A}_{b,K}^s$ and $\mathcal{B}_{b,K}^s$. Instead, we consider the pullbacks over $\mathcal{A}_{b,K}^s$ and $\mathcal{B}_{b,K}^s$ of the vanishing cycle sheaves associated with the Rapoport-Zink space of the same level. To explain such a choice, we observe that if appropriate integral abelian schemes existed (and indeed, Serre-Tate theorem implies that the datum of the universal Barsotti-Tate group over $\mathcal{M}_{b,K}$ formally locally defines such lifts), then the associated vanishing cycle sheaves over $\mathcal{A}_{b,K}^s$ and $\mathcal{B}_{b,K}^s$ would tautologically agree with pullbacks of those associated with the Rapoport-Zink spaces.

5.2 Proof. We deduce Theorem 4.6 from the following two results.

Proposition 5.1 *Maintaining the above notations. There is a system of* $W_{E_v} \times S_s$*-quasi-isomorphisms of complexes*

$$\pi_N^*(R\Psi_\eta^S Rf_{K_p\,*}(R\alpha_{K\,*}(\mathbb{Q}_l^{ac}))_{|\overline{S}(b)_{k^{ac}}}) \cong R\delta_{N\,*}\delta_N^*(q_2^* R\Psi_\eta^{\mathcal{M}} Rg_{K_p\,*}(\mathbb{Q}_l^{ac})),$$

indexed by quadruples (N, m, n, d), *such that, as* (N, m, n, d) *varies, they are compatible under the pullback by the projections among the Igusa varieties, the inclusion among the truncated Rapoport-Zink spaces, and the action of* $J_b(\mathbb{Q}_p)^\dagger$. *They are also compatible under pullback by the action of* $G(\mathbb{Q}_p)^\dagger$, *as the level* K *varies.*

Proposition 5.2 *Maintaining the above notations. There is a system of* $W_{E_v} \times S_s$*-quasi-isomorphisms of complexes*

$$R\delta_{N\,*}\delta_N^*(q_2^* R\Psi_\eta^{\mathcal{M}} Rg_{K_p\,*}(\mathbb{Q}_l^{ac})) \cong q_1^*(R\alpha_{K^p\,*}(\mathbb{Q}_l^{ac})_{|Ig_0}) \otimes q_2^*(R\Psi_\eta^{\mathcal{M}} Rg_{K_p\,*}(\mathbb{Q}_l^{ac})),$$

indexed by quadruples (N, m, n, d), *such that, as* (N, m, n, d) *varies, they are compatible under the pullback by the projections among the Igusa varieties, the inclusion among the truncated Rapoport-Zink spaces, and the action of* $J_b(\mathbb{Q}_p)^\dagger$. *They are also compatible under pullback by the action of* $G(\mathbb{Q}_p)^\dagger$, *as the level* K *varies.*

Proof of Theorem 4.6 The system of quasi-isomorphisms in the statement is the composition of those appearing in Propositions 5.1 and 5.2. □

Proof of proposition 5.1 It follows from the fact that the vanishing cycle functor commutes with derived pushforward ([1], Corollary 4.2, p.548; [2], Corollary 2.3, p. 373) that

$$\begin{aligned}\pi_N^*(R\Psi_\eta^S Rf_{K_p\,*}(R\alpha_{K\,*}(\mathbb{Q}_l^{\mathrm{ac}}))) &\cong \pi_N^*(R\Psi_\eta^S R\alpha_{K^p\,*}(Rf_{K_p\,*}(\mathbb{Q}_l^{\mathrm{ac}}))) \\ &\cong \pi_N^*(R\bar{\alpha}_{K^p\,*} R\Psi_\eta^A(Rf_{K_p\,*}(\mathbb{Q}_l^{\mathrm{ac}}))),\end{aligned}$$

and from the Proper Base Change Theorem ([14], Corollary 2.3, pp. 223-224) that

$$\pi_N^*(R\bar{\alpha}_{K^p\,*} R\Psi_\eta^A(Rf_{K_p\,*}(\mathbb{Q}_l^{\mathrm{ac}}))) \cong R\delta_{N,K^p\,*}(\pi_N^* R\Psi_\eta^A(Rf_{K_p\,*}(\mathbb{Q}_l^{\mathrm{ac}}))).$$

(Here above, and through out the rest of the proof, for $\mathcal{L}$ a sheaf defined over $\overline{S}_{K^p} \times k^{\mathrm{ac}}$, we write $\pi_N^*\mathcal{L}$ to mean $\pi_N^*(\mathcal{L}_{|\overline{S}(b)_{k^{\mathrm{ac}}}})$, where $\overline{S}(b)_{k^{\mathrm{ac}}} = \overline{S}_{K^p}(b) \times k^{\mathrm{ac}} \subset \overline{S}_{K^p} \times k^{\mathrm{ac}}$.) Let $q_2^* R\Psi_\eta^{\mathcal{M}}(Rg_{K_p\,*}(\mathbb{Q}_l^{\mathrm{ac}})) \cong \pi_N^* R\Psi_\eta^S(Rf_{K_p\,*}(\mathbb{Q}_l^{\mathrm{ac}}))$ denote the compatible system of W_{E_v}-quasi-isomorphisms in Theorem 3 ([13], Proposition 21, p. 605). It yields a system of quasi-isomorphisms

$$R\delta_{N\,*}\delta_N^*(q_2^* R\Psi_\eta^{\mathcal{M}} Rg_{K_p\,*}(\mathbb{Q}_l^{\mathrm{ac}})) \cong R\delta_{N\,*}\delta_N^*(\pi_N^* R\Psi_\eta^S(Rf_{K_p\,*}(\mathbb{Q}_l^{\mathrm{ac}}))).$$

On the other hand, the equality $\pi_N \circ \delta_N = \bar{\alpha} \circ \pi_N$ implies

$$R\delta_{N\,*}\delta_N^*(\pi_N^* R\Psi_\eta^S(Rf_{K_p\,*}(\mathbb{Q}_l^{\mathrm{ac}}))) \cong R\delta_{N\,*}(\pi_N^* \bar{\alpha}_{K_p}^* R\Psi_\eta^S(Rf_{K_p\,*}(\mathbb{Q}_l^{\mathrm{ac}}))).$$

Thus, in order to conclude, it suffices to prove

$$R\Psi_\eta^A(Rf_{K_p\,*}(\mathbb{Q}_l^{\mathrm{ac}})) \cong \bar{\alpha}_{K_p}^* R\Psi_\eta^S(Rf_{K_p\,*}(\mathbb{Q}_l^{\mathrm{ac}}))).$$

We recall that given a morphism of $W(k^{\mathrm{ac}})$-formal schemes which are locally of finite type, if the morphism is smooth then the associated map between the vanishing cycles of the trivial l-adic sheaf is a quasi-isomorphism ([12], Proposition 2.26, p. 235; [7], Lemma II.5.6, p.47). We apply this result to the (smooth) $\mathcal{O}_{E_v}$-morphisms

$\alpha_K : A^s_K \to S_K$, for all levels K (not just $K = K^p K_{p,0}$). By combining this with the pertinent instances of the Proper Base Change Theorem and the fact that the vanishing cycle functor commutes with derived pushforward, we obtain the following chain of quasi-isomorphisms

$$\begin{aligned} R\Psi^A_\eta(Rf_{K_p*}(\mathbb{Q}^{ac}_l)) &\cong Rf_{K_p*}(R\Psi^{A_K}_\eta(\mathbb{Q}^{ac}_l)) \cong Rf_{K_p*}(\bar{\alpha}^*_K R\Psi^{S_K}_\eta(\mathbb{Q}^{ac}_l)) \\ &\cong \bar{\alpha}^*_{K^p} Rf_{K_p*}(R\Psi^{S_K}_\eta(\mathbb{Q}^{ac}_l)) \cong \bar{\alpha}^*_{K^p} R\Psi^S_\eta(Rf_{K_p*}(\mathbb{Q}^{ac}_l)). \end{aligned}$$

Finally, it is an immediate consequence of the functoriality of the above constructions and of the compatibility among the quasi-isomorphisms in Theorem 3 that the resulting system of quasi-isomorphisms, as N, m, n, d and K vary, has the required properties. We remark that given the general statement of Theorem 3 ([13], Proposition 21, p. 605), this proof holds also for the vanishing cycle sheaves associated with the larger class of integral models considered in [13]. □

Proof of Proposition 5.2 We deduce from the Künneth formula, applied to the morphism $\gamma = \beta \times 1$, that

$$\begin{aligned} q_1^*(R\alpha_{K^p*}(\mathbb{Q}^{ac}_l)_{|\mathrm{Ig}_0}) \otimes q_2^*(R\Psi^{\mathcal{M}}_\eta Rg_{K_p*}(\mathbb{Q}^{ac}_l)) &\cong \\ \cong\ & q_1^*(R\beta_{K^p*}(\mathbb{Q}^{ac}_l)) \otimes q_2^*(R\Psi^{\mathcal{M}}_\eta Rg_{K_p*}(\mathbb{Q}^{ac}_l)) \\ \cong\ & R\gamma_{K^p*}\gamma^*_{K^p}(q_1^*(\mathbb{Q}^{ac}_l) \otimes q_2^*(R\Psi^{\mathcal{M}}_\eta Rg_{K_p*}(\mathbb{Q}^{ac}_l))) \\ \cong\ & R\gamma_{K^p*}\gamma^*_{K^p}(q_2^*(R\Psi^{\mathcal{M}}_\eta Rg_{K_p*}(\mathbb{Q}^{ac}_l))). \end{aligned}$$

Thus, it suffices to establish that, for any compatible system of lisse étale l-adic sheaves $\mathcal{F} = \mathcal{F}_{m,n,d}$, defined over the products $\mathrm{Ig}_{\mathbb{X},K^p,m} \times \overline{\mathcal{M}}^{n,d}_b$, there exists a compatible system of quasi-isomorphisms

$$R\delta_{N*}\delta^*_N\mathcal{F} \cong R\gamma_*\gamma^*\mathcal{F},$$

indexed by quadruples (N, m, n, d). Indeed, the existence of such a system of quasi-isomorphisms applied to the sheaves $\mathcal{F} = q_2^*(R\Psi^{\mathcal{M}}_\eta Rg_{K_p*}(\mathbb{Q}^{ac}_l))$, for each level K_p, implies the statement. (We remark that this holds also for the vanishing cycle sheaves associated with the larger class of integral models considered in [13].) We recall that for any K^p, m, n, d, N (with m, N sufficiently large with respect to d), $\gamma : \mathcal{A}^s \to \mathrm{Ig}_{K^p,m} \times \overline{\mathcal{M}}^{n,d}$ and $\delta_N : \mathcal{B}^s_N \to \mathrm{Ig}_{K^p,m} \times \overline{\mathcal{M}}^{n,d}$ are abelian schemes constructed via two different pullbacks from the universal abelian scheme over the reduction of the corresponding Shimura variety. We remark that while $\mathcal{A}^s$ depends only on the indexes K^p, m, n, d, the scheme $\mathcal{B}^s_N$ depends also on a choice of a sufficiently large integer N. In fact, $\mathcal{B}^s_N$ is defined via pullback under the morphism $\pi_N = \pi_{(N,m,n,d)}$. Furthermore, as we already remarked, the datum of the morphism $\pi_{(N,m,n,d)} : \mathrm{Ig}_{K^p,m} \times \overline{\mathcal{M}}^{n,d} \to \overline{S}_{K^p}(b)$ is equivalent to that of the abelian scheme $\mathcal{B}_N$ over $\mathrm{Ig}_{K^p,m} \times \overline{\mathcal{M}}^{n,d}$. We recall the definition of $\mathcal{B}_N$, for any quintuple (K^p, m, n, d, N). Let q denote the cardinality of the residue field k of E, and $F : \mathcal{A} \to \mathcal{A}^{(q)}$ the $(\log_p q)$-th power of the Frobenius morphism on $\mathcal{A}$. Let $\beta : \mathbb{X} \to \mathcal{H}$ denote (the pullback of) the universal quasi-isogeny over the Rapoport-Zink space $\overline{\mathcal{M}}$, and $j_m : \mathbb{X}[p^m] \to H^{sp}[p^m]$ denote (the pullback of) the universal Igusa structure of level m on the Barsotti-Tate group $H = \overline{A}[p^\infty]/C_{\mathbb{X}}$. It follows from the definition of the truncated Rapoport-Zink spaces that the restriction of β to $\overline{\mathcal{M}}^{n,d}$ satisfies the conditions that $p^n\beta$ is an isogeny and $\ker(p^n\beta) \subset \mathbb{X}[p^d]$. Further more, in [13] (Lemma 8, p. 591-592) we proved that, given N sufficiently large with respect to d and $m \geq d$, the Igusa structure j_m induces an isomorphisms of truncated Barsotti-Tate groups

$$j_m^{(q^N)} = f_N \circ j_{m\,|\mathbb{X}[p^d]} : \mathbb{X}[p^d] \cong H^{sp}[p^d] \cong H^{(q^N)}[p^d].$$

We define $\mathcal{B}_N = \pi_N^* \mathcal{A}$ as the quotient of $\mathcal{A}^{(q^N)}$ by the finite flat group scheme

$$j_m^{(q^N)}(\ker(p^n \beta)) \subset \mathcal{A}^{(q^N)}[p^d]$$

([13], Proposition 9, p. 592). It follows from the definition that $\mathcal{B}_N$ comes equipped with a p-power isogeny $\psi_N : \mathcal{A}^{(q^N)} \to \mathcal{B}_N$. We remark that it is an immediate consequence of the construction that, as the indexes K^p, m, d vary, the corresponding pairs $(\mathcal{B}_N, \psi_N)$ agree under pullback by the projections among Igusa varieties and the inclusions among truncated Rapoport-Zink spaces. As n varies, the pairs $(\mathcal{B}_N, \psi_N)$ agree under pullback only after composing ψ_N with the appropriate power of p, i.e. after replacing ψ_N by the quasi-isogeny $p^{-n}\psi_N$. Finally, as N vary, the data $(\mathcal{B}_N, \psi_N)$ agree under pullback by the appropriate power of Frobenius, i.e. for each N there is a canonical identification between $(\mathcal{B}_{N+1}, \psi_{N+1})$ and $(\mathcal{B}_N^{(q)}, \psi_N^{(q)})$. For any $s \geq 1$, we define $\phi_N = \psi_N^s \circ F^N : \mathcal{A}^s \to \mathcal{B}_N^s$. It is an easy observation that the morphisms ϕ_N are p-power isogenies of abelian schemes over $\mathrm{Ig}_{K^p,m} \times \overline{\mathcal{M}}^{n,d}$, for all indexes K^p, m, n, d, N, and that they are obviously S_s-equivariant. Let $\mathcal{F} = \mathcal{F}_{m,n,d}$ be a compatible system of lisse étale l-adic sheaves over the products $\mathrm{Ig}_{\mathbb{X},K^p,m} \times \overline{\mathcal{M}}_b^{n,d}$. Then, for each set of indexes, the isogeny ϕ_N induces a morphism of complexes

$$\phi_{N\,*} : R\gamma_* \gamma^* \mathcal{F} \to R\delta_{N\,*} \delta_N^* \mathcal{F},$$

which are quasi-isomorphisms since ϕ_N are p-power isogenies and $l \neq p$. Moreover, it follows from the above discussion that after composing each $\phi_{N\,*}$ with $\sigma^{-N \log_p q} p^{-n}$, we obtain a compatible system of quasi-isomorphisms $R\gamma_* \gamma^* \mathcal{F} \cong R\delta_{N\,*} \delta_N^* \mathcal{F}$ with the required properties (here, σ denotes the natural action of Frobenius of k on the complexes). □

5.3 Remark. We conclude this paper with a brief remark aimed to clarify the proof of Theorem 4.6 and its relation to the results in [13] (namely, to Theorem 3). To do so, we explain our approach from a different point of view. Let us consider a proof of Theorem 4.6 which deals with the special case of levels K which are maximal compact at p, i.e. $K = K^p K_{p,0}$, and the general case, i.e. $K = K^p K_p$ for $K_p \subset K_{p,0}$, separately. This new approach is not essentially different from our previous one. Indeed, in the case of levels $K = K^p K_{p,0}$, Theorem 4.6 simply states the existence of a compatible system of W_{E_v}-quasi-isomorphism of complexes

$$\pi_N^*(R\bar{\alpha}_{K^p\,*}(\mathbb{Q}_l^{\mathrm{ac}}))_{|\overline{S}(b)_{k^{\mathrm{ac}}}} \cong q_1^*(R\bar{\alpha}_{K^p\,*}(\mathbb{Q}_l^{\mathrm{ac}})_{|\mathrm{Ig}_0}) \otimes q_2^*(\mathbb{Q}_l^{\mathrm{ac}}).$$

This amounts to comparing the étale cohomology with constant l-adic coefficients of two abelian schemes in positive characteristic. The statement follows directly from the corresponding special case of proposition 5.2 and from the first half of the proof of proposition 5.1. In particular, for these levels, the pertinent vanishing cycle sheaves are trivial and the results do not rely on Theorem 3. Subsequently, given Theorem 4.6 in the special case, in order to extend that result to the general case, we need to establish the general statements of Propositions 5.1 and 5.2. In the case of proposition 5.2, the general statement follows immediately from its special case via Proper Base Change. The same it is not true for proposition 5.1, and in the second half of the proof of proposition 5.1 we rely on the theory of vanishing cycles to deduce the general statement of proposition 5.1 from Theorem 3. In particular, the more general version of Theorem 4.6 which is required to deduce the second statement in Theorem 4.5 (under the assumption that the Shimura varieties are proper) follows directly from the corresponding version of Theorem 3 ([13], Proposition 21, p. 605).

Acknowledgment. I would like to thank R. Taylor, M. Harris, and S.W. Shin for their interest in my work and for stimulating discussions, and the referee for helpful comments and useful suggestions.

References

[1] Berkovich, V. *Vanishing cycles for formal schemes.* Invent. Math. **115** (1994), no. 3, 539–571.
[2] Berkovich, V. *Vanishing cycles for formal schemes. II.* Invent. Math. **125** (1996), no. 2, 367–390.
[3] Berkovich, V. *Étale cohomology for non-Archimedean analytic spaces.* Inst. Hautes Études Sci. Publ. Math. No. **78** (1993), 5–161.
[4] Deligne, P. *Le formalisme des cycles évanescents.* Groupes de monodromie en géométrie algébrique. Séminaire de Géométrie Algébrique du Bois-Marie (SGA 7-II). Lecture notes in mathematics, Vol. **340**, Springer-Verlag, Berlin-New York, 1972-1973.
[5] Fulton, W.; Harris, J. *Representation Theory.* A first course. Graduate Texts in Mathematics, **129**. Readings in Mathematics. Springer-Verlag, New York, 1991.
[6] Grothendieck, A. *Groupes de Barsotti-Tate et cristaux de Dieudonné.* Sém. Math. Sup. Univ. Montréal. Presses Univ. Montréal, 1974.
[7] Harris, M.; Taylor, R. *On the geometry and cohomology of some simple Shimura varieties.* Volume **151**, Annals of Math. Studies, Princeton University Press, 2001.
[8] Igusa, J. *Kroneckerian model of fields of elliptic modular functions.* Amer. J. Math. **81**, 1959, 561–577.
[9] Katz, N.; Mazur, B. *Arithmetic moduli of elliptic curves.* Annals of Mathematics Studies, **108**, Princeton University Press, Princeton, NJ, 1985.
[10] Kottwitz, R. *Points on some Shimura varieties over finite fields.* J. Amer. Math. Soc. **5** (1992), no. 2, 373–444.
[11] Kottwitz, R. *Isocrystals with additional structure.* Compositio Math. **56** (1985), no. 2, 201–220.
[12] Mantovan, E. *On certain unitary group Shimura varieties.* Variétés de Shimura, espaces de Rapoport-Zink et correspondances de Langlands locales. Astérisque **291** (2004), 200–331.
[13] Mantovan, E. *On the cohomology of certain PEL-type Shimura varieties.* Duke Math. J. **129** (2005), no. 3, 573–610.
[14] Milne, J. S. *Étale Cohomology.* Princeton Mathematical Series, **33**, Princeton University Press, Princeton, NJ, 1980.
[15] Oort, F. *Foliations in moduli spaces of abelian varieties.* J. Amer. Math. Soc. **17** (2004), no. 2, 267–296.
[16] Oort, F.; Zink, Th. *Families of p-divisible groups with constant Newton polygon.* Documenta Mathematica **7** (2002), 183–201.
[17] Rapoport, M.; Richartz, M. *On the classification and specialization of F-isocrystals with additional structure.* Compositio Math. **103** (1996), no. 2, 153–181.
[18] Rapoport, M.; Zink, Th. *Period spaces for p-divisible groups.* Annals of Mathematics Studies, **141**, Princeton University Press, Princeton, NJ, 1996.
[19] Zink, Th. *On the slope filtration.* Duke Math. J. Vol. **109** (2001), 79–95.

Part II – Curves and Jacobians over Finite Fields

Fields Institute Communications
Volume **60**, 2011

Pairings on Hyperelliptic Curves

Jennifer Balakrishnan
Department of Mathematics
Massachusetts Institute of Technology
Cambridge, MA 02139 USA
jen@math.mit.edu

Juliana Belding
Department of Mathematics
Harvard University
Cambridge, MA 02138 USA
jbelding@math.harvard.edu

Sarah Chisholm
Department of Mathematics and Statistics
University of Calgary
Calgary, Alberta, Canada T2N 1N4
chisholm@math.ucalgary.ca

Kirsten Eisenträger
Department of Mathematics
The Pennsylvania State University
University Park, PA 16802 USA
eisentra@math.psu.eu

Katherine E. Stange
Department of Mathematics
Simon Fraser University
Burnaby, British Columbia, Canada V5A 1S6
kestange@sfu.ca

Edlyn Teske
Department of Combinatorics and Optimization
University of Waterloo
Waterloo, Ontario, Canada N2L 3G1
eteske@uwaterloo.ca

Dedicated to the memory of Isabelle Déchène (1974-2009)

2010 *Mathematics Subject Classification.* Primary 14G50; Secondary 94A60.

Abstract. We assemble and reorganize the recent work in the area of hyperelliptic pairings: We survey the research on constructing hyperelliptic curves suitable for pairing-based cryptography. We also showcase the hyperelliptic pairings proposed to date, and develop a unifying framework. We discuss the techniques used to optimize the pairing computation on hyperelliptic curves, and present many directions for further research.

1 Introduction

Numerous cryptographic protocols for secure key exchange and digital signatures are based on the computational infeasibility of the discrete logarithm problem in the underlying group. Here, the most common groups in use are multiplicative groups of finite fields and groups of points on elliptic curves over finite fields. Over the past years, many new and exciting cryptographic schemes based on pairings have been suggested, including one-round three-way key establishment, identity-based encryption, and short signatures [3, 4, 43, 64]. Originally, the Weil and Tate (-Lichtenbaum) pairings on supersingular elliptic curves were proposed for such applications, providing non-degenerate bilinear maps that are efficient to evaluate. Over time potentially more efficient pairings have been found, such as the eta [2], Ate [41] and R-ate [53] pairings. Computing any of these pairings involves finding functions with prescribed zeros and poles on the curve, and evaluating those functions at divisors.

As an alternative to elliptic curve groups, Koblitz [47] suggested Jacobians of hyperelliptic curves for use in cryptography. In particular, hyperelliptic curves of low genus represent a competitive choice. In 2007, Galbraith, Hess and Vercauteren [29] summarized the research on hyperelliptic pairings to date and compared the efficiency of pairing computations on elliptic and hyperelliptic curves. In this rapidly moving area, there have been several new developments since their survey: First, new pairings have been developed for the elliptic case, including so-called optimal pairings by Vercauteren [71] and a framework for elliptic pairings by Hess [40]. Second, several constructions of ordinary hyperelliptic curves suitable for pairing-based cryptography have been found [19, 22, 67, 20].

In this paper, we survey

- the constructions of hyperelliptic curves suitable for pairings, especially in the ordinary case,
- the hyperelliptic pairings proposed to date, and
- the techniques to optimize computations of hyperelliptic pairings.

We also

- give a unifying framework for hyperelliptic pairings which includes many of the recent variations of the Ate pairing, and
- present a host of potential further improvements.

In this paper, we do *not* provide any comparative implementation, or give recommendations on which pairings should be used to satisfy certain user-determined criteria; this is left for future work.

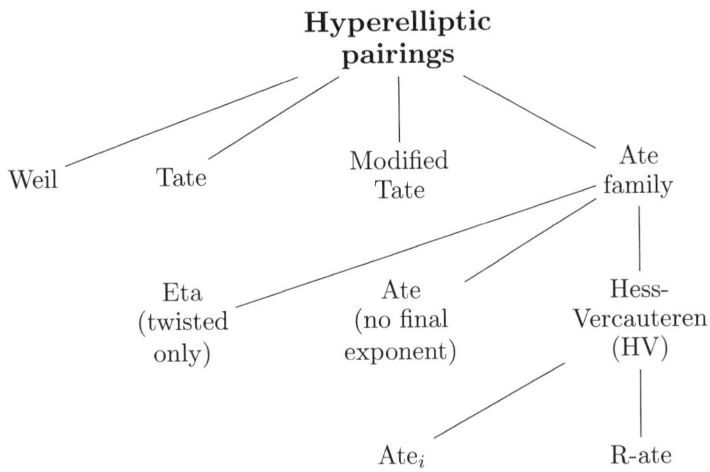

Figure 1 Classification of hyperelliptic pairings.

In our presentation, we focus on the case of genus 2 hyperelliptic curves and their Jacobians. Among all curves of higher genus, such curves are of primary interest for cryptographic applications: On the one hand, we find explicit formulae along with various optimizations (e.g., [50, 73]), providing for an arithmetic that is somewhat competitive with elliptic curves. On the other hand, the security is exactly the same as in the elliptic case, with the best attacks on the discrete logarithm problem in the Jacobian being square-root attacks based on the Pollard rho method (cf. [25]). However, Galbraith, Hess and Vercauteren [29, §10.1] argue that *pairing* computations on hyperelliptic curves will always be slow compared to elliptic curves: The most expensive part of a standard Tate pairing computation consists of repeatedly evaluating some function on a divisor and computing the product of the values obtained. Both in the elliptic and in the hyperelliptic case these divisors are defined over fields of the same size, but the functions in the hyperelliptic case are more complicated.

Figure 1 represents the collection of hyperelliptic pairings at a glance. For use in pairing-based applications, originally the Weil and Tate pairings were proposed. The Weil pairing is much more expensive to compute than the Tate pairing, so it is not used in practice. The pairings in the Ate family are potentially more efficient than the Tate pairing. Historically, the eta pairing was the first pairing to shorten the length of the Miller loop. It is defined on supersingular curves only and requires a final exponentiation. It gave rise to the Ate pairings which are defined for all curves. The hyperelliptic Ate pairing (which has a different definition than the elliptic Ate pairing!) has the advantage that its loop length is roughly half of the length of the Miller loop for the Tate pairing. It also is special in that it requires no final exponentiation (while the elliptic Ate pairing does require one). Other variations of the Ate pairing include the Hess-Vercauteren (HV) pairings. These are the pairings captured by our unifying framework, which generalizes work for the elliptic case by Hess [40] and Vercauteren [71]. HV pairings also have potentially shorter Miller loops than the Ate pairing, depending on the embedding degree of the Jacobian. All of the HV pairings involve a final exponentiation. Two examples of HV pairings are the R-ate and the Ate$_i$ pairings. Table 5.1 in Section 5 gives more details about the differences and merits of each pairing.

Our paper is organized as follows. In Section 2 we review some of the background on Jacobians of hyperelliptic curves. Section 3 discusses hyperelliptic curves of low embedding degree and what is known about constructing them. Section 4 gives an overview of the different pairings on hyperelliptic curves following the classification in Figure 1. We also introduce the HV pairing framework, give a direct proof of the non-degeneracy and bilinearity of the pairings captured by this framework and discuss how the Ate and R-ate pairings fit in. Section 5 describes the adaptation of Miller's algorithm to the hyperelliptic setting, presents common optimizations and compares all pairings according to their key characteristics of loop length and final exponentiation. Section 6 presents numerous problems for future work.

2 Jacobians of hyperelliptic curves

In this section, we fix some notation and terminology that will be used throughout the paper.

2.1 Hyperelliptic curves. A *hyperelliptic curve* C over a field K is a non-singular projective curve of the form

$$C : y^2 + H(x)y = F(x) \in K[x, y].$$

Let g be the genus of the curve. Throughout this paper, we restrict to the case where F is monic, $\deg F(x) = 2g + 1$, and $\deg H(x) \leq g$, so that C has one point at infinity, denoted P_∞. When $g = 1$, C is an *elliptic curve*. For significant parts of our discussion, we will consider the case where $g = 2$.

Although the points of a genus $g \geq 2$ hyperelliptic curve do not form a group, there is an involution of the curve taking $P = (x, y)$ to the point $(x, -y - H(x))$, which we will denote $-P$. Then, in accordance with the notation, $-(-P) = P$.

2.2 Divisors and abelian varieties. Let K be a field over which C is defined, and let $\overline{K}$ its algebraic closure. A *divisor* D on the curve C is a formal sum over all symbols (P), where P is a $\overline{K}$-point of the curve:

$$D = \sum_{P \in C(\overline{K})} n_P(P),$$

where all but finitely many of the coefficients $n_P \in \mathbb{Z}$ are zero. The collection of divisors forms an abelian group $\mathrm{Div}(C)$. The *degree* of a divisor is the sum

$$\sum_{P \in C(\overline{K})} n_P \in \mathbb{Z},$$

and the *support* of a divisor is the set of points of the divisor with non-zero coefficients n_P. For any rational function f on C, there is an associated divisor

$$\mathrm{div}(f) = \sum_{P \in C(\overline{K})} \mathrm{ord}_P(f)(P)$$

which encodes the number and location of its zeroes and poles. Any divisor which is the divisor of a function in this way is called a *principal divisor*.

An element σ in the Galois group of $\overline{K}$ over K, $\mathrm{Gal}(\overline{K}/K)$, acts on a divisor as follows:

$$\Big(\sum_{P\in C(\overline{K})} n_P(P)\Big)^{\sigma} = \sum_{P\in C(\overline{K})} n_P(P^{\sigma}).$$

In particular, let L be any intermediate field $K \subset L \subset \overline{K}$. Consider a function f defined over L; then $\mathrm{div}(f)$ is fixed by elements of $\mathrm{Gal}(\overline{K}/L)$. In fact, $\mathrm{div}(f)^{\sigma} = \mathrm{div}(f^{\sigma})$.

We give names to various sets of collections: $\mathrm{Div}(C)$ of divisors, $\mathrm{Div}^0(C)$ of degree zero divisors, $\mathrm{Ppl}(C)$ of principal divisors, $\mathrm{Div}_K(C)$ of divisors invariant under the action of $\mathrm{Gal}(\overline{K}/K)$, $\mathrm{Div}_K^0(C)$ of degree zero divisors invariant under the action of $\mathrm{Gal}(\overline{K}/K)$, and $\mathrm{Ppl}_K(C)$ of principal divisors invariant under the action of $\mathrm{Gal}(\overline{K}/K)$.

These are all abelian groups, which have the following subgroup relations:

$$\begin{array}{ccccc} \mathrm{Div}(C) & \supset & \mathrm{Div}^0(C) & \supset & \mathrm{Ppl}(C) \\ \cup & & \cup & & \cup \\ \mathrm{Div}_K(C) & \supset & \mathrm{Div}_K^0(C) & \supset & \mathrm{Ppl}_K(C). \end{array}$$

We make note of certain quotient groups:

$$\begin{aligned} \mathrm{Pic}(C) &:= \mathrm{Div}(C)/\,\mathrm{Ppl}(C), & \mathrm{Pic}^0(C) &:= \mathrm{Div}^0(C)/\,\mathrm{Ppl}(C), \\ \mathrm{Pic}_K(C) &:= \mathrm{Div}_K(C)/\,\mathrm{Ppl}_K(C), & \mathrm{Pic}_K^0(C) &:= \mathrm{Div}_K^0(C)/\,\mathrm{Ppl}_K(C). \end{aligned}$$

Elements of these quotient groups are equivalence classes of divisors. Divisors D_1 and D_2 of the same class are said to be *linearly equivalent*, and we write $D_1 \sim D_2$.

Recall that an elliptic curve is an example of an abelian variety. In general, an *abelian variety* A over K is a projective algebraic variety over K along with a group law $\varphi : A \times A \to A$ and an inverse map $\mathrm{Inv} : A \to A$ sending $x \mapsto x^{-1}$ such that φ and Inv are morphisms of varieties, both defined over K.

For an abelian variety A, a field K and an integer r, we let $A(K)[r]$ denote the set of r-torsion points of A defined over K, that is, the set of points in $A(K)$ of order dividing r. Now suppose A is an abelian variety over $\mathbb{F}_q$, with $q = p^m$. We say that A is *simple* if it is not isogenous over $\mathbb{F}_q$ to a product of lower dimensional abelian varieties. We call A *absolutely simple* if it is simple over $\overline{\mathbb{F}}_q$. We say A is *supersingular* if A is isogenous over $\overline{\mathbb{F}}_q$ to a power of a supersingular elliptic curve. (An elliptic curve E is supersingular if $E(\overline{\mathbb{F}}_q)$ has no points of order p.) An abelian variety A of dimension g over $\mathbb{F}_q$ is *ordinary* if $\#A(\overline{\mathbb{F}}_q)[p] = p^g$. Note that for dimension $g \geq 2$, there exist abelian varieties that are neither ordinary nor supersingular.

There is a natural isomorphism between the degree zero part of the Picard group $\mathrm{Pic}^0(C)$ of a hyperelliptic curve C and its *Jacobian* Jac_C, which is an abelian variety into which the curve embeds (cf. [26]). For the remainder of this paper, we will identify the Picard group $\mathrm{Pic}^0(C)$ with Jac_C.

2.3 Arithmetic in the Jacobian. We will work in the Jacobian Jac_C of a hyperelliptic curve C of genus g, whose elements are equivalence classes of degree-zero divisors. To do so, we choose a *reduced* representative in each such divisor class. A *reduced* divisor is one of the form

$$(P_1) + (P_2) + \cdots + (P_r) - r(P_\infty)$$

where $r \leq g$, P_∞ is the point at infinity on C, $P_i \neq -P_j$ for distinct i and j, and no P_i satisfying $P_i = -P_i$ appears more than once. Such a divisor is called *semi-reduced* if the condition $r \leq g$ is omitted. Each equivalence class contains exactly one reduced divisor. For a divisor D we will denote by $\rho(D)$ the reduced representative of its equivalence class. The action of Galois commutes with ρ, i.e. $\rho(D^\sigma) = \rho(D)^\sigma$, since the property of being reduced is preserved by the action of Galois.

To encode the reduced divisor in a convenient way, we write $(u(x), v(x))$ where $u(x)$ is a monic polynomial whose roots are the x-coordinates $x_1, \ldots, x_r$ of the r points

$$P_1 = (x_1, y_1), \ \ldots, \ P_r = (x_r, y_r),$$

and where $v(x_i) = y_i$ for $i = 1, \ldots, r$. This so-called *Mumford representation* [59] is uniquely determined by and uniquely determines the divisor. To find this representation, it suffices to find $u(x)$ and $v(x)$ satisfying the following conditions:

1. $u(x)$ is monic,
2. $\deg(v(x)) < \deg(u(x)) \leq g$, and
3. $u(x) \mid F(x) - v(x)H(x) - v(x)^2$,

where $F(x)$ and $H(x)$ are the polynomials defining the curve C (defined in Section 2.1). When we add two reduced divisors D_1 and D_2 the result $D_1 + D_2$ is not necessarily reduced. Beginning with two reduced divisors in Mumford representation, the algorithm to obtain the Mumford representation of the reduction of their sum can be explained in terms of the polynomials involved in the Mumford representation, without recourse to the divisor representation. This algorithm is originally due to Cantor [6], and in the form presented here to Koblitz [47]. The algorithm has two stages: in the first, we find a semi-reduced divisor $D \sim D_1 + D_2$, and in the second stage, we reduce D. Suppose that D_i has Mumford representation (u_i, v_i) for $i = 1, 2$.

STAGE 1:

1. Find $d(x) = \gcd(u_1(x), u_2(x), v_1(x) + v_2(x) + H(x))$. Finding this via the extended Euclidean algorithm gives $s_1(x)$, $s_2(x)$ and $s_3(x)$ such that
$$d = s_1 u_1 + s_2 u_2 + s_3(v_1 + v_2 + H).$$
2. Calculate the quantities
$$u = u_1 u_2 / d^2, \quad \text{and} \quad v = s_1 u_1 v_2 + s_2 u_2 v_1 + s_3(v_1 v_2 + F)/d \pmod{u(x)}.$$
(It is easily verified that the fraction on the right is defined since $d(x)$ is a divisor of the numerator.)

At this point, the result (u, v) is a semi-reduced divisor linearly equivalent to $D_1 + D_2$. This stage corresponds to simply adding D_1 and D_2 and canceling any points with their negatives if applicable. In fact, we obtain

$$D' = D_1 + D_2 - \operatorname{div}(d).$$

STAGE 2:

In this stage, if $\deg(u) > g$ we can replace (u, v) with a divisor (u', v') satisfying $\deg(u') < \deg(u)$. This replacement is as follows. Set

$$u' = (F - vH - v^2)/u, \quad \text{and} \quad v' = -H - v \pmod{u'}.$$

This stage corresponds to simplifying the divisor using the geometric group law nicely described for genus 2 by Lauter [51]. At each application of this loop to a divisor D_3, we obtain a divisor D'' satisfying[1]

$$D'' = D_3 - \operatorname{div}((F - vH - v^2)/u').$$

Applying this loop finitely many times, beginning with the result D' of stage one, we eventually obtain a reduced divisor D linearly equivalent to $D_1 + D_2$.

This algorithm has been optimized to avoid the use of the extended Euclidean algorithm and in this form it is much more efficient [29]. An enhanced version of Cantor's Algorithm is given as Algorithm 2 in this paper; see Section 5.1. If steps 5 and 8 through 13 are removed from Algorithm 2 one has the Cantor's Algorithm discussed here.

3 Hyperelliptic curves of low embedding degree

In this section we discuss hyperelliptic curves suitable for pairing-based cryptographic systems. The Jacobian varieties of such curves must have computable pairings, and computationally infeasible discrete logarithm problems. Specifically, we require low embedding degrees and large prime-order subgroups.

3.1 Embedding degree and ρ-value. Let r be a prime. Let C be a hyperelliptic curve over $\mathbb{F}_q$ of genus g with Jacobian variety $\mathrm{Jac}_C(\mathbb{F}_q)$ such that $r \mid \#\mathrm{Jac}_C(\mathbb{F}_q)$ and $\gcd(r, q) = 1$. The *embedding degree* of Jac_C with respect to r is the smallest integer k such that $r \mid (q^k - 1)$. Equivalently, the embedding degree of Jac_C with respect to r is the smallest integer k such that $\mathbb{F}_{q^k}^*$ contains the group of rth roots of unity μ_r. If Jac_C has embedding degree k with respect to r, then a pairing on C, such as the Weil pairing $e_r : \mathrm{Jac}_C(\mathbb{F}_q)[r] \times \mathrm{Jac}_C(\mathbb{F}_q)[r] \to \mu_r$, "embeds" $\mathrm{Jac}_C(\mathbb{F}_q)[r]$ (and any discrete logarithm problem in $\mathrm{Jac}_C(\mathbb{F}_q)[r]$) into $\mathbb{F}_{q^k}^*$, and $\mathbb{F}_{q^k}$ is the smallest-degree extension of $\mathbb{F}_q$ with this property; whence the name "embedding degree". Hitt [42] shows that if $q = p^m$ with $m > 1$, then $\mathrm{Jac}_C(\mathbb{F}_q)[r]$ may be embedded into a smaller field which is not an extension of $\mathbb{F}_q$ but only an extension of $\mathbb{F}_p$. The smallest such field is the so-called *minimal embedding field*, which is $\mathbb{F}_{p^{\mathrm{ord}_r p}}$.

We occasionally speak of the embedding degree of the hyperelliptic curve C, in which case we mean the embedding degree of its Jacobian.

Another important parameter is the *ρ-value*, which for a Jacobian variety of dimension g we define as $\rho = g \log q / \log r$. Since $\#\mathrm{Jac}_C(\mathbb{F}_q) = q^g + O(q^{g-1/2})$, the ρ-value measures the ratio of the bit-sizes of $\#\,\mathrm{Jac}_C(\mathbb{F}_q)$ and the subgroup order r. Jacobian varieties with a prime number of points have the smallest ρ-values: $\rho \approx 1$. We call a hyperelliptic curve, and its Jacobian variety, *pairing-friendly* if the Jacobian variety has small embedding degree and a large prime-order subgroup. In practice, we want $k \leq 60$ and $r > 2^{160}$.

Since the embedding degree k is the order of q in the multiplicative group $(\mathbb{Z}/r\mathbb{Z})^*$, and typically elements in $(\mathbb{Z}/r\mathbb{Z})^*$ have large order, we expect that for a random Jacobian over $\mathbb{F}_q$ with order-r subgroup, the embedding degree is approximately of the same size as r. (This reasoning has been made more precise

---

[1] In general, u' is a product of lines L_i whose divisors are $(P_i)+(-P_i)-2(P_\infty)$ for $i = 1, \ldots, r$ and $\operatorname{div}(F - vH - v^2)$ is the sum of the intersection points of C and a unique curve intersecting C at $3g$ points including $P_1, \ldots, P_r$.

for elliptic curves, by Balasubramanian and Koblitz [1] and Luca, Mireles and Shparlinski [57].) With $r > 2^{160}$, this means that evaluating a pairing for a *random* hyperelliptic curve becomes a computationally infeasible task. Just as in the case of elliptic curves, pairing-friendly hyperelliptic curves are rare and require special constructions.

3.2 Embedding degrees required for various security levels. For cryptographic applications, the discrete logarithm problems in $\mathrm{Jac}_C(\mathbb{F}_q)$ and in the multiplicative group $\mathbb{F}_{q^k}^*$ must both be computationally infeasible. For Jacobian varieties of hyperelliptic curves of genus 2 the best known discrete logarithm (DL) algorithm is the parallelized Pollard rho algorithm [70, 65], which has running time $O(\sqrt{r})$ where r is the size of the largest prime-order subgroup of $\mathrm{Jac}_C(\mathbb{F}_q)$. For Jacobian varieties of dimensions 3 and 4 there exist index calculus algorithms of complexities $O(q^{4/3+\varepsilon}) = O(|\,\mathrm{Jac}_C\,|^{4/9+\varepsilon})$ and $O(q^{3/2+\varepsilon}) = O(|\,\mathrm{Jac}_C\,|^{3/8+\varepsilon})$, respectively [35]. How this compares to the parallelized Pollard rho algorithm depends on the relative size of the subgroup order r – more precisely, only if $\rho < 9/8$ (genus 3 case) or $\rho < 4/3$ (genus 4 case) will the index calculus approach be superior to Pollard rho.

In any case, the best DL algorithms for genus 2, 3, and 4 are of exponential running time. On the other hand, the best algorithm for DL computation in finite fields is the index calculus attack (e.g., [62]) which has running time subexponential in the field size. Thus to achieve the same level of security in both groups, the size q^k of the extension field must be significantly larger than r. Table 3.1 shows sample subgroup sizes, extension field sizes, and embedding degrees with which to achieve common levels of security, for various cases $r \approx q^{g/\rho}$. The listed sizes for the prime-order subgroups and the extension fields (of large characteristic) follow the recommendations by NIST [61, Table 2].

Table 3.1 Embedding degrees for hyperelliptic curves of genus $g = 2$ required to obtain commonly desired levels of security.

Security level (bits)	Subgroup size (r)	Extension field size (q^k)	Embedding degree (k)					
			$\rho \approx 1$	$\rho \approx 2$	$\rho \approx 3$	$\rho \approx 4$	$\rho \approx 6$	$\rho \approx 8$
80	160	1024	$6g$	$3g$	$2g$	$1.5g$	g	$0.8g$
112	224	2048	$10g$	$5g$	$3.3g$	$2.5g$	$1.6g$	$1.3g$
128	256	3072	$12g$	$6g$	$4g$	$3g$	$2g$	$1.5g$
192	384	7680	$20g$	$10g$	$6.6g$	$5g$	$3.3g$	$2.5g$
256	512	15360	$30g$	$15g$	$10g$	$7.5g$	$5g$	$3.8g$

While Table 3.1 as such is for genus 2 only, it can easily be adapted to the cases of genus 3 and 4: Only in the case that the Jacobian has almost prime order ($\rho \approx 1$) we need to compensate for the aforementioned index-calculus algorithms in Jac_C. For this, if $g = 3$, multiply the second column entries by 9/8 and the fourth column entries by 8/9; if $g = 4$ multiply the second column entries by 4/3 and the fourth column entries by 3/4.

3.3 Ordinary hyperelliptic curves of low embedding degree. While there are numerous constructions for pairing-friendly elliptic curves – see e.g. the survey by Freeman, Scott and Teske [21] – there are not nearly as many constructions for hyperelliptic curves of low embedding degree and large prime-order subgroup. In this section, we discuss the case of ordinary Jacobians; see Section 3.4 for the supersingular case. We keep the discussion result-oriented, and refer the reader to the corresponding original papers for details on the specific constructions and the theory underneath.

Galbraith, McKee and Valença [32] argue that heuristically, for any fixed embedding degree k with $\varphi(k) \geq 4$ ($\varphi(k)$ = the Euler phi-function) and for any bound M on the field size q, there exist about as many genus 2 curves over $\mathbb{F}_q$ of embedding degree k (any ρ-value) as there exist elliptic curves over $\mathbb{F}_q$ of embedding degree k, namely $\Theta(M^{1/2}/\log M)$. For embedding degrees $k = 5, 10$, they identify several quadratic polynomials $q(x)$ parameterizing field sizes such that genus 2 curves over $\mathbb{F}_{q(x)}$ exist with embedding degree k (any ρ-value). (They also show that for $k = 8, 12$, such quadratic polynomials $q(x)$ do not exist.)

Freeman [18] was the first to actually construct ordinary genus 2 curves of low embedding degree. His construction is based on the Cocks-Pinch method [11][21, Theorem 4.1], which produces pairing-friendly *elliptic* curves over prime fields of any prescribed embedding degree and with $\rho \approx 2$. In the genus-2 case, Freeman obtains curves over prime fields $\mathbb{F}_q$ of any prescribed embedding degree k and ρ-value 8, that is, $r \approx q^{1/4}$ (where r denotes the prime subgroup order of the Jacobian).

Freeman [18, Proposition 2.3] further shows that the resulting Jacobian varieties have the property that $\mathrm{Jac}_C(\mathbb{F}_{q^k})$ always contains two linearly independent r-torsion points. For an elliptic curve $E/\mathbb{F}_q$, the corresponding result implies that the entire r-torsion group is contained in $E(\mathbb{F}_{q^k})$, but this is not necessarily the case for higher dimensional abelian varieties. (Recall that for genus g, the group $\mathrm{Jac}_C(\overline{\mathbb{F}}_q)[r]$ is $2g$-dimensional over $\mathbb{Z}/r\mathbb{Z}$.) This phenomenon gives rise to the notion of the *full* embedding degree, which is the smallest integer k such that all r-torsion points of Jac_C are defined over $\mathbb{F}_{q^k}$. Freeman [18, Algorithm 5.1] gives a construction of genus 2 curves of prescribed full embedding degree k (necessarily even), which may be useful in cryptographic applications that require more than two linearly independent r-torsion points (see Section 6.8). Again, this construction yields curves with ρ-value 8.

Note that an essential part of either construction [18] is the use of the complex multiplication (CM) method to compute the actual curve. In genus 2, this includes computation of the Igusa class polynomials (e.g., [72]) of the CM field $K = \mathrm{End}(\mathrm{Jac_C}) \otimes \mathbb{Q}$, which is currently feasible for CM fields K with class numbers less than 100 [49]. (Here, $\mathrm{End}(\mathrm{Jac_C})$ denotes the set of all endomorphisms of Jac_C defined over $\mathbb{F}_q$.)

Freeman, Stevenhagen and Streng [22, Algorithm 2.12] present a generalization of the Cocks-Pinch method, which, when coupled with complex multiplication methods, produces pairing-friendly abelian varieties over prime fields, of dimension g with ρ-values $\approx 2g^2$. This algorithm works for any prescribed embedding degree k, and applies to arbitrary genus $g \geq 2$. (However note that complex multiplication methods are available for special CM fields only if $g = 3$, and are completely undeveloped for $g \geq 4$.) In addition to explicit genus 2 examples with $\rho \approx 8$, a cryptographically interesting example is given for genus 3 ($k = 17$ and $\rho \approx 17.95$).

In the case of pairing-friendly *elliptic* curves, the method by Brezing and Weng [5] is a generalization of the Cocks-Pinch method [11] and produces elliptic curves over prime fields with $1 < \rho < 2$ for many embedding degrees. Freeman [19, Algorithm 3.8] combines the Brezing-Weng approach with the method from Freeman,

Stevenhagen and Streng [22] to construct so-called *families* of abelian varieties over prime fields with ρ-values strictly less than $2g^2$. An explicit construction for genus 2, embedding degree $k = 5$ and $\rho = 4$ is given – note that an instantiation with a 224-bit prime subgroup order r would exactly meet the 112-bit security level requirements (cf. Table 3.1). Other examples (for genus 2) include: $k = 6$, $\rho = 7.5$; $k = 8$, $\rho = 7.5$, and $k = 10$, $\rho = 6$ (able to exactly meet the 256-bit security level requirements) [19, 17]. In the case of genus 3, a construction yielding $k = 7$ and $\rho = 12$ is obtained.

All constructions mentioned so far in this section produce absolutely simple Jacobians. When considering simple abelian varieties A that are isogenous over some extension field $\mathbb{F}_{q^d}$ (q a prime) to a product of two elliptic curves, smaller ρ-values have been obtained:
Kawazoe and Takahashi [45] specialize to hyperelliptic curves with curve equation $y^2 = x^5 + ax$ over a prime field $\mathbb{F}_q$. For the cardinalities of the Jacobians of such curves, closed formulae exist. These formulae are exploited in adaptations of the Cocks-Pinch method (producing Jacobians with ρ-values around 4), and Brezing-Weng-type methods (for embedding degree divisible by 8, producing Jacobian varieties with $3 < \rho < 4$). The Jacobians split over $\mathbb{F}_{q^d}$, $d \in \{2, 4\}$.
Satoh [67] considers hyperelliptic curves C of the form $y^2 = x^5 + ax^3 + bx$ over $\mathbb{F}_q$, such that Jac_C splits over $\mathbb{F}_{q^2}$. This construction works for many embedding degrees and produces ρ-values < 4.
More generally, Freeman and Satoh [20] show that if E is defined over $\mathbb{F}_q$, and A is an abelian variety isogenous over $\mathbb{F}_{q^d}$ to $E \times E$, then A is isogenous over $\mathbb{F}_q$ to a primitive subvariety of the Weil restriction of E from $\mathbb{F}_{q^d}$ to $\mathbb{F}_q$. Thus, pairing-friendly abelian varieties of this type can be built from elliptic curves $E/\mathbb{F}_q$ that are not pairing-friendly over $\mathbb{F}_q$, but are pairing-friendly when base-extended to $\mathbb{F}_{q^d}$. The elliptic curves can be constructed via Cocks-Pinch or Brezing-Weng type methods. The generic ρ-value for Jacobians of genus 2 produced in this manner is 4. With the Brezing-Weng method, ρ-values between 2 and 4 can be obtained. This approach not only contains the constructions by Kawazoe and Takahashi [45] and Satoh [67] but also produces the lowest ever recorded ρ-values for ordinary genus 2 curves. Explicit examples of cryptographically interesting genus 2 curves are given, such as a $k = 9$, $\rho \approx 8/3$ curve and a $k = 27$, $\rho \approx 20/9$ curve.

In conclusion, to date, the best we can achieve for pairing-friendly ordinary genus 2 curves with arbitrary prescribed embedding degree k is a ρ-value of 4; and $\rho \approx 8$ if one insists on absolutely simple Jacobians. (Although to date, there is no apparent reason why Jacobians that split over small-degree extensions should be more vulnerable to DL attacks than the absolutely simple ones.) We have no constructions of ordinary hyperelliptic curves of genus $g \geq 2$ with ρ-values less than 2. In particular, we have no constructions of higher-dimensional pairing-friendly ordinary Jacobian varieties with a prime number of points. This is in sharp contrast to the elliptic case, where $\rho \approx 2$ can be achieved for any prescribed embedding degree, $1 < \rho < 2$ for selected embedding degrees, and constructions for prime-order elliptic curves exist for embedding degrees $k = 3, 4, 6, 10$, and 12 (cf. [21]).

3.4 Supersingular curves. Supersingular hyperelliptic curves over $\mathbb{F}_q$ are always pairing-friendly. In fact, Galbraith [28] shows that there exists a constant $k(g)$ such that the embedding degree of any supersingular abelian variety of dimension

g over any finite field $\mathbb{F}_q$ is bounded by $k(g)$. Rubin and Silverberg [66] prove that for simple supersingular abelian varieties, for $g \leq 6$ we have $k(g) \leq 7.5g$.

Specifically, for dimension $g = 2$, the embedding degree is bounded by 12, where $k = 12$ can only happen if $\mathbb{F}_q$ is a binary field $\mathbb{F}_{2^m}$ with m odd. If q is a square, or if $q = p^m$ with m odd and $p \neq 2, 3$, then the largest embedding degree is $k = 6$. If $\mathbb{F}_q = \mathbb{F}_{3^m}$ with m odd, the embedding degree is always bounded by 4. (In the case of dimension $g = 3$, the embedding degree is bounded by 18, and the bound for the dimension 4 case is 30. In both cases, this bound is achieved only in characteristic three. Over prime fields $\mathbb{F}_p$ with $p \geq 11$, there are no simple supersingular abelian varieties of dimension $g = 3$, while the largest embedding degree for dimension $g = 4$ is $k = 12$.)

As Rubin and Silverberg show [66, Corollaries 13,14], not all embedding degrees below these bounds are possible. For example, in the dimension 2 case and if $q = p^m$ with m odd, then for $p = 2$ we have $k \in \{1, 3, 6, 12\}$; if $p = 3$, we have $k \in \{1, 3, 4\}$; if $p = 5$ we have $k \in \{1, 3, 4, 5, 6\}$ and if $p \geq 7$ we have $k \in \{1, 3, 4, 6\}$.

Cryptographically interesting supersingular hyperelliptic curves can be explicitly constructed. For example, Galbraith et al. [33] give curve equations for various field characteristics that yield simple supersingular Jacobians of dimension $g = 2$ and of embedding degrees $k \in \{4, 5, 6, 12\}$. By carefully choosing the underlying fields, ρ-values close to 1 can be readily obtained.

3.5 Supersingular versus ordinary hyperelliptic curves. While the embedding degrees of supersingular abelian varieties are limited to a few, small values, their advantage is that they can achieve ρ-values significantly smaller than their ordinary counterparts. For example, let us consider the 112-bit security level (cf. Table 3.1). One could use the construction by Freeman and Satoh [20] of an ordinary absolutely simple hyperelliptic Jacobian of dimension 2, with embedding degree $k = 6$ and ρ-value 2.976, with a 230-bit prime-order subgroup, working over a finite field $\mathbb{F}_q$ with 342-bit q. Alternatively, one could use the embedding-degree 12 supersingular curve $y^2 + y = x^5 + x^3 + b$ ($b \in \{0, 1\}$) over $\mathbb{F}_{2^m}$ with $m \geq 250$ chosen such that its Jacobian contains a subgroup of prime order $r > 2^{224}$. (Note that Coppersmith' algorithm [12] for DL computation in finite fields of small characteristic requires to embed the Jacobian into a 3000-bit binary field $\mathbb{F}_{2^{12m}}$, to obtain roughly the same level of security provided by a 2048-bit field $\mathbb{F}_{q^{12}}$ with q large, cf. [55].) If m is chosen smaller than 342, this would result in bandwidth advantages for the supersingular Jacobian, given that in cryptographic applications the values that are transmitted are elements in $\mathrm{Jac}_C(\mathbb{F}_q)$. However, already at the 128-bit security level the advantage of supersingular curves disappears, in the light of the recent work by Freeman and Satoh [20]: this security level can be achieved with 256-bit prime-order subgroups either of an ordinary Jacobian over a 341-bit $\mathbb{F}_q$, with $k = 9$ and $\rho = 8/3$, or of a supersingular Jacobian over $\mathbb{F}_{2^m}$ with $m \geq 375$, of embedding degree 12 (again, m is chosen in response to Coppersmith' DL algorithm [12]: a 4500-bit binary field roughly provides the same security as a 3072-bit field of large characteristic). At high security levels ordinary curves are definitely preferable. For example, at the 256-bit level, a genus 2 curve with embedding degree $k = 27$ and (optimal to date) ρ-value of 20/9 (cf. [20]) requires a 568-bit field, while a binary supersingular curve of embedding degree 12 requires a 1875-bit field.

4 Pairings for hyperelliptic curves

In this section, we give an overview of the different pairings on hyperelliptic curves, as well as introduce the more general framework of *HV pairings* which unify the recent variations on the Ate pairing. In particular, we present a direct proof of bilinearity and non-degeneracy for these pairings and describe how the Ate_i and R-ate pairings fit into the framework.

We begin by introducing the historically most important pairings for hyperelliptic curves, the Tate-Lichtenbaum and Weil pairings. In what follows, let r be a positive integer and assume that C is defined over a finite field $\mathbb{F}_q$. Suppose that $K = \mathbb{F}_{q^k}$ is an extension of $\mathbb{F}_q$ such that $r \mid (q^k - 1)$. Throughout the section, we will use D to mean both a divisor and the divisor class represented by D.

For a positive integer s, a *Miller function* $f_{s,D}$ is a function with divisor

$$(f_{s,D}) = sD - \rho(sD),$$

uniquely defined up to scalar multiplication by elements of K^*. The *Miller loop length* of such a function is $\log_2 s$ and measures how quickly the function can be evaluated via Miller's algorithm (see Algorithm 1). The benefit of recent variations on the Tate-Lichtenbaum pairing is a reduction in Miller loop length, which is sometimes accomplished by combining several Miller functions (see Section 5).

4.1 Tate-Lichtenbaum pairing. For $D_1 \in \mathrm{Jac}_C(K)[r]$, the divisor rD_1 is linearly equivalent to zero, hence there is some function whose divisor is rD_1, namely the Miller function f_{r,D_1} defined above. Let D_2 be a divisor class, with representative $D_2 = \sum_P n_P(P)$ disjoint from D_1. We define a pairing called the *Tate-Lichtenbaum pairing* as follows

$$\begin{aligned} \tau : \mathrm{Jac}_C(K)[r] \times \mathrm{Jac}_C(K)/r\mathrm{Jac}_C(K) &\to K^*/(K^*)^r \\ (D_1, D_2) &\mapsto f_{r,D_1}(D_2) = \prod_P f_{r,D_1}(P)^{n_P}. \end{aligned}$$

This pairing is bilinear, non-degenerate and the result is independent of the choice of representatives of the divisor classes.

4.2 The Weil pairing. For $D_1, D_2 \in \mathrm{Jac}_C(\bar{K})[r]$, the *Weil pairing* is given by

$$\begin{aligned} e_r : \mathrm{Jac}_C(\bar{K})[r] \times \mathrm{Jac}_C(\bar{K})[r] &\to \mu_r \\ (D_1, D_2) &\mapsto \tau(D_1, D_2)\tau(D_2, D_1)^{-1} \end{aligned}$$

which can be computed via two Tate-Lichtenbaum pairings. It is bilinear, alternating, and non-degenerate.

4.3 The modified Tate-Lichtenbaum pairing. If $\mathrm{Jac}_C(K)$ contains no elements of order r^2, then there is an isomorphism

$$\mathrm{Jac}_C(K)[r] \cong \mathrm{Jac}_C(K)/r\mathrm{Jac}_C(K).$$

Under this identification, we define the *modified (or reduced) Tate-Lichtenbaum pairing* to be

$$\begin{aligned} t : \mathrm{Jac}_C(K)[r] \times \mathrm{Jac}_C(K)[r] &\to \mu_r \\ (D_1, D_2) &\mapsto \tau(D_1, D_2)^{(q^k-1)/r}. \end{aligned}$$

Since elements of K^* have order dividing $q^k - 1$ and $r \mid (q^k - 1)$, the rth powers which are the quotients of distinct representatives of the coset of $\tau(D_1, D_2)$ are removed by this *final exponentiation*, leaving a unique result lying in $\mu_r \subset K$.

Other powers of the Tate-Lichtenbaum pairing can also give non-degenerate bilinear pairings into μ_r which may yield shorter Miller loops (for example, with the use of efficiently computable automorphisms of C [16]; see Section 6.2).

4.4 Hyperelliptic Ate pairing. More generally, a *bilinear pairing* is a map

$$e : \mathbb{G}_1 \times \mathbb{G}_2 \to \mathbb{G}_3$$

where $\mathbb{G}_i$ are abelian groups, in additive notation, and $\mathbb{G}_3$ is a cyclic group, written multiplicatively, and for all $p_1, p_2 \in \mathbb{G}_1$, $q_1, q_2 \in \mathbb{G}_2$, we have

$$e(p_1 + p_2, q_1) = e(p_1, q_1)e(p_2, q_1),$$
$$e(p_1, q_1 + q_2) = e(p_1, q_1)e(p_1, q_2).$$

Let r be a prime dividing $\#\mathrm{Jac}_C(\mathbb{F}_q)$ and let k be the embedding degree of $\mathrm{Jac}_C(\mathbb{F}_q)$ with respect to r. We are interested in pairings where $\mathbb{G}_1$ and $\mathbb{G}_2$ are subgroups of $\mathrm{Jac}_C(K)$, where $K = \mathbb{F}_{q^k}$. In particular, a number of more convenient and faster pairings are known when

$$\begin{aligned} \mathbb{G}_1 &= \mathrm{Jac}_C(K)[r] \cap \ker(\pi - [1]), \\ \mathbb{G}_2 &= \mathrm{Jac}_C(K)[r] \cap \ker(\pi - [q]), \end{aligned} \tag{4.1}$$

where π is the qth power Frobenius automorphism. Since r divides $\#\mathrm{Jac}_C(\mathbb{F}_q)$, the group $\mathbb{G}_1$, being the eigenspace of 1, is at least 1-dimensional over $\mathbb{Z}/r\mathbb{Z}$. Since the eigenvalues of the Frobenius come in pairs $(\lambda, q/\lambda)$ [27, §5.2.3], q is also an eigenvalue of π on $\mathrm{Jac}_C[r]$, and thus there exists a divisor D such that $\pi(D) = qD$. This implies that $\pi^k D = q^k D = D$, since $r|(q^k - 1)$ and $rD = 0$. Consequently, $D \in \mathrm{Jac}_C(\mathbb{F}_{q^k})$, and the group $\mathbb{G}_2$ is also at least 1-dimensional over $\mathbb{Z}/r\mathbb{Z}$. If $k > 1$, then $\mathbb{G}_1 \neq \mathbb{G}_2$ and $\mathbb{G}_1 \times \mathbb{G}_2 \subset \mathrm{Jac}_C(\mathbb{F}_{q^k})[r]$ is at least 2-dimensional over $\mathbb{Z}/r\mathbb{Z}$.

In the remainder of this section, $\mathbb{G}_1$ and $\mathbb{G}_2$ always denote the groups defined in (4.1).

The most basic pairing defined for divisors in $\mathbb{G}_1$, $\mathbb{G}_2$ is the *hyperelliptic Ate pairing* [36]:

$$\begin{aligned} a : \mathbb{G}_2 \times \mathbb{G}_1 &\to \mu_r \\ (D_2, D_1) &\mapsto f_{q,\rho(D_2)}(D_1), \end{aligned}$$

where $\rho(D_2)$ is the reduced divisor class representative. Since the Frobenius π acts as $[q]$ on D_2, we have $f_{q,\rho(D_2)}(D_1) \in \mu_r$ and no final exponentiation is required [36, Lemma 2]. This is different from the elliptic Ate pairing [41], where a final exponentiation is always required. Another important difference of the hyperelliptic Ate pairing is that to obtain a well-defined value, one *must* use the reduced divisor $\rho(D_2)$, not simply any representative of the class D_2. The Miller loop length for the hyperelliptic Ate pairing is $\log_2 q$, in contrast to the elliptic case where the Miller loop length is $\log_2(t - 1)$ with t the trace of Frobenius.

4.5 The Hess-Vercauteren (HV) framework for pairings on Frobenius eigenspaces. Since 2007, several variations of the Ate pairing have been proposed for elliptic and hyperelliptic curves, exploiting the fact that products and ratios of bilinear, non-degenerate pairings on $\mathbb{G}_2 \times \mathbb{G}_1$ are also bilinear pairings, but not necessarily non-degenerate [75]. The key is to find combinations of pairings which are both non-degenerate and computable using shorter Miller loops. Following the work of Hess [40] and Vercauteren [71] in the elliptic curve case, we unify these various pairings on $\mathbb{G}_2 \times \mathbb{G}_1$ in a more general framework, which we call *HV pairings.* The main benefit of this framework is that the criteria for non-degeneracy are more straightforward to verify, giving a direct way to create new pairings. Further investigation of this framework and possible extensions seems likely to be fruitful (see Section 6.1 and (1) in Section 6.9).

Let D be any divisor in $\mathrm{Jac}_C(K)[r]$, and s an integer. Recall that any divisor D is equivalent to a unique reduced divisor which we denote $\rho(D)$. Let $h(x) \in \mathbb{Z}[x]$ be a polynomial of the form $h(x) = \sum_{i=0}^{n} h_i x^i$ satisfying $h(s) \equiv 0 \pmod r$. Define a *generalized Miller function* $f_{s,h,D}$ to be any function with divisor

$$\sum_{i=0}^{n} h_i \rho(s^i D). \tag{4.2}$$

To see that this divisor is principal, consider the principal divisor

$$\sum_{i=0}^{n} h_i (s^i D - \rho(s^i D)),$$

which differs by $(\sum_{i=0}^{n} h_i s^i)D$ from (4.2). Since $h(s) \equiv 0 \pmod r$, this is an integer multiple of rD, which is linearly equivalent to zero by assumption, and thus the divisor (4.2) is principal. As with the standard Miller function, the function $f_{s,h,D}$ is only defined up to scalar multiples. Also, we note that the Miller function $f_{r,D}$ for the Tate-Lichtenbaum pairing is equal to $f_{s,h,D}$ for the constant function $h(x) = r$ and arbitrary integer s.

Theorem 4.1 *Let $s \equiv q^j \pmod r$ for some $j \in \mathbb{Z}$. Let $h(x) \in \mathbb{Z}[x]$ with $h(s) \equiv 0 \pmod r$. Then*

$$\begin{aligned} a_{s,h} : \mathbb{G}_2 \times \mathbb{G}_1 &\to \mu_r \\ (D_2, D_1) &\mapsto f_{s,h,D_2}(D_1)^{(q^k-1)/r} \end{aligned}$$

is a bilinear pairing satisfying

$$a_{s,h}(D_2, D_1) = t(D_2, D_1)^{h(s)/r} \text{ and } a_{s,h}(D_2, D_1) = a(D_2, D_1)^{kq^{k-1}h(s)/r}$$

where t is the modified Tate-Lichtenbaum pairing and a is the hyperelliptic Ate pairing. The pairing $a_{s,h}$ is non-degenerate if and only if $h(s) \not\equiv 0 \pmod{r^2}$.

Remark 4.2 We note that since k is the embedding degree of $\mathrm{Jac}_C(\mathbb{F}_q)$ with respect to r, in Theorem 4.1 s will be a kth root of unity modulo r since q is a primitive k^{th} root. In Hess's framework, there is the additional condition that s be a primitive k^{th} root of unity modulo r^2. This requirement is necessary to show the existence of pairings such that the function $f_{s,h,D}$ is of "lowest degree" (see [40, §3]), but is not required for the result above.

Proof First we show that the pairing is well-defined on divisor classes. Suppose that $D_2' \sim D_2$. Then

$$\mathrm{div}(f_{s,h,D_2'}/f_{s,h,D_2}) = \sum_{i=1}^{n} h_i(\rho(s^i D_2') - \rho(s^i D_2)) = \emptyset.$$

This demonstrates well-definition in the factor $\mathbb{G}_2$. For the factor $\mathbb{G}_1$, it suffices to show that the pairing is trivial under the hypothesis that D_1 is a principal divisor.

Suppose $D_1 = \mathrm{div}(g)$. For any $D_2 \in \mathbb{G}_2$, by the hypothesis that $s \equiv q^j \mod r$ and $\rho(rD_2) = \emptyset$, it is the case that

$$\rho(s^i D_2) = \rho(q^{ij} D_2) = \rho(\pi^{ij} D_2) = \pi^{ij}\rho(D_2) = q^{ij}\rho(D_2) = s^i\rho(D_2) + rD'$$

for some divisor D' defined over $\mathbb{F}_q$. Therefore,

$$\sum_{i=1}^{n} h_i\rho(s^i D_2) = \sum_{i=1}^{n} h_i s^i \rho(D_2) + rD''$$

for some D'' defined over $\mathbb{F}_q$. Then by the hypothesis, this expression is an rth multiple of another divisor D''' defined over $\mathbb{F}_q$. By Weil reciprocity,

$$f_{s,h,D_2}(D_1)^{(q^k-1)/r} = g(rD''')^{(q^k-1)/r} = g(D''')^{q^k-1} = 1,$$

as required.

We show bilinearity and non-degeneracy directly, in contrast to Hess's more general approach in the elliptic curve case [40, Theorem 1].

Let $s = q^j + \ell r$, for $j, \ell \in \mathbb{Z}$. Linearity in the second coordinate follows from the definition of evaluation of a function on a divisor. To show linearity in the first coordinate, let $D_2, D_3 \in \mathbb{G}_2$ and $D_1 \in \mathbb{G}_1$ be non-trivial reduced divisors. Then

$$(f_{s,h,D_2+D_3}) = \sum_{i=0}^{n} h_i\rho(s^i D_2 + s^i D_3) = \sum_{i=0}^{n} h_i\rho(s^i D_2) + \sum_{i=0}^{n} h_i\rho(s^i D_3) + \sum_{i=0}^{n} h_i(g_i)$$

where

$$(g_i) = \rho(s^i D_2 + s^i D_3) - \rho(s^i D_2) - \rho(s^i D_3).$$

Since $rD_2 \sim 0$, $rD_3 \sim 0$ and $s = q^j + \ell r$, the function g_i has divisor

$$(g_i) = \rho(q^{ij} D_2 + q^{ij} D_3) - \rho(q^{ij} D_2) - \rho(q^{ij} D_3).$$

Since $D_2, D_3 \in \mathbb{G}_2$, the q-eigenspace of the Frobenius π, and since ρ commutes with π, we have

$$(g_i) = \rho(D_2 + D_3)^{\pi^{ij}} - \rho(D_2)^{\pi^{ij}} - \rho(D_3)^{\pi^{ij}}.$$

Then $(g_i) = (m)^{\pi^{ij}}$ where m is the function with divisor

$$(m) = \rho(D_2 + D_3) - \rho(D_2) - \rho(D_3).$$

As f_{s,h,D_2+D_3} is evaluated at the divisor $D_1 \in \mathbb{G}_1$, which is fixed by π, the value $g_i(D_1)$ equals $m(D_1)^{\pi^{ij}} = m(D_1)^{q^{ij}}$. Thus,

$$\prod_{i=0}^{n} g_i(D_1)^{h_i} = \prod_{i=0}^{n} m(D_1)^{h_i q^{ij}} = m(D_1)^{\sum_{i=0}^{n} h_i q^{ij}} = m(D_1)^{h(q^j)}.$$

Using the fact that $s = q^j + \ell r$ and $h(s) \equiv 0 \pmod r$, we see that this value is eliminated by the final exponentiation of $(q^k - 1)/r$. Since

$$f_{s,h,D_2+D_3}(D_1) = f_{s,h,D_2}(D_1) f_{s,h,D_3}(D_1) \prod_{i=0}^{n} g_i(D_1)^{h_i},$$

the pairing $a_{s,h}$ is linear with respect to the first coordinate.

We now show that

$$a_{s,h}(D_2, D_1) = t(D_2, D_1)^{h(s)/r}$$

using a similar argument. On the right, we have

$$t(D_2, D_1)^{h(s)/r} = \left(f_{r,D_2}(D_1)^{(q^k-1)/r}\right)^{h(s)/r}.$$

Since $D_2 \in \mathbb{G}_2$, we have $\rho(rD_2) = 0$, thus

$$(f_{r,D_2}^{h(s)/r}) = (h(s)/r)(rD_2 - \rho(rD_2)) = h(s)D_2 = \sum_{i=0}^{n} h_i s^i D_2.$$

On the left, we have

$$a_{s,h}(D_2, D_1) = f_{s,h,D_2}(D_1)^{(q^k-1)/r}.$$

where by definition

$$(f_{s,h,D_2}) = \sum_{i=0}^{n} h_i \rho(s^i D_2).$$

We can rewrite this as

$$(f_{s,h,D_2}) = \sum_{i=0}^{n} h_i s^i D_2 - \sum_{i=0}^{n} h_i (g_i),$$

where

$$(g_i) = s^i D_2 - \rho(s^i D_2).$$

Since we evaluate at $D_1 \in \mathbb{G}_1$ fixed by π and $s = q^j + \ell r$ for some $\ell \in \mathbb{Z}$, the contribution of the function with divisor $(\sum_{i=0}^{n} h_i(g_i))$ is eliminated by raising to the power $(q^k - 1)/r$. Furthermore, we may choose any functions f_{r,D_2} and f_{s,h,D_2} with the above divisors, as any discrepancy from scalar multiples will be canceled out when evaluating at the degree zero divisor D_1. Thus, $a_{s,h}(D_2, D_1) = t(D_2, D_1)^{h(s)/r}$.

We have that t is a non-degenerate pairing and $h(s) \equiv 0 \pmod{r}$. Therefore, by the relationship between $a_{s,h}$ and t, we conclude that $a_{s,h}$ is non-degenerate if and only if $h(s) \not\equiv 0 \pmod{r^2}$.

For the relationship with the hyperelliptic Ate pairing a, we use the fact that $t(D_2, D_1) = a(D_2, D_1)^{kq^{k-1}}$ [29, Theorem 2]. □

4.6 Examples of HV pairings. In this section, we describe how the pairings in the current literature fit into the HV framework. While these pairings can be expressed as $a_{s,h}$ for some $s \in \mathbb{Z}$ and $h(x) \in \mathbb{Z}[x]$, their actual computation takes an alternate form in order to make use of shorter Miller loops.

1. The *generalized Ate pairing* or *Ate$_i$ pairing*, was defined by Zhang [74] as the analogue of the Ate$_i$ pairing for elliptic curves [76]. For $s \equiv q^j \pmod{r}$,

$$\begin{aligned} a_s : \mathbb{G}_2 \times \mathbb{G}_1 &\to \mu_r \\ (D_2, D_1) &\mapsto f_{s,D_2}(D_1)^{(q^k-1)/r}. \end{aligned}$$

Since $r \mid (q^k - 1)$, we may assume $0 < j < k$. Note that if $s = q^j$ then no final exponentiation is needed, as is the case for the hyperelliptic Ate pairing. However, this choice of s is never an improvement over the Ate pairing as the Miller loop length is $i \log_2 q \geq \log_2 q$.

For $s \not\equiv q^j \pmod{r^2}$, it is straightforward to show this is the HV pairing $a_{s,h}$ where $h(x) = x - q^j$. Writing $s = q^j + \ell r$ for $\ell \in \mathbb{Z}$, we have $(f_{s,D}) = sD - \rho(sD) = (f_{s,h,D}) + \ell r D$. As $\ell r D \sim 0$, these functions differ only by a constant and thus give the same value after the final exponentiation.

2. The Ate pairings defined by Vercauteren [71, Theorem 1] for elliptic curves can be generalized directly to hyperelliptic curves. To define the pairing, we first choose an integer m relatively prime to r and express mr in base q as $mr = \sum_{i=0}^{n} h_i q^i$. We can decompose the mth power of the Tate-Lichtenbaum pairing as

$$t(D_2, D_1)^m = f_{\sum_{i=0}^{n} h_i q^i, D_2}(D_1)^{(q^k-1)/r} = \left(\prod_{i=0}^{n} f_{h_i q^i, D_2}(D_1) \cdot \prod_{j=0}^{n-1} g_j(D_1)\right)^{(q^k-1)/r} \quad (4.3)$$

where the g_j $(j = 0, \ldots, n-1)$ are auxiliary functions defined through

$$f_{\sum_{i=j}^{n} h_i q^i, D_2} = f_{\sum_{i=j+1}^{n} h_i q^i, D_2} f_{h_j q^j, D_2} g_j.$$

The pairing $a_{[h_0,\ldots,h_n]}$ is then defined as

$$\begin{aligned} a_{[h_0,\ldots,h_n]} : \mathbb{G}_2 \times \mathbb{G}_1 &\to \mu_r \\ (D_2, D_1) &\mapsto \left(\prod_{i=0}^{n} f_{h_i, D_2}(D_1)^{q^i} \cdot \prod_{j=0}^{n-1} g_j(D_1)\right)^{(q^k-1)/r}. \end{aligned}$$

It is easy to see that $a_{[h_0,\ldots,h_n]}(D_2, D_1)$ equals $t(D_2, D_1)^m$. Indeed, by definition of the Miller functions and the action of the Frobenius on D_1 and D_2, we have that

$$f_{h_i q^i, D_2}(D_1) = f_{q^i, D_2}(D_1)^{h_i} f_{h_i, q^i D_2}(D_1) = f_{q^i, D_2}(D_1)^{h_i} f_{h_i, D_2}(D_1)^{q^i}$$

as in the proof of [71, Theorem 1]. While not explicitly noted in that proof, it is also true that

$$\left(\prod f_{q^i, D_2}(D_1)^{h_i}\right)^{(q^k-1)/r} = 1,$$

by an argument similar to that of Theorem 4.1. Therefore $a_{[h_0,\ldots,h_n]}(D_2, D_1) = t(D_2, D_1)^m$. Thus, this pairing is simply the HV pairing $a_{q,h}$ where $h(x) = \sum_{i=0}^{n} h_i x^i$.

The pairing $a_{[h_0,\ldots,h_n]}$ is computed as a product of many Miller functions, as well as the auxiliary functions, and the total sum of the lengths of the Miller loops of the functions is $\sum_{i=0}^{n} \log_2 h_i$. Thus, for efficiency, this pairing is fastest if the coefficients of mr in base q expansion are small. Vercauteren gives an algorithm to find suitable multiples of r by searching for shortest vectors in a lattice spanned by vectors involving powers of q [71, §3.3]. This is the "lattice" idea which was further generalized by Hess [40]. See Section 6.1 for a discussion of the smallest loop length possible.

3. The *R-ate pairing*, introduced by Lee, Lee and Park in 2008 [53], was the first pairing defined as a ratio of generalized Ate pairings. We give a specific instantiation as an example (cf. [53, Corollary 3.3(3)]). Let $T_i \equiv q^i \pmod{r}$ and $T_j \equiv q^j \pmod{r}$, where $0 < i < j < k$, and write $T_i = aT_j + b$ for some $a, b \in \mathbb{Z}$. Then the R-ate pairing is

$$\begin{aligned} R : \mathbb{G}_2 \times \mathbb{G}_1 &\to \mu_r \\ (D_2, D_1) &\mapsto \left(f_{a, T_j D_2}(D_1) f_{b, D_2}(D_1) g(D_1)\right)^M, \end{aligned}$$

where g is an auxiliary function with divisor $aT_jD_2 + bD_2 - \rho(aT_jD_2 + bD_2)$ and $M \in \mathbb{N}$ is a final exponent. (The function g is the analogue of the ratio of a linear and vertical function for the elliptic curve case.) Although it is ambigous in the original paper, this pairing requires a final exponentiation to yield a unique value. The exponent $M = (q^k - 1)/r$ is sufficient, though a smaller exponent may also work, depending on the multiplicative orders of T_i and T_j modulo r (see [53, Corollary 3.3(3)] for details). It is easy to work out ([53, Theorem 3.2]) that

$$R(D_2, D_1) = \left(f_{T_i,D_2}(D_1)/f_{T_j,D_2}(D_1)^a\right)^M,$$

and thus R is in fact a ratio of generalized Ate pairings. Since $f_{a,T_jD_2}(D_1) = f_{a,D_2}(D_1)^{q^j}$ ([76, Theorem 1]), in practice, the R-ate pairing is computed as

$$R(D_2, D_1) = \left(f_{a,D_2}(D_1)^{q^j} f_{b,D_2}(D_1)g(D_1)\right)^M.$$

In this form, and with $M = (q^k - 1)/r$, it is a straightforward calculation to establish that R corresponds to the above Vercauteren pairing $a_{[h_0,..h_i,.,h_j]}$ with $h_0 = b, h_i = -1,\ h_j = a$ and all other coefficients equal to zero: let $\ell_i, \ell_j \in \mathbb{Z}$ such that $T_i = q^i + \ell_i r$ and $T_j = q^j + \ell_j r$, and express the r-multiple $(\ell_i - a\ell_j)r$ in base q, and use that f_{1,D_2} is a constant function and therefore eliminated by the final exponentiation. In other words, R is the HV-pairing $a_{s,h}$ where $s = q$ and $h(x) = ax^j - x^i + b$.

4.7 Twisted Ate pairing. In this section, we discuss the *twisted Ate pairing* $e : \mathbb{G}_1 \times \mathbb{G}_2 \to \mu_r$. The twisted Ate pairings use the fact that in certain situations, there is a "twist" of the Frobenius π which acts as $[q]$ on $\mathbb{G}_1$ and $[1]$ on $\mathbb{G}_2$, thereby reversing the roles of these groups in the Ate pairing. The main benefit of such pairings is that $D_1 \in \mathbb{G}_1$ is defined over $\mathbb{F}_q$, which means computing the Miller function f_{s,D_1} is simpler. An added benefit is that the points in $D_2 \in \mathbb{G}_2$ have x-coordinates in a subfield of $\mathbb{F}_{q^k}$ which also may simplify the evaluation, as explained in Section 5.3.

Let C be a curve over a finite field $K = \mathbb{F}_q$. A *twist* of C is a curve C' over $\mathbb{F}_q$ such that there exists an isomorphism $\phi : C' \to C$ defined over $\mathbb{F}_{q^\delta}$ for some $\delta \in \mathbb{Z}^+$. If δ is the minimal degree extension of $\mathbb{F}_q$ over which the isomorphism is defined, then the twist C' is of *degree* δ. For more on twists of curves, see Silverman [69, §10.2].

Let π be the Frobenius of C and let ϕ^π denote the isomorphism $C' \to C$ obtained by π acting on the coefficients of ϕ. Then $\phi^\pi \circ \phi^{-1}$ is an automorphism of C of order δ in $\mathrm{Aut}(C)$. Thus to look at twists of C, one needs to consider the automorphism group of C. For genus 2 hyperelliptic curves over $\mathbb{F}_q$, $\mathrm{Aut}(C)$ is isomorphic to one of the following groups [7, 8]:

$$C_2, C_{10}, C_2 \times S_3, V_4, D_8, D_{12}, 2D_{12}, \tilde{S_4}, \tilde{S_5}, M_{32}, \text{ or } M_{160},$$

where C_n is the cyclic group of order n, V_4 is the Klein 4-group, D_n is the dihedral group of order n, S_n is the symmetric group of order n, M_n is the group of order n arising from a certain exact sequence [8, Equation 6], and $2D_{12}, \tilde{S_4}, \tilde{S_5}$ are 2-coverings of D_{12}, S_4, and S_5, respectively. This implies that δ, as the order of an element in $\mathrm{Aut}(C)$, has to divide $\#\mathrm{Aut}(C)$ for one of the above automorphism groups.

If C has a twist of degree δ with $m = \gcd(k, \delta) > 1$, then it is possible to define a non-degenerate, bilinear pairing on $\mathbb{G}_1 \times \mathbb{G}_2$. For applications to cryptography, we are interested in using the highest degree twist available, because elements of $\mathbb{G}_2$ can then be represented as elements of the Jacobian of the twist C' defined over $\mathbb{F}_{q^{k/m}}$.

Given a curve C, let $r \mid \#\mathrm{Jac}_C(\mathbb{F}_q)$ be a large prime, k the embedding degree, and C' a degree δ twist of C. We have an injection

$$\begin{aligned}[\cdot] : \mu_\delta &\to \mathrm{Aut}(C)\\ \xi &\mapsto [\xi]\end{aligned}$$

where ξ is the automorphism defined by the twist. Then $\mathbb{G}_2 = \mathrm{Jac}_C(\mathbb{F}_q)[r] \cap \ker(\pi - [q]) = \mathrm{Jac}_C(\mathbb{F}_q)[r] \cap \ker([\xi]\pi^{k/m} - 1)$, and Zhang proved the following theorem ([74, Theorem 2]):

Theorem 4.3 *Let C be a hyperelliptic curve over $\mathbb{F}_q$ with a twist of degree δ. Let $m = \gcd(k, \delta)$ and $e = k/m$. Then*

$$\begin{aligned} a^{twist} : \mathbb{G}_1 \times \mathbb{G}_2 &\to \mu_r\\ (D_1, D_2) &\mapsto f_{q^e, D_1}(D_2),\end{aligned}$$

where the representatives of $D_1 \in \mathbb{G}_1$ and $D_2 \in \mathbb{G}_2$ have disjoint support, defines a non-degenerate bilinear pairing called the hyperelliptic twisted Ate pairing.

Remark 4.4 For C with $\gcd(k, \#\,\mathrm{Aut}(C)) \neq 1$, any pairing on $\mathbb{G}_1, \mathbb{G}_2 \subset \mathrm{Jac}_C(\mathbb{F}_q)$ in the HV framework has a twisted version, $a_{s,h}^{\mathrm{twist}} : \mathbb{G}_1 \times \mathbb{G}_2 \to \mu_r$ [40, Theorem 1].

We now define the *eta pairing*, which is essentially the twisted Ate pairing on supersingular curves, although historically it was introduced before the Ate pairing. The eta pairing makes use of a *distortion map* on C instead of a twist. Let $e(\cdot,\cdot)$ denote any bilinear, non-degenerate, Galois-invariant pairing on $\mathrm{Jac}_C(\mathbb{F}_q)[r]$. A non-degenerate pairing ensures that given a non-zero divisor class D_1 of order r, there exists D_2 such that $e(D_1, D_2) \neq 1$. However, there are certain instances where a specific D_1 and D_2 pair to 1, for example, where D_1, D_2 both are defined over $\mathbb{F}_q$ and the embedding degree $k > 1$. To remedy this, we introduce distortion maps.

Definition 4.5 Let e be a non-degenerate pairing and D_1 and D_2 non-zero divisor classes of prime order r on C. A *distortion map* is an endomorphism ψ of $\mathrm{Jac}_C(\mathbb{F}_q)$ such that $e(D_1, \psi(D_2)) \neq 1$.

Galbraith et al. [33] proved that distortion maps always exist for supersingular abelian varieties:

Theorem 4.6 *Let A be a supersingular abelian variety of dimension g over $\mathbb{F}_q$, and let r be a prime not equal to the characteristic of $\mathbb{F}_q$. For every two non-trivial elements D_1 and D_2 of $A(\mathbb{F}_q)[r]$, there exists an endomorphism ψ of A such that $e(D_1, \psi(D_2)) \neq 1$.*

The *eta pairing* has been introduced in 2007 by Barreto et al. [2] for supersingular curves. It provides a generalization of the results of Duursma and Lee [14] for a specific instance of supersingular curves. Consider a supersingular curve $C/\mathbb{F}_q$ (having one point at infinity) which has even embedding degree $k > 1$. Let D_1 and D_2 be reduced divisors of degree zero on C defined over $\mathbb{F}_q$ representing divisor classes with order r. Assume that there exists a distortion map ψ which allows for *denominator elimination* (see Section 5.3), meaning the x-coordinates of points in $\psi(D_2)$ lie in a subfield of $\mathbb{F}_{q^k}$.

Definition 4.7 For $T \in \mathbb{Z}$, the *eta pairing* η_T is given by

$$\begin{aligned} \eta_T : \mathbb{G}_1 \times \mathbb{G}_1 &\to \mu_r \\ (D_1, D_2) &\mapsto f_{T,D_1}(\psi(D_2))^{(q^k-1)/r}. \end{aligned}$$

Note that in the literature, the eta pairing is often defined without the final exponent, though it is necessary to obtain a unique value in μ_r. In general, this pairing is not a non-degenerate, bilinear pairing, but Barreto et al. [2, Theorem 1] give sufficient conditions on T under which $\eta_T(\cdot,\cdot)$ can be related to the modified Tate-Lichtenbaum pairing. In particular, this implies that for certain values of T, the eta pairing is indeed non-degenerate and bilinear. Moreover, the recent work of Lee, Lee and Lee [54, 52] allows us to compute the eta pairing on genus 2 curves for general divisors, which lifts an earlier restriction to the case of degenerate divisors (see Section 5.4).

5 Fast computation of hyperelliptic pairings

In this section, we summarize the state of the art for fast computation of pairings on hyperelliptic curves of genus 2.

5.1 Miller's algorithm. The algorithm used to compute Weil and Tate-Lichtenbaum pairings on elliptic curves was devised by Victor Miller in 1985 [58] and can be adapted to all pairings discussed in this paper [15]. Referring to the pairing definitions of Section 4 one sees that to compute a pairing, it is necessary to evaluate a Miller function at a divisor. Algorithm 1, further on referred to as "Miller's algorithm", computes such a value using the structure of an addition chain for s.

Usually, an addition chain takes the form of a double-and-add chain, as follows. Starting with the integer $k = 0$, at each step one performs one of two possible calculations to update the value of k: one either doubles to obtain $k \to 2k$ or doubles-and-adds to obtain $k \to 2k+1$. To determine the sequence of steps needed to obtain any desired integer s in this way, one reads the binary digits of s from left to right, doubling once for each '0' and doubling-and-adding for each '1.' (For example, $5 = 101_2$ is obtained as $0 \to 2(0)+1 = 1 \to 2(1) = 2 \to 2(2)+1 = 5$.) Starting from 0, this algorithm computes s in $\lfloor \log_2 s \rfloor + 1$ steps (each of which consists of either one or two additions).

Miller's Algorithm computes $f_{s,D}$ following this double-and-add process by computing the Miller function $f_{k,D}$ at each step along the way, obtaining $f_{s,D}$ at the end. A double step involves one addition, and a double-and-add step involves two. For each addition, we compute the new Miller function $f_{i+j,D}$ from the previously computed $f_{i,D}$ and $f_{j,D}$ via the relationship

$$f_{i+j,D} = f_{i,D} f_{j,D} h_{iD,jD}, \qquad i,j > 0,$$

where the auxiliary function $h_{D',D''}$ is a function with divisor

$$\rho(D') + \rho(D'') - \rho(D' + D'').$$

The computation of $h_{D',D''}$ is performed by an enhanced version of Cantor's Algorithm (cf. Section 2.3), here Algorithm 2. It is called under the name `Cantor()` once (if doubling) or twice (if doubling and adding) in each for-loop of Miller's

Algorithm. Using the result of Algorithm 2, one calculates $f_{2i,D}$ from $f_{i,D}$ ("double") or $f_{2i+1,D}$ from $f_{i,D}$ and $f_{1,D}$ ("double and add"), where $f_{1,D}$ is a constant function.

In order to compute the pairing value, the Miller function f_{s,D_2} must be evaluated at a divisor D_1, but this evaluation is not possible unless D_1 and D_2 have disjoint support, which is not the case if both are reduced. However, using reduced divisors and Mumford representation is too useful to dispense with, so the solution is the following. Let z be a uniformizer at P_∞ (for example, $z(x,y) = x^2/y$ is a convenient choice). Then, if f is a function with order $-r$ at P_∞, define the *leading coefficient* at P_∞ of f, denoted as $\mathrm{lc}_\infty(f)$, to be $(z^r f)(P_\infty)$. Then the *normalization* of f is the scalar multiple $f^{\text{norm}} = f/\mathrm{lc}_\infty(f)$ which has leading coefficient 1. For the hyperelliptic Ate pairing [36, Lemma 6], when z is $\mathbb{F}_q$-rational,

$$f_{q,\rho(D_2)}(D_1) = f^{\text{norm}}_{q,\rho(D_2)}(\epsilon(D_1)).$$

The right-hand expression requires computing the leading coefficient, but solves the problem of non-disjoint supports of D_1 and D_2 without losing the usefulness of Mumford representation.

For HV Pairings and the modified Tate-Lichtenbaum pairing, the same solution is possible. Consider the computation of $t(D_2, D_1) = f_{r,D_2}(D_1)^{(q^k-1)/r}$ where D_1, D_2 are reduced. Let $-b_i$ be the coefficient of P_∞ in D_i for $i = 1, 2$. (Note that $b_i = -1$ or -2, depending on whether or not the reduced divisor D_i is degenerate.) The function f_{r,D_2} has divisor rD_2 with order $-b_2 r$ at P_∞. Therefore, if z is an $\mathbb{F}_{q^k}$-rational uniformizer at P_∞,

$$f_{r,D_2}(D_1) = f^{\text{norm}}_{r,D_2}(\epsilon(D_1))/z(P_\infty)^{b_1 b_2 r}.$$

Since $b_1 b_2 r$ is a multiple of r, the contribution of $z(P_\infty)^{b_1 b_2 r(q^k-1)/r}$ is 1, and thus

$$f_{r,D_2}(D_1)^{(q^k-1)/r} = f^{\text{norm}}_{r,D_2}(\epsilon(D_1))^{(q^k-1)/r}.$$

As the HV pairing $a_{s,h}(D_2, D_1)$ is a simply a power of the modified Tate pairing $t(D_2, D_1)$ (see Theorem 4.1), in whichever form the pairing $a_{s,h}(D_2, D_1)$ is computed, evaluating normalized functions at effective divisors will give the pairing value.

In the elliptic curve case, it is more efficient to evaluate the Miller functions and the auxiliary functions $h_{D',D''}$ at the desired divisor (denoted D_2 in Miller's Algorithm) at each step, instead of reserving the evaluation for the end. In order to allow for this, D_2 is passed to Cantor's Algorithm. We now turn to a discussion of this aspect in the case of hyperelliptic curves.

In Miller's Algorithm, the current Miller function f is stored as two polynomials f_1 and f_2 such that $f = f_1/f_2$. Similarly, the auxiliary functions h are returned from Cantor's Algorithm as h_1 and h_2. It remains to explain how to evaluate a polynomial function $g(x, y)$ on C at the effective part of a divisor given in Mumford representation $(u(x), v(x))$ (we need only the effective part because of the preceeding discussion and the computation of the leading coefficient). We need to evaluate $G(x) = g(x, v(x))$ at the zeroes of $u(x)$. This is the same as computing the resultant $\mathrm{Res}(G(x), u(x))$. Performing a resultant calculation is sufficiently costly that it is best left to the end of Miller's Algorithm, as long as the size of the Miller functions can be kept low in the meantime. Fortunately, in preparation for the eventual final resultant, it suffices to compute the Miller functions in x and y modulo $u(x)$, while substituting $y = v(x)$, effectively capping their degrees.

Algorithm 1 Miller's Algorithm

Input: $D_1 = (u_1, v_1)$, $D_2 = (u_2, v_2)$, d, $s = \sum_{i=0}^{N} s_i 2^i$
Output: $f_{s,D_1}^{\text{norm}}(\epsilon(D_2))^d$

$D \leftarrow D_1$
$f_1 \leftarrow 1, f_2 \leftarrow 1, f_3 \leftarrow 1$
for $i = N - 1$ down to 0 **do**
  $f_1 \leftarrow f_1^2 \pmod{u_2}, f_2 \leftarrow f_2^2 \pmod{u_2}, f_3 \leftarrow f_3^2$
  $(D, h_1, h_2, h_3) \leftarrow \text{Cantor}(D, D, D_2)$
  $f_1 \leftarrow f_1 \cdot h_1 \pmod{u_2}, f_2 \leftarrow f_2 \cdot h_2 \pmod{u_2}, f_3 \leftarrow f_3 \cdot h_3$
  if $s_i = 1$ **then**
    $(D, h_1, h_2, h_3) \leftarrow \text{Cantor}(D, D_1, D_2)$
    $f_1 \leftarrow f_1 \cdot h_1 \pmod{u_2}, f_2 \leftarrow f_2 \cdot h_2 \pmod{u_2}, f_3 \leftarrow f_3 \cdot h_3$
  end if
end for
return $\left(\text{Res}(f_1, u_2)/(f_3^{\deg(u_2)} \cdot \text{Res}(f_2, u_2))\right)^d$

Algorithm 2 Cantor's Algorithm

Input: $D_1 = (u_1, v_1)$, $D_2 = (u_2, v_2)$, $D' = (u, v)$
Output: $\rho(D_1 + D_2), f(x, v(x)) \pmod{u}, g(x, v(x)) \pmod{u}, \text{lc}_\infty(h_{D_1,D_2})$ where $h_{D_1,D_2} = f/g$

1: compute (d_1, e_1, e_2) such that $d_1 = e_1u_1 + e_2u_2 = \gcd(u_1, u_2)$
2: compute (d, c_1, c_2) such that $d = c_1d_1 + c_2(v_1 + v_2 + H) = \gcd(d_1, v_1 + v_2 + H)$
3: $s_1 \leftarrow c_1e_1, s_2 \leftarrow c_1e_2, s_3 \leftarrow c_2$
4: $U \leftarrow (u_1u_2)/d^2, V \leftarrow (s_1u_1v_2 + s_2u_2v_1 + s_3(v_1v_2 + F))/d \pmod{U}$
5: $f \leftarrow d \pmod{u}, g \leftarrow 1, h \leftarrow 1$
6: **while** $\deg(U) > g$ **do**
7:   $U' \leftarrow (F - VH - V^2)/U, V' \leftarrow (-H - V) \pmod{U'}$
8:   $f \leftarrow f \cdot (v - V) \pmod{u}$
9:   $g \leftarrow g \cdot U' \pmod{u}$
10:   **if** $\deg(V) > g$ **then**
11:     $h \leftarrow -\text{leadingcoeff}(V) \cdot h$
12:   **end if**
13:   $U \leftarrow U', V \leftarrow V'$
14: **end while**
15: **return** $(U, V), f, g, h$

If Steps 5 and 8 through 13 are removed from Cantor's Algorithm and only (U, V) is returned, the algorithm computes $\rho(D_1 + D_2)$ for any divisors D_1 and D_2 in Mumford representation (this is the usual meaning of "Cantor's Algorithm" as in Section 2.3). If these steps are included, then Cantor's Algorithm can also return $f, g \pmod{u}$ such that $f/g = h_{D_1,D_2}(x, v(x))$ for some specified divisor (u, v). This is the form in which it is used in Miller's Algorithm.

In the case that we are pairing degenerate divisors (see Section 5.4), a norm computation may be preferred to the resultant method [29].

5.2 Using effective divisors and the leading coefficient. The leading coefficient of $f_{s,D}$ is an element of the field of definition of the function. Therefore, in the case of twisted pairings, the leading coefficient of f_{s,D_1} is defined over $\mathbb{F}_q$. Therefore, if the pairing includes a final exponentiation, the leading coefficient will be eliminated and thus may be ignored in the computation of the pairing.

5.3 Final exponentiation. As described in Section 4, most of the hyperelliptic pairings involve a *final exponentiation* of a Miller function $f_{s,D}(D')$ by $(q^k-1)/r$, where $D \in \mathrm{Jac}_C(\mathbb{F}_q)[r]$ and D' is an arbitrary divisor in $\mathrm{Jac}_C(\mathbb{F}_{q^k})$. As has been widely reported, this extra computation has its benefits, in particular when k is even. Many of these are described by Scott [68] and Galbraith, Hess, and Vercauteren [29]; we summarize the main ones here.

When k is even, the field $\mathbb{F}_{q^k}$ can be constructed as a degree two extension of $\mathbb{F}_{q^\ell}$, where $2\ell = k$. We can represent elements as $a + ib$ with $a, b \in \mathbb{F}_{q^\ell}$ and γ^2 a quadratic non-residue over $\mathbb{F}_{q^\ell}$. It is straightforward to check that

$$(1/(a+\gamma b))^{q^\ell - 1} = (a - \gamma b)^{q^\ell - 1}$$

which means inversion can be replaced by conjugation since the result is the same after final exponentiation. In particular, this applies to any denominators of computations in Miller's algorithm.

There is a further optimization, *denominator elimination*, which in fact allows one to ignore all denominators in Miller's algorithm. In computing $f_{s,D}(D')$ where D is a divisor defined over the base field $\mathbb{F}_q$, one computes the numerator and denominator values separately (see Algorithm 1). If $D' = (u(x), v(x))$ has $u(x)$ defined over $\mathbb{F}_{q^\ell}$, then the computation of the denominator involves only D and $u(x)$ and therefore becomes trivial after final exponentiation. In the case of supersingular curves, for example, a suitable evaluation divisor can be found using a distortion map ψ (see Section 4.7) such that $\psi(D')$ has x-coordinates in $\mathbb{F}_{q^\ell}$ [33].

The final exponentiation is generally computed in multiple steps by writing $(q^k - 1)/r$ as a product of polynomials in base q expansion and exploiting finite field constructions, in particular the qth power of Frobenius, which speeds up computation [29]. Other methods for faster computation include signed sliding window methods [37], as well as trace and tori methods [34, 38].

Remark 5.1 As the Ate pairing does not require final exponentiation, these techniques are unavailable. Furthermore, as stated by Granger et al., there are also possible security implications; namely, the problem of *pairing inversion* (given γ and D_1, find D_2 such that $a(D_1, D_2) = \gamma$) may not be as hard (see [36, Intro.]). However, we remark that if $r^2 \nmid (q^k - 1)$ and r is prime, a superfluous final exponentiation of the Ate pairing still gives a non-degenerate result.

5.4 Degenerate divisors. For a genus 2 curve, a general reduced divisor D is of the form $D = (P_1) + (P_2) - 2(\infty)$ and a degenerate divisor is of the form $D = (P) - (\infty)$. As there are fewer points in the support, the arithmetic is faster when adding a general divisor to a degenerate divisor than when adding two general divisors. This speeds up the computation of the Miller function $f_{s,D}$ where D is degenerate. Furthermore, the evaluation of a Miller function on a degenerate divisor is also faster by at least half, since there is only one affine point. Many of the fastest hyperelliptic pairing computations use degenerate divisors, including the examples noted with [a], [b] and [c] in the Table 5.1. We summarize here when it is possible to use degenerate divisors as either the first or second argument of a pairing.

Should $\mathrm{Jac}_C(\mathbb{F}_q)$ be of prime order r, then for any $P \in C(\mathbb{F}_q)$, the divisor $D = (P) - (\infty)$ can be used as the first argument, regardless of the pairing. Furthermore, if C is supersingular, then using a distortion map ψ (see Section 4.7), we have that $\psi(D)$ is also degenerate and pairs non-trivially with D. Hence, for supersingular curves with prime-order $\mathrm{Jac}_C(\mathbb{F}_q)$, we can use degenerate divisors as both arguments of the Tate-Lichtenbaum pairing. This fact was originally exploited in the definition of the η_T pairing by Duursma and Lee [14]. In the more general situation where $\#\mathrm{Jac}_C(\mathbb{F}_q)$ is not prime and/or the curve C is not supersingular, using degenerate divisors is not as straightforward, as noted by Frey and Lange [24]. If $\#\mathrm{Jac}_C(\mathbb{F}_q) = nr$ where $\gcd(n, r) = 1$, there is no guarantee that there exists a degenerate divisor D of order r. The probability that a reduced divisor is of order r is $1/n$ and the probability that a divisor is degenerate is roughly $1/q$, by the Hasse-Weil bounds on $C(\mathbb{F}_q)$ and $\mathrm{Jac}_C(\mathbb{F}_q)$. Therefore, assuming independence, a heuristic argument gives that the probability a divisor is degenerate and order r is $1/qn$. This implies that using a degenerate divisor for the first argument is not necessarily possible.

However, Frey and Lange [24] show that for q large enough (as in a cryptographic setting), it is possible to use a degenerate divisor as the second argument. In other words, there exists $D_2 = (P) - (\infty) \in \mathrm{Jac}_C(\mathbb{F}_{q^k})$ such that for any $D_1 \in \mathrm{Jac}_C(\mathbb{F}_q)[r]$, the Tate-Lichtenbaum pairing $\tau(D_1, D_2)$ is non-trivial. The probability that $P \in C(\mathbb{F}_{q^k})$ yields such a divisor D_2 has a lower bound of $1/k \log_2 q$. Moreover, if $k = 2d$ is even, it is possible to choose $P = (x, y)$ with $x \in \mathbb{F}_{q^d}$ and $y \in \mathbb{F}_{q^k}$, using a degenerate divisor on the quadratic twist of $C/\mathbb{F}_{q^d}$. This technique is used for example by Fan, Gong and Jao [16] and allows for denominator elimination.

Remark 5.2 As remarked by Galbraith, Hess and Vercauteren [29, §7], there are potential security implications with using degenerate divisors, depending on the application. While the discrete logarithm problem with a degenerate divisor as a base point is no easier than that with a general divisor [44], other hardness assumptions such as pairing inversion (see Remark 5.1) are potentially compromised, as Granger et al. have noted [36]. To our knowledge, the topic remains unresolved.

We also remark that there are protocols in which it may not always be possible to use degenerate divisors, for example, when computing a pairing where one input is required to be a random multiple of a divisor D.

5.5 Rubin-Silverberg point compression. Another method available to us in genus 2 is the point compression technique of Rubin and Silverberg [66], who note that supersingular abelian varieties can be identified with subvarieties of Weil restrictions of supersingular elliptic curves.

Recall that a *supersingular q-Weil number* is a complex number of the form $\sqrt{q}\zeta$, where ζ is a root of unity and $\sqrt{q}$ denotes the positive square root. Let m be the order of ζ.

The following theorem allows us to define a useful invariant:

Theorem 5.3 ([66]) *Suppose A is a simple supersingular abelian variety of dimension g over $\mathbb{F}_q$, where q is a power of a prime p, and $P(x)$ is the characteristic polynomial of the Frobenius endomorphism of A. Then $P(x) = G(x)^e$, where $G(x) \in \mathbb{Z}[x]$ is a monic irreducible polynomial with $e = 1$ or 2. All of the roots of G are supersingular q-Weil numbers.*

We call the roots of G the *q-Weil numbers* for A.

Definition 5.4 The *cryptographic exponent* of A is defined by

$$c_A = \begin{cases} \dfrac{m}{2} & \text{, if } q \text{ is a square} \\ \dfrac{m}{\gcd(2,m)}, & \text{if } q \text{ is not a square.} \end{cases}$$

Let $\alpha_A = c_A/g$; it is the *security parameter* of A.

Now let $\mathbb{F} \subset \mathbb{F}'$ be finite fields, E an elliptic curve over $\mathbb{F}$, and let $Q \in E(\mathbb{F}')$. Recall that the trace from $\mathbb{F}'$ to $\mathbb{F}$ is given by

$$\mathrm{Tr}_{\mathbb{F}'/\mathbb{F}}(Q) = \sum_{\sigma \in \mathrm{Gal}(\mathbb{F}'/\mathbb{F})} \sigma(Q).$$

Rubin and Silverberg prove the following result:

Theorem 5.5 ([66]) *Let E be a supersingular elliptic curve over $\mathbb{F}_q$, π a q-Weil number for E ($\pi \notin \mathbb{Q}$). Fix $r \in \mathbb{N}$ with $\gcd(r, 2pc_E) = 1$. Then there is a simple supersingular abelian variety A over $\mathbb{F}_q$ having the following properties.*

1. $\dim A = \varphi(r)$.
2. *For every primitive r^{th} root of unity ζ, $\pi\zeta$ is a q-Weil number for A.*
3. $c_A = rc_E$.
4. $\alpha_A = (r/\phi(r))\alpha_E$.
5. *There is a natural identification of $A(\mathbb{F}_q)$ with the following subgroup of $E(\mathbb{F}_{q^r})$:*

$$\{Q \in E(\mathbb{F}_{q^r}) : \mathrm{Tr}_{\mathbb{F}_{q^r}/\mathbb{F}_{q^{r/l}}}(Q) = 0 \textit{ for every prime } l \mid r\}.$$

This theorem can be thought of as a form of point compression for supersingular elliptic curves. More concretely, the theorem allows us to replace the Jacobian of a hyperelliptic curve C over $\mathbb{F}$ with an elliptic curve E over an extension $\mathbb{F}'$ of $\mathbb{F}$, while still exploiting the per-bit security gain of higher genus hyperelliptic curves. From a security standpoint, there is no difference between working with $E(\mathbb{F}')$ and working with $\mathrm{Jac}_C(\mathbb{F})$. On the other hand, one needs fewer bits to represent divisors with support in $C(\mathbb{F})$ than to represent points in $E(\mathbb{F}')$.

As noted by Galbraith [28], recent implementations [2] indicate that pairings on elliptic curves with the Rubin-Silverberg compression are, in general, more efficient than using the pairings on Jacobians of hyperelliptic curves. However, it seems that Rubin and Silverberg have initiated a promising investigation into the arithmetic geometry of abelian varieties and its applications to pairings. Much work remains to be done, in particular with respect to the torsion structure of these varieties.

5.6 A comparison of pairings. We conclude this section by summarizing in Table 5.1 all known variants of the Tate-Lichtenbaum pairing defined in Section 4, in terms of their loop length and whether or not there is a final exponent of $(q^k - 1)/r$. Note that if there is a final exponent, in the case of even embedding degree k, this allows for the optimizations described in Section 5.3. The last column gives references to specific examples of curves of genus 2 in the literature for which the efficiency of the pairing has been analyzed, either theoretically, via implementation or both.

All pairings in Table 5.1 except the Tate-Lichtenbaum pairing and the modified Tate-Lichtenbaum pairing are defined on $\mathbb{G}_2 \times \mathbb{G}_1$, but if $\gcd(k, \#\,\mathrm{Aut}(C)) \neq 1$,

Table 5.1 A comparison of pairings.

Pairing	Curves	Final Exponent	Loop Length	Examples for $g = 2$
Modified Tate	All	Yes	$\log_2 r$	[16][a], [39, §5], [10],
Ate [36]	All	No	$\log_2 q$	
Eta [2]	Super-singular	Yes	Varies ($\log_2 q$) possible	[2][b]
HV [40, 71]	All	Yes	Varies $(\log_2 r)/\varphi(k)$ possible	[71, §4][c]
Ate$_i$ [74]	All	Yes	$\log_2(q^i \pmod r)$ $(\log_2 r)/\varphi(k)$ possible	[74, §5][d]
R-ate [53]	All	Yes	Varies	[53, §5][e], [31, §4,5][f]

[a] Fan, Gong and Jao use efficiently computable automorphisms to compute a power of the modified Tate-Lichtenbaum pairing on two Kawazoe-Takahashi families of non-supersingular curves over prime fields. This algorithm allows for a theoretical reduction of up to one fourth in the length of the Miller loop ($\log_2 r$). They implement this on curves over $\mathbb{F}_p$ where p is a 329-bit prime and $k = 4$ and compare this with pairings on a supersingular curve defined over $\mathbb{F}_p$ with p a 256-bit prime and $k = 4$. Using all known optimizations (degenerate divisors, encapsulated group operations, final exponentiation, fast field arithmetic), the pairing computation on the non-supersingular curve is about 55.8% faster.

[b] This is one of the fastest known pairing implementations on a hyperelliptic curve and makes use of many optimizations including degenerate divisors and a special octupling formula.

[c] Vercauteren gives an example of a family of supersingular curves with $k = 12$ such that the loop length is approximately $\log_2 r/\varphi(k)$.

[d] Zhang gives examples of Kawazoe-Takahashi curves with $k = 8, 24$ such that the twisted Ate$_i$ pairing has loop length approximately $\log_2 r/\varphi(k)$.

[e] Lee, Lee and Park show that for supersingular curves the loop length can theoretically be approximately $(\log_2 q)/2$. They also compute an example on a Duursma-Lee curve with $k = 5$, achieving a loop length 21% shorter than the Ate.

[f] Galbraith, Lin and Mireles Morales [31] describe how to use the R-ate pairing on a real model of a hyperelliptic curve of genus 2 over $\mathbb{F}_p$ with $k = 6$. By using a distortion map ψ on $\mathrm{Jac}_C(\mathbb{F}_p)[r]$ such that the image of $\mathbb{G}_1$ is in the p-eigenspace, $\mathbb{G}_2$, they are also able to make use of denominator elimination. They conclude that such pairings are theoretically competitive with both pairings on certain elliptic curves with $k = 3$ and with hyperelliptic curves in the imaginary model with $k = 4$.

then there exist the twisted versions on $\mathbb{G}_1 \times \mathbb{G}_2$ which have the same final exponent and loop length.

6 Future work on hyperelliptic pairings

In this section, we present possible areas for future work, expanding upon the list in the 2007 survey paper of Galbraith, Hess and Vercauteren [29]. We list some newer problems, mention some recent advancements in the elliptic curve case which may find generalizations in pairings for $g \geq 2$, and conclude by revisiting the 2007 list [29].

6.1 Achieving optimal loop length. Since 2007, there has been a flurry of new work to reduce the loop length in Miller's algorithm using variants of the Ate pairing. In particular, the Ate pairing on hyperelliptic curves of genus g already reduces the loop length by up to a factor of g when compared to the Tate-Lichtenbaum pairing [36]. Vercauteren [71] uses the following definition to characterize pairings with certain loop lengths.

Definition 6.1 [71] Let $e : \mathbb{G}_1 \times \mathbb{G}_2 \mapsto \mu_r \subset \mathbb{F}_{q^k}^*$ be a non-degenerate, bilinear pairing defined using a combination of Miller functions. We call $e(\cdot, \cdot)$ an *optimal* pairing if it can be computed using $(\log_2 r)/\varphi(k) + \varepsilon(k)$ Miller iterations, where φ is the Euler phi function and $\varepsilon(k) \leq \log_2 k$.

Note that this means a pairing is optimal if the total sum of all the loop lengths of the Miller functions is approximately $(\log_2 r)/\varphi(k)$.

For an HV pairing $a_{s,h(x)}$ with $h(x) = \sum_{i=0}^{n} h_i x^i$, the total sum of loop lengths is $\sum_{i=0}^{n} \log_2 h_i$. Thus to be optimal, it is necessary but not sufficient that the coefficients of h are bounded by $r^{\varphi(k)}$. This can be achieved by finding the shortest vectors in a lattice spanned by vectors involving powers of s [71, §3.3]. Vercauteren and Zhang both give examples of genus 2 HV pairings (see Table 5.1) where the polynomial $h(x)$ satisfies this bound and has only one coefficient which is not ± 1, therefore providing examples of optimal hyperelliptic pairings. It remains open whether given a hyperelliptic curve it is always possible to construct an optimal HV pairing. One direction would be to look at extending the method of Vercauteren [71] which constructs optimal pairings on *parameterized* families of elliptic curves.

Vercauteren also conjectures that for elliptic curves without efficiently computable automorphisms other than the Frobenius, no pairing can be better than optimal [71, §2]. More specifically, he conjectures that for such a curve, any non-degenerate pairing requires at least $(1 - \delta) \log_2 r/\varphi(k)$ Miller iterations where $0 < \delta < 1/4$. For a curve with a set of efficiently computable endomorphisms $\mathcal{E} \subset \mathrm{End(E)}$, Vercauteren defines a *superoptimal* pairing as one which can be computed using $(\log_2 r)/\#\mathcal{E} + \varepsilon(k)$ Miller iterations. It remains to examine what is the best possible for genus 2 curves, both with and without the existence of efficiently computable endomorphisms (see also Section 6.2). Furthermore, it is not known whether there are other non-degenerate, bilinear hyperelliptic pairings on $\mathbb{G}_1 \times \mathbb{G}_2$ which are not part of the HV framework.

Lastly, we remark that the computation of an HV pairing cannot be measured solely by the sum of loop lengths. There is also the cost of computing the auxiliary functions (see (2),(3) in Section 4.6). It remains to formally compare the cost of these additional computations with the benefit of a shorter total sum of Miller loop lengths.

6.2 Using efficiently computable automorphisms. One newer method to speed up computations is to use efficiently computable automorphisms of the curve C (beyond the Frobenius). For example, Fan, Gong and Jao use efficiently computable automorphisms in computing a power of the modified Tate-Lichtenbaum pairing on some specific non-supersingular genus 2 curves over prime fields [16]. An open task is to explore how far can this be generalized to other genus 2 curves.

Furthermore, Hess [40] extends his pairing framework for ordinary elliptic curves to exploit efficiently computable automorphisms. This does not generally give an improved loop length since $\#\mathcal{E} \leq \varphi(k)$ for most ordinary elliptic curves. However, as hyperelliptic curves have a greater variety of $\mathrm{Aut}(C)$, it would be worthwhile to examine what improvements in loop length can be made by extending the HV framework to exploit these automorphisms.

6.3 Fast arithmetic and the embedding degree. In the case of even embedding degree k, it is traditional to exploit the degree two subfield, as explained in Section 5.3. In fact, Koblitz and Menezes define *pairing friendly fields* to be finite fields of the form $\mathbb{F}_{q^k}$ such that $k = 2^i 3^j$ for $0 \leq i, j \in \mathbb{Z}$ and $q \equiv 1 \pmod{12}$ [48, §5]. (If k is strictly a power of 2 then it is only required that $q \equiv 1 \pmod 4$.) By a theorem of Lidl and Niederreiter [56, Theoreom 3.75] and more particularly, by a specific instance of this theorem given by Koblitz and Menezes [48, Theorem 2], we can construct the extension $\mathbb{F}_{q^k}$ for k of this form using a tower of quadratic and cubic extensions. There are thus certain advantages we can make use of for $k = 2^i 3^j$. For instance, there exist fast arithmetic methods for degree 2 and 3 subextensions; namely, the Karatsuba method for quadratic subextensions and the Toom-Cook method for cubic subextensions [46, §4.3.3]. These methods are used to economize the arithmetic in the smaller fields which reduce the number of field multiplications. However, there are embedding degrees not of this form, particularly among recent constructions of non-supersingular curves, and hence it would be worthwhile to see if these ideas can be extended to embedding degrees k containing other prime factors.

6.4 Degenerate divisors. As discussed in Section 5.4, one common optimization is to use degenerate divisors. Frey and Lange [24] give a lower bound on the probability that $P \in C(\mathbb{F}_{q^k})$ gives a non-trivial pairing value when used as a degenerate divisor in the second argument of the Tate-Lichtenbaum pairing. However, to our knowledge, there is no method to efficiently find such points beyond simple trial and error.

We also consider using degenerate divisors with Ate-type pairings a on $\mathbb{G}_2 \times \mathbb{G}_1$ (or twisted Ate on $\mathbb{G}_1 \times \mathbb{G}_2$). While a heuristic argument shows that the likelihood that a divisor of $\mathbb{G}_1$ is degenerate is small, it would be useful to know if there are particular curves where this is more likely and if so, how to find such divisors. It also remains to analyze the likelihood that an element of $\mathbb{G}_2$ is degenerate. We note that for $D \in \mathbb{G}_2$, if $D = (P) - (\infty)$, then $\pi(D) = (\pi(P)) - (\infty)$ implies that the divisor class qD is also degenerate.

6.5 Ignoring the last bit. In the case of the modified Tate-Lichtenbaum pairing on elliptic curves, when computing $f_{r,D_1}(D_2)$, it is possible to ignore the last bit in the expansion of r. This follows from the fact that since r is odd, the last iteration of the Miller loop of the Tate-Lichtenbaum pairing is the evaluation at D_2 of the line function corresponding to the line through $(r-1)P$ and P. This is a vertical line and so by the choice of divisor D_2 with x-coordinates lying over $\mathbb{F}_{q^d}$, this is eliminated by the final exponentiation. While this does not give a large improvement compared to other loop length reductions, it is worth verifying whether this trick might be used in the case of hyperelliptic curves.

6.6 Compression and higher degree twists. Galbraith and Lin [30] give explicit formula to compute the Weil pairing on elliptic curves given only x-coordinates, and the Tate-Lichtenbaum and Ate pairings given both x-coordinates but at most one y-coordinate. This form of point compression is advantageous for elliptic curve pairings with small embedding degree, where one would be working over a field of large order (and consequently, taking a square root to recover y could be expensive). The compression makes use of explicit recurrence formulas for elliptic curve point multiplication and for Miller functions in the case of embedding degree $k = 2$. As these recurrences are given solely in terms of the x-coordinate of the point, the pairings are also computed in terms of the x-coordinate of the points involved. Note, however, that neglecting the value of y introduces a sign ambiguity, but this is resolved by taking the trace of the pairing, which is independent of the sign of y. It is perhaps worth investigating if the analogous results may be obtained for hyperelliptic pairings (for curves of the form $y^2 = F(x)$) of small embedding degree.

Another form of compression involves algebraic tori, which are d-dimensional generalizations of the multiplicative group $\mathbb{G}_m$. Naehrig, Barreto and Schwabe [60] use algebraic tori to compress computations, not just in the final exponentiation but also in the Miller loop of elliptic curve pairings. Their methods rely on explicit formulas for multiplication and squaring of torus elements and also exploit degree 6 twists. One might want to try similar methods for certain twists of hyperelliptic curves.

Another benefit of twists, as explained in Section 4.7, is that curves with a twist of degree d allow one to use the twisted versions of Ate-type pairings. This means one computes the Miller function $f_{s,D_1}(D_2)$ for $D_1 \in \mathbb{G}_1$ and the divisor $D_2 = (u(x), v(x)) \in \mathbb{G}_2$ with $u(x)$ defined over the subfield $\mathbb{F}_{q^{k/(d,k)}}$, as opposed to computing $f_{s,D_2}(D_1)$. Furthermore, the points of $\mathbb{G}_2$ can be represented as points on the Jacobian of the twist C' which allows for faster computations in the group $\mathbb{G}_2$. The example of Zhang [74] uses a twist of degree 8; to our knowledge, pairings on curves with twists of degree 10 have not been implemented.

6.7 Trace zero subvarieties. For a hyperelliptic curve C of genus g defined over $\mathbb{F}_q$, a *trace zero subvariety* of C is a subgroup of the Jacobian of C whose construction is connected to the Weil restriction of scalars. The use of trace zero varieties for cryptographic applications was first suggested by Frey [23]. The trace zero subvariety of C over a field extension of degree ℓ is a subgroup of $\mathrm{Jac}_C(\mathbb{F}_{q^\ell})$, which is isomorphic to the quotient $\mathrm{Jac}_C(\mathbb{F}_{q^\ell})/\mathrm{Jac}_C(\mathbb{F}_q)$.

It can also be defined concretely as follows: Let π be the qth power Frobenius. Let ℓ be a prime and assume that $\ell \nmid \#\mathrm{Jac}_C(\mathbb{F}_q)$. We define the trace zero subvariety G_ℓ of $\mathrm{Jac}_C(\mathbb{F}_{q^\ell})$ to be the set of elements of trace zero. I.e.,

$$G_\ell(\mathbb{F}_q) := \{D \in \mathrm{Jac}_C(\mathbb{F}_{q^\ell}) : D + \pi(D) + \cdots + \pi^{\ell-1}(D) = \mathcal{O}\}.$$

Since $G_\ell(\mathbb{F}_q)$ is the kernel of the trace map, it is a subgroup of $\mathrm{Jac}_C(\mathbb{F}_{q^\ell})$. To perform arithmetic in a trace zero subvariety one can use the algorithms that work in the whole Jacobian. So far, no specific algorithms for the group law are known that make use of the subgroup properties.

Since $G_\ell(\mathbb{F}_q)$ is a subgroup of $\mathrm{Jac}_C(\mathbb{F}_{q^\ell})$, we can define a Tate-Lichtenbaum pairing on it by restriction: suppose the order of $G_\ell(\mathbb{F}_q)$ is divisible by a large prime factor r, but not by r^2. Let $\mathbb{G}_1 := G_\ell[r] \cap \ker(\pi^\ell - [1])$ and $\mathbb{G}_2 := G_\ell[r] \cap \ker(\pi^\ell - [q^\ell])$.

Then the Tate-Lichtenbaum pairing on G_ℓ is a map

$$t : \mathbb{G}_1 \times \mathbb{G}_2 \to \mu_r.$$

On the points of $\mathbb{G}_1$, π acts as multiplication by an integer s ([13]), and the same is true for the action of π on $\mathbb{G}_2$ ([9, Proposition 3]). Cesena [9] gives a new algorithm for computing the Tate-Lichtenbaum pairing over trace zero subvarieties of supersingular elliptic curves by exploiting the action of the q-Frobenius. He uses the fact that the q-Frobenius π is an efficient endomorphism (rather than just the q^r-Frobenius), together with the fact that for particular supersingular elliptic curves the action of the Frobenius can be computed more efficiently [9, Lemmas 1–3]. For these curves, the action of π is (close to being) multiplication of a power of q.

Experimentally, Cesena's algorithm is as efficient as the Tate-Lichtenbaum pairing on supersingular elliptic curves, though less efficient than the eta pairing η_T or the optimal Ate pairing of Vercauteren. It remains to explore whether Cesena's algorithm generalizes to supersingular hyperelliptic curves or non-supersingular trace zero varieties.

6.8 Exploiting torsion groups of dimension > 2**.** If r is coprime with the characteristic of $\mathbb{F}_q$, the r-torsion group of a Jacobian variety of dimension g is isomorphic to $(\mathbb{Z}/r\mathbb{Z})^{2g}$. With the exception of the recent work by Okamoto and Takashima [63], all known pairing-based cryptographic applications require only two linearly independent torsion points and thus can be realized in the elliptic curve setting; in fact, also the Okamoto-Takashima protocols can as well be implemented using a product of two (supersingular) elliptic curves. It is an open problem to find a cryptographic application that uses curves of genus 2 (or larger) and that does *not* work using elliptic curves. Both for the ordinary and the supersingular case, constructions of Jacobians of dimension 2 with low *full* embedding degree (cf. Section 3.3) are available ([18, 63]).

6.9 More problems. For completeness, we include the problems posed by Galbraith, Hess and Vercauteren [29], making note of any recent advancements:

1. *Construct pairing-friendly ordinary hyperelliptic curves with smaller ρ-values.* At this point in time, the smallest ρ-value obtained for an ordinary hyperelliptic curve of small embedding degree is $\rho \approx 20/9$ (for $g = 2$, $k = 27$; cf. Section 3). It is highly desirable to have curves with ρ-value < 2.
2. *Curves with $g \geq 3$.* For curves with $g \geq 3$, is it possible to develop efficient pairing-based cryptosystems which are also secure against the index calculus attacks available for these curves?
3. *Pairings on real models of hyperelliptic curves.* There have been recent examples [31] of efficient pairing computations on real models of hyperelliptic curves, as remarked in Section 5.6. Are real models competitive with the imaginary models in general? Furthermore, are there efficient pairings on non-hyperelliptic curves?
4. *Torsion structure.* Is there an efficient method for selecting divisors from $\mathrm{Jac}_C(\mathbb{F}_{q^k})[r]$ for pairing computations? (See also Section 6.4.) Furthermore, if this group has more than two generators, what cryptographic applications are possible? (See also Section 6.8.)

5. *Rubin-Silverberg point compression and Weil restriction.* Can the Rubin-Silverberg method (see Section 5.5) be made more efficient in the elliptic curve case and/or generalized to Jacobians of curves of genus $g \geq 2$?
6. *Weil restriction.* As in Rubin-Silverberg point compression, certain abelian varieties can be identified with subvarieties of the Weil restriction of supersingular elliptic curves. When the abelian variety is a Jacobian, are there explicitly computable homomorphisms between the elliptic curve and the Jacobian representation?

Acknowledgments. The authors are most grateful to David Freeman for his elaborate feedback on earlier versions of this paper, which significantly improved our work. The authors further thank Paulo S.L.M. Barreto, Anja Becker, Felix Fontein, Steven Galbraith, and Alfred Menezes for helpful discussions and comments on an earlier draft of the paper. Thanks to Alan Silvester for conscientious technical editing. Thanks to Jonathan Wise. The work of the first author has been supported by a National Science Foundation Graduate Research Fellowship. The third author has been supported by the Informatics Circle of Research Excellence Chair in Algorithmic Number Theory and Cryptography. The fourth author is partially supported by National Science Foundation grant DMS-0801123 and a Sloan Research Fellowship. The work of the fifth author has been supported by a National Science Foundation Fellowship (#0802915) and a National Science and Engineering Research Council Fellowship (#373333). The fifth author is also grateful to Harvard University and the Pacific Institute for the Mathematical Sciences UBC, where some of this work was conducted. Lastly, the sixth author acknowledges support by the National Science and Engineering Research Council of Canada, and by the Premier's Research Excellence Award of the province of Ontario.

The idea for this paper was conceived during the workshop "Women in Numbers" at the Banff International Research Station in November 2008, where the authors formed a project group studying hyperelliptic pairings.

References

1. R. Balasubramanian and N. Koblitz, *The improbability that an elliptic curve has subexponential discrete log problem under the Menezes-Okamoto-Vanstone algorithm*, J. Cryptology **11** (1998), no. 2, 141–145.
2. P.S.L.M. Barreto, S. Galbraith, C. Ó hÉigeartaigh, and M. Scott, *Efficient pairing computation on supersingular abelian varieties*, Des. Codes Cryptogr. **42** (2007), 239–271.
3. D. Boneh and M. Franklin, *Identity-based encryption from the Weil pairing*, SIAM J. Comput. **32** (2003), no. 3, 586–615.
4. D. Boneh, B. Lynn, and H. Shacham, *Short signatures from the Weil pairing*, J. Cryptology **17** (2004), no. 4, 297–319.
5. F. Brezing and A. Weng, *Elliptic curves suitable for pairing based cryptography*, Des. Codes Cryptogr. **37** (2005), 133–141.
6. D. Cantor, *Computing in the Jacobian of a hyperelliptic curve*, Math. Comp. **48** (1987), no. 177, 95–101.
7. G. Cardona, *On the number of curves of genus 2 over a finite field*, Finite Fields Appl. **9** (2003), no. 4, 505–526.
8. G. Cardona, E. Nart, and J. Pujolàs, *Curves of genus two over fields of even characteristic*, Math. Z. **250** (2005), no. 1, 177–201.
9. E. Cesena, *Pairing with supersingular trace zero varieties revisited*, Cryptology ePrint Archive Report 2008/404, `http://eprint.iacr.org/2008/404/`.
10. Y.-J. Choie and E. Lee, *Implementation of Tate pairing on hyperelliptic curves of genus 2*, ICISC 2003, LNCS, vol. 2971, Springer-Verlag, 2004, pp. 97–111.

11. C. Cocks and R.G.E. Pinch, *Identity-based cryptosystems based on the Weil pairing*, unpublished manuscript, 2001.
12. D. Coppersmith, *Fast evaluation of logarithms in fields of characteristic two*, IEEE Trans. Inform. Theory **30** (1984), 587–594.
13. D. Doche and T. Lange, *Handbook of Elliptic and Hyperelliptic Curve Cryptography*, ch. 15 (Arithmetic of special curves), Chapman & Hall/CRC, 2006.
14. I. Duursma and H.-S. Lee, *Tate pairing implementation for hyperelliptic curves* $y^2 = x^p - x + d$, Adv. Cryptology - Asiacrypt 2003, LNCS, vol. 2894, Springer-Verlag, 2003, pp. 111–123.
15. K. Eisenträger, K. Lauter, and P. Montgomery, *Improved Weil and Tate pairings for elliptic and hyperelliptic curves*, Algorithmic Number Theory ANTS-VI, LNCS, vol. 3076, Springer-Verlag, 2004, pp. 169–183.
16. X. Fan, G. Gong, and D. Jao, *Speeding up pairing computations on genus 2 hyperelliptic curves with efficiently computable automorphisms*, Pairing 2008, LNCS, vol. 5209, Springer-Verlag, 2008, pp. 243–264.
17. D. Freeman, *A generalized Brezing-Weng method for constructing pairing-friendly ordinary abelian varieties: Additional examples*, `http://theory.stanford.edu/~dfreeman/papers/gen-bw-examples.pdf`.
18. ———, *Constructing pairing-friendly genus 2 curves with ordinary Jacobians*, Pairing 2007, LNCS, vol. 4575, Springer-Verlag, 2007, pp. 152–176.
19. ———, *A generalized Brezing-Weng algorithm for constructing pairing-friendly ordinary abelian varieties*, Pairing 2008, LNCS, vol. 5209, Springer-Verlag, 2008, pp. 146–163.
20. D. Freeman and T. Satoh, *Constructing pairing-friendly hyperelliptic curves using Weil restriction*, Preprint August 2009, 28 pages.
21. D. Freeman, M. Scott, and E. Teske, *A taxonomy of pairing-friendly elliptic curves*, J. Cryptology, **23** (2010), 224–280.
22. D. Freeman, P. Stevenhagen, and M. Streng, *Abelian varieties with prescribed embedding degree*, Algorithmic Number Theory ANTS-VIII, LNCS, vol. 5011, Springer-Verlag, 2008, pp. 60–73.
23. G. Frey, *Applications of arithmetical geometry to cryptographic constructions*, Finite Fields and Applications – Proceedings of the Fifth International Conference on Finite Fields and Applications F_q5, Augsburg, August 2-6, 1999, Springer, 2001, pp. 128–161.
24. G. Frey and T. Lange, *Fast bilinear maps from the Tate-Lichtenbaum pairing on hyperelliptic curves*, Algorithmic Number Theory ANTS-VII, LNCS, vol. 4076, Springer-Verlag, 2006, pp. 466–479.
25. ———, *Handbook of Elliptic and Hyperelliptic Curve Cryptography*, ch. 23 (Algebraic realizations of DL systems), Chapman & Hall/CRC, 2006.
26. ———, *Handbook of Elliptic and Hyperelliptic Curve Cryptography*, ch. 4 (Background on curves and Jacobians), Chapman & Hall/CRC, 2006.
27. ———, *Handbook of Elliptic and Hyperelliptic Curve Cryptography*, ch. 5 (Varieties over special fields), Chapman & Hall/CRC, 2006.
28. S. Galbraith, *Supersingular curves in cryptography*, Adv. Cryptology – ASIACRYPT 2001, LNCS, vol. 2248, Springer-Verlag, 2001, pp. 495–513.
29. S. Galbraith, F. Hess, and F. Vercauteren, *Hyperelliptic pairings*, Pairing 2007, LNCS, vol. 4575, Springer-Verlag, 2007, pp. 108–131.
30. S. Galbraith and X. Lin, *Computing pairings using x-coordinates only*, Des. Codes Cryptogr. **50** (2009), 305–324.
31. S. Galbraith, X. Lin, and D. Mireles Morales, *Pairings on hyperelliptic curves with a real model*, Pairing 2008, LNCS, vol. 5209, Springer-Verlag, 2008, pp. 265–281.
32. S. Galbraith, J. McKee, and P. Valença, *Ordinary abelian varieties having small embedding degree*, Proc. Workshop on Mathematical Problems and Techniques in Cryptology, CRM, Barcelona, 2005, pp. 29–45.
33. S. Galbraith, J. Pujolàs, C. Ritzenthaler, and B. Smith, *Distortion maps for genus two curves*, J. Math. Cryptology **3** (2009), 1–18.
34. S. Galbraith and M. Scott, *Exponentiation in pairing-friendly groups using homomorphisms*, Pairing 2008, LNCS, vol. 5209, Springer-Verlag, 2008, pp. 211–224.
35. P. Gaudry, E. Thomé, N. Thériault, and C. Diem, *A double large prime variation for small genus hyperelliptic index calculus*, Math. Comp. **76** (2007), 475–492.

36. R. Granger, F. Hess, R. Oyono, N. Thériault, and F. Vercauteren, *Ate pairing on hyperelliptic curves*, Adv. Cryptology - Eurocrypt 2007, LNCS, vol. 4515, Springer-Verlag, 2007, pp. 419–436.
37. R. Granger, D. Page, and N. P. Smart, *High security pairing-based cryptography revisited*, Algorithmic Number Theory ANTS-VII, LNCS, vol. 4076, Springer-Verlag, 2006, pp. 480–494.
38. R. Granger, D. Page, and M. Stam, *On small characteristic algebraic tori in pairing-based cryptography*, LMS J. Comput. Math. **9** (2006), 64–85.
39. C. Ó hÉigeartaigh and M. Scott, *Pairing calculation on supersingular genus 2 curves*, Selected Areas in Cryptography - SAC 2006, LNCS, vol. 4356, Springer-Verlag, 2007, pp. 302–316.
40. F. Hess, *Pairing lattices*, Pairing 2008, LNCS, vol. 5209, Springer-Verlag, 2008, pp. 18–38.
41. F. Hess, N. Smart, and F. Vercauteren, *The Eta pairing revisited*, IEEE Trans. Inform. Theory **52** (2006), 4595–4602.
42. L. Hitt, *On the minimal embedding field*, Pairing 2007, LNCS, vol. 4575, Springer-Verlag, 2007, pp. 294–301.
43. A. Joux, *A one round protocol for tripartite Diffie-Hellman*, J. Cryptology **17** (2004), no. 4, 263–276.
44. M. Katagi, I. Kitamura, T. Akishita, and T. Takagi, *Novel efficient implementations of hyperelliptic curve cryptosystems using degenerate divisors*, WISA 2004, LNCS, vol. 3325, Springer-Verlag, 2005, pp. 345–359.
45. M. Kawazoe and T. Takahashi, *Pairing-friendly hyperelliptic curves of type* $y^2 = x^5 + ax$, Pairing 2008, LNCS, vol. 5209, Springer-Verlag, 2008, pp. 164–177.
46. D. Knuth, *The Art of Computer Programming*, 3rd ed., vol. 2, Addison-Wesley, 1997.
47. N. Koblitz, *Hyperelliptic cryptosystems*, J. Cryptology **1** (1989), no. 3, 139–150.
48. N. Koblitz and A. Menezes, *Pairing-based cryptography at high security levels*, Cryptography and Coding 2005, LNCS, vol. 3796, Springer-Verlag, 2005, pp. 13–36.
49. D. Kohel, *Quartic CM fields database*, `http://echidna.maths.usyd.edu.au/kohel/dbs/complex_multiplication2.html`.
50. T. Lange, *Formulae for arithmetic on genus 2 hyperelliptic curves*, Appl. Algebra Engrg. Comm. Comput. **15** (2005), 295–328.
51. K. Lauter, *The equivalence of the geometric and algebraic group laws for Jacobians of genus 2 curves*, Topics in Algebraic and Noncommutative Geometry, Contemp. Math., vol. 324, Amer. Math. Soc., 2003, pp. 165–171.
52. E. Lee, H.-S. Lee, and Y. Lee, *Eta pairing computation on general divisors over hyperelliptic curves* $y^2 = x^p - x + d$, J. Symbolic Comput. **43** (2008), 452–474.
53. E. Lee, H.-S. Lee, and C.-M. Park, *Efficient and generalized pairing computations on abelian varieties*, IEEE Trans. Inform. Theory **55** (2009), 1793–1803.
54. E. Lee and Y. Lee, *Tate pairing computation on the divisors of hyperelliptic curves of genus 2*, J. Korean Math. Soc. **45** (2008), no. 4, 1057–1073.
55. A. K. Lenstra, *Unbelievable security. Matching AES security using public key cryptosystems*, Adv. Cryptology - ASIACRYPT 2001, LNCS, vol. 2248, Springer-Verlag, 2001, pp. 67–86.
56. R. Lidl and H. Niederreiter, *Finite Fields*, 2nd ed., Cambridge University Press, 1997.
57. F. Luca, D. J. Mireles, and I. E. Shparlinski, *MOV attack in various subgroups on elliptic curves*, Illinois J. Math. **48** (2004), 1041–1052.
58. V. Miller, *The Weil pairing and its efficient calculation*, J. Cryptology **17** (2004), 235–261.
59. D. Mumford, *Tata Lectures on Theta I, II*, Birkhäuser, Boston, 1983/84.
60. M. Naehrig, P. Barreto, and P. Schwabe, *On compressible pairings and their computation*, AFRICACRYPT 2008, LNCS, vol. 5023, Springer-Verlag, 2008, pp. 371–388.
61. NIST, *Recommendation for key management – Part 1: General (revised)*, NIST Special Publication 800-57, May 2006, `http://csrc.nist.gov/publications/nistpubs/800-57/SP800-57-Part1.pdf`.
62. A. Odlyzko, *Discrete logarithms in finite fields and their cryptographic significance*, Adv. Cryptology - Eurocrypt 1984, LNCS, vol. 209, Springer-Verlag, 1985, pp. 224–314.
63. T. Okamoto and K. Takashima, *Homomorphic encryption and signatures from vector decomposition*, Pairing 2008, LNCS, vol. 5209, Springer-Verlag, 2008, pp. 57–74.
64. K. Paterson, *Advances in Elliptic Curve Cryptography*, ch. X (Cryptography from pairings), Cambridge University Press, 2005.

65. J. M. Pollard, *Monte Carlo methods for index computation (mod p)*, Math. Comp. **32** (1978), no. 143, 918–924.
66. K. Rubin and A. Silverberg, *Supersingular abelian varieties in cryptology*, Adv. Cryptology – CRYPTO 2002, LNCS, vol. 2442, Springer-Verlag, 2002, pp. 336–353.
67. T. Satoh, *The Brezing-Weng-Freeman method for certain genus 2 hyperelliptic curves*, Adv. Cryptology – Eurocrypt 2009, LNCS, vol. 5479, Springer-Verlag, 2009, pp. 536–553.
68. M. Scott, *Implementing cryptographic pairings*, Pairing 2007, LNCS, vol. 4575, Springer-Verlag, 2007, pp. 177–196.
69. J.H. Silverman, *The Arithmetic of Elliptic Curves*, Graduate Texts in Mathematics, vol. 106, Springer-Verlag, New York, 1992, Corrected reprint of the 1986 original.
70. P. C. van Oorschot and M. J. Wiener, *Parallel collision search with cryptanalytic applications*, J. Cryptology **12** (1999), 1–28.
71. F. Vercauteren, *Optimal pairings*, IEEE Trans. Inform. Theory **56** (2010), 455–461.
72. A. Weng, *Constructing hyperelliptic curves of genus 2 suitable for cryptography*, Math. Comp. **72** (2003), 435–458.
73. T. Wollinger, J. Pelzl, and C. Paar, *Cantor versus Harley: Optimization and analysis of explicit formulae for hyperelliptic curve cryptosystem*, IEEE Trans. Comput. **54** (2005), 861–872.
74. F. Zhang, *Twisted Ate pairing on hyperelliptic curves and applications*, Cryptology ePrint Archive Report 2008/274, `http://eprint.iacr.org/2008/274/`.
75. C. Zhao, C. Zhang, and J. Huang, *All pairings are in a group*, IEICE Trans. Fundament. Electronics, Communications and Computer Sciences **E91-A** (2008), 3084–3087.
76. ———, *A note on the Ate pairing*, Int. J. Inform. Security **7** (2008), 379–382.

Fields Institute Communications
Volume **60**, 2011

Biased Statistics for Traces of Cyclic p-Fold Covers over Finite Fields

Alina Bucur
Department of Mathematics, University of California San Diego (UCSD)
La Jolla, CA 92093-0112 USA
Current address: Institute for Advanced Study, Einstein Drive, Princeton, NJ 08540 USA
alina@math.ucsd.edu

Chantal David
Department of Mathematics, Concordia University,
Montreal, QC Canada H3G 1M8
Current address: Institute for Advanced Study, Einstein Drive, Princeton, NJ 08540 USA
cdavid@mathstat.concordia.ca

Brooke Feigon
School of Mathematics, University of East Anglia
Norwich, UK, NR4 7TJ
Current address: Department of Mathematics, University of Toronto, 40 St. George Street, Toronto, ON Canada M5S 2E4
bfeigon@math.toronto.edu

Matilde Lalín
Département de mathématiques et de statistique, Université de Montréal
Montréal, QC Canada H3T 1J4
Current address: Department of Mathematical and Statistical Sciences, University of Alberta, CAB 632, Edmonton, AB Canada T6G 2G1
mlalin@dms.umontreal.ca

To epsilon

Abstract. In this paper, we discuss in more detail some of the results on the statistics of the trace of the Frobenius endomorphism associated to cyclic p-fold covers of the projective line that were presented in [1]. We also show new findings regarding statistics associated to such curves where we fix the number of zeros in some of the factors of the equation in the affine model.

2010 *Mathematics Subject Classification.* Primary 11G20; Secondary 11T55, 11G25.

The first author was supported by NSF DMS-0652529 and DMS-0635607.

The second author was supported by NSERC Discovery Grant 155635-2008 and a grant to the Institute for Advanced Study from the Minerva Research Foundation.

The third author was supported with NSERC funds.

The fourth author was supported by NSERC Discovery Grant 355412-2008 and a University of Alberta Faculty of Science Start-up grant.

1 Introduction

Let p be a prime and fix a prime power q such that $q \equiv 1 \pmod p$. In [1], we discussed the statistics for the distribution of the trace of the Frobenius endomorphism of curves C when C varies over irreducible components of the moduli space of cyclic p-fold covers of $\mathbb{P}^1(\mathbb{F}_q)$. To do so, we first consider all curves with affine models

$$Y^p = F(X), \qquad F \in \mathcal{F}_{(d_1,\ldots,d_{p-1})} \subseteq \mathbb{F}_q[X], \tag{1.1}$$

where

$$\begin{aligned}\mathcal{F}_{(d_1,\ldots,d_{p-1})} \;=\; & \{F = F_1F_2^2 \ldots F_{p-1}^{p-1} : F_1,\ldots,F_{p-1} \in \mathbb{F}_q[X] \\ & \text{are monic, square-free, pairwise coprime,} \\ & \text{and } \deg F_i = d_i \text{ for } 1 \le i \le p-1\}.\end{aligned}$$

Unless otherwise mentioned, all polynomials will be monic.

Roughly speaking, varying over all curves of an irreducible component of the moduli space of cyclic p-fold covers means to vary over models (1.1) with F in certain unions of sets of the type $\mathcal{F}_{(d_1,\ldots,d_{p-1})}$, and statistics for the trace of Frobenius over the components of the moduli space can be deduced from the statistics associated to these sets (see Section 5).

Let $\mu_p \subseteq \mathbb{C}^*$ be the set of pth roots of unity, let $\mu_p^0 = \mu_p \cup \{0\}$, let ξ_p be a primitive pth root of unity, and let χ_p be a non-trivial character of order p of $\mathbb{F}_q$. Let C be the cyclic p-fold cover with affine model (1.1). Then, the number of affine points of C is

$$\sum_{x \in \mathbb{F}_q} 1 + \chi_p(F(x)) = q + S_p(F),$$

and the affine trace is

$$-S_p(F) = -\sum_{x \in \mathbb{F}_q} \chi_p(F(x)).$$

Theorem 1.1 [1, Theorem 7.3] *Let $\varepsilon_1,\ldots,\varepsilon_{p-1} \in \mu_p^0$ such that m of the ε_i are zero. Then, as $d_1,\ldots,d_{p-1} \to \infty$,*

$$\frac{\left|\{F \in \mathcal{F}_{(d_1,\ldots,d_{p-1})} : \chi_p(F(x_i)) = \varepsilon_i,\ 1 \le i \le q\}\right|}{\left|\mathcal{F}_{(d_1,\ldots,d_{p-1})}\right|} \sim \left(\frac{p-1}{q+p-1}\right)^m \left(\frac{q}{p(q+p-1)}\right)^{q-m}.$$

Furthermore, let $X_1,\ldots,X_q$ be q i.i.d. random variables taking the value 0 with probability $(p-1)/(q+p-1)$ and each of the values in μ_p with equal probability $q/(p(q+p-1))$. Then, as $d_1,\ldots,d_{p-1} \to \infty$,

$$\frac{\left|\{F \in \mathcal{F}_{(d_1,\ldots,d_{p-1})} \;:\; S_p(F) = t\}\right|}{\left|\mathcal{F}_{(d_1,\ldots,d_{p-1})}\right|} \sim \operatorname{Prob}\left(\sum_{i=1}^{q} X_i = t\right)$$

for any $t \in \mathbb{Z}[\xi_p] \subset \mathbb{C}$.

The case $p = 2$ was proven in [2], and the case $p = 3$ was proven in [1]. The proof of the general case was sketched in [1], and we give more details in Section 3 of this paper. It may not be clear a priori where the random variables of Theorem 1.1 come from, but they can be explained by a simple heuristic. See [1] for more details.

As an intermediate step in the proof of Theorem 1.1, we have to consider the sets of polynomials

$$\mathcal{F}_{(d_1,\dots,d_{p-1})}^{(k_1,\dots,k_{p-1})} = \left\{F \in \mathcal{F}_{(d_1,\dots,d_{p-1})} : F_i \text{ has } k_i \text{ roots over } \mathbb{F}_q, 1 \le i \le p-1\right\}.$$

Those spaces do not have a natural geometric interpretation as the sets of polynomials $\mathcal{F}_{(d_1,\dots,d_{p-1})}$ which parametrize irreducible components of moduli spaces, but they also lead to interesting results involving natural probabilities. We present those results in this paper. We first concentrate on the case $p = 3$ where the results are easier to explain, and then move on to the general case for any odd prime p.

Theorem 1.2 *Fix $0 \le k \le q$ and let*

$$\mathcal{F}_{(d_1,d_2)}^{k} = \left\{F = F_1F_2^2 \in \mathcal{F}_{(d_1,d_2)} : F_2 \textit{ has } k \textit{ roots over } \mathbb{F}_q\right\}.$$

Let $\varepsilon_1, \dots, \varepsilon_q \in \{0, 1, \xi_3, \xi_3^2\}$ such that m of the ε_i are zero. Then, as $d_1, d_2 \to \infty$,

$$\frac{\left|\left\{F \in \mathcal{F}_{(d_1,d_2)}^{k} : \chi_3(F(x_i)) = \varepsilon_i,\ 1 \le i \le q\right\}\right|}{\left|\mathcal{F}_{(d_1,d_2)}^{k}\right|} \sim \frac{\binom{m}{k}}{\binom{q}{k}} \left(\frac{1}{q+1}\right)^{m-k} \left(\frac{q}{3(q+1)}\right)^{q-m}.$$

Let $X_1, \dots, X_q$ be random variables taking the value 0 with probability $1/(q+1)$ and any of the values $1, \xi_3, \xi_3^2$ with equal probability $q/(3(q+1))$ together with a bias counting the number of k-tuples of roots taken by the variables (as described in Section 2). Then, as $d_1, d_2 \to \infty$,

$$\frac{\left|\left\{F \in \mathcal{F}_{(d_1,d_2)}^{k} \ :\ S_3(F) = t\right\}\right|}{\left|\mathcal{F}_{(d_1,d_2)}^{k}\right|} \sim \operatorname{Prob}\left(\sum_{i=1}^{q} X_i = t\right)$$

for any $t \in \mathbb{Z}[\xi_3] \subset \mathbb{C}$.

The proof of Theorem 1.2 is given in Section 2. We can interpret those results in the following way: Let $\mathcal{F}_d$ be the set of monic square-free polynomials F of degree d. It follows by exactly the same steps as Theorem 1.1 for $p = 2$ that, as $d \to \infty$,

$$\frac{|\{F \in \mathcal{F}_d \ :\ S_3(F) = t\}|}{|\mathcal{F}_d|} \sim \operatorname{Prob}\left(\sum_{i=1}^{q} X_i = t\right), \tag{1.2}$$

where $X_1, \dots, X_q$ are i.i.d. random variables variables taking the value 0 with probability $1/(q+1)$ and any of the values $1, \xi_3, \xi_3^2$ with equal probability $q/(3(q+1))$.[1] Then, if F_2 has no roots over $\mathbb{F}_q$ ($k = 0$), the polynomials $F \in \mathcal{F}_{(d_1,d_2)}^{k}$ lead to the same probability as the square-free polynomials. If F_2 has k roots over $\mathbb{F}_q$ ($1 \le k \le q$), then the random variables $X_1, \dots, X_q$ of Theorem 1.2 are distributed in such a way that the probability that $X_i = \varepsilon_i$ for $1 \le i \le q$ depends on m, the number of zeros among $\varepsilon_1, \dots, \varepsilon_q$ (and the X_i are not i.i.d. in this case). More precisely, $\operatorname{Prob}(X_i = \varepsilon_i, 1 \le i \le q)$ is $\binom{m}{k}T^{-1}$ times the probability associated with square-free polynomials, where $\binom{m}{k}$ is the number of k-tuples of zeros among $\varepsilon_1, \dots, \varepsilon_q$ and T is a normalizing factor insuring that the sum of the

[1] The difference between (1.2) and Theorem 1.1 with $p = 2$ is that in the former case, we consider the trace $S_3(F)$ and in the latter case, the trace $S_2(F)$. Both results follow directly from a count on how many polynomials in $\mathcal{F}_d$ take a prescribed set of values.

probabilities is 1 (see Section 2 for more details). We then say the probability $\mathrm{Prob}(X_i = \varepsilon_i, 1 \le i \le q)$ in this case is the probability associated with the family of square-free polynomials biased by the number of zeros of $\varepsilon_1, \dots, \varepsilon_q$.

We now study the general case, where we take polynomials $F = F_1 F_2^2 \dots F_{p-1}^{p-1}$ in $\mathcal{F}_{(d_1,\dots,d_{p-1})}$ where some of the F_i have a prescribed number of roots.

Theorem 1.3 *Fix $1 \le v \le p-1$, and let $k_{v+1}, \dots, k_{p-1}$ be non-negative integers with $k = k_{v+1} + \dots + k_{p-1}$. We also suppose that $0 \le k \le q$. Let*

$$\mathcal{F}^{(*,k_{v+1},\dots,k_{p-1})}_{(d_1,\dots,d_{p-1})} = \left\{F \in \mathcal{F}_{d_1,\dots,d_{p-1}} : F_i \text{ has } k_i \text{ roots over } \mathbb{F}_q,\ v+1 \le i \le q\right\}.$$

Let $\varepsilon_1, \dots, \varepsilon_q \in \mu_p^0$ such that m of the ε_i are zero. Then, as $d_1, \dots, d_{p-1} \to \infty$,

$$\frac{\left|\left\{F \in \mathcal{F}^{(*,k_{v+1},\dots,k_{p-1})}_{(d_1,\dots,d_{p-1})} : \chi_p(F(x_i)) = \varepsilon_i,\ 1 \le i \le q\right\}\right|}{\left|\mathcal{F}^{(*,k_{v+1},\dots,k_{p-1})}_{(d_1,\dots,d_{p-1})}\right|} \sim \frac{\binom{m}{k}}{\binom{q}{k}} \left(\frac{v}{q+v}\right)^{m-k} \left(\frac{q}{p(q+v)}\right)^{q-m}.$$

Let $X_1, \dots, X_q$ be q random variables taking the value 0 with probability $v/(q+v)$ and any of the values in μ_p with equal probability $q/(p(q+v))$ together with a bias counting the number of k-tuples of roots taken by the variables (as described in Section 3). Then, as $d_1, \dots, d_{p-1} \to \infty$,

$$\frac{\left|\left\{F \in \mathcal{F}^{(*,k_{v+1},\dots,k_{p-1})}_{(d_1,\dots,d_{p-1})} : S_p(F) = t\right\}\right|}{\left|\mathcal{F}^{(*,k_{v+1},\dots,k_{p-1})}_{(d_1,\dots,d_{p-1})}\right|} \sim \mathrm{Prob}\left(\sum_{i=1}^{q} X_i = t\right)$$

for any $t \in \mathbb{Z}[\xi_p] \subset \mathbb{C}$.

The proof of Theorem 1.3 is given in Section 4. We can interpret the results as we did for the case of $p = 3$. If $v = 1$ and $k = 0$, then the probability for the set of polynomials $F_1 F_2^2 \dots F_{p-1}^{p-1}$ is the same as for the square-free polynomials. If $v = 1$ and $1 \le k \le q$, then the probability $\mathrm{Prob}(X_i = \varepsilon_i, 1 \le i \le q)$ in this case is the probability associated with the family of square-free polynomials biased by the number of zeros of $\varepsilon_1, \dots, \varepsilon_q$. When $v = 2$ and $k = 0$, then the probability $\mathrm{Prob}(X_i = \varepsilon_i, 1 \le i \le q)$ in this case is the probability associated with the family of polynomials $F_1 F_2^2$ with F_1, F_2 monic, square-free and co-prime. If $v = 2$ and $1 \le k \le q$, the probability $\mathrm{Prob}(X_i = \varepsilon_i, 1 \le i \le q)$ in this case is the probability associated with the family of polynomials of the form $F_1 F_2^2$ biased by the number of zeros of $\varepsilon_1, \dots, \varepsilon_q$. The general case follows similarly.

We can give a more geometric version of Theorem 1.1 in terms of the moduli space of cyclic p-fold covers. For any prime p, let $\mathcal{H}_{g,p}$ denote the moduli space of cyclic p-fold covers of genus g. It breaks into a disjoint union of irreducible components $\mathcal{H}^{(d_1,\dots,d_{p-1})}$ indexed by the inertia type of the branch points of $F(X)$, and

$$\mathcal{H}_{g,p} = \bigcup_{\substack{d_1 + 2d_2 + \dots + (p-1)d_{p-1} \equiv 0 \pmod p \\ 2g = (p-1)(d_1 + \dots + d_{p-1} - 2)}} \mathcal{H}^{(d_1,\dots,d_{p-1})}, \tag{1.3}$$

where the union is disjoint and each component $\mathcal{H}^{(d_1,\ldots,d_{p-1})}$ is irreducible (see Section 5 for more details). We remark that for $p = 2$, there is just one component in the above decomposition and the moduli space is irreducible.

Consider the first étale cohomology group with $\mathbb{Q}_\ell$ coefficients, where ℓ is a prime with $\ell \equiv 1 \bmod p$ (so that $\mathbb{Q}_\ell$ contains the pth roots of unity). The group of order p generated by the cyclic automorphism acts on the curve C and therefore on the first cohomology group, which gives us a representation of the aforementioned cyclic group on this $\mathbb{Q}_\ell$-vector space. Since the group is abelian and we have enough roots of unity in $\mathbb{Q}_\ell$, the representation splits into a direct sum of 1-dimensional representations. Since these are 1-dimensional representations, they correspond to multiplication by some scalar. In order to find the scalar, one can use the Lefschetz–Verdier fixed point formula from which it follows that our cyclic automorphism acts on these subspaces by multiplication by different powers of χ_p and the dimensions of isotypical subspaces are equal. For more details, see [3]. However, the Riemann–Hurwitz formula shows that the trivial character appears with multiplicity 0. This shows that the cyclic automorphism of order p splits the first cohomology group of C into $p-1$ subspaces $H^1_{\chi_p}, H^1_{\chi_p^2}, \ldots, H^1_{\chi_p^{p-1}}$ on which the automorphism acts by multiplication by $\chi_p, \chi_p^2, \ldots, \chi_p^{p-1}$ respectively. Futhermore, the action of the cyclic automorphism is defined over the base field $\mathbb{F}_q$ (since this contains the pth roots of unity) and Frobenius fixes $\mathbb{F}_q$. Thus the two actions (of Frobenius and of the cyclic automorphism) commute and it suffices to study the trace of the Frobenius on one of these subspaces, say $\mathrm{Tr}(\mathrm{Frob}_C|_{H^1_{\chi_p}})$. Moving to another subspace corresponds to a new choice of the character χ_p.

Theorem 1.4 [1, Theorem 7.4] *If q is fixed and $d_1, \ldots, d_{p-1} \to \infty$, the distribution of the trace of the Frobenius endomorphism associated to C as C ranges over the component $\mathcal{H}^{(d_1,\ldots,d_{p-1})}$ of cyclic p-fold covers of $\mathbb{P}^1(\mathbb{F}_q)$ is that of the sum of $q+1$ i.i.d. random variables $X_1, \ldots, X_{q+1}$, where each X_i takes the value 0 with probability $(p-1)/(q+p-1)$ and each value in μ_p with probability $q/(p(q+p-1))$. More precisely, for any $t \in \mathbb{Z}[\xi_p] \subset \mathbb{C}$ with $|t| \le q+1$ and any $1 > \varepsilon > 0$, we have, as $d_1, \ldots, d_{p-1} \to \infty$,*

$$\frac{\left|\left\{C \in \mathcal{H}^{(d_1,\ldots,d_{p-1})} : \mathrm{Tr}(\mathrm{Frob}_C|_{H^1_{\chi_p}}) = -t\right\}\right|'}{\left|\mathcal{H}^{(d_1,\ldots,d_{p-1})}\right|'} \sim \mathrm{Prob}\left(\sum_{i=1}^{q+1} X_i = t\right).$$

In the last theorem, and in the rest of the paper, the $'$ notation, applied both to summation and cardinality, means that curves C on the moduli spaces are counted with the usual weights $1/|\mathrm{Aut}(C)|$ (as in the mass formula). The proof of Theorem 1.4 is given in Section 5.

2 Cyclic trigonal curves and proof of Theorem 1.2

For $p = 3$, a cyclic p-fold cover of $\mathbb{P}^1(\mathbb{F}_q)$ is called a cyclic trigonal curve, and every cyclic trigonal curve has an affine model of the form $Y^3 = F(X)$ where $F(X) = F_1(X)F_2(X)^2 \in \mathcal{F}_{(d_1,d_2)}$.

Let $0 \le k \le q$. We consider in this section the sets of polynomials

$$\mathcal{F}^k_{(d_1,d_2)} = \left\{F = F_1F_2^2 \in \mathcal{F}_{(d_1,d_2)} : F_2 \text{ has } k \text{ roots over } \mathbb{F}_q\right\}.$$

Fix $q \equiv 1 \pmod 3$. throughout this section, ξ_3 denotes a non-trivial third root of unity in $\mathbb{C}^*$, and χ_3 a non-trivial cubic character of $\mathbb{F}_q$. We will denote by ζ_q the (incomplete) zeta function of the rational function field $\mathbb{F}_q[X]$ given by

$$\zeta_q(s) = \sum_F |F|^{-s} = \prod_P \left(1 - |P|^{-s}\right)^{-1} = (1 - q^{1-s})^{-1}.$$

Proposition 2.1 [1, Proposition 4.3] *Let $0 \le \ell \le q$, let $x_1, \dots, x_\ell$ be distinct elements of $\mathbb{F}_q$, and $a_1, \dots, a_\ell \in \mathbb{F}_q^*$. Then for any $1 > \varepsilon > 0$, we have*

$$\begin{aligned} &|\{F \in \mathcal{F}_{(d_1,d_2)} : F(x_i) = a_i, 1 \le i \le \ell\}| \\ &= \frac{K q^{d_1+d_2}}{\zeta_q(2)^2} \left(\frac{q}{(q+2)(q-1)}\right)^\ell \left(1 + O\left(q^{-(1-\varepsilon)d_2 + \varepsilon\ell} + q^{-d_1/2+\ell}\right)\right), \end{aligned}$$

where

$$K = \prod_P \left(1 - \frac{1}{(|P|+1)^2}\right), \tag{2.1}$$

and the product runs over all monic irreducible polynomials of $\mathbb{F}_q[X]$.

In particular, taking $\ell = 0$, we have

$$\left|\mathcal{F}_{(d_1,d_2)}\right| = \frac{K q^{d_1+d_2}}{\zeta_q(2)^2} \left(1 + O\left(q^{-(1-\varepsilon)d_2} + q^{-d_1/2}\right)\right). \tag{2.2}$$

The statistics for the number of polynomials $F \in \mathcal{F}^k_{(d_1,d_2)}$ taking prescribed values then follow easily from the previous proposition.

Corollary 2.2 [1, Corollary 4.4] *Let $x_1, \dots, x_q$ be an enumeration of elements in $\mathbb{F}_q$. Let $a_1 = \dots = a_m = 0$, and $a_{m+1}, \dots, a_q \in \mathbb{F}_q^*$. Then, for $1 > \varepsilon > 0$,*

$$\begin{aligned} &\left|\left\{F \in \mathcal{F}^k_{(d_1,d_2)} \;:\; F(x_i) = a_i,\ 1 \le i \le q\right\}\right| \\ &= \binom{m}{k} \frac{K q^{d_1+d_2}}{\zeta_q(2)^2} \left(\frac{1}{q+2}\right)^m \left(\frac{q}{(q+2)(q-1)}\right)^{q-m} \\ &\quad \times \left(1 + O\left(q^{-(1-\varepsilon)(d_2-k)+\varepsilon q} + q^{-(d_1+k-m)/2+q}\right)\right). \end{aligned}$$

Corollary 2.3 *Let $x_1, \dots, x_q$ be an enumeration of elements in $\mathbb{F}_q$. Let $a_1 = \dots = a_m = 0$, and $a_{m+1}, \dots, a_q \in \mathbb{F}_q^*$. Then, for $1 > \varepsilon > 0$,*

$$\begin{aligned} &\frac{\left|\left\{F \in \mathcal{F}^k_{(d_1,d_2)} : F(x_i) = a_i, 1 \le i \le q\right\}\right|}{\left|\mathcal{F}^k_{(d_1,d_2)}\right|} \\ &= \frac{\binom{m}{k}}{\binom{q}{k}} \left(\frac{1}{q+1}\right)^{m-k} \left(\frac{q}{(q-1)(q+1)}\right)^{q-m} \\ &\quad \times \left(1 + O\left(q^{-(1-\varepsilon)(d_2-k)+\varepsilon q} + q^{-(d_1+k-3q)/2}\right)\right), \end{aligned}$$

and

$$\begin{aligned} &\frac{\left|\left\{F \in \mathcal{F}^k_{(d_1,d_2)} : \chi(F(x_i)) = \varepsilon_i, 1 \le i \le q\right\}\right|}{\left|\mathcal{F}^k_{(d_1,d_2)}\right|} \\ &= \frac{\binom{m}{k}}{\binom{q}{k}} \left(\frac{1}{q+1}\right)^{m-k} \left(\frac{q}{3(q+1)}\right)^{q-m} \\ &\quad \times \left(1 + O\left(q^{-(1-\varepsilon)(d_2-k)+\varepsilon q} + q^{-(d_1+k-3q)/2}\right)\right). \end{aligned}$$

Proof We first use Corollary 2.2 to compute $\left|\mathcal{F}^k_{(d_1,d_2)}\right|$. Let $M(a_1,\dots,a_q)$ be the number of values of a_i which are zero. Then

$$\left|\mathcal{F}^k_{(d_1,d_2)}\right| = \sum_{m=k}^{q} \sum_{\substack{(a_1,\dots,a_q)\in(\mathbb{F}_q)^q \\ M(a_1,\dots,a_q)=m}} \left|\left\{F\in\mathcal{F}^k_{(d_1,d_2)} : F(x_i)=a_i, 1\le i\le q\right\}\right|.$$

By Corollary 2.2, the main term of the above sum equals

$$\begin{aligned}
&\frac{Kq^{d_1+d_2}}{\zeta_q(2)^2}\left(\frac{q}{q+2}\right)^q \sum_{m=k}^{q}\binom{q}{m}\binom{m}{k}q^{-m} \\
&= \frac{Kq^{d_1+d_2}}{\zeta_q(2)^2}\left(\frac{q}{q+2}\right)^q \binom{q}{k}\sum_{m=k}^{q}\binom{q-k}{m-k}q^{-m} \\
&= \binom{q}{k}\frac{Kq^{d_1+d_2-k}}{\zeta_q(2)^2(1+q^{-1})^k}\left(\frac{q+1}{q+2}\right)^q.
\end{aligned}$$

Taking the maximal value of the error term for m between k and q, this gives

$$\begin{aligned}
\left|\mathcal{F}^k_{(d_1,d_2)}\right| &= \binom{q}{k}\frac{Kq^{d_1+d_2-k}}{\zeta_q(2)^2(1+q^{-1})^k}\left(\frac{q+1}{q+2}\right)^q \\
&\quad\times\left(1+O\left(q^{-(1-\varepsilon)(d_2-k)+\varepsilon q}+q^{-(d_1+k-3q)/2}\right)\right). \qquad (2.3)
\end{aligned}$$

The first assertion follows by dividing the result of Corollary 2.2 by $\left|\mathcal{F}^k_{(d_1,d_2)}\right|$, and the second assertion by remarking that for $\varepsilon_i=0$, then $\chi(F(x_i))=\varepsilon_i$ if and only if $F(x_i)=0$, and for $\varepsilon_i=1$, then $\chi(F(x_i))=\varepsilon_i$ if and only if $F(x_i)$ is one of the $(q-1)/3$ cubes in $\mathbb{F}_q^*$, and similarly for $\varepsilon_i=\xi_3,\xi_3^2$. □

Taking $d_1, d_2\to\infty$ in Corollary 2.3, this proves the first statement of Theorem 1.2.

We can also describe the asymptotic of Corollary 2.3 in terms of a natural probability.

Let X be the random variable taking the value 0 with probability $1/(q+1)$, and any value $\in\{1,\xi_3,\xi_3^2\}$ with probability $q/(3(q+1))$, as in Theorem 1.2. For each q-tuple $(\varepsilon_1,\dots,\varepsilon_q)\in\{0,1,\xi_3,\xi_3^2\}^q$, let m be the number of i such that $\varepsilon_i=0$. Let $X_1,\dots,X_q$ be random variables distributed as X with a bias counting the (unordered) k-tuples $\{i_1,\dots,i_k\}\subseteq\{1,\dots,q\}$ such that $\varepsilon_{i_1}=\cdots=\varepsilon_{i_k}=0$. More precisely, let

$$\mathrm{Prob}\left(X_i=\varepsilon_i : 1\le i\le q\right) = \binom{m}{k}\frac{1}{T}\left(\frac{1}{q+1}\right)^m\left(\frac{q}{3(q+1)}\right)^{q-m}, \qquad (2.4)$$

where

$$\begin{aligned} T &= \sum_{(\varepsilon_1,\dots,\varepsilon_q)\in\{0,1,\xi_3,\xi_3^2\}^q} \binom{m}{k}\left(\frac{1}{q+1}\right)^m \left(\frac{q}{3(q+1)}\right)^{q-m} \\ &= \sum_{m=k}^{q} \binom{m}{k} 3^{q-m}\binom{q}{m}\left(\frac{1}{q+1}\right)^m \left(\frac{q}{3(q+1)}\right)^{q-m} \\ &= \binom{q}{k}\sum_{m=k}^{q}\binom{q-k}{q-m}\left(\frac{1}{q+1}\right)^m \left(\frac{q}{q+1}\right)^{q-m} \\ &= \binom{q}{k}\left(\frac{1}{q+1}\right)^k. \end{aligned}$$

Using the value of T in (2.4), we get

$$\operatorname{Prob}\left(X_i=\varepsilon_i : 1\le i\le q\right) = \frac{\binom{m}{k}}{\binom{q}{k}}\left(\frac{1}{q+1}\right)^{m-k}\left(\frac{q}{3(q+1)}\right)^{q-m}, \tag{2.5}$$

which is the probability appearing in Corollary 2.3.

The second statement of Theorem 1.2 follows by summing the probabilities for all tuples $(\varepsilon_1,\dots,\varepsilon_q)$ such that $\varepsilon_1+\dots+\varepsilon_q=t$.

3 General p-fold covers and proof of Theorem 1.1

We recall that

$$\mathcal{F}_{(d_1,\dots,d_{p-1})} = \left\{ \begin{array}{c} F=F_1F_2^2\dots F_{p-1}^{p-1} : F_i \text{ monic, square-free,} \\ \text{pairwise coprime, } \deg F_i = d_i, 1\le i\le p-1 \end{array}\right\}.$$

The proof of the following proposition was sketched in [1]. We give more details here. For $F,G\in\mathbb{F}_q[X]$, let $\gcd(F,G)$ denote their greatest common divisor.

Proposition 3.1 [1, Proposition 7.1] *Fix $0\le \ell\le q$, $x_1,\dots,x_\ell$ distinct points in $\mathbb{F}_q$ and $a_1,\dots,a_\ell$ nonzero elements of $\mathbb{F}_q$. Then, for each $r\ge 1$ and $\varepsilon>0$,*

$$\begin{aligned} &\left|\left\{F\in\mathcal{F}_{(d_1,\dots,d_r)} : F(x_i)=a_i, 1\le i\le \ell\right\}\right| \\ &= \frac{L_{r-1}q^{d_1+\dots+d_r}}{\zeta_q(2)^r}\left(\frac{q}{(q+r)(q-1)}\right)^\ell \\ &\quad\times\left(1+O\left(q^{\varepsilon\ell}\sum_{h=2}^{r} q^{\varepsilon(d_h+\dots+d_r)-d_h}+q^{-d_1/2+\ell}\right)\right), \end{aligned}$$

where

$$L_{r-1} := K_1\dots K_{r-1},$$

with

$$K_j := \prod_P\left(1-\frac{j}{(|P|+1)(|P|+j)}\right),\quad j\ge 1,$$

and $K_0=L_0=1$. Furthermore, taking $\ell=0$, this gives

$$\begin{aligned} &\left|\mathcal{F}_{(d_1,\dots,d_r)}\right| \\ &= \frac{L_{r-1}q^{d_1+\dots+d_r}}{\zeta_q(2)^r}\left(1+O\left(\sum_{h=2}^{r} q^{\varepsilon(d_h+\dots+d_r)-d_h}+q^{-d_1/2}\right)\right). \end{aligned} \tag{3.1}$$

Proof If $r = 1$, $\mathcal{F}_{(d_1)}$ is the set of square-free polynomials, and this is Lemma 5 in [2], where the empty sum $q^{-d_2} + \cdots + q^{-d_r}$ of the error term is understood to be 0. We then suppose that $r \geq 2$. Let

$$\mathcal{G}_{d_1,\ldots,d_r} = \left\{ \begin{array}{l} (F_1, \ldots, F_r) \in \mathbb{F}_q[X]^r : F_i \text{ monic, square-free,} \\ \text{pairwise coprime, } \deg(F_i) = d_i, 1 \leq i \leq r \end{array} \right\}.$$

By Lemma 4.2 in [1],

$$\begin{aligned}
&\left|\left\{F \in \mathcal{F}_{(d_1,\ldots,d_r)} : F(x_i) = a_i, 1 \leq i \leq \ell\right\}\right| \\
&= \sum_{\substack{(F_2,\ldots,F_r) \in \mathcal{G}_{d_2,\ldots,d_r} \\ \prod_{j=2}^r F_j(x_i) \neq 0, 1 \leq i \leq \ell}} S_{d_1}^{F_2\ldots F_r}(\ell) \\
&= \frac{q^{d_1-\ell}}{\zeta_q(2)(1-q^{-2})^\ell} \sum_{\substack{(F_2,\ldots,F_r) \in \mathcal{G}_{d_2,\ldots,d_r} \\ \prod_{j=2}^r F_j(x_i) \neq 0, 1 \leq i \leq \ell}} \prod_{P \mid F_2 \ldots F_r} (1 + |P|^{-1})^{-1} \\
&+ \sum_{\substack{(F_2,\ldots,F_r) \in \mathcal{G}_{d_2,\ldots,d_r} \\ \prod_{j=2}^r F_j(x_i) \neq 0, 1 \leq i \leq \ell}} O\left(q^{d_1/2}\right) \\
&= \frac{q^{d_1-\ell}}{\zeta_q(2)(1-q^{-2})^\ell} \sum_{\deg F_r = d_r} \cdots \sum_{\deg F_2 = d_2} b(F_2 \ldots F_r) + O\left(q^{d_1/2 + d_2 + \ldots + d_r}\right),
\end{aligned} \tag{3.2}$$

where, following the notation from Lemma 4.2 in [1],

$$S_d^U(\ell) = \left|\{F \in \mathcal{F}_d : (F, U) = 1, F(x_i) = a_i, 1 \leq i \leq \ell\}\right|,$$

and we define

$$b(F) = \begin{cases} \mu^2(F) \prod_{P \mid F} (1 + |P|^{-1})^{-1} & F(x_i) \neq 0, 1 \leq i \leq \ell, \\ 0 & \text{otherwise,} \end{cases}$$

for any polynomial $F \in \mathbb{F}_q[X]$. (Here the product is over all the monic irreducible polynomials $P \in \mathbb{F}_q[X]$ dividing F.)

We now evaluate

$$M_r := \sum_{\deg F_r = d_r} \cdots \sum_{\deg F_2 = d_2} b(F_2 \ldots F_r).$$

We notice that b is multiplicative and that $b(F_2 \ldots F_r) = 0$ if the F_i are not relatively prime in pairs. Then we have

$$M_r = \sum_{\deg(F_r) = d_r} \sum_{\substack{\deg(F_{r-1}) = d_{r-1} \\ \gcd(F_{r-1}, F_r) = 1}} \cdots \sum_{\substack{\deg(F_2) = d_2 \\ \gcd(F_2, F_r) = 1, \ldots, \gcd(F_2, F_3) = 1}} b(F_2) \ldots b(F_r).$$

For any $j \geq 1$, let

$$c_j^U(F) = \begin{cases} \mu^2(F) \prod_{P \mid F} (1 + j|P|^{-1})^{-1} & F(x_i) \neq 0, 1 \leq i \leq \ell, \ \gcd(F, U) = 1, \\ 0 & \text{otherwise.} \end{cases} \tag{3.3}$$

Notice that in particular, $b(F) = c_1^1(F)$. Then we can write

$$M_r = \sum_{\deg(F_r) = d_r} c_1^1(F_r) \sum_{\deg(F_{r-1}) = d_{r-1}} c_1^{F_r}(F_{r-1}) \ldots \sum_{\deg(F_2) = d_2} c_1^{F_3 \ldots F_r}(F_2).$$

Using Lemma 3.2 with $j = 1$, we have

$$\begin{aligned} &\sum_{\deg(F_2)=d_2} c_1^{F_3\dots F_r}(F_2) \\ &= \frac{K_1 q^{d_2}}{\zeta_q(2)} \left(\frac{q+1}{q+2}\right)^{\ell} \left(\prod_{P|F_3\dots F_r} \left(\frac{|P|+1}{|P|+2}\right)\right) \left(1 + O\left(q^{\varepsilon(d_2+\dots+d_r+\ell)-d_2}\right)\right). \end{aligned} \tag{3.4}$$

Using (3.4) in (3.2), this proves the theorem for $r = 2$. For $r \geq 3$, we first have to evaluate

$$\begin{aligned} &\sum_{\deg(F_3)=d_3} c_1^{F_4\dots F_r}(F_3) \sum_{\deg(F_2)=d_2} c_1^{F_3\dots F_r}(F_2) \\ &= \frac{K_1 q^{d_2}}{\zeta_q(2)} \left(\frac{q+1}{q+2}\right)^{\ell} \sum_{\deg(F_3)=d_3} c_1^{F_4\dots F_r}(F_3) \left(\prod_{P|F_3\dots F_r} \left(\frac{|P|+1}{|P|+2}\right)\right) \\ &\quad \times \left(1 + O\left(q^{\varepsilon(d_2+\dots+d_r+\ell)-d_2}\right)\right) \\ &= \frac{K_1 q^{d_2}}{\zeta_q(2)} \left(\frac{q+1}{q+2}\right)^{\ell} \left(\prod_{P|F_4\dots F_r} \left(\frac{|P|+1}{|P|+2}\right)\right) \\ &\quad \times \sum_{\substack{\deg(F_3)=d_3 \\ F_3(x_i)\neq 0, 1\leq i\leq \ell \\ \gcd(F_3, F_4\dots F_r)=1}} \mu^2(F_3) \prod_{P|F_3} \frac{|P|}{|P|+2} \left(1 + O\left(q^{\varepsilon(d_2+\dots+d_r+\ell)-d_2}\right)\right) \\ &= \frac{K_1 q^{d_2}}{\zeta_q(2)} \left(\frac{q+1}{q+2}\right)^{\ell} \left(\prod_{P|F_4\dots F_r} \left(\frac{|P|+1}{|P|+2}\right)\right) \\ &\quad \times \sum_{\deg(F_3)=d_3} c_2^{F_4\dots F_r}(F_3) \left(1 + O\left(q^{\varepsilon(d_2+\dots+d_r+\ell)-d_2}\right)\right). \end{aligned}$$

Using Lemma 3.2 with $j = 2$, this gives

$$\begin{aligned} &\sum_{\deg(F_3)=d_3} c_1^{F_4\dots F_r}(F_3) \sum_{\deg(F_2)=d_2} c_1^{F_3\dots F_r}(F_2) \\ &= \frac{K_1 K_2 q^{d_2+d_3}}{\zeta_q(2)^2} \left(\frac{q+1}{q+3}\right)^{\ell} \prod_{P|F_4\dots F_r} \left(\frac{|P|+1}{|P|+3}\right) \\ &\quad \times \left(1 + O\left(q^{\varepsilon(d_2+\dots+d_r+\ell)-d_2} + q^{\varepsilon(d_3+\dots+d_r+\ell)-d_3}\right)\right). \end{aligned}$$

Using the last equality in (3.2), this proves the proposition with $r = 3$. In general, continuing in this way up to the last sum $\sum_{\deg F_r = d_r}$, we obtain the result. $\square$

Lemma 3.2 *Assume the hypotheses of Proposition 3.1 and let U be a polynomial of degree u with $U(x_i) \neq 0$ for $1 \leq i \leq \ell$. Let $j \geq 1$, and let $c_j^U(F)$ be defined* *as in* (3.3). *Then for any* $1 > \varepsilon > 0$,

$$\sum_{\deg(F)=d} c_j^U(F)$$

$$= \frac{K_j\, q^d}{\zeta_q(2)} \left(\frac{q+j}{q+j+1}\right)^{\ell} \left(\prod_{P|U} \left(\frac{|P|+j}{|P|+j+1}\right)\right) \left(1+O\left(q^{\varepsilon(d+u+\ell)-d}\right)\right).$$

Proof We will use the function field version of the Wiener-Ikehara Tauberian Theorem. This is Theorem 17.1 in [5]. For our application, it is important to get an error term which is independent of U and q and for this we need a more precise statement of the Tauberian Theorem than in [5] so we will go through the proof here.

First, we consider the Dirichlet series

$$G_j(s) = \sum_F \frac{c_j^U(F)}{|F|^s} = \prod_{\substack{P \\ P(x_i)\neq 0, 1\leq i\leq \ell, P\nmid U}} \left(1 + \frac{1}{|P|^s} \cdot \frac{|P|}{|P|+j}\right)$$

$$= \frac{\zeta_q(s)}{\zeta_q(2s)} H_j(s) \left(1+\frac{1}{q^{s-1}(q+j)}\right)^{-\ell} \prod_{P|U} \left(1+\frac{1}{|P|^s}\cdot\frac{|P|}{|P|+j}\right)^{-1},$$

where

$$H_j(s) = \prod_P \left(1 - \frac{j}{(|P|^s+1)(|P|+j)}\right).$$

In the above, we have used the hypothesis that $U(x_i) \neq 0$, and therefore the primes $P|U$ are different from the primes $X - x_i$ for $1 \leq i \leq \ell$. Notice that $H_j(s)$ converges absolutely for $\mathrm{Re}(s) > 0$, and $G_j(s)$ is meromorphic for $\mathrm{Re}(s) > 0$ with simple poles at the points s where $\zeta_q(s) = (1-q^{1-s})^{-1}$ has poles, that is, $s_n = 1 + i\frac{2\pi n}{\log q}$, with $n \in \mathbb{Z}$. Notice that $H_j(1) = K_j$, and $\mathrm{Res}_{s=1}\zeta_q(s) = \frac{1}{\log q}$. Thus $G_j(s)$ has a simple pole at $s = 1$ with residue

$$\frac{K_j}{\zeta_q(2)\log q} \left(\frac{q+j}{q+j+1}\right)^{\ell} \left(\prod_{P|U}\left(\frac{|P|+j}{|P|+j+1}\right)\right). \tag{3.5}$$

Define $Z(u)$ by $Z(q^{-s}) = G_j(s)$. Thus

$$Z(u) = \frac{1-qu^2}{1-qu}\left(1+\frac{uq}{q+j}\right)^{-\ell} \prod_P \left(1 - \frac{j}{(u^{-\deg P}+1)(q^{\deg P}+j)}\right)$$
$$\times \prod_{P|U}\left(1+\frac{u^{\deg P}q^{\deg P}}{q^{\deg P}+j}\right)^{-1}.$$

Fix any ε such that $1 > \varepsilon > 0$. Then $Z(u)$ is a meromorphic function on the disk $\{u \mid |u| \leq q^{-\varepsilon}\}$ with a simple pole at $u = q^{-1}$. Let $C = \{u \in \mathbb{C} \mid |u| = q^{-\varepsilon}\}$, oriented counterclockwise. For any $0 < \delta < q^{-1}$, let $C_\delta = \{u \in \mathbb{C} \mid |u| = \delta\}$, oriented clockwise. Notice that $\frac{Z(u)}{u^{d+1}}$ is a meromorphic function between the two circles, with a simple pole at $u = q^{-1}$ with residue

$$\begin{aligned}\operatorname{Res}_{u=q^{-1}} \frac{Z(u)}{u^{d+1}} &= \lim_{u \to q^{-1}} (u - q^{-1}) \frac{Z(u)}{u^{d+1}} \\ &= -\frac{K_j}{\zeta_q(2)} \left(\frac{q+j}{q+j+1} \right)^\ell \left(\prod_{P|U} \left(\frac{|P|+j}{|P|+j+1} \right) \right) q^d.\end{aligned}$$

Thus by the Cauchy integral formula,

$$\frac{1}{2\pi i} \oint_{C_\delta + C} \frac{Z(u)}{u^{d+1}} du = -\frac{K_j}{\zeta_q(2)} \left(\frac{q+j}{q+j+1} \right)^\ell \left(\prod_{P|U} \left(\frac{|P|+j}{|P|+j+1} \right) \right) q^d. \tag{3.6}$$

Now we observe that

$$Z(u) = \sum_{n=0}^{\infty} \left(\sum_{\deg F = n} c_j^U(F) \right) u^n.$$

Thus

$$\frac{1}{2\pi i} \oint_{C_\delta} \frac{Z(u)}{u^{d+1}} du = - \sum_{\deg F = d} c_j^U(F). \tag{3.7}$$

Combining (3.6) and (3.7) we see that

$$\sum_{\deg F = d} c_j^U(F) = \frac{K_j}{\zeta_q(2)} \left(\frac{q+j}{q+j+1} \right)^\ell \left(\prod_{P|U} \left(\frac{|P|+j}{|P|+j+1} \right) \right) q^d + \frac{1}{2\pi i} \oint_C \frac{Z(u)}{u^{d+1}} du.$$

Let M be the maximum value of $|Z(u)|$ on C. Then clearly

$$\left| \frac{1}{2\pi i} \oint_C \frac{Z(u)}{u^{d+1}} du \right| \le M q^{\varepsilon d}$$

so

$$\sum_{\deg(F) = d} c_j^U(F) = \frac{K_j}{\zeta_q(2)} \left(\frac{q+j}{q+j+1} \right)^\ell \left(\prod_{P|U} \left(\frac{|P|+j}{|P|+j+1} \right) \right) q^d + O(M q^{\varepsilon d}). \tag{3.8}$$

To conclude the proof we note that

$$\begin{aligned}M &= \max_{|q^{-s}| = q^{-\varepsilon}} |H_j(s)| \left| \left(1 + \frac{1}{q^{s-1}(q+j)} \right)^{-\ell} \prod_{P|U} \left(1 + \frac{1}{|P|^s} \cdot \frac{|P|}{|P|+j} \right)^{-1} \right| \\ &\ll (1 - q^{-\varepsilon})^{-\ell} \prod_{P|U} (1 - |P|^{-\varepsilon})^{-1} \ll q^{\varepsilon \ell} \prod_{P|U} |P|^{\varepsilon} \le q^{\varepsilon \ell} |U|^{\varepsilon} = q^{\varepsilon(u+\ell)}.\end{aligned} \tag{3.9}$$

By (3.9), the absolute error term in (3.8) is $O\left(q^{\varepsilon(d+u+\ell)}\right)$ and we get the result. □

Proposition 3.1 will be used with $r = p - 1$.

Denote

$$\mathcal{F}^{(k_1, \dots, k_{p-1})}_{(d_1, \dots, d_{p-1})} = \left\{ F = F_1 \dots F_{p-1}^{p-1} \in \mathcal{F}_{(d_1, \dots, d_{p-1})} : F_i \text{ has } k_i \text{ roots in } \mathbb{F}_q, 1 \le i \le p-1 \right\}.$$

Corollary 3.3 *Fix $0 \le m \le q$. Choose $x_1, \ldots, x_q$ an enumeration of the points of $\mathbb{F}_q$, and values $a_1 = \ldots = a_m = 0$, $a_{m+1}, \ldots, a_q \in \mathbb{F}_q^*$. Pick a partition $m = k_1 + \ldots + k_{p-1}$. Then, for any $\varepsilon > 0$,*

$$\begin{aligned}
&\left|\left\{F \in \mathcal{F}_{(d_1,\ldots,d_{p-1})}^{(k_1,\ldots,k_{p-1})} : F(x_i) = a_i, 1 \le i \le q\right\}\right| \\
&= \binom{m}{k_1, \ldots, k_{p-1}} \frac{L_{p-2} q^{d_1+\cdots+d_{p-1}}}{\zeta_q(2)^{p-1}} \left(\frac{1}{q+p-1}\right)^m \left(\frac{q}{(q+p-1)(q-1)}\right)^{q-m} \\
&\times \left(1 + O\left(q^{\varepsilon q} \sum_{h=2}^{p-1} q^{\varepsilon(d_h+\cdots+d_{p-1}-(k_h+\cdots+k_{p-1}))-(d_h-k_h)} + q^{-(d_1-k_1)/2+q}\right)\right).
\end{aligned}$$

Proof Let F be a polynomial in $F_{(d_1,\ldots,d_{p-1})}^{(k_1,\ldots,k_{p-1})}$ such that $F(x_i) = a_i$ for $1 \le i \le q$. Then, F can be written as

$$F(X) = \prod_{j=1}^{m} (X - x_j)^{b_j} G(X),$$

where the b_j are positive integers with the property that for any $1 \le i \le p-1$, the number of b_j's such that $b_j = i$ is k_i and $G \in F_{(d_1-k_1,\ldots,d_{p-1}-k_{p-1})}$ is such that $G(x_i) \ne 0$ for $1 \le i \le q$. There are then $\binom{m}{k_1,\ldots,k_{p-1}}$ choices for the b_j's. After this choice of b_j's, we choose the values $\alpha_i = G(x_i)$ for $1 \le i \le m$. This gives,

$$\begin{aligned}
&\left|\left\{F \in \mathcal{F}_{(d_1,\ldots,d_{p-1})}^{(k_1,\ldots,k_{p-1})} : F(x_i) = a_i, 1 \le i \le q\right\}\right| \\
&= \binom{m}{k_1, \ldots, k_{p-1}} \sum_{\substack{(\alpha_1,\ldots,\alpha_m) \\ \in (\mathbb{F}_q^*)^m}} \Bigg|\Bigg\{G \in \mathcal{F}_{(d_1-k_1,d_2-k_2,\ldots,d_{p-1}-k_{p-1})} : G(x_i) \\
&= a_i \prod_{j=1}^{m} (x_i - x_j)^{-b_j} \text{for } m+1 \le i \le q, \text{ and } G(x_i) = \alpha_i \text{ for } 1 \le i \le m\Bigg\}\Bigg|.
\end{aligned}$$

The result is proved by using Proposition 3.1 with $r = p-1$ in the last expression. □

Corollary 3.4 *Choose $x_1, \ldots, x_q$ an enumeration of the points of $\mathbb{F}_q$. Fix $0 \le m \le q$, $a_1 = \ldots = a_m = 0$ and $a_{m+1}, \ldots, a_q \in \mathbb{F}_q^*$. Then for any $\varepsilon > 0$,*

$$\begin{aligned}
&\left|\left\{F \in \mathcal{F}_{(d_1,\ldots,d_{p-1})} : F(x_i) = a_i, 1 \le i \le q\right\}\right| \\
&= \frac{L_{p-2} q^{d_1+\ldots+d_{p-1}}}{\zeta_q(2)^{p-1}} \left(\frac{p-1}{q+p-1}\right)^m \left(\frac{q}{(q+p-1)(q-1)}\right)^{q-m} \\
&\times \left(1 + O\left(q^{\varepsilon q+(1-\varepsilon)m} \sum_{h=2}^{p-1} q^{\varepsilon(d_h+\cdots+d_{p-1})-d_h} + q^{-(d_1-m)/2+q}\right)\right).
\end{aligned}$$

Let $\varepsilon_1, \dots, \varepsilon_q \in \mu_p^0$ *and let* m *be the number of values of* ε_i *which are* 0. *Then for any* $\varepsilon > 0$,

$$\begin{aligned}&\left|\left\{F \in \mathcal{F}_{(d_1,\dots,d_{p-1})} : \chi_p(F(x_i)) = \varepsilon_i, 1 \le i \le q\right\}\right| \\ &= \frac{L_{p-2}q^{d_1+\dots+d_{p-1}}}{\zeta_q(2)^{p-1}} \left(\frac{p-1}{q+p-1}\right)^m \left(\frac{q}{p(q+p-1)}\right)^{q-m} \\ &\times \left(1 + O\left(q^{\varepsilon q+(1-\varepsilon)m} \sum_{h=2}^{p-1} q^{\varepsilon(d_h+\dots+d_{p-1})-d_h} + q^{-(d_1-m)/2+q}\right)\right).\end{aligned}$$

Proof Adding over all the $p-1$-partitions of m and using Corollary 3.3 and the identity

$$\sum_{k_1+\dots+k_{p-1}=m} \binom{m}{k_1, \dots, k_{p-1}} = (p-1)^m,$$

which follows from the Multinomial Theorem, we get the first assertion. For the second assertion, we note that if $\varepsilon \in \mu_p$, there are $\frac{q-1}{p}$ elements $\alpha \in \mathbb{F}_q^*$ such that $\chi_p(\alpha) = \varepsilon$. □

Finally, Theorem 1.1 follows by dividing

$$\left|\left\{F \in \mathcal{F}_{(d_1,\dots,d_{p-1})} : \chi_p(F(x_i)) = \varepsilon_i, 1 \le i \le q\right\}\right|$$

(Corollary 3.4) by $\left|\mathcal{F}_{(d_1,\dots,d_{p-1})}\right|$ (Proposition 3.1). This gives

$$\begin{aligned}&\frac{\left|\left\{F \in \mathcal{F}_{(d_1,\dots,d_{p-1})} : \chi_p(F(x_i)) = \varepsilon_i,\ 1 \le i \le q\right\}\right|}{\left|\mathcal{F}_{(d_1,\dots,d_{p-1})}\right|} \\ &= \left(\frac{p-1}{q+p-1}\right)^m \left(\frac{q}{p(q+p-1)}\right)^{q-m} \\ &\times \left(1 + O\left(q^{\varepsilon q+(1-\varepsilon)m} \sum_{h=2}^{p-1} q^{\varepsilon(d_h+\dots+d_{p-1})-d_h} + q^{-(d_1-m)/2+q}\right)\right).\end{aligned}$$

Notice that the order in which we perform the nested sum in equation (3.2) is arbitrary. Thus, we can assume, without loss of generality, that $d_2 \ge \dots \ge d_{p-1}$ while proving Proposition 3.1. As a result, the error terms in all the statements of this section go to zero for $d_1, \dots, d_{p-1} \to \infty$ as long as ε is small enough, for example, $0 < \varepsilon < 1/r$ in Proposition 3.1 and $0 < \varepsilon < 1/(p-1)$ in the other statements.

4 General p-fold covers and proof of Theorem 1.3

Let v be a fixed integer such that $0 \le v \le p-1$. We now study the distribution of the traces $S_p(F)$ when F varies over polynomials in

$$\mathcal{F}_{(d_1,\dots,d_{p-1})}^{(*,k_{v+1},\dots,k_{p-1})} = \left\{F \in \mathcal{F}_{d_1,\dots,d_{p-1}} : F_i \text{ has } k_i \text{ roots over } \mathbb{F}_q, v+1 \le i \le p-1\right\}.$$

Then, if $F \in \mathcal{F}_{(d_1,\dots,d_{p-1})}^{(*,k_{v+1},\dots,k_{p-1})}$, the number of roots is fixed for $F_{v+1}, \dots, F_{p-1}$ only. We will always write $k = k_{v+1} + \dots + k_{p-1}$ for the total number of fixed roots. Notice that for $v = 0$ we necessarily have $k = m$.

Using Corollary 3.3, we obtain the following three results.

Lemma 4.1 *Let $1 \le v \le p-1$. Fix m such that $0 \le k \le m \le q$. Choose $x_1, \dots, x_q$ an enumeration of the points of $\mathbb{F}_q$, and values $a_1 = \dots = a_m = 0$, $a_{m+1}, \dots, a_q \in \mathbb{F}_q^*$. Then, for any $\varepsilon > 0$, we have*

$$\begin{aligned}
&\left|\left\{F \in \mathcal{F}_{(d_1,\dots,d_{p-1})}^{(*,k_{v+1},\dots,k_{p-1})} : F(x_i) = a_i, 1 \le i \le q\right\}\right| \\
&= \binom{m}{m-k, k_{v+1}, \dots, k_{p-1}} v^{m-k} \frac{L_{p-2} q^{d_1+\cdots+d_{p-1}}}{\zeta_q(2)^{p-1}} \\
&\times \left(\frac{1}{q+p-1}\right)^m \left(\frac{q}{(q+p-1)(q-1)}\right)^{q-m} \\
&\times \left(1 + O\left(q^{\varepsilon q + (1-\varepsilon)m} \sum_{h=2}^{p-1} q^{\varepsilon(d_h + \cdots + d_{p-1}) - d_h} + q^{-(d_1+k-m)/2+q}\right)\right).
\end{aligned}$$

Notice that the case $v = 0$ was already covered in Corollary 3.3.

Proof We first write

$$\begin{aligned}
&\left|\left\{F \in \mathcal{F}_{(d_1,\dots,d_{p-1})}^{(*,k_{v+1},\dots,k_{p-1})} : F(x_i) = a_i, 1 \le i \le q\right\}\right| \\
&= \sum_{k_1+\cdots+k_v = m-k} \left|\left\{F \in \mathcal{F}_{(d_1,\dots,d_{p-1})}^{(k_1,\dots,k_{p-1})} : F(x_i) = a_i, 1 \le i \le q\right\}\right|. \qquad (4.1)
\end{aligned}$$

By Corollary 3.3, the main term of (4.1) is

$$\begin{aligned}
&\frac{L_{p-2} q^{d_1+\cdots+d_{p-1}}}{\zeta_q(2)^{p-1}} \left(\frac{1}{q+p-1}\right)^m \left(\frac{q}{(q+p-1)(q-1)}\right)^{q-m} \\
&\times \sum_{k_1+\cdots+k_v = m-k} \binom{m}{k_1, \dots, k_{p-1}},
\end{aligned}$$

and then we use the fact that

$$\begin{aligned}
&\sum_{k_1+\cdots+k_v = m-k} \binom{m}{k_1, \dots, k_{p-1}} \\
&= \sum_{k_1+\cdots+k_v = m-k} \binom{m-k}{k_1, \dots, k_v} \binom{m}{m-k, k_{v+1}, \dots, k_{p-1}} \\
&= v^{m-k} \binom{m}{m-k, k_{v+1}, \dots, k_{p-1}}.
\end{aligned}$$

By using a bound on the error term of Corollary 3.3 which is valid for all $k_1, \dots, k_v$ such that $k_1 + \cdots + k_v = m-k$, the result follows. □

Lemma 4.2 *As defined above, let $1 \le v \le p-1$ and $k = k_{v+1} + \cdots + k_{p-1}$ with $0 \le k \le q$. Then, for any $\varepsilon > 0$,*

$$\begin{aligned}
\left|\mathcal{F}_{(d_1,\dots,d_{p-1})}^{(*,k_{v+1},\dots,k_{p-1})}\right| &= \binom{q}{q-k, k_{v+1}, \dots, k_{p-1}} \frac{L_{p-2} q^{d_1+\cdots+d_{p-1}}}{\zeta_q(2)^{p-1} (q+v)^k} \left(\frac{q+v}{q+p-1}\right)^q \\
&\times \left(1 + O\left(q^q \sum_{h=2}^{p-1} q^{\varepsilon(d_h+\cdots+d_{p-1}) - d_h} + q^{-(d_1+k-3q)/2}\right)\right).
\end{aligned}$$

For $v=0$, we obtain

$$\left|\mathcal{F}^{(k_1,\dots,k_{p-1})}_{(d_1,\dots,d_{p-1})}\right| = \binom{q}{q-k,k_1,\dots,k_{p-1}} \frac{L_{p-2}q^{d_1+\dots+d_{p-1}-k}}{\zeta_q(2)^{p-1}} \left(\frac{q}{q+p-1}\right)^q \times \left(1+O\left(q^{\varepsilon q}\sum_{h=2}^{p-1} q^{\varepsilon(d_h+\dots+d_{p-1}-(k_h+\dots+k_{p-1}))-(d_h-k_h)} + q^{-(d_1-k_1)/2+q}\right)\right).$$

Proof We have

$$\left|\mathcal{F}^{(*,k_{v+1},\dots,k_{p-1})}_{(d_1,\dots,d_{p-1})}\right| = \sum_{m=k}^{q}\binom{q}{m}(q-1)^{q-m}\left|\left\{F\in\mathcal{F}^{(*,k_{v+1},\dots,k_{p-1})}_{(d_1,\dots,d_{p-1})} : F(x_i)=a_i, 1\le i\le q\right\}\right|. \tag{4.2}$$

For $v\ge 1$ we use Lemma 4.1 and the chicken, chicken, chicken identity

$$\binom{q}{m}\binom{m}{m-k,k_{v+1},\dots,k_{p-1}} = \binom{q}{q-k,k_{v+1},\dots,k_{p-1}}\binom{q-k}{q-m}.$$

Then, the main term of $\left|\mathcal{F}^{(*,k_{v+1},\dots,k_{p-1})}_{(d_1,\dots,d_{p-1})}\right|$ is

$$\begin{aligned}
&\frac{L_{p-2}q^{d_1+\dots+d_{p-1}}}{\zeta_q(2)^{p-1}}\left(\frac{q}{q+p-1}\right)^q v^{-k}\sum_{m=k}^{q}\binom{q}{m}\binom{m}{m-k,k_{v+1},\dots,k_{p-1}} v^m q^{-m}\\
&= \binom{q}{q-k,k_{v+1},\dots,k_{p-1}}\frac{L_{p-2}q^{d_1+\dots+d_{p-1}}}{\zeta_q(2)^{p-1}}\left(\frac{q}{q+p-1}\right)^q v^{-k}\sum_{m=k}^{q}\binom{q-k}{q-m}v^m q^{-m}\\
&= \binom{q}{q-k,k_{v+1},\dots,k_{p-1}}\frac{L_{p-2}q^{d_1+\dots+d_{p-1}}}{\zeta_q(2)^{p-1}}\left(\frac{q+v}{q+p-1}\right)^q (q+v)^{-k}.
\end{aligned}$$

Replacing m by the maximal value $m=q$ in the error term of Lemma 4.1, the result follows for $v\neq 0$.

If $v=0$ we use Corollary 3.3 in equation (4.2) (in this case, the sum in equation (4.2) has one term for $m=k$). The main term of $\left|\mathcal{F}^{(k_1,\dots,k_{p-1})}_{(d_1,\dots,d_{p-1})}\right|$ is

$$\begin{aligned}
&= \binom{q}{k}(q-1)^{q-k}\binom{k}{k_1,\dots,k_{p-1}}\frac{L_{p-2}q^{d_1+\dots+d_{p-1}}}{\zeta_q(2)^{p-1}}\left(\frac{1}{q+p-1}\right)^k\\
&\quad\times\left(\frac{q}{(q+p-1)(q-1)}\right)^{q-k}\\
&= \binom{q}{q-k,k_1,\dots,k_{p-1}}\frac{L_{p-2}q^{d_1+\dots+d_{p-1}}}{\zeta_q(2)^{p-1}}\left(\frac{1}{q+p-1}\right)^k\left(\frac{q}{q+p-1}\right)^{q-k},
\end{aligned}$$

and the error term is the same as in Corollary 3.3.

□

Proposition 4.3 *As defined above, let $1\le v\le p-1$ and $k=k_{v+1}+\dots+k_{p-1}$ with $0\le k\le q$. Fix m such that $0\le k\le m\le q$. Choose $x_1,\dots,x_q$ an enumeration* *of the points of* $\mathbb{F}_q$, *and values* $a_1 = \ldots = a_m = 0$, $a_{m+1}, \ldots, a_q \in \mathbb{F}_q^*$. *Then for any* $\varepsilon > 0$,

$$\frac{\left|\left\{F \in \mathcal{F}_{(d_1,\ldots,d_{p-1})}^{(*,k_{v+1},\ldots,k_{p-1})} : F(x_i) = a_i, 1 \le i \le q\right\}\right|}{\left|\mathcal{F}_{(d_1,\ldots,d_{p-1})}^{(*,k_{v+1},\ldots,k_{p-1})}\right|} = \frac{\binom{m}{k}}{\binom{q}{k}} \left(\frac{v}{q+v}\right)^{m-k} \left(\frac{q}{(q-1)(q+v)}\right)^{q-m} \times \left(1 + O\left(q^q \sum_{h=2}^{p-1} q^{\varepsilon(d_h+\cdots+d_{p-1})-d_h} + q^{-(d_1+k-3q)/2}\right)\right),$$

and

$$\frac{\left|\left\{F \in \mathcal{F}_{(d_1,\ldots,d_{p-1})}^{(*,k_{v+1},\ldots,k_{p-1})} : \chi_p(F(x_i)) = \varepsilon_i, 1 \le i \le q\right\}\right|}{\left|\mathcal{F}_{(d_1,\ldots,d_{p-1})}^{(*,k_{v+1},\ldots,k_{p-1})}\right|} = \frac{\binom{m}{k}}{\binom{q}{k}} \left(\frac{v}{q+v}\right)^{m-k} \left(\frac{q}{p(q+v)}\right)^{q-m} \times \left(1 + O\left(q^q \sum_{h=2}^{p-1} q^{\varepsilon(d_h+\cdots+d_{p-1})-d_h} + q^{-(d_1+k-3q)/2}\right)\right).$$

If $v = 0$, *we need* $m = k$, *and in this case*

$$\frac{\left|\left\{F \in \mathcal{F}_{(d_1,\ldots,d_{p-1})}^{(k_1,\ldots,k_{p-1})} : F(x_i) = a_i, 1 \le i \le q\right\}\right|}{\left|\mathcal{F}_{(d_1,\ldots,d_{p-1})}^{(k_1,\ldots,k_{p-1})}\right|} = \frac{1}{\binom{q}{k}(q-1)^{q-k}} \times \left(1 + O\left(q^{\varepsilon q} \sum_{h=2}^{p-1} q^{\varepsilon(d_h+\cdots+d_{p-1}-(k_h+\cdots+k_{p-1}))-(d_h-k_h)} + q^{-(d_1-k_1)/2+q}\right)\right),$$

and

$$\frac{\left|\left\{F \in \mathcal{F}_{(d_1,\ldots,d_{p-1})}^{(k_1,\ldots,k_{p-1})} : \chi_p(F(x_i)) = \varepsilon_i, 1 \le i \le q\right\}\right|}{\left|\mathcal{F}_{(d_1,\ldots,d_{p-1})}^{(k_1,\ldots,k_{p-1})}\right|} = \frac{1}{\binom{q}{k}p^{q-k}} \times \left(1 + O\left(q^{\varepsilon q} \sum_{h=2}^{p-1} q^{\varepsilon(d_h+\cdots+d_{p-1}-(k_h+\cdots+k_{p-1}))-(d_h-k_h)} + q^{-(d_1-k_1)/2+q}\right)\right).$$

Proof For $v > 0$, the first assertion follows by dividing the result of Lemma 4.1 by the result of Lemma 4.2, and the second assertion by observing that $\chi_p(F(a_i)) = \xi_p$ if and only if $F(a_i)$ takes $(q-1)/p$ possible values in $\mathbb{F}_q^*$. For $v = 0$ we use Corollary 3.3 instead of Lemma 4.1. □

The result of Theorem 1.3 then follows by taking $d_1, \ldots, d_{p-1} \to \infty$ in Proposition 4.3.

We now compare the result of Proposition 4.3 with a probabilistic model. Let X be the random variable taking value 0 with probability $\frac{v}{q+v}$, and any value in

μ_p with probability $\frac{q}{p(q+v)}$. For each q-tuple $(\varepsilon_1, \ldots, \varepsilon_q) \in \mu_p^0$, let m be the number of i such that $\varepsilon_i = 0$. Let $X_1, \ldots, X_q$ be random variables distributed as X with a bias counting the (unordered) k-sets $\{i_1, \ldots, i_k\}$ where $\varepsilon_{i_j} = 0$ for $j = 1, \ldots, k$, i.e.,

$$\operatorname{Prob}(X_i = \varepsilon_i \ : \ 1 \le i \le q) = \binom{m}{k} \frac{1}{T} \left(\frac{v}{q+v}\right)^m \left(\frac{q}{p(q+v)}\right)^{q-m},$$

where

$$\begin{aligned} T &= \sum_{(\varepsilon_1,\ldots,\varepsilon_q)\in(\mu_p^0)^q} \binom{m}{k} \left(\frac{v}{q+v}\right)^m \left(\frac{q}{p(q+v)}\right)^{q-m} \\ &= \sum_{m=k}^{q} \binom{q}{m} p^{q-m} \binom{m}{k} \left(\frac{v}{q+v}\right)^m \left(\frac{q}{p(q+v)}\right)^{q-m} \\ &= \binom{q}{k} \sum_{m=k}^{q} \binom{q-k}{q-m} \left(\frac{v}{q+v}\right)^m \left(\frac{q}{q+v}\right)^{q-m} \\ &= \binom{q}{k} \left(\frac{v}{q+v}\right)^k . \end{aligned}$$

Thus, we obtain

$$\operatorname{Prob}(X_i = \varepsilon_i \ : \ 1 \le i \le q) \ = \ \frac{\binom{m}{k}}{\binom{q}{k}} \left(\frac{v}{q+v}\right)^{m-k} \left(\frac{q}{p(q+v)}\right)^{q-m}, \qquad (4.3)$$

which are the probabilities in Proposition 4.3 for $v \neq 0$. The second statement of Theorem 1.3 then follows from Proposition 4.3 by summing the probabilities for all tuples $(\varepsilon_1, \ldots, \varepsilon_q)$ such that $\varepsilon_1 + \cdots + \varepsilon_q = t$.

Finally, we remark that the probability for $\mathcal{F}_{(d_1,\ldots,d_{p-1})}$ can be written as the mixed probability involving the probabilities for all the $\mathcal{F}_{(d_1,\ldots,d_{p-1})}^{(*,k_{v+1},\ldots,k_{p-1})}$ as

$$\begin{aligned} &\frac{\left|\{F \in \mathcal{F}_{(d_1,\ldots,d_{p-1})} : \chi(F(x_i)) = \varepsilon_i, 1 \le i \le q\}\right|}{\left|\mathcal{F}_{(d_1,\ldots,d_{p-1})}\right|} \\ &= \sum_{\substack{0 \le k_i \le d_i \\ v < i \le p-1}} \frac{\left|\left\{F \in \mathcal{F}_{(d_1,\ldots,d_{p-1})}^{(*,k_{v+1},\ldots,k_{p-1})} : \chi(F(x_i)) = \varepsilon_i, 1 \le i \le q\right\}\right|}{\left|\mathcal{F}_{(d_1,\ldots,d_{p-1})}^{(*,k_{v+1},\ldots,k_{p-1})}\right|} \frac{\left|\mathcal{F}_{(d_1,\ldots,d_{p-1})}^{(*,k_{v+1},\ldots,k_{p-1})}\right|}{\left|\mathcal{F}_{(d_1,\ldots,d_{p-1})}\right|}. \end{aligned}$$

5 The geometric point of view

We prove in this section that Theorem 1.4 is a consequence of Theorem 1.1.

Let C be a cyclic p-fold cover of $\mathbb{P}^1(\mathbb{F}_q)$. Then, C has an affine model of the form $Y^p = F(X)$, where $F(X)$ is a polynomial in $\mathbb{F}_q[X]$. If $G(X) = H(X)^p F(X)$, then $Y^p = F(X)$ and $Y^p = G(X)$ are isomorphic over $\mathbb{F}_q$, so it suffices to consider curves $Y^p = F(X)$ where $F(X)$ is a polynomial which is pth-power free.

Let $F \in \mathbb{F}_q[X]$ be pth-power free and monic. Recall that pth-power free over $\mathbb{F}_q$ is the same as pth-power free over $\overline{\mathbb{F}}_q$. So F factors in $\overline{\mathbb{F}}_q[X]$ as

$$F(X) = \prod_{i=1}^{d_1}(X - a_{1,i}) \prod_{i=1}^{d_2}(X - a_{2,i})^2 \cdots \prod_{i=1}^{d_{p-1}} (X - a_{p-1,i})^{p-1}$$

where the $a_{i,j}$ are distinct elements of $\overline{\mathbb{F}}_q$.

Let C_F be the cyclic p-fold cover given by

$$Y^p = F(X) = F_1(X)F_2(X)^2 \dots F_{p-1}(X)^{p-1}$$

where the F_i are square-free, relatively prime, and $\deg F_i = d_i$ for $1 \le i \le p-1$ and $d = \deg F = d_1 + 2d_2 + \dots + (p-1)d_{p-1}$. The number of branch points on C_F is $R = d_1 + \dots + d_{p-1}$ if $d \equiv 0 \pmod{p}$ or $R = d_1 + \dots + d_{p-1} + 1$ otherwise (as the point at infinity is a branch point in the latter case), and the Riemann-Hurwitz formula implies that the genus is $g = (p-1)(R-2)/2$. Then, the curve C_F has genus g if $d = d_1 + 2d_2 + \dots + (p-1)d_{p-1} \equiv 0 \pmod{p}$ and $2g = (p-1)(d_1 + \dots + d_{p-1} - 2)$, or if $d = d_1 + 2d_2 + \dots + (p-1)d_{p-1} \not\equiv 0 \pmod{p}$ and $2g = (p-1)(d_1 + \dots + d_{p-1} - 1)$.

Over $\overline{\mathbb{F}}_q$, one can reparametrize and choose an affine model for any cyclic p-fold cover with $d_1 + 2d_2 + \dots + (p-1)d_{p-1} \equiv 0 \pmod{p}$. Furthermore, the moduli space $\mathcal{H}_{g,p}$ of cyclic p-fold covers of $\mathbb{P}^1(\mathbb{F}_q)$ of a fixed genus g splits into irreducible subspaces indexed by equivalence classes of $(p-1)$-tuples of nonnegative integers $(d_1, \dots, d_{p-1})$ with the property that $d_1 + 2d_2 + \dots + (p-1)d_{p-1} \equiv 0 \pmod{p}$. We will not expand on the equivalence relations here, but we would like the reader to note that in the $p = 3$ case, they amount to interchanging d_1 and d_2. The moduli space can be written as a disjoint union over its connected components,

$$\mathcal{H}_{g,p} = \bigcup \mathcal{H}^{(d_1,\dots,d_{p-1})}, \tag{5.1}$$

where each component $\mathcal{H}^{(d_1,\dots,d_{p-1})}$ is irreducible. For more details about these Hurwitz spaces, see [4].

From now on, we assume that $d_1 + 2d_2 + \dots + (p-1)d_{p-1} \equiv 0 \mod p$, and we define

$$\begin{aligned}
\mathcal{F}^j_{(d_1,\dots,d_{p-1})} &= \{F = F_1 F_2^2 \dots F_{p-1}^{p-1} \in \mathcal{F}_{(d_1,\dots,d_{j-1},d_j-1,d_{j+1},\dots,d_{p-1})}\} \\
&\qquad \text{for } 1 \le j \le p-1, \\
\mathcal{F}^0_{(d_1,\dots,d_{p-1})} &= \mathcal{F}_{(d_1,\dots,d_{p-1})}, \\
\mathcal{F}_{[d_1,\dots,d_{p-1}]} &= \cup_{j=0}^{p-1} \mathcal{F}^j_{(d_1,\dots,d_{p-1})}.
\end{aligned}$$

For any set $\mathcal{F}$ of monic polynomials in $\mathbb{F}_q[X]$, we denote by $\widehat{\mathcal{F}}$ the set of polynomials αF where $\alpha \in \mathbb{F}_q$ and $F \in \mathcal{F}$. This defines the sets $\widehat{\mathcal{F}}_{(d_1,\dots,d_{p-1})}$, $\widehat{\mathcal{F}}^j_{(d_1,\dots,d_{p-1})}$ and $\widehat{\mathcal{F}}_{[d_1,\dots,d_{p-1}]}$ which are used in this section.

When we write a cyclic p-fold cover of $\mathbb{P}^1$ as

$$C_F : Y^p = F(X) \tag{5.2}$$

where $F(X)$ is pth-power free, we choose an affine model of the curve. To compute the statistics for the components $\mathcal{H}^{(d_1,\dots,d_{p-1})}$ of the moduli space $\mathcal{H}_{g,p}$, we need to work with families of models where we count each curve, seen as a projective variety of dimension 1, up to isomorphism, with the same multiplicity. To do so, we have to consider all pth-power free polynomials in $\mathbb{F}_q[X]$, and not only monic ones. We fix a genus g, and a component $\mathcal{H}^{(d_1,\dots,d_{p-1})}$ for this genus as in the decomposition (5.1). For each curve of this component, we want to count its different affine models $C' : Y^p = G(X)$. Since C' is obtained from an automorphism of $\mathbb{P}^1(\mathbb{F}_q)$, this means that $G \in \widehat{\mathcal{F}}_{[d_1,\dots,d_{p-1}]}$ (since $G \in \widehat{\mathcal{F}}_{(d_1,\dots,d_{p-1})}$ if the roots of F are sent to the roots of G, and $G \in \widehat{\mathcal{F}}^j_{(d_1,\dots,d_{p-1})}$ if a root of F_j is sent to the point at infinity).

Assume that C has genus $g > (p-1)^2$ and there are two ways of writing C as a cyclic p-fold cover, i.e. two maps $\phi_{1,2} : C \to \mathbb{P}^1$. They induce a map $\phi : C \to \mathbb{P}^1 \times \mathbb{P}^1$, $\phi = (\phi_1, \phi_2)$. The image of ϕ is a curve rationally equivalent to n_1 times the horizontal fiber plus n_2 times the vertical fiber, where $n_i \in \{1, p\}$ since it has to divide the degree of the projections. The adjunction formula says that the arithmetic genus of the image is equal to $(n_1 - 1)(n_2 - 1) \leq (p-1)^2$. Hence the geometric genus is at most $(p-1)^2$, which is strictly less then the genus of C itself. Hence the map ψ from C to the image of ϕ cannot be an isomorphism, in fact it must have degree > 1. But both ϕ_i factor through ψ, hence the degree of ψ can be only either 1 or p. Since we already excluded 1, it follows that the degree is p, which in turn implies that $n_1 = n_2 = 1$. Hence $\operatorname{im}(\phi)$ must be the graph of an automorphism $\mathbb{P}^1 \to \mathbb{P}^1$, and ϕ_1 and ϕ_2 are therefore related by this automorphism.

Hence all curves C' isomorphic to C are obtained from the automorphisms of $\mathbb{P}^1(\mathbb{F}_q)$, namely the $q(q^2-1)$ elements of $\mathrm{PGL}_2(\mathbb{F}_q)$. By running over the elements of $\mathrm{PGL}_2(\mathbb{F}_q)$, we obtain $q(q^2-1)/|\mathrm{Aut}(C)|$ different models $C' : Y^p = G(X)$ where $G \in \widehat{\mathcal{F}}_{[d_1,\dots,d_{p-1}]}$. This shows that

$$\left|\mathcal{H}^{(d_1,\dots,d_{p-1})}\right|' = \sum_{C \in \mathcal{H}^{(d_1,\dots,d_{p-1})}} \frac{1}{|\mathrm{Aut}(C)|} = \frac{\left|\widehat{\mathcal{F}}_{[d_1,\dots,d_{p-1}]}\right|}{q(q^2-1)}, \tag{5.3}$$

where, as before, the $'$ notation means that the curves C on the moduli space are counted with the usual weights $1/|\mathrm{Aut}(C)|$.

For $1 \leq j \leq p-1$, we denote

$$\widehat{S}_p^j(F) = \sum_{x \in \mathbb{P}^1(\mathbb{F}_q)} \chi_p^j(F(x)),$$

where the value of F at the point at infinity is given by the value at zero of $X^{d_1+2d_2+\cdots+(p-1)d_{p-1}} F(1/X)$. Fix an enumeration of the points on $\mathbb{P}^1(\mathbb{F}_q)$, $x_1, \dots, x_{q+1}$, such that x_{q+1} denotes the point at infinity. Then

$$F(x_{q+1}) = \begin{cases} \text{leading coefficient of } F & F \in \widehat{\mathcal{F}}_{(d_1,\dots,d_{p-1})}, \\ 0 & F \in \cup_{j=1}^{p-1} \widehat{\mathcal{F}}^j_{(d_1,\dots,d_{p-1})}. \end{cases}$$

The number of points on the projective curve C_F with affine model (5.2) is given by

$$\sum_{x \in \mathbb{P}^1(\mathbb{F}_q)} \left(1 + \sum_{j=1}^{q-1} \chi_p^j(F(x))\right) = q + 1 + \sum_{j=1}^{p-1} \widehat{S}_p^j(F)$$

and

$$\mathrm{Tr}(\mathrm{Frob}_C|_{H^1_{\chi_p^j}}) = -\widehat{S}_p^j(F), \quad 1 \leq j \leq p-1. \tag{5.4}$$

It follows easily from the definitions above that

$$\sum_{j=1}^{p-1} \widehat{S}_p^j(F) = \sum_{j=1}^{p-1} S_p^j(F) + \begin{cases} p-1 & F \in \widehat{\mathcal{F}}_{(d_1,\dots,d_{p-1})} \text{ and leading} \\ & \text{coefficient of } F \text{ is a } p\text{th-power}, \\ -1 & F \in \widehat{\mathcal{F}}_{(d_1,\dots,d_{p-1})} \text{ and leading} \\ & \text{coefficient of } F \text{ is not a } p\text{th-power}, \\ 0 & F \in \cup_{j=1}^{p-1} \widehat{\mathcal{F}}^j_{(d_1,\dots,d_{p-1})}. \end{cases}$$

As in (5.3), we write

$$\left|\left\{C \in \mathcal{H}^{(d_1,\ldots,d_{p-1})} : \mathrm{Tr}(\mathrm{Frob}_C|_{H^1_{\chi_p}}) = -t\right\}\right|' = \sum_{\substack{C\in\mathcal{H}^{(d_1,\ldots,d_{p-1})}\\ \mathrm{Tr}(\mathrm{Frob}_C|_{H^1_{\chi_p}})=-t}} \frac{1}{|\mathrm{Aut}(C)|}. \tag{5.5}$$

It then follows from (5.3), (5.4) and (5.5) that

$$\frac{\left|\left\{C \in \mathcal{H}^{(d_1,\ldots,d_{p-1})} : \mathrm{Tr}(\mathrm{Frob}_C|_{H^1_{\chi_p}}) = -t\right\}\right|'}{\left|\mathcal{H}^{(d_1,\ldots,d_{p-1})}\right|'} = \frac{\left|\left\{F \in \widehat{\mathcal{F}}_{[d_1,\ldots,d_{p-1}]} : \widehat{S}_p(F) = t\right\}\right|}{\left|\widehat{\mathcal{F}}_{[d_1,\ldots,d_{p-1}]}\right|}. \tag{5.6}$$

We first compute

$$\begin{aligned}\left|\widehat{\mathcal{F}}_{[d_1,\ldots,d_{p-1}]}\right| &= (q-1)\sum_{j=0}^{p-1}\left|\mathcal{F}^j_{(d_1,\ldots,d_{p-1})}\right| \\ &= \frac{(q-1)(q+p-1)}{q}\,\frac{L_{p-2}q^{d_1+\cdots+d_{p-1}}}{\zeta_q(2)^{p-1}} \\ &\quad\times\left(1+O\left(\sum_{h=2}^{p-1} q^{\varepsilon(d_h+\cdots+d_{p-1})-d_h+1} + q^{-(d_1-1)/2}\right)\right)\end{aligned} \tag{5.7}$$

by Proposition 3.1.

Fix a $(q+1)$-tuple $(\varepsilon_1,\ldots,\varepsilon_{q+1})$ where $\varepsilon_i \in \mu_p^0$ for $1 \le i \le q+1$. Denote by m the number of i such that $\varepsilon_i = 0$. We want to evaluate the probability that the character χ_p takes exactly these values at the points $F(x_1),\ldots,F(x_{q+1})$ where x_{q+1} is the point at infinity of $\mathbb{P}^1(\mathbb{F}_q)$, as F ranges over $\widehat{\mathcal{F}}_{[d_1,\ldots,d_{p-1}]}$.

Case 1: $\varepsilon_{q+1} = 0$.

In this case, only polynomials from $\bigcup_{j=1}^{p-1}\widehat{\mathcal{F}}^j_{(d_1,\ldots,d_{p-1})}$ can have $\chi_p(F(x_{q+1})) = \varepsilon_{q+1}$. Also, the number of zeros among $\varepsilon_1,\ldots,\varepsilon_q$ is now $m-1$. Thus, using Corollary 3.4,

$$\begin{aligned}&\left|\left\{F \in \widehat{\mathcal{F}}_{[d_1,\ldots,d_{p-1}]} : \chi_p(F(x_i)) = \varepsilon_i, 1 \le i \le q+1\right\}\right| \\ &= \sum_{\alpha\in\mathbb{F}_q^*}\left|\left\{F \in \cup_{j=1}^{p-1}\mathcal{F}^j_{(d_1,\ldots,d_{p-1})} : \chi_p(F(x_i)) = \varepsilon_i\chi_p^{-1}(\alpha), 1 \le i \le q\right\}\right| \\ &= \frac{(q-1)(p-1)}{q}\left(\frac{L_{p-2}q^{d_1+\cdots+d_{p-1}}}{\zeta_q(2)^{p-1}}\left(\frac{p-1}{q+p-1}\right)^{m-1}\left(\frac{q}{p(q+p-1)}\right)^{q-m+1}\right) \\ &\quad\times\left(1+O\left(q^{\varepsilon q+(1-\varepsilon)m}\sum_{h=2}^{p-1} q^{\varepsilon(d_h+\cdots+d_{p-1})-d_h} + q^{-(d_1-m)/2+q}\right)\right).\end{aligned} \tag{5.8}$$

Case 2: $\varepsilon_{q+1} \in \mu_p$.

In this case, only polynomials from $\widehat{\mathcal{F}}_{(d_1,\dots,d_{p-1})}$ can have $\chi_p(F(x_{q+1})) = \varepsilon_{q+1}$, and there are m values of $\varepsilon_1, \dots, \varepsilon_q$ which are zero. Thus,

$$\begin{aligned}
&\left|\left\{F \in \widehat{\mathcal{F}}_{[d_1,\dots,d_{p-1}]} : \chi_p(F(x_i)) = \varepsilon_i, 1 \le i \le q+1\right\}\right| \\
&= \sum_{\substack{\alpha \in \mathbb{F}_q^* \\ \chi_p(\alpha)=\varepsilon_{q+1}}} \left|\left\{F \in \mathcal{F}_{(d_1,\dots,d_{p-1})} : \chi_p(F(x_i)) = \varepsilon_i \varepsilon_{q+1}^{-1}, 1 \le i \le q\right\}\right| \\
&= \frac{q-1}{p} \frac{L_{p-2} q^{d_1+\dots+d_{p-1}}}{\zeta_q(2)^{p-1}} \left(\frac{p-1}{q+p-1}\right)^m \left(\frac{q}{p(q+p-1)}\right)^{q-m} \\
&\quad \times \left(1 + O\left(q^{\varepsilon q+(1-\varepsilon)m} \sum_{h=2}^{p-1} q^{\varepsilon(d_h+\dots+d_{p-1})-d_h} + q^{-(d_1-m)/2+q}\right)\right),
\end{aligned} \tag{5.9}$$

which is the same as (5.8).

Then, it follows from (5.7), (5.8) and (5.9) that

$$\begin{aligned}
&\frac{\left|\left\{F \in \widehat{\mathcal{F}}_{[d_1,\dots,d_{p-1}]} \ : \ \chi_p(F(x_i)) = \varepsilon_i, 1 \le i \le q+1\right\}\right|}{\left|\widehat{\mathcal{F}}_{[d_1,\dots,d_{p-1}]}\right|} \\
&= \left(\frac{p-1}{q+p-1}\right)^m \left(\frac{q}{p(q+p-1)}\right)^{q+1-m} \\
&\quad \times \left(1 + O\left(q^{\varepsilon q+(1-\varepsilon)m+1} \sum_{h=2}^{p-1} q^{\varepsilon(d_h+\dots+d_{p-1})-d_h} + q^{-(d_1-m)/2+q}\right)\right).
\end{aligned}$$

Putting everything together, we obtain

$$\begin{aligned}
&\frac{\left|\left\{C \in \mathcal{H}^{(d_1,\dots,d_{p-1})} \ : \ \mathrm{Tr}(\mathrm{Frob}_C|_{H^1_{\chi_p}}) = -t\right\}\right|'}{\left|\mathcal{H}^{(d_1,\dots,d_{p-1})}\right|'} \\
&= \frac{\left|\left\{F \in \widehat{\mathcal{F}}_{[d_1,\dots,d_{p-1}]} \ : \ \widehat{S}_p(F) = t\right\}\right|}{\left|\widehat{\mathcal{F}}_{[d_1,\dots,d_{p-1}]}\right|} \\
&= \sum_{\substack{(\varepsilon_1,\dots,\varepsilon_{q+1}) \\ \varepsilon_1+\dots+\varepsilon_{q+1}=t}} \frac{\left|\left\{F \in \widehat{\mathcal{F}}_{[d_1,\dots,d_{p-1}]} \ : \ \chi_p(F(x_i)) = \varepsilon_i, 1 \le i \le q+1\right\}\right|}{\left|\widehat{\mathcal{F}}_{[d_1,\dots,d_{p-1}]}\right|} \\
&= \sum_{\substack{(\varepsilon_1,\dots,\varepsilon_{q+1}) \\ \varepsilon_1+\dots+\varepsilon_{q+1}=t}} \left(\frac{p-1}{q+p-1}\right)^m \left(\frac{q}{p(q+p-1)}\right)^{q+1-m} \\
&\quad \times \left(1 + O\left(q^{\varepsilon q+(1-\varepsilon)m+1} \sum_{h=2}^{p-1} q^{\varepsilon(d_h+\dots+d_{p-1})-d_h} + q^{-(d_1-m)/2+q}\right)\right) \\
&= \mathrm{Prob}\left(\sum_{i=1}^{q+1} X_i = t\right) \left(1 + O\left(q^{q+1} \sum_{h=2}^{p-1} q^{\varepsilon(d_h+\dots+d_{p-1})-d_h} + q^{-(d_1-3q)/2}\right)\right)
\end{aligned}$$

where $X_1, \ldots, X_{q+1}$ are i.i.d. random variables that take the value 0 with probability $(p-1)/(q+p-1)$ and any value in μ_p with probability $q/(p(q+p-1))$. Taking $d_1, \ldots, d_{p-1} \to \infty$, this proves Theorem 1.4.

Acknowledgments. This work was initiated at the Banff workshop "Women in Numbers" organized by Kristin Lauter, Rachel Pries, and Renate Scheidler in November 2008, and the authors would like to thank the organizers and BIRS for a great workshop and excellent working conditions. The authors also wish to thank Eduardo Dueñez, Nicholas Katz, Kiran Kedlaya, Pär Kurlberg, Lea Popovic, Rachel Pries, and Zeév Rudnick for helpful discussions related to this work.

References

[1] Bucur A., David C., Feigon B., and Lalín M. *Statistics for traces of cyclic trigonal curves over finite fields.* Int. Math. Res. Not. IMRN **2010**, no. 5, 932–967

[2] Kurlberg P. and Rudnick Z. *The fluctuations in the number of points on a hyperelliptic curve over a finite field.* J. Number Theory **129** (2009), no. 3, 580–587.

[3] Raynaud M. *Caractéristique d'Euler-Poincaré d'un faisceaux et cohomologie des variétés abeliennes.* Seminaire Bourbaki. vol. 9. Exp. 286 (1995), 129–147.

[4] Romagny M. and Wewers S. *Hurwitz Spaces.* Groupes de Galois arithmétiques et différentiels, Sémin. Congr. vol. 13. Soc. Math. France, Paris, 2006, pp. 313–341.

[5] Rosen M. *Number Theory in Function Fields.* Graduate Texts in Mathematics. vol. 210. Springer-Verlag, New York, 2002. pp. xii+358.

Fields Institute Communications
Volume **60**, 2011

The ℓ-Rank Structure of a Global Function Field

Lisa Berger
Mathematics Department
Stony Brook University
Stony Brook, NY 11794-3651, USA
lbrgr@math.sunysb.edu

Jing Long Hoelscher
Department of Mathematics, Statistics, and Computer Science
University of Illinois at Chicago
851 S. Morgan Street
Chicago, IL 60607-7045, USA
jlong@math.uic.edu

Yoonjin Lee
Department of Mathematics
Ewha Womans University
11-1 Daehyun-Dong
Seodaemun-Gu
Seoul 120-750, South Korea
yoonjinl@ewha.ac.kr

Jennifer Paulhus
Department of Mathematical Sciences
Villanova University
800 Lancaster Avenue
Villanova, PA 19085-1699, USA
jennifer.paulhus@villanova.edu

Renate Scheidler
Department of Mathematics and Statistics
University of Calgary
2500 University Drive NW
Calgary, AB T2N 3Z4, Canada
rscheidl@ucalgary.ca

2010 *Mathematics Subject Classification.* Primary 11R58; Secondary 11R65, 11Y40, 14Q05, 14H05.

The third author is the corresponding author and is supported by Priority Research Centers Program through the National Research Foundation of Korea (NRF), funded by the Ministry of Education, Science and Technology (2009-0093827).

The last author is supported by NSERC of Canada.

Abstract. For any prime ℓ, it is possible to construct global function fields whose Jacobians have high ℓ-rank by moving to a sufficiently large constant field extension. This was investigated in some detail by Bauer et al. in [2]. The two main results of [2] are an upper bound on the size of the field of definition of the ℓ-torsion $\mathcal{J}[\ell]$ of the Jacobian, and a lower bound on the increase in the base field size that guarantees a strict increase in ℓ-rank. Here, we provide improvements to both these results, and give examples which illustrate that our techniques have the potential to yield the correct ℓ-rank over any intermediate field of the field of definition of $\mathcal{J}[\ell]$, including base fields that might be too large to be handled directly by computer algebra packages.

1 Introduction

Let q be any prime power and ℓ a prime not dividing q. Then the Jacobian of any function field $\mathbb{K}/\mathbb{F}_q$ is a finite Abelian group that can have ℓ-rank as large as $2g$. However, Jacobians of large ℓ-rank, and even just positive ℓ-rank, tend to be rare. This statement was made precise by Achter [1], who gave for any r with $0 \le r \le 2g$ and any sufficiently general family of curves over $\mathbb{F}_q$ an explicit formula for the proportion of curves whose Jacobians have ℓ-rank r over $\mathbb{F}_q$. For example, if $q \equiv 1 \pmod{\ell}$, then an elliptic curve has ℓ-rank 0, 1, or 2 over $\mathbb{F}_q$ with approximate probabilities $1 - \ell/(\ell^2-1)$, $1/\ell$, and $1/\ell(\ell^2-1)$, respectively. Similar behaviour holds for curves of higher genus.

It is thus clear that algebraic function fields of high ℓ-rank require special construction. For genus 2 hyperelliptic curves $y^2 = f(x)$ with $f(x) \in \mathbb{F}_q[x]$ square-free and $\deg(f) = 5$, the 2-rank can simply be read from the factorization of $f(x)$ into irreducibles over $\mathbb{F}_q$ [12]. In fact, the 2-rank of a quadratic function field is generally well understood [18]. A number of constructions for hyperelliptic curves of high 3-rank were presented in [2]; see also Chapter 7 of [16] for a somewhat different approach to generating such fields via cubic extensions. In a sequence of papers, Pacelli et al. found infinite families of quadratic [10, 11] and higher degree [13, 14] function fields of large 3-rank, and more generally, n-rank.

In this paper, we focus our attention on a method that constructs arbitrary function fields of large ℓ-rank by enlarging the base field $\mathbb{F}_q$. This procedure was first presented in Section 5 of [2], where it was described in the context of hyperelliptic function fields, but it is applicable to any type of function field $\mathbb{K}/\mathbb{F}_q$ — and in fact, any Jacobian variety — and any prime $\ell \nmid q$. The technique has the advantage that one can start with a function field $\mathbb{K}$ of any genus g over a very small base field $\mathbb{F}_q$. It requires the computation of the L-polynomial of $\mathbb{K}$ over this small field $\mathbb{F}_q$, which can be accomplished with a computer algebra package such as Magma [4]. Using only the factorization modulo ℓ of the reciprocal polynomial $F(t) \equiv t^{2g}L(t^{-1}) \pmod{\ell}$ of this L-polynomial, it is possible to find an upper bound on the degree n_ℓ over $\mathbb{F}_q$ of the field of definition of the full ℓ-torsion of the Jacobian of $\mathbb{K}/\mathbb{F}_q$. This factorization also provides lower bounds on the minimal extension degree over any base field that guarantees an increase in ℓ-rank, as well as a lower bound on that ℓ-rank increase.

The results in this article represent a number of improvements to the work of [2, Section 5]. We provide a simpler means for finding good upper bounds on n_ℓ and analyze the possible ℓ-ranks over all intermediate base fields $\mathbb{F}_q \subset \mathbb{F}_{q^n} \subset \mathbb{F}_{q^{n_\ell}}$.

In many cases, our results make it possible to find all these intermediate ℓ-ranks, requiring the computation of the L-polynomial $L(t)$ of $\mathbb{K}$ over the base field $\mathbb{F}_q$ only. Sometimes, the ℓ-rank of the Jacobian over an intermediate field $\mathbb{F}_{q^n}$ for very small n may also be needed. Apart from these ingredients, our methods involve only basic algebra and linear algebra.

We begin with an overview of the techniques of [2] in Section 3 and describe our improvements in Sections 4, 5 and 7, illustrating them with a number of examples. In the process, we rediscover and improve on known results for curves of genus 1 and 2 in Section 6.

2 Background and notation

For introductory reading on algebraic function fields, the reader is encouraged to consult [15, 17]. Throughout this paper, q will denote a prime power, $\mathbb{F}_q$ a finite field of order q, $\overline{\mathbb{F}}_q$ an algebraic closure of $\mathbb{F}_q$, and $\mathbb{F}_q^* = \mathbb{F}_q \setminus \{0\}$. Furthermore, $\mathbb{F}_q[x]$ and $\mathbb{F}_q(x)$ are the ring of polynomials and the field of rational functions, respectively, in the indeterminate x with coefficients in $\mathbb{F}_q$. For $F \in \mathbb{F}_q[x]$, $\deg(F)$ denotes the degree of F.

An *algebraic function field* $\mathbb{K}/\mathbb{F}_q$ (of one variable over $\mathbb{F}_q$) is a finite algebraic extension of $\mathbb{F}_q(x)$. If $[\mathbb{K} : \mathbb{F}_q(x)]$ denotes the degree of the extension $\mathbb{K}/\mathbb{F}_q(x)$, then $\mathbb{K} = \mathbb{F}_q(x, y)$ where y is the root of some polynomial $F(x, Y) \in \mathbb{F}_q[x][Y]$ that is monic with respect to Y, has degree $[\mathbb{K} : \mathbb{F}_q(x)]$ in Y, and is irreducible over $\mathbb{F}_q(x)$. We will always assume that $\mathbb{F}_q$ is the *(full) constant field* of $\mathbb{K}$; that is, $\mathbb{F}_q$ is algebraically closed in $\mathbb{K}$. Equivalently, the polynomial $F(x, Y)$ is absolutely irreducible, i.e. irreducible over $\overline{\mathbb{F}}_q$.

Recall that a *place* $\mathfrak{p}$ of $\mathbb{K}/\mathbb{F}_q$ is the unique (principal) maximal ideal of some discrete valuation ring $\mathcal{O} = \mathcal{O}_{\mathfrak{p}}$ of $\mathbb{K}$. We denote the set of places of $\mathbb{K}/\mathbb{F}_q$ by $\mathbb{P}_{\mathbb{K}}$. A *divisor* D of $\mathbb{K}/\mathbb{F}_q$ is a formal sum of places, i.e. $D = \sum_{\mathfrak{p} \in \mathbb{P}_{\mathbb{K}}} a_{\mathfrak{p}} \mathfrak{p}$ where $a_{\mathfrak{p}} = 0$ for all but finitely many $\mathfrak{p} \in \mathbb{P}_{\mathbb{K}}$. The *degree* of D is the integer $\deg(D) = \sum_{\mathfrak{p} \in \mathbb{P}_{\mathbb{K}}} a_{\mathfrak{p}}$. Clearly, the divisors form an infinite Abelian group, the *divisor group* $\mathcal{D}_{\mathbb{K}}(\mathbb{F}_q)$ of $\mathbb{K}/\mathbb{F}_q$, and the divisors of degree zero form a subgroup of $\mathcal{D}_{\mathbb{K}}(\mathbb{F}_q)$, the *degree zero divisor group* $\mathcal{D}^0_{\mathbb{K}}(\mathbb{F}_q)$. A divisor is *principal* if there exists a non-zero element $z \in \mathbb{K}$ such that $D = \sum_{\mathfrak{p} \in \mathbb{P}_{\mathbb{K}}} v_{\mathfrak{p}}(z)\mathfrak{p}$, where $v_{\mathfrak{p}}$ is the discrete valuation on $\mathbb{K}$ associated to the place $\mathfrak{p} \in \mathbb{P}_{\mathbb{K}}$. Since every principal divisor has degree zero, the set $\mathcal{P}_{\mathbb{K}}(\mathbb{F}_q)$ of principal divisors is a subgroup of $\mathcal{D}^0_{\mathbb{K}}(\mathbb{F}_q)$.

The quotient group $\mathrm{Jac}_{\mathbb{K}}(\mathbb{F}_q) = \mathcal{D}^0_{\mathbb{K}}(\mathbb{F}_q)/\mathcal{P}_{\mathbb{K}}(\mathbb{F}_q)$ is the *(degree zero) divisor class group* of $\mathbb{K}/\mathbb{F}_q$; it is a finite Abelian group. The notation "Jac" stems from the following fact. There exists a unique non-singular, projective, absolutely irreducible, algebraic curve C so that $\mathbb{K}/\mathbb{F}_q$ is the function field of C. Then $\mathrm{Jac}_{\mathbb{K}}(\mathbb{F}_q)$ is isomorphic to the group of $\mathbb{F}_q$-rational points on the *Jacobian variety* of C. Hence, $\mathrm{Jac}_{\mathbb{K}}(\mathbb{F}_q)$ is sometimes simply referred to as the Jacobian of $\mathbb{K}/\mathbb{F}_q$. We denote by g the *genus* of $\mathbb{K}/\mathbb{F}_q$, or equivalently, the genus of C.

When $\mathbb{K}/\mathbb{F}_q$ has genus $g = 1$ and the set $C(\mathbb{F}_q)$ of points on C with coordinates in $\mathbb{F}_q$ is non-empty, C is an *elliptic curve*. In this case $C(\mathbb{F}_q)$ is an abelian group, where the addition is defined geometrically via the "chord and tangent" addition law, described most easily via the property "any three collinear points on the curve sum to zero". With these conditions, $\mathrm{Jac}_{\mathbb{K}}(\mathbb{F}_q)$ is isomorphic to $C(\mathbb{F}_q)$. Indeed, as an algebraic variety, an elliptic curve is isomorphic to its Jacobian, while for curves of higher genus this is no longer the case.

The divisor group of $\mathbb{K}/\mathbb{F}_q$ is closely related to the *zeta function* of $\mathbb{K}/\mathbb{F}_q$, i.e. the function $\zeta(s) = \sum_{D \in \mathcal{D}_{\mathbb{K}}(\mathbb{F}_q)} q^{-s \deg(D)}$. Here, s is a complex variable. If we set $t = q^{-s}$, then for all $s \in \mathbb{C}$ with $\mathrm{Re}(s) > 1$,

$$\zeta(s) = \sum_{D \in \mathcal{D}_{\mathbb{K}}(\mathbb{F}_q)} t^{\deg(D)} = \frac{L(t)}{(1-t)(1-qt)} ,$$

where $L(t) = a_0 + a_1 t + \cdots + a_{2g} t^{2g}$ is called the *L-polynomial* of $\mathbb{K}/\mathbb{F}_q$. It is a polynomial of degree $2g$ with integer coefficients satisfying $a_0 = 1$ and $q^{g-i} a_i = a_{2g-i}$ for $0 \leq i \leq g-1$. The reciprocals of the roots of $L(t)$ are algebraic integers of absolute value $\sqrt{q}$ — this is generally referred to as the *Riemann hypothesis for function fields*, although it was in fact proved by Weil — and we have $|\operatorname{Jac}_{\mathbb{K}}(\mathbb{F}_q)| = L(1)$. This implies the *Hasse-Weil bounds*

$$(\sqrt{q}-1)^{2g} \leq |\operatorname{Jac}_{\mathbb{K}}(\mathbb{F}_q)| \leq (\sqrt{q}+1)^{2g} ,$$

which show that $\operatorname{Jac}_{\mathbb{K}}(\mathbb{F}_q)$ is a very large group even for function fields of modest size. As a result, computing the structure of $\operatorname{Jac}_{\mathbb{K}}(\mathbb{F}_q)$, or even just its order, generally tends to be a difficult problem. For small function fields, however, it is possible to compute the zeta function, L-polynomial, and possibly even the Jacobian, using for example a computer algebra package such as Magma.

For any prime ℓ, we denote the *ℓ-rank* of $\operatorname{Jac}_{\mathbb{K}}(\mathbb{F}_q)$ by $\ell\text{-rank}(\operatorname{Jac}_{\mathbb{K}}(\mathbb{F}_q))$. It is the dimension of the $\mathbb{F}_\ell$-module $\operatorname{Jac}_{\mathbb{K}}(\mathbb{F}_q)/\ell \operatorname{Jac}_{\mathbb{K}}(\mathbb{F}_q)$, which is also the minimum number of generators of the ℓ-Sylow subgroup of $\operatorname{Jac}_{\mathbb{K}}(\mathbb{F}_q)$ when viewed as a finite Abelian group. An upper bound on the ℓ-rank of $\operatorname{Jac}_{\mathbb{K}}(\mathbb{F}_q)$ is $2g$ if $\ell \nmid q$ and g if $\ell \mid q$. Here, we will only consider the former case and henceforth exclude the scenario where ℓ is the characteristic of $\mathbb{F}_q$.

3 Increasing the ℓ-rank via enlarging the base field

For completeness, we provide an overview of the aforementioned method given in Section 5 of [2]; for details and proofs, the reader is referred to that source. We continue to let $\mathbb{K}/\mathbb{F}_q$ be an algebraic function field and $\operatorname{Jac}_{\mathbb{K}}(\mathbb{F}_q)$ its Jacobian. Consider the field $\overline{\mathbb{K}} = \mathbb{K}\overline{\mathbb{F}}_q$, the compositum of $\mathbb{K}$ and $\overline{\mathbb{F}}_q$. Then $\overline{\mathbb{K}}/\overline{\mathbb{F}}_q$ is an unramified function field extension of $\mathbb{K}/\mathbb{F}_q$ of the same genus as $\mathbb{K}/\mathbb{F}_q$. Since $\mathbb{F}_q$ is algebraically closed in $\mathbb{K}$, we see that if we write $\mathbb{K} = \mathbb{F}_q(x,y)$ with $y \in \mathbb{K}$, then $\overline{\mathbb{K}} = \overline{\mathbb{F}}_q(x,y)$ and $[\overline{\mathbb{K}} : \overline{\mathbb{F}}_q(x)] = [\mathbb{K} : \mathbb{F}_q(x)]$.

Similarly, for any $n \in \mathbb{N}$, set $\mathbb{K}_n = \mathbb{K}\mathbb{F}_{q^n} = \mathbb{F}_{q^n}(x,y) \subset \overline{\mathbb{K}}$; then $[\mathbb{K}_n : \mathbb{F}_{q^n}(x)] = [\mathbb{K} : \mathbb{F}_q(x)]$. If $L(t) = \prod_{i=1}^{2g}(1-\omega_i t)$ is the L-polynomial of $\mathbb{K}/\mathbb{F}_q$, with $\omega_i \in \mathbb{C}$ for $1 \leq i \leq 2g$, then the L-polynomial of $\mathbb{K}_n/\mathbb{F}_{q^n}$ is $\prod_{i=1}^{2g}(1-\omega_i^n t)$.

For brevity, set $\mathcal{J} = \operatorname{Jac}_{\overline{\mathbb{K}}}(\overline{\mathbb{F}}_q)$ and $\mathcal{J}_n = \operatorname{Jac}_{\mathbb{K}_n}(\mathbb{F}_{q^n})$. As before, let ℓ be any prime not dividing q, and denote the ℓ-torsion of $\mathcal{J}$ and $\mathcal{J}_n$ by $\mathcal{J}[\ell]$ and $\mathcal{J}_n[\ell]$, respectively. Then $\mathcal{J}_n[\ell] \subset \mathcal{J}_{kn}[\ell]$ for all $k \in \mathbb{N}$, and $\mathcal{J}[\ell] \cong (\mathbb{Z}/\ell)^{2g}$. In fact, there exists a smallest field $\mathbb{F}_{q^{n_\ell}}$ such that $\mathcal{J}[\ell] \subseteq \mathcal{J}_{n_\ell}[\ell]$; this field is the *field of rationality* or *field of definition* of $\mathcal{J}[\ell]$. It follows that $\mathcal{J}_{n_\ell}[\ell] \cong (\mathbb{Z}/\ell)^{2g}$, and hence $\ell\text{-rank}(\mathcal{J}_{n_\ell}) = 2g$ is maximal. The method in [2, Section 5] provided partial answers to the following questions:

1. What is the exact value of n_ℓ? Is it at least possible to find an upper bound b on n_ℓ?

2. For any $n \in \mathbb{N}$, is it possible to ascertain if $\ell\text{-rank}(\mathcal{J}_n) > \ell\text{-rank}(\mathcal{J}_1)$? If yes, is it possible to find (lower bounds on) such a value of n as well as the increase from $\ell\text{-rank}(\mathcal{J}_1)$ to $\ell\text{-rank}(\mathcal{J}_n)$?

The upper bound b on n_ℓ given in [2] requires a potentially tedious search, which is eliminated by the results of this paper. Moreover, we extend the answer given in [2] to Question 2 above by considering a more general scenario. In particular, this will allow us to deduce ℓ-ranks over large base fields, which might be too big for computers to handle, from those over smaller base fields.

The Galois group $\mathrm{Gal}(\overline{\mathbb{F}}_q/\mathbb{F}_q)$ acts on the ℓ-torsion $\mathcal{J}[\ell]$ of $\mathcal{J}$, whence we obtain a representation $\rho_\ell : \mathrm{Gal}(\overline{\mathbb{F}}_q/\mathbb{F}_q) \to \mathrm{Aut}(J[\ell])$. This representation factors through $\mathrm{Gal}(\mathbb{F}_{q^{n_\ell}}/\mathbb{F}_q)$. For any choice of $\mathbb{F}_\ell$-basis of $\mathcal{J}[\ell]$, we thus obtain

$$\rho_\ell : \mathrm{Gal}(\mathbb{F}_{q^{n_\ell}}/\mathbb{F}_q) \hookrightarrow \mathrm{GL}_{2g}(\mathbb{F}_\ell) \ .$$

The Galois group $\mathrm{Gal}(\mathbb{F}_{q^{n_\ell}}/\mathbb{F}_q)$ above is generated by the restriction π_{q,n_ℓ} of the Frobenius automorphism π_q on $\overline{\mathbb{F}}_q$ to $\mathbb{F}_{q^{n_\ell}}$. This is the automorphism that sends every element in $\mathbb{F}_{q^{n_\ell}}$ to its q-th power. Thus, the image of π_{q,n_ℓ} under ρ_ℓ is a matrix $M_\ell \in \mathrm{GL}_{2g}(\mathbb{F}_\ell)$ of order n_ℓ. In order to determine $n_\ell = \mathrm{ord}(M_\ell)$, we define ρ_ℓ in terms of a basis for which M_ℓ is in *primary rational canonical form.*

The explanation of the primary rational canonical form requires a brief excursion into linear algebra. Let V be a vector space over some field K and ϕ a linear transformation on V. In our context, we will have $K = \mathbb{F}_\ell$, $V = \mathcal{J}_{n_\ell}[\ell] \cong \mathbb{F}_\ell^{2g}$, and $\phi = \pi_{q,n_\ell}$; note that since π_{q,n_ℓ} acts on the places of $\mathbb{K}_{n_\ell}/\mathbb{F}_{q^{n_\ell}}$, this action extends naturally to $\mathcal{J}_{n_\ell}$ and hence to $\mathcal{J}_{n_\ell}[\ell]$.

Recall that the *minimal polynomial* of ϕ is the unique monic polynomial $G_\phi(t) \in K[t]$ of minimal degree with $G_\phi(\phi) = 0$; it divides all other polynomials $F(t) \in K[t]$ with $F(\phi) = 0$, including the *characteristic polynomial* $F_\phi(t)$ of ϕ whose roots are the eigenvalues of ϕ with appropriate multiplicities. In fact, $F_\phi(t)$ and $G_\phi(t)$ have the same roots, but potentially with different multiplicities.

For any monic polynomial $F(t) = t^m + a_{m-1}t^{m-1} + \cdots + a_0 \in K[t]$, the *companion matrix* of $F(t)$ is the $m \times m$ matrix

$$A_F = \begin{pmatrix} 0 & 1 & 0 & 0 & \cdots & 0 & 0 \\ 0 & 0 & 1 & 0 & \cdots & 0 & 0 \\ \vdots & & & & & & \vdots \\ 0 & 0 & 0 & 0 & \cdots & 0 & 1 \\ -a_0 & -a_1 & -a_2 & -a_3 & \cdots & -a_{m-2} & -a_{m-1} \end{pmatrix} .$$

Note that $F(t)$ is both the minimal and the characteristic polynomial of its own companion matrix A_F.

A subspace W of V is said to be *ϕ-cyclic* if it spanned by the vectors $\phi^i(v)$, $i \geq 0$, for some $v \in V$. Then W is ϕ-cyclic if and only if W has an ordered basis relative to which the matrix associated to the restriction $\phi|_W$ of ϕ to W is the companion matrix of the minimal polynomial of $\phi|_W$.

Finally, for any finite collection of square matrices $A_1, A_2, \ldots, A_r$, we denote by $\mathrm{diag}(A_1, A_2, \ldots, A_r)$ the matrix that has the sub-matrices $A_1, A_2, \ldots, A_r$ along its diagonal and zeros everywhere else; it is a square matrix whose size is the sum of the sizes of the A_i, $1 \leq i \leq r$. We make use of the following decomposition theorem:

Theorem 3.1 (Theorem 5.1 of [2], based on results of [8]) *Let $\phi : V \to V$ be a linear transformation on a vector space V over a field K. Then there exist distinct monic irreducible polynomials $P_1, P_2, \dots, P_s \in K[t]$ with the following properties:*

1. *For each i with $1 \le i \le s$, there exist $k_i \in \mathbb{N}$ and ϕ-cyclic subspaces $V_{i1}, V_{i2}, \dots, V_{ik_i}$ such that V is the direct sum of the V_{ij} $(1 \le i \le s, 1 \le j \le k_i)$.*
2. *For each i with $1 \le i \le s$ and each j with $1 \le j \le k_i$, the minimal polynomial of the restriction of ϕ to V_{ij} is of the form $P_i^{m_{ij}}$ where the integers m_{ij} satisfy $m_{i1} \ge m_{i2} \ge \cdots \ge m_{ik_i} \ge 1$. In particular, $\dim(V_{ij}) = m_{ij} \deg(P_i)$.*
3. *V has a basis relative to which the matrix of ϕ is of the form*
$$A_\phi = \mathrm{diag}(A_{P_1^{m_{11}}}, \dots, A_{P_1^{m_{1k_1}}}, \dots, A_{P_s^{m_{s1}}}, \dots, A_{P_s^{m_{sk_s}}}) , \tag{3.1}$$
where $A_{P_i^{m_{ij}}}$ is the companion matrix of $P_i^{m_{ij}}$ for $1 \le i \le s$ and $1 \le j \le k_i$.
4. *The minimal polynomial of ϕ is $G_\phi = P_1^{m_{11}} P_2^{m_{21}} \cdots P_s^{m_{s1}}$.*
5. *The characteristic polynomial of ϕ is $F_\phi = P_1^{m_1} P_2^{m_2} \cdots P_s^{m_s}$ where $m_i = \sum_{j=1}^{k_i} m_{ij}$ for $1 \le i \le s$.*

The polynomials $P_i^{m_{ij}}$ $(1 \le i \le s, 1 \le j \le k_i)$ are uniquely determined by V and ϕ and are called the *elementary divisors* of ϕ (or of any matrix representing ϕ). The matrix A_ϕ is said to be the *primary rational canonical form* of ϕ (or of any matrix representing ϕ).

If A_ϕ has finite order in $\mathrm{GL}_{\dim(V)}(K)$ — this is the case in our context — then
$$\mathrm{ord}(A_\phi) = \mathrm{lcm}\left\{\mathrm{ord}(A_{P_i^{m_{ij}}}) \mid 1 \le i \le s, 1 \le j \le k_i\right\} ,$$
which is easy to compute if $\dim(V)$ is not too large and each companion matrix has sufficiently small order. Thus, if we were able to obtain the elementary divisors of π_{q,n_ℓ}, then we could easily obtain $n_\ell = \mathrm{ord}(A_{\pi_{q,n_\ell}})$. Unfortunately, it is unclear how to obtain these elementary divisors. It is however possible to obtain the characteristic polynomial $F_{\pi_{q,n_\ell}}(t) \in \mathbb{F}_\ell[t]$ from the L-polynomial of $\mathbb{K}/\mathbb{F}_q$ via $F_{\pi_{q,n_\ell}}(t) \equiv t^{2g} L(t^{-1}) \pmod{\ell}$. In other words, if we compute the L-polynomial of $\mathbb{K}/\mathbb{F}_q$ — using Magma, for example — and factor its reciprocal modulo ℓ to obtain $F_{\pi_{q,n_\ell}}(t) = P_1^{m_1} P_2^{m_2} \cdots P_s^{m_s}$ with $P_1, P_2, \dots, P_s \in \mathbb{F}_\ell[t]$ monic and irreducible, then we can construct all possible matrices $A_{\pi_{q,n_\ell}}$ as given in (3.1) (with $\phi = \pi_{q,n_\ell}$) for all $1 \le i \le s$ and $1 \le j \le m_i$. One of these is the primary rational canonical form of π_{q,n_ℓ}, and although it is unknown which candidate matrix is the correct one, the maximum of their orders is an upper bound on n_ℓ. Moreover, in the case when $F_{\pi_{q,n_\ell}}(t)$ is square-free, i.e. $m_i = 1$ for $1 \le i \le s$, we see that $A_{\pi_{q,n_\ell}} = \mathrm{diag}(A_{P_1}, A_{P_2}, \dots, A_{P_s})$ is uniquely determined. This is actually the case for many curves, and particularly for most hyperelliptic curves; see the work of Chavdarov [5] and Kowalski [9, Section 6].

We now summarize the main results of [2, Section 5]. The first provides a partial answer to Question 1 above.

Proposition 3.2 (Theorem 5.2 of [2]) *Let $L(t)$ be the L-polynomial of a function field $\mathbb{K}/\mathbb{F}_q$ of genus g, and set $F(t) \equiv t^{2g} L(t^{-1}) \pmod{\ell}$, $F(t) \in \mathbb{F}_\ell[t]$. Let $F = P_1^{m_1} P_2^{m_2} \cdots P_s^{m_s}$ be the factorization of $F(t)$ into powers of distinct monic* *irreducibles $P_i(t) \in \mathbb{F}_\ell[t]$ and define a set $\mathcal{S}$ as follows:*

$$\mathcal{S} = \Big\{(P_i^{m_{ij}}) \mid 1 \le i \le s,\ 1 \le j \le k_i,\ m_{i1} \ge \cdots \ge m_{ik_i} \ge 1, \text{ and } \sum_{j=1}^{k_i} m_{ij} = m_i \text{ for } 1 \le i \le s\Big\} .$$

For any tuple $\mathbf{P} = (P_i^{m_{ij}}) \in \mathcal{S}$, define the matrix

$$A_{\mathbf{P}} = \operatorname{diag}(A_{P_1^{m_{11}}}, \ldots, A_{P_1^{m_{1k_1}}} \ldots , A_{P_s^{m_{s1}}}, \ldots, A_{P_s^{m_{sk_s}}}) ,$$

where $A_{P_i^{m_j}}$ is the companion matrix of $P_i^{m_{ij}}$. Set

$$b = \max\{\operatorname{ord}(A_{\mathbf{P}}) \mid \mathbf{P} \in \mathcal{S}\} ,$$

and let $\mathbb{F}_{q^{n_\ell}}$ be the field of rationality of $\mathcal{J}[\ell]$. Then $n_\ell \le b$, with equality if $F(t)$ is square-free.

We will improve this result in Theorem 4.3 below; in particular, we will note that in order to determine the maximum of all the $\operatorname{ord}(A_{\mathbf{P}})$ with $\mathbf{P} \in \mathcal{S}$, it suffices to consider only the trivial partitions consisting of just one summand $m_{i1} = m_i$ $(1 \le i \le s)$.

The following appears as a short discussion on pp. 521-522 of [2]. It can be useful if a lower bound on the ℓ-rank of $\mathcal{J}_{\mathbb{K}}(\mathbb{F}_q)$ is a priori known; for example, if $\mathbb{K}$ was obtained via one of the special constructions referred to in Section 1. For any two polynomials $P(t), G(t) \in \mathbb{F}_\ell[t]$ with $P(t)$ irreducible, we let $v_P(G)$ denote the exact power of $P(t)$ that divides $G(t)$.

Proposition 3.3 *Let $L(t)$ be the L-polynomial of a function field $\mathbb{K}/\mathbb{F}_q$ of genus g, and set $F(t) \equiv t^{2g}L(t^{-1}) \pmod{\ell}$, $F(t) \in \mathbb{F}_\ell[t]$. Suppose that $\ell\text{-rank}(\mathcal{J}_{\mathbb{K}}(\mathbb{F}_q)) \ge r$. Then $(t-1)^r$ divides $F(t)$, and if $v_{t-1}(F) = r$, then $\ell\text{-rank}(\mathcal{J}_{\mathbb{K}}(\mathbb{F}_q)) = r$.*

If $\ell\text{-rank}(\mathcal{J}_{\mathbb{K}}(\mathbb{F}_q)) \ge r$, then each polynomial $P_i^{m_i}$ $(1 \le i \le s)$ is a product of at least r elementary divisors $P^{m_{ij}}$ of π_{q,n_ℓ}.

Finally, [2] provided the following partial answer to Question 2 above.

Proposition 3.4 (Theorem 5.6 of [2]) *Let $L(t)$ be the L-polynomial of a function field $\mathbb{K}/\mathbb{F}_q$ of genus g, and set $F(t) \equiv t^{2g}L(t^{-1}) \pmod{\ell}$, $F(t) \in \mathbb{F}_\ell[t]$. Suppose $F(t)$ has an irreducible factor $P(t) \in \mathbb{F}_\ell[t]$ different from $t-1$. Let $A_P \in \mathrm{GL}_{\deg(P)}(\mathbb{F}_\ell)$ be the companion matrix of P, and set $n = \operatorname{ord}(A_P)$. Then $\ell\text{-rank}(\mathcal{J}_n) \ge \ell\text{-rank}(\mathcal{J}_k) + \deg(P)$ for any proper divisor k of n.*

The next four sections contain new results and improvements on the propositions from [2] presented above. Specifically, Section 4 gives a more explicit expression for the bound b on n_ℓ of Proposition 3.2 that eliminates the need to compute the maximum of the orders of all the matrix candidates. Section 5 provides a simple upper bound on (and multiple of) b. Section 6 applies our results to curves of genus 1 and 2, and Section 7 adds to Proposition 3.3 and improves on Proposition 3.4.

4 An explicit expression for the bound b

In order to further investigate the bound b on the degree n_ℓ of the field of rationality of $\mathcal{J}[\ell]$ over $\mathbb{F}_q$, we require some additional linear algebra. Recall that an *elementary Jordan matrix* over some field K is an $m \times m$ matrix of the form

$$J_{m,\lambda} = \begin{pmatrix} \lambda & 1 & 0 & \cdots & 0 & 0 \\ 0 & \lambda & 1 & \cdots & 0 & 0 \\ 0 & 0 & \lambda & \cdots & 0 & 0 \\ \vdots & \vdots & & \ddots & & \vdots \\ 0 & 0 & 0 & \cdots & \lambda & 1 \\ 0 & 0 & 0 & \cdots & 0 & \lambda \end{pmatrix} . \tag{4.1}$$

Let $A \in \mathrm{GL}_m(K)$. Over an algebraic closure $\overline{K}$ of K, the elementary divisors of A are of the form $(t - \lambda_i)^{m_{ij}}$ with $\lambda_i \in \overline{K}$ and $1 \leq i \leq s, 1 \leq j \leq k_i$, and the *Jordan canonical form* of A (or of the linear transformation represented by A) is the matrix

$$J_A = \mathrm{diag}(J_{m_{ij},\lambda_i})_{1 \leq i \leq s,\, 1 \leq j \leq k_i} \ .$$

In particular, the companion matrix A_F of a polynomial $F(t) \in K[t]$ with roots $\lambda_1, \lambda_2, \ldots, \lambda_r \in \overline{K}$ of respective multiplicities $m_1, m_2, \ldots, m_r$ has Jordan canonical form $J_{A_F} = \mathrm{diag}(J_{m_1,\lambda_1}, \ldots, J_{m_r,\lambda_r})$.

We begin with two basic results on elementary Jordan matrices and companion matrices. As usual, for any real number R, let $\lceil R \rceil$ denote the ceiling of R, i.e. the smallest integer that is greater than or equal to R. Furthermore, for any $\gamma \in \overline{\mathbb{F}}_\ell^*$, let $\mathrm{ord}_\ell(\gamma)$ denote the order of γ in $\overline{\mathbb{F}}_\ell^*$; note that since $\gamma \in \mathbb{F}_{\ell^d}^*$ for some $d \in \mathbb{N}$, $\mathrm{ord}_\ell(\gamma)$ divides $\ell^d - 1$ and is hence finite and not divisible by ℓ.

Lemma 4.1 *Let ℓ be a prime and $\lambda \in \overline{\mathbb{F}}_\ell^*$. Then* $\mathrm{ord}(J_{m,\lambda}) = \ell^{\lceil \log_\ell(m) \rceil} \mathrm{ord}_\ell(\lambda)$.

Proof For brevity, set $n = \lceil \log_\ell(m) \rceil$. Then $\ell^{n-1} < m \leq \ell^n$. Induction yields

$$J_{m,\lambda}^i = \begin{pmatrix} \lambda^i & \binom{i}{1}\lambda^{i-1} & \binom{i}{2}\lambda^{i-2} & \cdots & \binom{i}{i-1}\lambda & 1 & 0 & 0 & \cdots & 0 \\ 0 & \lambda^i & \binom{i}{1}\lambda^{i-1} & \cdots & \binom{i}{i-2}\lambda^2 & \binom{i}{i-1}\lambda & 1 & 0 & \cdots & 0 \\ \vdots & \vdots & \vdots & & & & & \vdots & & \vdots \\ 0 & 0 & 0 & \cdots & 0 & 0 & 0 & 0 & \cdots & \lambda^i \end{pmatrix}$$

for $i \in \mathbb{N}$. Note that for $i \geq m$, all the entries on and above the main diagonal of $J_{m,\lambda}^i$ have the form $\binom{i}{j}\lambda^{i-j}$ with $0 \leq j \leq m-1$.

Recall that the exact power of ℓ dividing a binomial coefficient $\binom{i}{j}$ is the number of carries that occur when j is added to $i - j$ in base ℓ. For $i = \ell^n$ and $1 < j < \ell^n$, at least one such carry occurs, so $\binom{\ell^n}{j} \equiv 0 \pmod{\ell}$. Since $\ell^n \geq m$, we have $J_{m,\lambda}^{\ell^n} = \lambda^{\ell^n} I_m$, where I_m is the $m \times m$ identity matrix.

Since $\mathrm{ord}_\ell(\lambda)$ is not divisible by ℓ, it follows that

$$\mathrm{ord}(J_{m,\lambda}^{\ell^n}) = \mathrm{ord}_\ell(\lambda^{\ell^n}) = \frac{\mathrm{ord}_\ell(\lambda)}{\gcd(\mathrm{ord}_\ell(\lambda), \ell^n)} = \mathrm{ord}_\ell(\lambda) \ .$$

On the other hand, $\mathrm{ord}(J_{m,\lambda}^{\ell^n}) = \mathrm{ord}(J_{m,\lambda}) / \gcd(\ell^n, \mathrm{ord}(J_{m,\lambda}))$, so it suffices to show that $\ell^n \mid \mathrm{ord}(J_{m,\lambda})$ to complete the proof.

To that end, let r be the exact power of ℓ dividing $\mathrm{ord}(J_{m,\lambda})$, and write $\mathrm{ord}(J_{m,\lambda})/\ell^r = a\ell + b$ with $a \geq 0$ and $0 < b < \ell$. Then $\mathrm{ord}(J_{m,\lambda}) - \ell^r = a\ell^{r+1} + (b-1)\ell^r$ with $0 \leq b \leq \ell - 1$. Thus, adding ℓ^r to $\mathrm{ord}(J_{m,\lambda}) - \ell^r$ in base ℓ produces no carries, so $\binom{\mathrm{ord}(J_{m,\lambda})}{\ell^r} \not\equiv 0 \pmod{\ell}$. If $r < n$, then $1 \leq \ell^r \leq \ell^{n-1} \leq m - 1$, so this would result in a non-zero entry above the main diagonal of

$J_{m,\lambda}^{\operatorname{ord}(J_{m,\lambda})} = I_m$, a contradiction. Hence $r \geq n$. It follows that $\ell^n \mid \operatorname{ord}(J_{m,\lambda})$, and hence $\operatorname{ord}(J_{m,\lambda}) = \ell^n \operatorname{ord}_\ell(\lambda)$. $\square$

Lemma 4.2 *Let ℓ be a prime, $P(t) \in \mathbb{F}_\ell[t]$ an irreducible polynomial, and A_P the companion matrix of P. Then $\operatorname{ord}(A_{P^m}) = \ell^{\lceil \log_\ell(m) \rceil} \operatorname{ord}(A_P)$ for all $m \in \mathbb{N}$.*

Proof Let $d = \deg(P)$. Then P decomposes into linear factors over $\mathbb{F}_{\ell^d}^*$, say

$$P(t) = (t - \lambda_1)(t - \lambda_2) \cdots (t - \lambda_d) ,$$

with $\lambda_1, \lambda_2, \ldots, \lambda_d \in \mathbb{F}_{\ell^d}^*$ distinct. The Jordan canonical forms of the companion matrices A_P and A_{P^m} are

$$\begin{aligned} J_{A_P} &= \operatorname{diag}(\lambda_1, \lambda_2, \ldots, \lambda_d) , \\ J_{A_{P^m}} &= \operatorname{diag}(J_{m,\lambda_1}, J_{m,\lambda_2}, \ldots, J_{m,\lambda_d}) , \end{aligned}$$

respectively. Since $\lambda_1, \lambda_2, \ldots, \lambda_d \in \mathbb{F}_{\ell^d}^*$ are Galois conjugates, they all have the same order in $\mathbb{F}_{\ell^d}^*$; call this order r. Thus, $\operatorname{ord}(J_{A_P}) = \operatorname{lcm}(\operatorname{ord}_\ell(\lambda_1), \ldots, \operatorname{ord}_\ell(\lambda_d)) = r$. Furthermore, by Lemma 4.1, $\operatorname{ord}(J_{m,\lambda_i}) = \ell^n \operatorname{ord}_\ell(\lambda_i) = \ell^n r$ for $1 \leq i \leq d$, where as before, $n = \lceil \log_\ell(m) \rceil$. Since similar matrices have the same order, we have

$$\begin{aligned} \operatorname{ord}(A_{P^m}) &= \operatorname{ord}(J_{A_{P^m}}) = \operatorname{lcm}(\operatorname{ord}(J_{m,\lambda_i}) | 1 \leq i \leq s) \\ &= \ell^n r = \ell^n \operatorname{ord}(J_{A_P}) = \ell^n \operatorname{ord}(A_P) \end{aligned}$$

as claimed. $\square$

We can now give a simpler expression for the bound b of Proposition 3.2 that can be gleaned directly from the L-polynomial of $\mathbb{K}/\mathbb{F}_q$ and avoids computing $\operatorname{ord}(A_\mathbf{P})$ for the entire collection of $\mathbf{P} \in \mathcal{S}$. We also observe that b is in fact a multiple of n_ℓ.

Theorem 4.3 *With the notation of Proposition 3.2, we have*

$$n_\ell \mid b = \operatorname{ord}(A_F) = \ell^{\lceil \log_\ell(\max\{m_i | 1 \leq i \leq s\}) \rceil} \cdot \operatorname{lcm}\left(\operatorname{ord}(A_{P_i}) \mid 1 \leq i \leq s \right) ,$$

where A_F is the companion matrix of $F(t)$.

Proof Note that for any integers μ, ν, M, N with $\mu, \nu \geq 0$ and M, N not divisible by ℓ, we have $\operatorname{lcm}(\ell^\mu M, \ell^\nu N) = \ell^{\max\{\mu,\nu\}} \operatorname{lcm}(M, N)$.

Recall that $\operatorname{ord}(A_{P_i})$ is equal to the order of each of the roots of P_i in $\mathbb{F}_{\ell^{\deg(P_i)}}$ and is thus coprime to any power of ℓ. Let $\mathbf{P} = (P_i^{m_{ij}}) \in \mathcal{S}$. Then by Lemma 4.2,

$$\begin{aligned} \operatorname{ord}(A_\mathbf{P}) &= \operatorname{lcm}\left(\operatorname{ord}(A_{P_i^{m_{ij}}}) \right) = \operatorname{lcm}\left(\ell^{\lceil \log_\ell(m_{ij}) \rceil} \operatorname{ord}(A_{P_i}) \right) \\ &= \ell^{\max\{\lceil \log_\ell(m_{ij}) \rceil\}} \operatorname{lcm}\left(\operatorname{ord}(A_{P_i}) \right) = \ell^{\lceil \log_\ell(m_{i1}) \rceil} \operatorname{lcm}\left(\operatorname{ord}(A_{P_i}) \right) , \end{aligned}$$

where the last equality above follows from the fact that $m_{i1} \geq m_{i2} \geq \ldots m_{ik_i}$ for all i with $1 \leq i \leq s$. Hence, in order to find b, we need to maximize the above expression over all $\mathbf{P} \in \mathcal{S}$. Since $\operatorname{lcm}(\operatorname{ord}(A_{P_i}))$ is fixed for each $\mathbf{P} \in \mathcal{S}$, this amounts to finding for each i with $1 \leq i \leq s$ the maximum value m_{i1} for all partitions of m_i. Clearly, this maximum is attained for the one-term partition $m_i = m_{i1}$, in which case $\mathbf{P} = (P_i^{m_i})$ corresponds to $F(t)$.

It is also clear that $\operatorname{ord}(A_\mathbf{P})$ divides b for all $\mathbf{P} \in \mathcal{S}$; in particular, $n_\ell \mid b$. Finally, we note that the matrices A_F and $\operatorname{diag}(A_{P_1^{m_1}}, \ldots, A_{P_s^{m_s}})$ have the same Jordan canonical form, and hence the same order b. $\square$

To illustrate the above bound, we revisit Example 5.5 of [2].

Example 4.4 Consider $q = 179$, the hyperelliptic function field $\mathbb{K} = \mathbb{F}_{179}(x, y)$ of genus 4 with

$$y^2 = x^9 + 151\,x^8 + 168\,x^7 + 10\,x^6 + 32\,x^5 + 141\,x^4 + 110\,x^3 + 35\,x^2 + 160\,x + 2\ ,$$

and $\ell = 3$. With the aid of Magma, we found that the zeta function of $\mathbb{K}$ is $\zeta(t) = L(t)/(179\,t^2 - 180\,t + 1)$ with

$$\begin{aligned} L(t) = {} & 179^4\,t^8 - 17 \cdot 179^3\,t^7 + 315 \cdot 179^2\,t^6 - 3041 \cdot 179\,t^5 + 56275\,t^4 \\ & - 3041\,t^3 + 315\,t^2 - 17\,t + 1\ . \end{aligned}$$

Reducing $t^8 L(t^{-1})$ modulo 3, we obtain $F(t) = t^8 + t^7 + t^5 + t^4 + 2\,t^3 + 2\,t + 1 \in \mathbb{F}_3[t]$. Over $\mathbb{F}_3$, $F(t)$ factors as $F = P_1 P_2 P_3^2 P_4$ where

$$P_1(t) = t + 1, \quad P_2(t) = t + 2, \quad P_3(t) = t^2 + 1, \quad P_4(t) = t^2 + t + 2$$

are all irreducible over $\mathbb{F}_3$. Using Proposition 3.2, it was necessary to compute the maximum of the orders of the two matrices

$$\mathrm{diag}(A_{P_1}, A_{P_2}, A_{P_3}, A_{P_3}, A_{P_4}) \quad \text{and} \quad \mathrm{diag}(A_{P_1}, A_{P_2}, A_{P_3^2}, A_{P_4})\ .$$

Using Theorem 4.3, we simply compute $\lceil \log_3(2) \rceil = 1$, and $\mathrm{ord}(A_{P_1}) = 2$, $\mathrm{ord}(A_{P_2}) = 1$, $\mathrm{ord}(A_{P_3}) = 4$, $\mathrm{ord}(A_{P_4}) = 8$. Thus, n_3 is a divisor of $b = 3 \cdot \mathrm{lcm}(2, 1, 4, 8) = 24$. Alternatively, we could have computed $\mathrm{ord}(A_F) = 24$.

We state some simple special cases of Theorem 4.3:

Corollary 4.5 *Let $L(t)$ be the L-polynomial of a function field $\mathbb{K}/\mathbb{F}_q$ of genus g, and set $F(t) \equiv t^{2g} L(t^{-1}) \pmod{\ell}$, $F(t) \in \mathbb{F}_\ell[t]$. Let $b = \mathrm{ord}(A_F)$, where A_F is the companion matrix of F. Then the following hold:*

1. *If $F(t)$ splits into distinct irreducible factors in $\mathbb{F}_\ell[t]$, all of which have degrees dividing $d \in \mathbb{N}$, then*

$$n_\ell = b = \mathrm{lcm}(\mathrm{ord}_\ell(\alpha_i)) \mid \ell^d - 1\ ,$$

where the $\alpha_i \in \mathbb{F}_{\ell^d}$ run through all the roots of F.

2. *If $F(t)$ is a product of powers of linear factors in $\mathbb{F}_\ell[t]$, and the largest exponent of any such linear factor is $m > 1$, then*

$$n_\ell \mid b = \ell^{\lceil \log_\ell(m) \rceil}\, \mathrm{lcm}(\mathrm{ord}_\ell(\alpha_i)) \mid \ell^{\lceil \log_\ell(m) \rceil}(\ell - 1)\ ,$$

where the $\alpha_i \in \mathbb{F}_{\ell^d}$ run through all the roots of F.

Proof Recall that $n_\ell = b$ if $F(t)$ is square-free by Proposition 3.2. The corollary now follows from Theorem 4.3 and the fact that for any irreducible polynomial $P \in \mathbb{F}_\ell[t]$ and any root α of $P(t)$, $\mathrm{ord}(A_P) = \mathrm{ord}_\ell(\alpha) \mid \ell^{\deg(P)} - 1$. □

5 A simple bound on $b = \mathrm{ord}(A_F)$

It is at times desirable to have a bound on n_l that does not require any matrix order computation. In order to present such a bound, we further investigate the irreducible factors of the L-polynomial.

Lemma 5.1 *Let $P(t) \in \mathbb{F}_\ell[t]$ be an irreducible polynomial, and suppose there exists $\alpha \in \overline{\mathbb{F}}_\ell$ such that $\alpha^2 \neq q$ and $P(\alpha) = P(q/\alpha) = 0$. Then the map $\psi : \overline{\mathbb{F}}_\ell \to \overline{\mathbb{F}}_\ell$ via $\psi(\gamma) = q/\gamma$ is a permutation of order 2 on the roots of $P(t)$, so $\deg(P)$ is even.*

Proof Let $d = \deg(P)$. Note that since $P(t)$ has at least two distinct roots α and q/α, we have $d \geq 2$.

The Frobenius on $\overline{\mathbb{F}}_\ell$ sends elements to their ℓ-th powers. Since this Frobenius acts transitively on the roots of $P(t)$, these roots are of the form $\alpha_i = \alpha^{\ell^i}$ for $0 \leq i \leq d-1$. Let $q/\alpha = \alpha^{\ell^j}$. Then $j \neq 0$ as otherwise $\alpha^2 = q$. Hence $1 \leq j \leq d-1$.

Now for all i with $0 \leq i \leq d-1$,

$$\psi(\alpha_i) = \frac{q}{\alpha_i} = \frac{q}{\alpha^{\ell^i}} = \left(\frac{q}{\alpha}\right)^{\ell^i} = (\alpha^{\ell^j})^{\ell^i} = \alpha^{\ell^{j+i}} = \alpha_{i+j \ (\mathrm{mod}\ d)} \ ,$$

so ψ maps any root α_i of $P(t)$ to another root of $P(t)$. It follows that ψ is a permutation on the roots of $P(t)$ whose order divides 2. Moreover, since $0 < j < d$ implies $i \not\equiv i + j \pmod{d}$, the order of ψ is exactly 2. In particular, for $i = 0$, we obtain $\alpha_0 = \psi(\psi(\alpha_0)) = \alpha_{2j \ (\mathrm{mod}\ d)}$, so $d \mid 2j$. Since $1 \leq j \leq d-1$, we see that $d = 2j$ is even. □

Corollary 5.2 *With the notation and assumptions of Lemma 5.1, if A_P is the companion matrix of P, then* $\mathrm{ord}(A_P)$ *divides* $(\ell^{\deg(P)/2} + 1)\,\mathrm{ord}_\ell(q)$.

Proof From the proof of Lemma 5.1, we see that $q/\alpha = \alpha^{\ell^j}$ with $2j = d = \deg(P)$. Thus, $\alpha^{\ell^j+1} = q$, and hence $\alpha^{(\ell^j+1)\,\mathrm{ord}_\ell(q)} = 1$. Since $\mathrm{ord}(A_P) = \mathrm{ord}_\ell(\alpha)$, the claim follows. □

Definition 5.3 *Let $P(t) \in \mathbb{F}_\ell[t]$ be monic and irreducible. Then $P(t)$ is said to be of*

- type 1 *if $P(t)$ is the minimal polynomial of a square root of q;*
- type 2 *if there exists $\alpha \in \overline{\mathbb{F}}_\ell$ such that $\alpha^2 \neq q$ and $P(\alpha) = P(q/\alpha) = 0$;*
- type 3 *otherwise.*

Proposition 5.4 *Let $G(t) \in \mathbb{F}_\ell[t]$ be a monic polynomial of even degree with $G(0) \neq 0$, such that for every root α of $G(t)$, q/α is also a root of $G(t)$. Then $G(t) = G_1(t)G_2(t)G_3(t)$ where*

- *$G_1(t)$ is a (possibly empty) product of powers of type 1 irreducibles; specifically:*
  - *if q is a square modulo ℓ, say $q = \alpha^2$ with $\alpha \in \overline{\mathbb{F}}_\ell^*$, then $G_1(t) = (t-\alpha)^{2j_-}(t+\alpha)^{2j_+}$ for some $j_+, j_- \geq 0$;*
  - *if q is a non-square modulo ℓ, then $G_1(t) = (t^2 - q)^j$ for some $j \geq 0$;*
- *G_2 is a (possibly empty) product of powers of type 2 irreducibles;*
- *G_3 is a (possibly empty) product of powers of polynomials of the form $P(t)Q(t)$ where $P(t)$ and $Q(t)$ are type 3 irreducibles of the same degree with disjoint root sets such that q/α is a root of $Q(t)$ whenever α is a root of $P(t)$.*

Proof For any root α of $G(t)$, let $P_\alpha(t)$ denote the minimal polynomial of α. Let $G_1(t)$ and $G_2(t)$ denote the product of all type 1 and type 2 irreducible factors of $G(t)$, respectively, and set $G_3(t) = G(t)/G_1(t)G_2(t)$. Then $G_3(t)$ consists of type 3 irreducible factors of $G(t)$ only. Moreover, every root α of $G_1(t)$ satisfies $\alpha = q/\alpha$, every root of $G_2(t)$ satisfies $\alpha \neq q/\alpha$ and $P_\alpha(t) = P_{q/\alpha}(t)$, and every root of $G_3(t)$ satisfies $\alpha \neq q/\alpha$ and $P_\alpha(t) \neq P_{q/\alpha}(t)$.

Let α be any root of $G_3(t)$, and set $d = \deg(P_\alpha)$. Then $\alpha_i = \alpha^{\ell^i}$, $0 \leq i \leq d-1$, are the roots of $P_\alpha(t)$; see the proof of Lemma 5.1. Therefore, $q/\alpha_i = (q/\alpha)^{\ell^i}$, $0 \leq i \leq d-1$, are all the roots of $P_{q/\alpha}(t)$, so $\deg(P_{q/\alpha}) = d$. Moreover, no α_i is a root of $P_{q/\alpha}(t)$, otherwise $P_\alpha(t) = P_{\alpha_i}(t) = P_{q/\alpha}(t)$. Hence $G_3(t)$ is of the form described above.

Now $\deg(G_3)$ is even, and by Lemma 5.1, $\deg(G_2)$ is also even. Thus, $\deg(G_1)$ must be even. Every root of $G_1(t)$ is a square root of q. If q is a non-square modulo ℓ, then $P_\alpha(t) = t^2 - q$, so $G_1(t)$ is as specified above. If q is a square modulo ℓ, say $q = \alpha^2$ with $\alpha \in \mathbb{F}_{\ell^d}^*$, then $P_\alpha(t) = t \pm a$. Thus, $G_1(t) = (t-a)^{n_-}(t+a)^{n_+}$ with $n_-, n_+ \geq 0$. Then $n_- + n_+ = \deg(G_1)$ is even. Now

$$G_1(0) = \frac{G(0)}{G_2(0)G_3(0)} = q^{(\deg(G)-\deg(G_2)-\deg(G_3))/2} = q^{\deg(G_1)/2} = \alpha^{\deg(G_1)} .$$

On the other hand, $G_1(0) = (-\alpha)^{n_-}\alpha^{n_+} = (-1)^{n_-}\alpha^{n_-+n_+} = (-1)^{n_-}\alpha^{\deg(G_1)}$. It follows that n_- is even, and hence n_+ is also even. □

Note that it is easy, given the irreducible factors of $G(t)$, to determine which are of type 2, as any such factor $P(t) = t^d + a_{d-1}t^{d-1} + \cdots + a_1 t + a_0$ satisfies $a_0 = q^d$ and $a_i = a_{d-i}q^{d-i}$ for $1 \leq i \leq d/2 - 1$.

The above results lead to a bound on b, with b as given in Proposition 3.2 and Theorem 4.3, and hence on n_l, that can be read solely from the factorization of $F(t)$ over $\mathbb{F}_\ell$. Hence, one can avoid computing any matrix orders.

Theorem 5.5 *Let $L(t)$ be the L-polynomial of a function field $\mathbb{K}/\mathbb{F}_q$ of genus g, and set $F(t) \equiv t^{2g}L(t^{-1}) \pmod{\ell}$, $F(t) \in \mathbb{F}_\ell[t]$. Set $b = \operatorname{ord}(A_F)$, where A_F is the companion matrix of $F(t)$. For $i = 1, 2, 3$, put $l_i = 1$ whenever $F(t)$ has no type i irreducible factor; otherwise, put $l_i = \ell^{\lceil \log_\ell(m_i) \rceil} N_i$, where*

$$\begin{aligned}
m_i &= \max\{v_P(F) \mid P \textit{ is a type } i \textit{ irreducible factor of } F\} \quad (i = 1,2,3), \\
N_1 &= \begin{cases} \operatorname{ord}_\ell(\alpha) & \textit{if } q = \alpha^2 \textit{ is a square modulo } \ell \\ & \textit{and } v_{t+\alpha}(F) = 0, \\ \operatorname{ord}_\ell(-\alpha) & \textit{if } q = \alpha^2 \textit{ is a square modulo } \ell \\ & \textit{and } v_{t-\alpha}(F) = 0, \\ \max\{\operatorname{ord}_\ell(\alpha), \operatorname{ord}_\ell(-\alpha)\} & \textit{if } q = \alpha^2 \textit{ is a square modulo } \ell \\ & \textit{and } v_{t+\alpha}(F)\cdot v_{t-\alpha}(F) \neq 0, \\ 2\operatorname{ord}_\ell(q) & \textit{if } q \textit{ is a non-square modulo } \ell, \end{cases} \\
N_2 &= \operatorname{ord}_\ell(q)\operatorname{lcm}(\ell^{\deg(P)/2} + 1 \mid P \textit{ is a type 2 irreducible factor of } F), \\
N_3 &= \operatorname{lcm}(\ell^{\deg(P)} - 1 \mid P \textit{ is a type 3 irreducible factor of } F).
\end{aligned}$$

Then b divides $\operatorname{lcm}(l_1, l_2, l_3)$.

Proof Assume $l_1, l_2, l_3 > 1$. From the proof of Theorem 4.3, we obtained $b = \operatorname{lcm}(\ell^{\lceil \log_\ell(v_P(F)) \rceil}\operatorname{ord}(A_P))$ where the lcm runs over all the irreducible factors P of F. Thus, $b = \operatorname{lcm}(b_1, b_2, b_3)$ where

$$b_i = \operatorname{lcm}(\ell^{\lceil \log_\ell(v_P(F)) \rceil}\operatorname{ord}(A_P) \mid P \text{ is a type } i \text{ irreducible factor of } F)$$

for $i = 1, 2, 3$. As before, the power of ℓ in the lcm is just the largest power of ℓ occurring in any term. Furthermore, for any irreducible polynomial $P(t)$ and any root α of $P(t)$, $\operatorname{ord}(A_P) = \operatorname{ord}_\ell(\alpha)$ divides $|\mathbb{F}_{\ell^{\deg(P)}}^*| = \ell^{\deg(P)} - 1$ if $P(t)$ is of type 3, and $\operatorname{ord}(A_P) \mid (\ell-1)(\ell^{\deg(P)/2}+1)$ if $P(t)$ is of type 2 by Corollary 5.2. It follows that $b_3 \mid l_3$ and $b_2 \mid l_2$.

We use Proposition 5.4 to analyze b_1. Suppose first that $q = \alpha^2$ is a square in $\mathbb{F}_\ell^*$. If $v_{t+\alpha}(F) = 0$, then the only type 1 irreducible factor of $F(t)$ is $P(t) = t - \alpha$ with companion matrix $A_P = (\alpha)$. So by Lemma 4.2, $b_1 = \ell^{\lceil \log_\ell(v_{t-\alpha}(F)) \rceil} \operatorname{ord}_\ell(\alpha) = l_1$. If $v_{t+\alpha}(F) \neq 0$, then set $n_\alpha = \operatorname{ord}_\ell(\alpha)$, $n_{-\alpha} = \operatorname{ord}_\ell(-\alpha)$, and assume without loss of generality that $n_{-\alpha} \geq n_\alpha$. Then $n_{-\alpha} = n_\alpha$ or $n_{-\alpha} = 2n_\alpha$, so $\operatorname{lcm}(n_{-\alpha}, n_\alpha) = n_{-\alpha} = \max\{n_\alpha, n_{-\alpha}\}$. It follows again by Lemma 4.2 that

$$b_1 = \operatorname{lcm}(\ell^{\lceil \log_\ell(v_{t-\alpha}(F)) \rceil} n_\alpha\, , \ell^{\lceil \log_\ell(v_{t+\alpha}(F)) \rceil} n_{-\alpha}) = \ell^{m_1} \max\{n_\alpha, n_{-\alpha}\} = l_1 \ .$$

Finally, if q is a non-square modulo ℓ, then A_{t^2-q} has eigenvalues $\pm\sqrt{q}$, both of order $2 \operatorname{ord}_\ell(q)$. Thus, again $b_1 = l_1$. □

Corollary 5.6 *With the notation of Theorem 5.5, b divides $\operatorname{lcm}(l_1', l_2', l_3)$, where for $i = 1, 2$, $l_i' = 1$ if $F(t)$ has no type i irreducible factors; else*

$$l_1' = \begin{cases} \ell^{\lceil \log_\ell(m_1) \rceil}(\ell - 1) & \text{if } q \text{ is a square modulo } \ell\,, \\ 2\,\ell^{\lceil \log_\ell(m_1) \rceil}(\ell - 1) & \text{if } q \text{ is a non-square modulo } \ell\,, \end{cases}$$

$$l_2' = \ell^{\lceil \log_\ell(m_2) \rceil}(\ell - 1) \operatorname{lcm}(\ell^{\deg(P)/2} + 1 \mid P \text{ is a type 2 irreducible factor of } F)\,.$$

Proof This follows from the simple fact that the order of every element in $\mathbb{F}_\ell^*$ divides $\ell - 1$. □

The bound on b in Theorem 5.5 can be sharp:

Example 5.7 We revisit Example 4.4. We have $\ell = 3$ and $q = 179 \equiv -1 \pmod 3$, so q is a non-square modulo ℓ that has order 2. $P_3(t) = t^2 + 1$ is the only type 1 factor of $F(t)$ and $k = v_{P_3}(F) = 2$. So $l_1 = l_1' = 2 \cdot 3 \cdot 2 = 12$. $P_4(t)$ is the only type 2 factor of $F(t)$, so $l_2 = l_2' = 2 \cdot (3^1 + 1) = 8$. Finally, $P_1(t) = t + 1$ and $P_2(t) = t + 2$ form a pair of type 3 factors, so $l_3 = 3 - 1 = 2$. Hence, $b \mid \operatorname{lcm}(12, 8, 2) = 24$, and in fact $b = 24$ from Example 4.4.

Once again, we state a simple special case:

Corollary 5.8 *Let $L(t)$ be the L-polynomial of a function field $\mathbb{K}/\mathbb{F}_q$ of genus g, and set $F(t) \equiv t^{2g}L(t^{-1}) \pmod \ell$, $F(t) \in \mathbb{F}_\ell[t]$. Suppose that $F(t)$ is irreducible over $\mathbb{F}_\ell$, and let A_F be the companion matrix of $F(t)$. Then*

$$\operatorname{ord}(A_F) \mid \operatorname{ord}_\ell(q)(\ell^g + 1) \mid (\ell - 1)(\ell^g + 1) \ .$$

Proof This follows immediately from Theorem 5.5 or Corollary 5.2, since $F(t)$ is a type 2 polynomial. □

6 Genus 1 and 2 curves

In the case of elliptic (i.e. genus 1) curves, it is well-known that $\operatorname{ord}_\ell(q) \mid n_\ell$, which gives a lower bound on n_ℓ. In fact, for $q \not\equiv 1 \pmod \ell$, the Balasubramanian-Koblitz Theorem [3] states that $\mathcal{J}[\ell] \subseteq \mathbb{F}_{q^n}$ if and only if $\operatorname{ord}_\ell(q) \mid n$. Proposition 6.1 below is a slightly more precise statement than Lemma 2.1 of [6].

Proposition 6.1 *Let $L(t)$ be the L-polynomial of an elliptic function field $\mathbb{K}/\mathbb{F}_q$, and set $F(t) = t^2L(t^{-1}) \pmod \ell$, $F(t) \in \mathbb{F}_\ell[t]$. Let n_ℓ be the degree of the field of rationality of $\mathcal{J}[\ell]$ over $\mathbb{F}_q$. Then the following hold:*

1. *If $F(t) = (t - \alpha)(t - q/\alpha)$ splits into two distinct linear factors in $\mathbb{F}_\ell[t]$, then $n_\ell = \operatorname{ord}_\ell(\alpha) \mid \ell - 1$.*
2. *If $F(t)$ is irreducible in $\mathbb{F}_\ell[t]$, then $n_\ell \mid \ell^2 - 1$.*

3. *If* $F(t) = (t-\alpha)^2$ *is the square of a linear factor in* $\mathbb{F}_\ell[t]$, *then* $n_\ell \mid \ell \operatorname{ord}_\ell(\alpha) \mid \ell(\ell-1)$. *This can only occur if* q *is a square modulo* ℓ.

Proof This is immediate from Corollaries 4.5 and 5.8. Note also that if $F(t) = (t-\alpha)^2$ with $\alpha \in \mathbb{F}_\ell$, then $F(0) = \alpha^2 = q$, so q must be a square modulo ℓ. □

We point out that $F(t)$ is a product of 2 (linear) type 3 polynomials in case 1, a type 2 polynomial in case 2, and a square of a type 1 polynomial in case 3. So the above results could also have been obtained through Theorem 5.5 and Corollary 5.6. We also note that every factorization described above can happen, as evidenced by the example below. Here, $q = 5$ and $\ell = 11$. The first column in the table below lists a cubic polynomial $D(x) \in \mathbb{F}_5[x]$ so that $y^2 = D(x)$ is an elliptic curve over $\mathbb{F}_5$. The second column provides the factorization of $F(t) \equiv t^2 L(t^{-1}) \pmod{11}$, $F(t) \in \mathbb{F}_{11}[t]$, into monic irreducibles over $\mathbb{F}_{11}$. The third column specifies which of the cases in Proposition 6.1 this factorization corresponds to.

$D(x) \in \mathbb{F}_5[x]$	Factorization of $F(t)$ over $\mathbb{F}_{11}$	Case in Prop. 6.1
$x^3 + 2x + 1$	$(t+3)(t+9)$	1
$x^3 + 1$	$t^2 + 5$	2
$x^3 + x + 1$	$(t+7)^2$	3

For genus 2 curves, bounds on the field of rationality of the ℓ-torsion can be found in [7, Proposition 6.2]. That source assumes that the function field has complex multiplication by the ring of integers of a quartic CM field. Our result has no such restriction and is an improvement on [7] in some cases.

Proposition 6.2 *Let* $L(t)$ *be the L-polynomial of a function field* $\mathbb{K}/\mathbb{F}_q$ *of genus* 2, *and set* $F(t) = t^4 L(t^{-1}) \pmod{\ell}$, $F(t) \in \mathbb{F}_\ell[t]$. *Let* n_ℓ *be the degree of the field of rationality of* $\mathcal{J}[\ell]$ *over* $\mathbb{F}_q$.

1. *Suppose first that* $F(t)$ *is square-free.*
   (a) *If* $F(t)$ *splits into four (distinct) linear factors in* $\mathbb{F}_\ell[t]$, *then*
   $$n_\ell = \operatorname{lcm}(\operatorname{ord}_\ell(\alpha_i)) \mid \ell - 1 \ ,$$
   where the $\alpha_i \in \mathbb{F}_\ell$ *run through all the roots of* F.
   (b) *If* $F(t)$ *splits into either two or three distinct irreducible factors in* $\mathbb{F}_\ell[t]$, *then* $n_\ell \mid \ell^2 - 1$.
   (c) *If* $F(t)$ *is irreducible in* $\mathbb{F}_\ell[t]$, *then* $n_\ell \mid \operatorname{ord}_\ell(q)(\ell^2+1) \mid (\ell-1)(\ell^2+1)$.
2. *Suppose now that* $F(t)$ *is not square-free.*
   (a) *If* $F(t)$ *has a quadratic irreducible factor in* $\mathbb{F}_\ell[t]$, *then* $n_\ell \mid \ell(\ell^2-1)$.
   (b) *If* $F(t)$ *is the fourth power of a linear factor in* $\mathbb{F}_\ell[t]$, *then for the root* α *of* $F(t)$,
   $$n_\ell \mid \ell^{\lceil \log_\ell(4) \rceil} \operatorname{ord}_\ell(\alpha) \mid \begin{cases} \ell^2(\ell-1) & \text{if } \ell = 2 \text{ or } \ell = 3\,, \\ \ell(\ell-1) & \text{if } \ell \geq 5\,. \end{cases}$$
   This can only occur when q *is a square modulo* ℓ.
   (c) *In every other case, we have*
   $$n_\ell \mid \operatorname{lcm}(\operatorname{ord}_\ell(\alpha_i)) \mid \ell(\ell-1),$$
   where the $\alpha_i \in \mathbb{F}_\ell$ *run through all the roots of* F.

Proof First note that $F(t)$ cannot have an irreducible factor of degree 3, as such a factor would be a type 2 factor, which cannot have odd degree by Lemma

5.1. Neither can $F(t)$ split into a cube of a linear factor and a second different linear factor. To see this, suppose that three of the roots of $F(t)$ in $\mathbb{F}_\ell^*$ are identical, say $\alpha = \beta = q/\alpha$. Then $\alpha = \beta$ implies $q/\beta = q/\alpha$, so all four roots must be identical. Hence the list above exhausts all possible cases.

The claim for irreducible $F(t)$ follows from Corollary 5.8. The other results are a consequence of Theorem 4.3, or of Theorem 5.5 and Corollary 5.6. Note again that if $F(t) = (t-\alpha)^4$ with $\alpha \in \mathbb{F}_\ell$, then the coefficient of t in $F(t)$ is equal to q times that of t^3. Thus, $-4\alpha q = -4\alpha^3$, which implies that $q = \alpha^2$ must be a square modulo ℓ. □

Once again, each of the above factorizations can occur: in the example below, $q = 11$ and $\ell = 7$. The first column lists a polynomial $D(x) \in \mathbb{F}_{11}[x]$ of degree 5 so that $y^2 = D(x)$ is a hyperelliptic curve of genus 2 over $\mathbb{F}_{11}$. The second and third columns are analogous to those of the previous table, with column 3 specifying which of the cases in Proposition 6.2 the factorization case of $F(t)$ over $\mathbb{F}_7$ corresponds to.

$D(x) \in \mathbb{F}_{11}[x]$	Factorization of $F(t)$ over $\mathbb{F}_7$	Case in Prop. 6.2
$x^5+x^4+x^3+x^2+6x+10$	$(t+1)(t+3)(t+4)(t+6)$	1 (a)
$x^5+x^4+x^3+x^2+x+6$	$(t^2+t+3)(t^2+6t+3)$	1 (b)
$x^5+x^4+x^3+x^2+x+3$	$(t+3)(t+6)(t^2+6t+4)$	1 (b)
$x^5+x^4+x^3+x^2+x+1$	$t^4+4t^3+6t^2+2t+2$	1 (c)
$x^5+x^4+x^3+x^2+x+2$	$(t+5)^2(t^2+6t+4)$	2 (a)
$x^5+x^4+x^3+x^2+3x+8$	$(t^2+t+4)^2$	2 (a)
$x^5+x^4+x^3+x^2+x$	$(t+2)^2(t+5)^2$	2 (b)
$x^5+x^4+x^3+x^2+x+10$	$(t+2)^2(t+3)(t+6)$	2 (b)
$x^5+x^4+x^3+x^2+x+4$	$(t+5)^4$	2 (c)

We point out that the last line of the above table represents an example where Theorem 4.3 (and hence Theorem 5.5 as well) do not give a sharp bound on n_ℓ. Both theorems yield the multiple 42 of n_7, whereas computations of the ℓ-rank of the Jacobian in Magma reveal that $n_7 = 21$.

7 Incremental increases in ℓ-rank

Rather than achieving full ℓ-rank $2g$, we now turn to the question of incremental increases in ℓ-rank that was already addressed in Proposition 3.4. We provide an improvement to that proposition. In particular, we will see how to achieve multiple such increases over intermediate base fields $\mathbb{F}_q \subseteq \mathbb{F}_{q^n} \subseteq \mathbb{F}_{q^{n\ell}}$. We begin with some useful preliminary results.

Lemma 7.1 *Let ℓ be a prime, $\lambda \in \overline{\mathbb{F}}_\ell^*$, and $m \in \mathbb{N}$ with $m \geq 2$. Then for any $n \in \mathbb{N}$, the eigenspace of the Jordan matrix $J_{m,\lambda}^n \in \mathrm{GL}_m(\overline{\mathbb{F}}_\ell)$ has dimension 1 if and only if $\ell \nmid n$.*

Proof The matrix $J_{m,\lambda}^n$ has λ^n as its only eigenvalue, and the corresponding eigenspace is the null space of the matrix $J_{m,\lambda}^n - \lambda^n I_m$. This matrix is of the form $\begin{pmatrix} 0 & A \\ 0 & 0 \end{pmatrix}$, where A is an $(m-1)\times(m-1)$ matrix whose i-th row $(1 \leq i \leq m-1)$ has the form

$$\begin{pmatrix} 0 & 0 & \dots & 0 & \binom{n}{1}\lambda^{n-1} & \binom{n}{2}\lambda^{n-2} & \cdots & \binom{n}{m-i}\lambda^{n-m+i} \end{pmatrix}.$$

Here, for ease of notation, we adopt the convention $\binom{n}{j} = 0$ for $j > n$. Thus, A is an upper triangular matrix of determinant $(n\lambda^{n-1})^{m-1}$. So A has maximal rank (i.e. trivial null space) if and only if $\ell \nmid n$, and this holds exactly when the null space of $J^n_{m,\lambda} - \lambda^n I_m$ has dimension 1. □

Corollary 7.2 *With the notation of Lemma 7.1, if $\ell \nmid n$ and $\mathrm{ord}_\ell(\lambda) \mid n$, then $J^n_{m,\lambda}$ is similar to $J_{m,1}$.*

Lemma 7.3 *Let $P \in \mathbb{F}_\ell[t]$ be an irreducible polynomial, $m \in \mathbb{N}$, and A_P, A_{P^m} the companion matrices of P and P^m, respectively. Then for any multiple n of $\mathrm{ord}(A_P)$, the matrix $A^n_{P^m}$ has 1 as its only eigenvalue. The corresponding eigenspace contains a direct sum of $\deg(P)$ one-dimensional spaces, and is equal to this direct sum if $m = 1$ or $\ell \nmid n$.*

Proof For brevity, set $d = \deg(P)$. Then over $\overline{\mathbb{F}}_\ell$, A^n_P is similar to the $d \times d$ identity matrix I_d, and thus has 1 as its only eigenvalue, with d linearly independent eigenvectors. Since A^n_P and I_d have entries in $\mathbb{F}_\ell$, so does the kernel of their difference $\ker(A^n_P - I_d)$. Thus, $\dim_{\mathbb{F}_\ell} \ker(A^n_P - I_d) = d$ and the result holds for $m = 1$.

Now suppose $m \geq 2$. Then over $\overline{\mathbb{F}}_\ell$, A_{P^m} is similar to its Jordan canonical form $J_{A_{P^m}} = \mathrm{diag}(J_{m,\lambda_1}, J_{m,\lambda_2}, \dots, J_{m,\lambda_d})$, where $\lambda_1, \lambda_2, \dots, \lambda_d \in \overline{\mathbb{F}}_\ell$ are the roots of $P(t)$. So A_{P^m} has d distinct eigenvalues $\lambda_j \in \overline{\mathbb{F}}_\ell$, $1 \leq j \leq d$, each of which corresponds to a one-dimensional eigenspace W_j over $\overline{\mathbb{F}}_\ell$. Since $\mathrm{ord}_\ell(\lambda_j) = \mathrm{ord}(A_P)$ divides n for $1 \leq j \leq d$, each power J^n_{m,λ_j} has 1 as its only eigenvalue. It follows that the only eigenvalue of $A^n_{P^m}$ is 1, and $A^n_{P^m}$ acts trivially on each W_j for $1 \leq j \leq d$. Therefore, $\dim_{\overline{\mathbb{F}}_\ell} \ker(A^n_{P^m} - I_{md}) \geq d$, where I_{md} is the $md \times md$ identity matrix. Since $A^n_{P^m}$ and I_{md} have entries in $\mathbb{F}_\ell$, so does $\ker(A^n_{P^m} - I_{md})$. Thus, $\dim_{\mathbb{F}_\ell} \ker(A^n_{P^m} - I_{md}) \geq d$, and the eigenspace of $A^n_{P^m}$ corresponding to 1 contains the direct sum $W = \bigoplus_{j=1}^d W_j$.

If $\ell \nmid n$, then Corollary 7.2 implies that J^n_{m,λ_j} is similar to $J_{m,1}$ for all $1 \leq j \leq d$. So over $\overline{\mathbb{F}}_\ell$, $A^n_{P^m}$ is similar to the matrix $\mathrm{diag}(J_{m,1}, J_{m,1}, \dots, J_{m,1})$, and thus has 1 as its only eigenvalue, with d Jordan blocks. Therefore, $\dim_{\overline{\mathbb{F}}_\ell} \ker(A^n_{P^m} - I_{md}) = d$. Again, since $A^n_{P^m}$ and I_{md} both have entries in $\mathbb{F}_\ell$, we have $\dim_{\mathbb{F}_\ell} \ker(A^n_{P^m} - I_{md}) = d$, and the eigenspace of $A^n_{P^m}$ corresponding to the only eigenvalue 1 is equal to W. □

We can now ascertain lower bounds on the ℓ-rank of $\mathcal{J}_n = \mathrm{Jac}_{\mathbb{K}_n}(\mathbb{F}_{q^n})$ for any $n \in \{1, 2, \dots, n_\ell\}$, and sometimes even the exact ℓ-rank.

Theorem 7.4 *Let $L(t)$ be the L-polynomial of a function field $\mathbb{K}/\mathbb{F}_q$ of genus g, and set $F(t) \equiv t^{2g} L(t^{-1}) \pmod{\ell}$, $F(t) \in \mathbb{F}_\ell[t]$. Let $P_1, P_2, \cdots, P_s \in \mathbb{F}_\ell[t]$ be the collection of distinct monic irreducible factors of $F(t)$, with respective companion matrices A_{P_i}, $1 \leq i \leq s$. Let $n \in \mathbb{N}$, and set*

$$\mathcal{I}_n = \{i \mid 1 \leq i \leq s \text{ and } \mathrm{ord}(A_{P_i}) \text{ divides } n\} \ .$$

Then the following hold:

1. ℓ-rank$(\mathcal{J}_n) \geq \sum_{i \in \mathcal{I}_n} \deg(P_i)$, *with equality if $v_{P_i}(F) = 1$ for all $i \in \mathcal{I}_n$.*
2. ℓ-rank$(\mathcal{J}_1) \geq 1$ *if and only if* $t - 1 \mid F(t)$. *Moreover,* ℓ-rank$(\mathcal{J}_1) = 1$ *if* $v_{t-1}(F) = 1$.

3. *Suppose* $\mathrm{ord}(A_{P_i}) \nmid n$ *for all* $1 \le i \le s$ *such that* $P_i(t) \ne t-1$. *If* $v_{t-1}(F) \le 1$ *or* $\ell \nmid n$, *then* $\ell\text{-rank}(\mathcal{J}_n) = \ell\text{-rank}(\mathcal{J}_1)$.

Proof We note that an element of $\mathcal{J}[\ell]$ is defined over $\mathbb{F}_{q^n}$, i.e. belongs to $\mathcal{J}_n[\ell]$, if and only if it is fixed by the Frobenius π_{q,n_ℓ}^n acting on $\mathcal{J}[\ell]$.

Let $P_i^{m_{ij}}$ ($1 \le i \le s$, $1 \le j \le k_i$) be the elementary divisors of π_{q,n_ℓ}. Then the primary rational canonical form of π_{q,n_ℓ} is $A_{\pi_{q,n_\ell}} = \mathrm{diag}(A_{P_i^{m_{ij}}})$. Thus, the n-th power π_{q,n_ℓ}^n of π_{q,n_ℓ} has a matrix representation $A_{\pi_{q,n_\ell}}^n = \mathrm{diag}(A_{P_i^{m_{ij}}}^n)$. By Lemma 7.3, for $i \in \mathcal{I}_n$, $A_{P_i^{m_{ij}}}^n$ has 1 as its only eigenvalue, and the corresponding eigenspace W_{ij} contains a subspace that is isomorphic to $\mathbb{F}_\ell^{\deg(P_i)}$. It follows that π_{q,n_ℓ}^n has 1 as an eigenvalue, and the corresponding eigenspace $W = \bigoplus_{i,j} W_{ij}$ contains a subspace that is $\mathbb{F}_\ell$-isomorphic to $\mathbb{F}_\ell^{d_n}$, where $d_n = \sum_{i\in\mathcal{I}_n} \deg(P_i)$. Moreover, the eigenvalue 1 results in a trivial action of π_{q,n_ℓ}^n on W, so $w^{q^n} = w$ for all $w \in W$. Hence, elements in W must be defined over $\mathbb{F}_{q^n}$, implying that $W \subseteq \mathcal{J}_n[\ell]$. Thus, $\ell\text{-rank}(\mathcal{J}_n) \ge d_n$, yielding the inequality of part 1.

To obtain equality in the case when $F(t)$ is square-free, note that $k_i = m_{i1} = 1$ for $1 \le i \le s$. Thus, each W_{i1} is isomorphic to $\mathbb{F}_\ell^{\deg(P_i)}$ by Lemma 7.3, and hence $W \cong \mathbb{F}_\ell^{d_n}$. Moreover, for $i \notin \mathcal{I}_n$, the matrix $A_{P_i^{m_{i1}}}^n$ has only eigenvalues distinct from 1, so π_{q,n_ℓ}^n does not act trivially on any element outside W. It follows that $\ell\text{-rank}(\mathcal{J}_n) = d_n$.

For part 2 of Theorem 7.4, if $t-1 \mid F(t)$, then $\ell\text{-rank}(\mathcal{J}_1) \ge \deg(t-1) = 1$ by part 1 of the theorem applied to $n = 1$. Conversely, if $\ell\text{-rank}(\mathcal{J}_1) \ge 1$, then $t-1 \mid F(t)$ by Proposition 3.3 (with $r = 1$). Similarly, $v_{t-1}(F) = 1$ implies $t-1 \mid F(t)$, and hence $\ell\text{-rank}(\mathcal{J}_1) \ge 1$. Then the same proposition yields $\ell\text{-rank}(\mathcal{J}_1) = 1$.

For part 3, suppose first that $t-1 \nmid F(t)$. Then neither π_{q,n_ℓ} nor π_{q,n_ℓ}^n has 1 as an eigenvalue, so neither map acts trivially on any non-zero element of $\mathcal{J}[\ell]$. Thus, $\ell\text{-rank}(\mathcal{J}_1) = \ell\text{-rank}(\mathcal{J}_n) = 0$.

Now assume that $t-1 \mid F(t)$. Then $\mathcal{I}_n = \mathcal{I}_1$. If $v_{t-1}(F) = 1$, then by parts 1 and 2 of the theorem, $\ell\text{-rank}(\mathcal{J}_n) = \ell\text{-rank}(\mathcal{J}_1) = 1$. If $(t-1)^2 \mid F(t)$, then let $(t-1)^{m_j}$, $1 \le j \le k$, be the elementary divisors of π_{q,n_ℓ} corresponding to the factor $t-1$ of $F(t)$. Then $A_{\pi_{q,n_\ell}}$ is similar to $\mathrm{diag}(J_{m_1,1}, \ldots, J_{m_k,1}, B)$, where B is a matrix whose eigenvalues are all distinct from 1. By Corollary 7.2, A_{π_{q,n_ℓ}^n} is similar to $\mathrm{diag}(J_{m_1,1}, \ldots, J_{m_k,1}, C)$, where C also has only eigenvalues distinct from 1. It follows that π_{q,n_ℓ} and π_{q,n_ℓ}^n act trivially on k-dimensional subspaces of $\mathcal{J}_1[\ell]$ and $\mathcal{J}_n[\ell]$, respectively, but do not act trivially on any elements outside these respective subspaces. So $\ell\text{-rank}(\mathcal{J}_n) = \ell\text{-rank}(\mathcal{J}_1) = k$. □

For brevity, we henceforth refer to the tuple of ℓ-ranks of the Jacobians $\mathcal{J}_n$, $1 \le n \le n_\ell$, as the *ℓ-rank structure* of the extension $\mathbb{K}/\mathbb{F}_q$. The results of Theorem 7.4, combined with Theorem 4.3 (or Theorem 5.5) can sometimes yield the entire ℓ-rank structure of a given function field. Before we illustrate this, we discuss a particularly simple case.

Corollary 7.5 *Let* $L(t)$ *be the* L*-polynomial of a function field* $\mathbb{K}/\mathbb{F}_q$ *of genus* g, *set* $F(t) \equiv t^{2g}L(t^{-1}) \pmod{\ell}$, $F(t) \in \mathbb{F}_\ell[t]$, *and let* A_F *be the companion matrix of* $F(t)$. *If* $F(t)$ *is irreducible, then* $\ell\text{-rank}(\mathcal{J}_{\mathrm{ord}(A_F)}) = 2g$ *and* $\ell\text{-rank}(\mathcal{J}_k) = 0$ *for any* $k < \mathrm{ord}(A_F)$.

Proof ℓ-rank$(\mathcal{J}_{\mathrm{ord}(A_F)}) = \deg(F) = 2g$ by part 1 of Theorem 7.4. Since $F(t)$ is irreducible and $\deg(F)$ is even, we have $t-1 \nmid F(t)$, so ℓ-rank$(\mathcal{J}_k) = \ell$-rank$(\mathcal{J}_1) = 0$ for any $k < \mathrm{ord}(A_F)$ by parts 3 and 2 of Theorem 7.4. □

We revisit Example 5.4 of [2] to apply this corollary:

Example 7.6 Consider $q = 373$, the hyperelliptic function field $\mathbb{K} = \mathbb{F}_{373}(x, y)$ of genus 4 with

$$y^2 = x^9 + 245\,x^8 + 175\,x^7 + 340\,x^6 + 122\,x^5 + 70\,x^4 + 196\,x^3 + 210\,x^2 + 316\,x + 337\ ,$$

and $\ell = 3$. With the aid of Magma, we found that $\zeta(t) = L(t)/(373\,t^2 - 374\,t + 1)$ is the zeta function of $\mathbb{K}$ with

$$\begin{aligned} L(t) = {} & 373^4\,t^8 + 33 \cdot 373^3\,t^7 + 347 \cdot 373^2\,t^6 - 3785 \cdot 373\,t^5 - 188703\,t^4 \\ & - 3785\,t^3 + 347\,t^2 + 33\,t + 1\ . \end{aligned}$$

Then $F(t) = t^8 + 2\,t^6 + t^5 + t^3 + 2\,t^2 + 1$ is irreducible over $\mathbb{F}_3$, and $\mathrm{ord}(A_F) = 41$. By Corollary 7.5, ℓ-rank$(\mathcal{J}_{41}) = 8$, and ℓ-rank$(\mathcal{J}_k) = 0$ for $1 \le k \le 40$.

We now illustrate with two examples how our previous results can be employed to obtain the complete ℓ-rank structure of a function field $\mathbb{K}/\mathbb{F}_q$. In fact, knowing the L-polynomial of $\mathbb{K}$ will often yield this entire structure; at times, it is also necessary to find the ℓ-rank of $\mathcal{J}_n$ for one or a few very small values of n via direct computation, for example, using Magma. Generally, it is even possible to deduce the elementary divisors of π_{q,n_ℓ} using these techniques.

Example 7.7 We first revisit Example 5.5 of [2], which was already discussed in Examples 4.4 and 5.7. Here, $\mathbb{K} = \mathbb{F}(x, y)$ with y as given in Example 4.4 is a genus 4 hyperelliptic function field over $\mathbb{F}_{179}$. We wish to determine the 3-rank structure of $\mathbb{K}/\mathbb{F}_{179}$, i.e. the 3-rank of $\mathcal{J}_n$ with $1 \le n \le n_3$. Theorems 4.3 and 5.5 yield $n_3 \mid 24$.

Recall that $F = P_1 P_2 P_3^2 P_4 \in \mathbb{F}_3[t]$ where

$$P_1(t) = t + 1, \quad P_2(t) = t + 2, \quad P_3(t) = t^2 + 1, \quad P_4(t) = t^2 + t + 2$$

are all irreducible over $\mathbb{F}_3$, and $\mathrm{ord}(A_{P_1}) = 2$, $\mathrm{ord}(A_{P_2}) = 1$, $\mathrm{ord}(A_{P_3}) = 4$, $\mathrm{ord}(A_{P_4}) = 8$.

By part 2 of Theorem 7.4, 3-rank$(\mathcal{J}_1) = 1$, and by part 3 of the same theorem, 3-rank$(\mathcal{J}_n) = 1$ for all odd n. Using the notation of Theorem 7.4, we obtain the following sets $\mathcal{I}_n$:

n	1 (mod 2)	2 (mod 4)	4 (mod 8)	0 (mod 8)
$\mathcal{I}_n$	$\{2\}$	$\{1, 2\}$	$\{1, 2, 3\}$	$\{1, 2, 3, 4\}$

By part 1 of Theorem 7.4, 3-rank$(\mathcal{J}_n) = \deg(P_2) = 1$ for n odd (which we already observed), 3-rank$(\mathcal{J}_n) = \deg(P_1) + \deg(P_2) = 2$ for $n \equiv 2 \pmod 4$, 3-rank$(\mathcal{J}_n) \ge 4$ for $n \in \{4, 12, 20\}$, and 3-rank$(\mathcal{J}_n) \ge 6$ for $n \in \{8, 16, 24\}$. From Theorems 4.3 and 5.5, 3-rank$(\mathcal{J}_{24}) = 8$. This leaves the 3-ranks of $\mathcal{J}_n$ for $n = 4, 8, 12, 16$ and 20 only partially determined.

We first analyze $n = 12$; this will resolve all the ambiguous cases. The primary rational canonical form of π_{179,n_3} is one of

$$A_1 = \mathrm{diag}(A_{P_1}, A_{P_2}, A_{P_3}, A_{P_3}, A_{P_4}) \ \text{ or } \ A_2 = \mathrm{diag}(A_{P_1}, A_{P_2}, A_{P_3^2}, A_{P_4})\ .$$

By considering the orders of the diagonal sub-matrices in A_1 and A_2, is easy to see that $A_1^{12} = A_2^{12} = \mathrm{diag}(I_6, -I_2)$. It follows that π_{179,n_3}^{12} acts trivially on a subspace of $\mathcal{J}[3]$ of dimension exactly 6. So 3-rank$(\mathcal{J}_{12}) = 6$.

Note that $\mathbb{F}_{179^{12}}$ is the largest proper subfield of $\mathbb{F}_{179^{24}}$, so this implies that $n_3 = 24$. Since A_1 has order 8, it follows that the primary rational canonical form of π_{179,n_3} must be A_2.

Now $A_{P_1}^n = A_{P_2}^n = (1)$ for all n even, and both these matrices have a one-dimensional eigenspace. By Lemma 7.3, $A_{P_3^2}^n$ has 1 as its only eigenvalue, with a 2-dimensional eigenspace, for $n = 4, 8, 16, 20$. Since $\mathrm{ord}(A_{P_4}) = 8$, $A_{P_4}^4 = A_{P_4}^{20}$ has a double eigenvalue -1. A_2^4 and A_2^{20} have 1 as a 6-fold eigenvalue with a 4-dimensional eigenspace, and -1 as a double eigenvalue. Hence, π_{179,n_3}^4 acts trivially on a subspace of $\mathcal{J}[3]$ of dimension 4. Thus, 3-rank$(\mathcal{J}_4) =$ 3-rank$(\mathcal{J}_{20}) = 4$. Similarly, $A_{P_4}^8 = A_{P_4}^{16} = I_2$, so A_2^8 and A_2^{16} each have 1 as their only eigenvalue, with a 6-dimensional eigenspace. Thus, 3-rank$(\mathcal{J}_8) =$ 3-rank$(\mathcal{J}_{16}) = 6$. This yields the following 3-rank structure for $\mathbb{K}/\mathbb{F}_{179}$:

n	3-rank
odd	1
2, 6, 10, 14, 18, 22	2
4, 20	4
8, 12, 16	6
24	8

To check our results, we used Magma to compute the 3-Sylow subgroups of $\mathcal{J}_n$ for all proper divisors n of 24, and for a few other values of n, and found

$$\mathrm{Syl}_3(\mathcal{J}_1) \cong \mathrm{Syl}_3(\mathcal{J}_3) \cong \mathbb{Z}/9,$$
$$\mathrm{Syl}_3(\mathcal{J}_2) \cong \mathrm{Syl}_3(\mathcal{J}_{10}) \cong (\mathbb{Z}/9)^2, \quad \mathrm{Syl}_3(\mathcal{J}_6) \cong (\mathbb{Z}/27)^2,$$
$$\mathrm{Syl}_3(\mathcal{J}_4) \cong \mathrm{Syl}_3(\mathcal{J}_{20}) \cong (\mathbb{Z}/9)^4,$$
$$\mathrm{Syl}_3(\mathcal{J}_8) \cong \mathrm{Syl}_3(\mathcal{J}_{16}) \cong (\mathbb{Z}/3)^2 \times (\mathbb{Z}/9)^4, \quad \mathrm{Syl}_3(\mathcal{J}_{12}) \cong (\mathbb{Z}/3)^2 \times (\mathbb{Z}/27)^4.$$

We were unable to compute $\mathrm{Syl}_3(J_{24})$ with Magma due to limited computer memory.

Example 7.8 Consider $q = 149$, the hyperelliptic function field $\mathbb{K} = \mathbb{F}_{149}(x, y)$ of genus 4 with

$$y^2 = x^9 + 43\,x^8 + 35\,x^7 + 11\,x^6 + 22\,x^5 + 38\,x^4 + 75\,x^3 + 28\,x^2 + 61\,x + 5\ ,$$

and $\ell = 5$. With the aid of Magma, we computed the L-polynomial $L(t)$ of $\mathbb{K}/\mathbb{F}_{149}$ and obtained $F(t) \equiv t^8 L(t^{-1}) \pmod 5$. We found that

$$F(t) = t^8 + 4t^7 + 3t^6 + 2t^5 + 2t^4 + 3t^3 + 3t^2 + t + 1 \in \mathbb{F}_5[t]$$

factors as $F = P_1 P_2 P_3^2 P_4 P_5$, where

$$P_1(t) = t + 4,\ P_2(t) = t + 1,\ P_3(t) = t + 3,\ P_4(t) = t^2 + t + 2,\ P_5 = t^2 + 2t + 3$$

are all irreducible over $\mathbb{F}_5$. One easily verifies that $\mathrm{ord}(A_{P_1}) = 1$, $\mathrm{ord}(A_{P_2}) = 2$, $\mathrm{ord}(A_{P_3}) = 4$, and $\mathrm{ord}(A_{P_4}) = \mathrm{ord}(A_{P_5}) = 24$. Both Theorems 4.3 and 5.5 imply $n_5 \mid 120$.

As in the previous example, we deduce that 5-rank$(\mathcal{J}_n) = 1$ for n odd and obtain

n	1 (mod 2)	2 (mod 4)	4, 8, 12, 16, 20 (mod 24)	0 (mod 24)
$\mathcal{I}_n$	$\{1\}$	$\{1,2\}$	$\{1,2,3\}$	$\{1,2,3,4,5\}$

By part 1 of Theorem 7.4, 5-rank$(\mathcal{J}_n) = 1$ for n odd, 5-rank$(\mathcal{J}_n) = 2$ for $n \equiv 2 \pmod 4$, 5-rank$(\mathcal{J}_n) \geq 3$ for $n \equiv 4, 8, 16, 20 \pmod{24}$, and 5-rank$(\mathcal{J}_n) \geq 7$ for $n \equiv 0 \pmod{24}$. Also, from Theorems 4.3 and 5.5, 5-rank$(\mathcal{J}_{120}) = 8$. This leaves the 5-ranks of $\mathcal{J}_n$ for $n \equiv 0 \pmod 4$ with $n < 120$ only partially determined.

The primary rational canonical form $A_{\pi_{149,n_5}}$ of π_{149,n_5} is one of

$$\begin{aligned} A_1 &= \operatorname{diag}(A_{P_1}, A_{P_2}, A_{P_3}, A_{P_3}, A_{P_4}, A_{P_5}) \text{ or} \\ A_2 &= \operatorname{diag}(A_{P_1}, A_{P_2}, A_{P_3^2}, A_{P_4}, A_{P_5}) \ . \end{aligned}$$

Hence, $n_5 = \operatorname{ord}(A_1) = 24$ or $n_5 = \operatorname{ord}(A_2) = 120$.

We consider multiples n of 20 less than 120; note that $\operatorname{ord}(A_{P_3^2}) = 20$ by Lemma 4.1. For such n, we have $A_1^n = A_2^n = \operatorname{diag}(I_4, A_{P_4}^n, A_{P_5}^n)$, and $A_{P_4}^n$ and $A_{P_5}^n$ do not have 1 as an eigenvalue. It follows that π_{149,n_5}^n acts trivially on a subspace of $\mathcal{J}[5]$ of dimension exactly 4. So 5-rank$(\mathcal{J}_n) = 4$ for these n.

Note that by Lemma 7.3, $A_{P_3^2}^4$ has 1 as its only eigenvalue, with a one-dimensional eigenspace. Also, $A_{P_4}^4$ and $A_{P_5}^4$ do not have 1 as an eigenvalue. Hence, 5-rank$(\mathcal{J}_4) = 4$ if $A_{\pi_{149,n_5}} = A_1$ and 5-rank$(\mathcal{J}_4) = 3$ if $A_{\pi_{149,n_5}} = A_2$. At this point, it is unclear how to resolve this ambiguity, so we resort to Magma to determine 5-rank$(\mathcal{J}_4)$. We obtain 5-rank$(\mathcal{J}_4) = 3$, so $A_{\pi_{149,n_5}} = A_2$ and $n_5 = 120$. This resolves the rest of the 5-rank structure completely as follows.

Since $7 \leq$ 5-rank$(\mathcal{J}_n) <$ 5-rank$(\mathcal{J}_{120}) = 8$ for $n \equiv 0 \pmod{24}$ and $n < 120$, we have 5-rank$(\mathcal{J}_n) = 7$ for such n. Also, for any multiple n of 4 not divisible by 20, $A_{P_3^2}^n$ has 1 as its only eigenvalue, with a one-dimensional eigenspace. So 5-rank$(\mathcal{J}_n) = 3$ for $n \equiv 0 \pmod 4$, $n \not\equiv 0 \pmod{20}$, $n \not\equiv 0 \pmod{24}$. This yields the following 5-rank structure for $\mathbb{K}/\mathbb{F}_{149}$:

Congruence class of n, $1 \leq n \leq 120$	5-rank
$n \equiv 1 \pmod 2$	1
$n \equiv 2 \pmod 4$	2
$n \equiv 0 \pmod 4$, $n \not\equiv 0 \pmod{20}$, $n \not\equiv 0 \pmod{24}$	3
$n \equiv 0 \pmod{20}$, $n < 120$	4
$n \equiv 0 \pmod{24}$, $n < 120$	7
$n = 120$	8

Again, we used Magma to compute $\mathcal{J}_n[5]$ for most of the divisors n of 120, and found

$$\begin{aligned} &\text{5-rank}(\mathcal{J}_1) = \text{5-rank}(\mathcal{J}_3) = \text{5-rank}(\mathcal{J}_5) = \text{5-rank}(\mathcal{J}_{15}) = 1, \\ &\text{5-rank}(\mathcal{J}_2) = \text{5-rank}(\mathcal{J}_6) = \text{5-rank}(\mathcal{J}_{10}) = \text{5-rank}(\mathcal{J}_{30}) = 2, \\ &\text{5-rank}(\mathcal{J}_4) = \text{5-rank}(\mathcal{J}_8) = \text{5-rank}(\mathcal{J}_{12}) = 3, \\ &\text{5-rank}(\mathcal{J}_{20}) = 4, \\ &\text{5-rank}(\mathcal{J}_{24}) = 7. \end{aligned}$$

We again could not compute 5-rank$(\mathcal{J}_n)$ for $n = 40, 60, 120$ using Magma because of limited memory.

8 Conclusion

For any function field $\mathbb{K}/\mathbb{F}_q$ of genus g, we have provided several tools for analyzing the ℓ-ranks of the Jacobians of constant field extensions $\mathbb{K}_n/\mathbb{F}_{q^n}$, with $n \in \mathbb{N}$ and ℓ a prime not dividing q. This includes the determination of the field of definition of the absolute ℓ-torsion, or at least an upper bound on the extension degree of this field over $\mathbb{F}_q$.

Our investigation extends the previous results of [2]. Of key importance is the factorization of $F(t) \in \mathbb{F}_\ell[t]$ into monic irreducibles over $\mathbb{F}_\ell$, where $F(t) \equiv t^{2g}L(t^{-1}) \pmod{\ell}$ and $L(t)$ is the L-polynomial $L(t)$ of $\mathbb{K}/\mathbb{F}_q$. Assuming that $L(t)$ can be computed, our method can often determine the entire ℓ-rank structure of $\mathbb{K}/\mathbb{F}_q$. This is always true if $F(t)$ is square-free, and also in some other cases when $F(t)$ only contains small and few powers. In fact, oftentimes, our approach is able to identify the elementary divisors of the q-th power Frobenius, which in turn yield the complete ℓ-rank structure. At times, when $F(t)$ is not square-free, it may additionally be necessary to determine the ℓ-rank of $\mathcal{J}_n$ for one or a few very small values of n.

Apart from the L-polynomial, all the ingredients of our method stem from basic algebra and linear algebra. Obtaining $L(t)$ can be difficult and represents the main practical and computational obstacle to this approach. Magma computes the zeta function, and hence the L-polynomial, of a hyperelliptic function field of moderate size reasonably efficiently. For small ℓ, it can also find the ℓ-Sylow subgroups of an Abelian group quite quickly. Thus, our approach is very suitable for not too large base fields $\mathbb{F}_q$ and small values of ℓ, and is especially fruitful for hyperelliptic curves. Unfortunately, it is unclear whether it is possible to obtain $F(t)$ without computing $L(t)$, or to ascertain the elementary divisors of the Frobenius via some other means.

Acknowledgments. This work was begun at the *WIN — Women in Numbers* workshop, held at the Banff International Research Station (BIRS) in Banff, Alberta (Canada) November 2-7, 2008. The authors are grateful to BIRS for hosting this workshop at this beautiful venue and providing a stimulating work environment. The first four authors are also indebted to the University of Calgary's Department of Mathematics and Statistics for hosting them during a visit to finish this work. Finally, we owe thanks to an anonymous referee for her or his careful proof-reading and constructive comments that lead to considerable improvements of this paper.

References

[1] Achter, J. *The distribution of class groups of function fields.* J. Pure Appl. Algebra **204** (2006), 316–333.

[2] Bauer, M., Jacobson Jr, J. M., Lee, Y. and Scheidler, R. *Construction of Hyperelliptic Function Fields of High Three-Rank.* Math. Comp. **77**(261) (2008), 503–530.

[3] Balasubramanian, R. and Koblitz, N. *The improbability that an elliptic curve has subexponential discrete log problem under the Menezes-Okamoto-Vanstone algorithm.* J. Cryptology **11** (1998), 141–145.

[4] Bosma, W., Cannon, J. and Playoust, C. *The Magma algebra system. I. The user language.* J. Symbolic Comput. **24** (1997), 235–265.

[5] Chavdarov, N. *The generic irreducibility of the numerator of the zeta function in a family of curves with large monodromy.* Duke Math. J. **87** (1997), 151–180.

[6] Charles, D. and Lauter, K. *Computing modular polynomials.* LMS J. Comput. Math. **8** (2005), 195–204.

[7] Freeman D. and Lauter, K. *Computing Endomorphism Rings of Jacobians of Genus 2 Curves over Finite Fields.* Symp. Algebraic Geometry and its Applications, World Scientific, 2008, pp. 29–66.

[8] Hungerford, T. W. *Algebra.* Springer, New York 1974.

[9] Kowalski, E. *The large sieve, monodromy and zeta functions of curves.* J. Reine Angew. Math. **601** (2006), 29–69.

[10] Lee, Y. and Pacelli, A. M. *Class groups of imaginary function fields: the inert case.* Proc. Amer. Math. Soc. **133** (2005), 2883–2889.
[11] Lee Y. and Pacelli, A. M. *Higher rank subgroups in the class groups of imaginary function fields.* J. Pure Applied Algebra **207** (2006), 51–62.
[12] Miret, J., Pujolàs, J. and Rio, A. *Bisection for genus 2 curves in odd characteristic.* Proc. Japan Academy – Series A **85** (2009), 55–61.
[13] Pacelli, A. M., *Abelian subgroups of any order in class groups of global function fields.* J. Number Theory **106** (2004), 26–49.
[14] Pacelli, A. M. *The prime at infinity and the rank of the class group of global function fields.* J. Number Theory **116** (2006), 311–323.
[15] Rosen, M. *Number Theory in Function Fields.* Springer, Berlin 2002.
[16] Rozenhart, P. *Fast Tabulation of Cubic Function Fields.* Doctoral Dissertation, University of Calgary (Canada) 2009.
[17] Stichtenoth, H. *Algebraic Function Fields and Codes.* Second ed., Springer, Berlin 2009.
[18] Zhang, X.-K. *Ambiguous classes and 2-rank of class group of quadratic function field* (Chinese summary). J. China Univ. Sci. Tech. **17** (1987), 425–431.

Part III – Galois Covers of Function Fields in Positive Characteristic

Fields Institute Communications
Volume **60**, 2011

A Survey of Galois Theory of Curves in Characteristic p

Rachel Pries
Mathematics Department
Colorado State University
Fort Collins, CO 80523-1874 USA
pries@math.colostate.edu

Katherine Stevenson
Department of Mathematics
Cal State Northridge
18111 Nordhoff St.
Northridge, CA 91330-8313 USA
katherine.stevenson@csun.edu

Abstract. This survey is about Galois theory of curves in characteristic p, a topic which has inspired major research in algebraic geometry and number theory and which contains many open questions. We illustrate important phenomena which occur for covers of curves in characteristic p. We explain key results on the structure of fundamental groups. We end by describing areas of active research and giving two new results about the genus and p-rank of certain covers of the affine line.

1 Introduction

The purpose of this paper is to introduce the reader to the topic of Galois theory of curves in characteristic p. Since topological methods are no longer applicable, this topic has inspired major research in algebraic geometry and number theory, to recapture some information on the structure of fundamental groups. In spite of these advances, there are still many fascinating open questions on this topic.

In section 2, we recall some Galois theoretic facts for complex curves which are meaningless or false for curves in characteristic p. Section 3 contains some crucial examples of Kummer and Artin-Schreier covers of curves in characteristic p. The main algebraic definitions of objects such as the fundamental group, higher

2010 *Mathematics Subject Classification.* Primary 14H30, 14H40, 14G20; Secondary 14-06, 14-02.

This project was initiated at the workshop WIN Women in Numbers in November 2008. The authors would like to thank the Banff International Research Station for hosting the workshop and the National Security Agency, the Fields Institute, the Pacific Institute for the Mathematical Sciences, Microsoft Research, and University of Calgary for their financial support. The first author was partially supported by NSF grant 07-01303.

ramification groups, and Jacobians, can be found in Section 4. In Section 5, we outline the proofs of several major results, including some of the contributions of Grothendieck, Harbater, Pop, Raynaud, Serre, and Tamagawa.

Finally, in Section 6, we describe a few areas of active research involving embedding problems and arithmetic invariants of Galois covers defined over an algebraically closed field of characteristic $p > 0$. We prove two new results on these topics. To describe the results, let ℓ be a prime distinct from p and let a be the order of ℓ modulo p. Let L be an ℓ-group whose maximal elementary abelian quotient is $(\mathbb{Z}/\ell)^a$. Let G be a semi-direct product $L \rtimes \mathbb{Z}/p$. In Proposition 6.7, we prove that the smallest genus which occurs for a (wildly ramified) G-Galois cover $\phi : W \to \mathbb{P}^1_k$ branched only at ∞ is $g_W = 1+|L|(p-3)/2$. This result can be viewed as the solution to an embedding problem with prescribed ramification conditions. In Proposition 6.9, when $L \simeq (\mathbb{Z}/\ell)^a$, we prove that W can be chosen such that its Jacobian J_W has p-rank $s_W = (\ell^a - 1)(p-3)/2$ and furthermore such that the p-torsion $J_W[p]$ decomposes completely into s_W copies of $\mathbb{Z}/p \oplus \mu_p$ and $(p-1)/2$ copies of $E_{ss}[p]$, the p-torsion group scheme of a supersingular elliptic curve. In particular, the Newton polygon of J_W only has slopes 0, 1/2, and 1. The result is interesting because this combination of arithmetic invariants is somewhat unusual.

2 Facts about Galois covers of complex curves

Here are some of the basic properties of Galois covers of complex curves that are false for covers of curves defined over a field of characteristic $p > 0$. Suppose $\mathcal{X}$ is a smooth connected projective complex curve, i.e., a Riemann surface, of genus g. Suppose $\mathcal{B} \subset \mathcal{X}$ is a finite set of $r \geq 0$ points and $x \in \mathcal{X} - \mathcal{B}$ is a point. Suppose G is a finite group.

2.1 The fundamental group $\pi_1(\mathcal{X} - \mathcal{B}, x)$. The complex curve $\mathcal{X}$ is homeomorphic to the quotient of a polygon with $4g$ sides, where the quotient is determined by identifying the sides with the consecutive labels α_i, β_i, α_i^{-1}, β_i^{-1} for $1 \leq i \leq g$. Also the point x can be identified with a corner of the polygon. Let γ_i be a loop in $\mathcal{X}$, starting at x, that circles around the ith point of $\mathcal{B}$. The topological fundamental group $\pi_1(\mathcal{X} - \mathcal{B}, x)$ is generated by the homotopy classes of the loops $\alpha_1, \beta_1, \ldots, \alpha_g, \beta_g, \gamma_1, \ldots, \gamma_r$ with the sole relation $\prod_{i=1}^g [\alpha_i, \beta_i] \prod_{j=1}^r \gamma_j = 1$. This statement about the fundamental group implies the following facts:

(i) If $r > 0$, then $\pi_1(\mathcal{X} - \mathcal{B}, x)$ is a free group on $2g + r - 1$ generators.

(ii) The structure of $\pi_1(\mathcal{X} - \mathcal{B}, x)$ depends only on the genus of $\mathcal{X}$ and the cardinality of $\mathcal{B}$.

Because there is a bijection between finite quotients of $\pi_1(\mathcal{X} - \mathcal{B}, x)$ and finite Galois covers of $\mathcal{X}$ with branch locus in $\mathcal{B}$, one immediately deduces the following:

(iii) A finite group G is the Galois group of a cover of $\mathcal{X}$ with branch locus in $\mathcal{B}$ if and only if G can be generated by $2g + r - 1$ elements.

(iv) In particular, there are no nontrivial Galois covers of the complex affine line $\mathbb{A}^1_{\mathbb{C}}$ (i.e., the complex plane is simply connected).

(v) Given $\mathcal{X}$, $\mathcal{B}$, and G, the number of isomorphism classes of Galois covers of $\mathcal{X}$ with branch locus in $\mathcal{B}$ and with Galois group G is finite.

2.2 Ramification of complex covers. Suppose $\varphi : \mathcal{Y} \to \mathcal{X}$ is a Galois cover of complex curves with branch locus in $\mathcal{B}$ and Galois group G. Consider a point $Q \in \varphi^{-1}(\mathcal{B})$. The decomposition group D_Q consists of the automorphisms

$\sigma \in \mathrm{Gal}(\mathcal{Y}/\mathcal{X})$ such that $\sigma(Q) = Q$. The image under φ of a loop in $\mathcal{Y}$ around Q will be a loop in $\mathcal{X}$ around $\varphi(Q)$ traversed $|D_Q|$ times. By triangulating $\mathcal{X}$ and $\mathcal{Y}$ appropriately and computing their Euler characteristics, one can determine the genus of $\mathcal{Y}$. This yields some more facts:

(vi) The decomposition groups of a Galois cover $\varphi : \mathcal{Y} \to \mathcal{X}$ of complex curves are cyclic.

(vii) If $g_{\mathcal{X}}$ is the genus of $\mathcal{X}$, the genus $g_{\mathcal{Y}}$ of $\mathcal{Y}$ is given by the Riemann-Hurwitz formula to be

$$2g_{\mathcal{Y}} - 2 = |G|(2g_{\mathcal{X}} - 2) + \sum_{Q \in \varphi^{-1}(\mathcal{B})} (|D_Q| - 1).$$

Thus, $g_{\mathcal{Y}}$ is determined by $g_{\mathcal{X}}$, $|G|$, $|\mathcal{B}|$ and the orders of the decomposition groups.

2.3 Jacobians and torsion points. The definition of the Jacobian $J_{\mathcal{X}}$ of a complex curve $\mathcal{X}$ can be found in [21, VIII]. Recall that Ω^1 is the vector space of holomorphic 1-forms ω on $\mathcal{X}$. If γ is a loop in $\mathcal{X}$, there is a linear functional $\int_\gamma : \Omega^1 \to \mathbb{C}$. The value of the integral $\int_\gamma \omega$ depends only on the equivalence class $[\gamma]$ of γ in the homology group $H_1(\mathcal{X}, \mathbb{Z})$, which is the abelianization of the fundamental group. The dual space $(\Omega^1)^*$ is the vector space of linear functionals $\lambda : \Omega^1 \to \mathbb{C}$. A *period* is a linear functional which equals $\int_{[\gamma]}$ for some equivalence class $[\gamma]$ in $H_1(\mathcal{X}, \mathbb{Z})$. The set Λ of periods is a subgroup of $(\Omega^1)^*$.

The Jacobian of $\mathcal{X}$ is $J_{\mathcal{X}} := (\Omega^1)^*/\Lambda$. If $\mathcal{X}$ has genus g, then $\dim(\Omega^1) = g$. Also $H_1(\mathcal{X}, \mathbb{Z})$ is a $\mathbb{Z}$-module of rank $2g$. Thus $J_{\mathcal{X}} \simeq \mathbb{C}^g/\Lambda$ is a complex torus of dimension g. In fact, it is an abelian variety of dimension g [11, Chapter 6].

If ℓ is a prime, consider the multiplication-by-ℓ map m_ℓ on $J_{\mathcal{X}}$. The kernel $J_{\mathcal{X}}[\ell]$ of m_ℓ is the subgroup of ℓ-torsion points of the Jacobian. As an abelian group, $J_{\mathcal{X}}[\ell] \simeq (1/\ell)\Lambda/\Lambda$, thus:

(viii) The subgroup of ℓ-torsion points of the Jacobian satisfies $J_{\mathcal{X}}[\ell] \simeq (\mathbb{Z}/\ell)^{2g}$. In particular, there are ℓ^{2g} points of the Jacobian that are ℓ-torsion points.

2.4 Transition to characteristic $p > 0$. We will consider covers of curves defined over an algebraically closed field k of characteristic $p > 0$. The topological tools used above, such as loops, are meaningless for k-curves. For this reason, new algebraic definitions are needed for objects such as the fundamental group or Jacobian of a k-curve. Surprisingly, many attributes of fundamental groups and covers will remain the same in characteristic p. Most importantly, Grothendieck proved that fact (iii) holds for finite groups G that are prime-to-p (see Section 5.1). However, there are some substantial differences between the characteristic p and characteristic 0 settings. In particular, we will see that statements (i)-(viii) are each false for covers of k-curves. In each case, the statement must be revised in characteristic p to cope with the appearance of new ramified p-group covers and the disappearance of unramified p-group covers.

3 Examples of covers of curves in characteristic $p > 0$

Let k be an algebraically closed field of characteristic $p > 0$, e.g., $k = \bar{\mathbb{F}}_p$. Before developing the theory, we provide some examples of Galois extensions of $k(x)$. While the constructions are simple, these examples are crucial for understanding the fundamental group in characteristic $p > 0$.

For a field extension L/K, let $\mathrm{Gal}(L/K)$ be the set of automorphisms of L fixing every element of K. A field extension is Galois if and only if $|\mathrm{Gal}(L/K)| = [L : K]$.

3.1 Kummer extensions. Let ℓ be a prime. As long as ℓ is distinct from p, Kummer extensions still yield Galois extensions of $k(x)$ with Galois group $\mathbb{Z}/\ell$:

$$k(x) \hookrightarrow k(x)[y]/(y^\ell - x) \cong k(y). \tag{1}$$

This is an extension of degree ℓ. Since $p^\alpha \equiv 1 \bmod \ell$ for some positive integer α, there is an ℓth root of unity $\zeta_\ell \in \mathbb{F}_{p^\alpha} - \{1\} \subset k$. Then $\sigma : y \mapsto \zeta_\ell y$ is an automorphism of degree ℓ. Thus $\mathrm{Gal}(k(y)/k(x)) = \langle \sigma \rangle \simeq \mathbb{Z}/\ell$ and the extension is Galois.

The only places of $k(x)$ over which $k(y)$ is ramified are x and $1/x$. To see that x is the only affine place over which the extension is ramified, note that $0 = \partial(y^\ell - x)/\partial y = \ell y^{\ell-1}$ if and only if $y = 0$. For more information about this example, see [41, III.7.3].

If $\ell = p$, then the polynomial $t^p - 1 \equiv (t-1)^p \bmod p$ has only one root in k. So extension (1) has degree p, but $\mathrm{Gal}(k(y)/k(x))$ is trivial, and thus extension (1) is not Galois.

3.2 Artin-Schreier extensions. A new equation is needed in order to produce the group $\mathbb{Z}/p$ as the Galois group of an extension of $k(x)$. For $f(x) \in k[x]$ with $d = \deg(f(x))$ prime-to-p, consider the degree p extension:

$$k(x) \hookrightarrow L := k(x)[y]/(y^p - y - f(x)). \tag{2}$$

Then $\tau : y \mapsto y + 1$ is an automorphism of L of order p because $(y+1)^p \equiv y^p + 1 \bmod p$. Thus $\mathrm{Gal}(L/k(x)) = \langle \tau \rangle \simeq \mathbb{Z}/p$ and the extension is Galois.

There is no affine place of $k(x)$ over which L is ramified because

$$\partial(y^p - y - f(x))/\partial y = py^{p-1} - 1 \equiv -1 \bmod p \neq 0.$$

The only place of $k(x)$ over which L is ramified is the infinite place. For more information about this example, see [41, III.7.8].

3.3 New phenomena in characteristic $p > 0$. Artin-Schreier extensions can be used to give counterexamples to some of the facts from Section 2 for covers of k-curves. First, consider the affine line $\mathbb{A}^1_k = \mathbb{P}^1_k - \infty$, so that $2g + r - 1 = 0$. The Artin-Schreier extension (2) with equation $y^p - y = f(x)$ yields a nontrivial Galois cover $\phi : Y \to \mathbb{P}^1_k$ branched only at ∞. The decomposition group above ∞ has order p. This shows that facts (iii) and (iv) are false for k-curves. Moreover, by changing either the degree or the coefficients of $f(x)$, one sees that these covers occur in infinite families, and thus fact (v) is false for k-curves as well. It turns out that the genus of Y is $(p-1)(d-1)/2$ (see Section 4.2). This depends on a new invariant $d = \deg(f(x))$ which shows that fact (vii) is false for k-curves.

To construct a counterexample to fact (vi) for k-curves, consider a tower of a Kummer and Artin-Schreier extension with equations $x_1^\ell = x$ and $y^p - y = x_1^d$ where $\ell \mid (p-1)$, $p \nmid d$, and $\ell \nmid d$. This yields an extension $M/k(x)$ of degree ℓp. Consider the following automorphisms in $\mathrm{Gal}(M/k(x))$ where ζ_ℓ is a primitive ℓth root of unity:

$$\tau : x_1 \mapsto x_1, y \mapsto y+1, \text{ and } \sigma : x_1 \mapsto \zeta_\ell x_1, y \mapsto \zeta_\ell^d y.$$

Then $\sigma\tau\sigma^{-1}(y) = y + \zeta_\ell^{-d} \neq \tau(y)$. This shows that the extension is Galois with Galois group G a non-abelian semi-direct product of the form $\mathbb{Z}/p \rtimes \mathbb{Z}/\ell$. The extension is totally ramified above ∞ and so the decomposition group equals G which is not cyclic.

For a counterexample to fact (viii) for k-curves, suppose $p = 2$, and consider the k-curve E defined by the Artin-Schreier equation $y^2 - y = x^3$, which is an elliptic curve. Then E is supersingular [40, V, # 5.7] and thus the Jacobian of E has no 2-torsion points other than the identity [40, V, Thm. 3.1].

It is harder to show, but the same phenomena contradicting facts (iii)-(vii) hold for Galois covers of an arbitrary affine k-curve $X - B$ with Galois group G under the basic condition that p divides $|G|$. The same phenomenon contradicting fact (viii) occurs for any smooth projective k-curve of positive genus. These will be major themes of the next sections. More theory about Galois covers for curves defined over k is needed in order to define the fundamental group of a k-curve. Having done so, we show that facts (i)-(ii) are also false for k-curves in Sections 4.7, 5.2, and 5.4.

4 Algebraic definitions

Here we provide the basic definitions required to make sense of covers in arithmetic geometry. This section is meant to be a reference for the following sections. The reader may find it easier to skip this section and refer back to it as necessary. The idea is to mimic the construction of covering spaces in topology and analysis, where $U \to C$ is a covering if locally the inverse function theorem holds. In the algebraic context, the comparable concept is that of an étale or unramified morphism.

Let K be an algebraically closed field; (the material in this section is valid in any characteristic). Unless stated otherwise, all curves in this section are smooth connected K-curves. Let X be a projective K-curve. The genus of X is the dimension of $H^0(X, \Omega^1)$. Let $B \subset X$ be a finite (possibly empty) set of points and let $C = X - B$.

4.1 Terminology for Galois covers. An algebraic field extension L of F is a *separable* F-algebra if for every element $y \in L$ the minimal polynomial of y over F factors into distinct linear factors in its splitting field. The extension is *inseparable* otherwise. For example, extension (1) is purely inseparable when $\ell = p$. If R is an integral domain and $R \subset S$ is a ring extension, then S is *generically separable* as an R-algebra if $\text{frac}(S)$ is a separable $\text{frac}(R)$-algebra. A morphism of K-curves $\phi : Y \to X$ is *generically separable* if X can be covered by affine open subsets $U = \text{Spec}(R)$ such that the ring extension $R \subset \mathcal{O}(\phi^{-1}(U))$ is generically separable. A *cover* is a morphism $\phi : Y \to X$ which is finite and generically separable.

If $\phi : Y \to X$ is a cover, then the *Galois group* $\text{Gal}(Y/X)$ consists of the automorphisms σ of Y satisfying $\phi \circ \sigma = \phi$. If G is a finite group, then a *G-Galois cover* is a cover $\phi : Y \to X$ *together* with an inclusion $\rho : G \hookrightarrow \text{Gal}(Y/X)$ such that $\mathcal{O}_Y^G = f^*(\mathcal{O}_X)$ (where the left side denotes the sheaf of G-invariants). If Y is irreducible, this forces ρ to be an isomorphism. As X is a smooth curve, this condition is equivalent to saying that G acts simply transitively on a generic geometric fibre of $\phi : Y \to X$, so that $|\,\text{Gal}(Y/X)| = deg(f)$. Given an abstract finite group G, there could be many inclusions ρ with this property. If the inclusion ρ is not fixed then $\phi : Y \to X$ is called a *Galois cover with Galois group G*.

For example, extension (2) is a Galois cover with group $\mathbb{Z}/p$ and extension (2) together with the choice of automorphism $\tau : y \mapsto y + 1$ is a $\mathbb{Z}/p$-Galois extension.

4.2 Ramification: Wild, tame and p-tame. Let $\phi : Y \to X$ be a G-Galois cover. The cover ϕ is *prime-to-p* if $|G|$ is prime-to-p.

Let Q be a point of Y and let $P = \phi(Q) \in X$. The *decomposition group D_Q* at Q is the subgroup of G consisting of automorphisms that fix the point Q. The number of points in the fibre $\phi^{-1}(P)$ equals $|G|/|D_Q|$. The *inertia group I_Q* at Q is the subgroup of D_Q that induces the identity automorphism on the residue field at Q. Since K is algebraically closed, the inertia group equals the decomposition group. The cover is *ramified* at Q if I_Q is non-trivial, and it is *totally ramified* at Q if $I_Q = G$. The *branch locus* of ϕ is the set of points $P \in X$ for which there exists a ramified point $Q \in \phi^{-1}(P)$. The phrase *with branch locus in B* means that the branch locus is contained in B.

When K has characteristic $p > 0$, the cover ϕ is *wildly ramified* at Q if p divides $|I_Q|$ and is *tame* otherwise. The cover ϕ is tame if it is tame at all ramification points and is wild otherwise. When K has characteristic 0, a ramified point Q is *p-tame* if p does not divide $|I_Q|$.

For example, if $\ell \neq p$, then equation (1) yields a cover $\phi : Y \to \mathbb{P}^1_K$ with branch locus $B = \{0, \infty\}$, which is tame because the inertia group is $\mathbb{Z}/\ell$ above both branch points. Equation (2) yields a cover $\phi : Y \to \mathbb{P}^1_K$ branched only at ∞, which is wild because the inertia group is $\mathbb{Z}/p$.

4.3 The fundamental group. If $Z \to X$ is a G-Galois cover with branch locus in $\mathcal{B}$ and $\pi : G \to H$ is a surjection of finite groups, then $Z \to X$ must factor through an H-Galois cover $Y \to X$ with branch locus in B. Consider the set of Galois groups of finite Galois covers of X with branch locus in B. Consider also the collection of surjections $\pi : G \to H$ when a G-Galois cover $Z \to X$ with branch locus in B factors through an H-Galois cover with branch locus in B. This set of groups and collection of surjections forms an inverse system. The inverse limit of this system is the algebraic fundamental group $\pi_1(C)$ where $C = X - B$. The isomorphism class of $\pi_1(C)$ does not depend on the choice of the base point so we eliminate the base point from the notation. A more precise and complete definition of the fundamental group can be found in [19, Section 2].

By definition, the finite quotients of $\pi_1(C)$ correspond to finite Galois covers of C. Thus to understand $\pi_1(C)$ one needs to understand:

1. What are the finite quotients of $\pi_1(C)$?
2. How do the finite quotients fit together into an inverse system?

Answering the first question is called "the inverse Galois problem" for C. The second question is more subtle and is related to embedding problems. Roughly speaking, question (2) asks: Given an H-Galois cover $\psi : V \to C$ and a surjection $G \twoheadrightarrow H$, what G-Galois covers of C exist which factor through ψ? (See Section 5.5.)

For an algebraically closed field K of characteristic 0, Grothendieck showed that the fundamental group of a K-curve $X - B$ of genus g with $r = |B|$ punctures is isomorphic to the profinite completion of the topological fundamental group of a Riemann surface of genus g with r punctures [1, XIII, Cor. 2.12]. Thus, $\pi_1(X - B)$ is the group obtained by taking the profinite group on generators $\alpha_1, \beta_1, ..., \alpha_g, \beta_g, \gamma_1, ..., \gamma_r$ and imposing the sole relation $\prod_{i=1}^{g}[\alpha_i, \beta_i] \prod_{j=1}^{r} \gamma_j = 1$. In particular, if $r > 0$, then $\pi_1(X - B)$ is a free profinite group on $2g + r - 1$ generators. This implies that every group generated by $2g + r - 1$ elements is a quotient of $\pi_1(X - B)$. Moreover, the freeness implies that: Given an H-Galois cover $\psi : V \to C$, a group G generated by $2g + r - 1$ elements, and a surjection $G \twoheadrightarrow H$, there exists a G-Galois cover of C that factors through ψ.

For example, if $\mathrm{char}(K) = 0$ and $X = \mathbb{P}^1_K$ and $B = \{0, \infty\}$, then $\pi_1(X - B)$ is the profinite group $\hat{\mathbb{Z}}$ on one generator. This implies that, for each $\ell \in \mathbb{N}$, there is exactly one isomorphism class of $\mathbb{Z}/\ell$-Galois cover $\phi_\ell : Y \to \mathbb{P}^1_K$ branched at $B = \{0, \infty\}$. The Kummer Equation (1) in Section 3.1 is an equation representing this isomorphism class. If $\ell_1 \mid \ell_2$, then ϕ_{ℓ_2} factors through ϕ_{ℓ_1}.

In summary, for a curve defined over an algebraically closed field K of characteristic 0, the fundamental group is finitely generated as a profinite group; in particular, this implies that the answer to question (2) above is completely determined by the answer to question (1). This is because a finitely generated profinite group is determined by its finite quotients [10, Prop. 15.4].

The fundamental group of a curve defined over an algebraically closed field k of characteristic $p > 0$ is known in only two cases: (1) when X is the projective line $\mathbb{P}^1_k$ and $B = \emptyset$ then $\pi_1(X)$ is trivial; and (2) when X is an elliptic curve E and $B = \emptyset$ then $\pi_1(E)$ is a finitely generated abelian profinite group, (see subsection 4.7). Section 5 contains some of the major results obtained about the fundamental group and its finite quotients in characteristic $p > 0$. In particular, when $B \neq \emptyset$, the fundamental group is not finitely generated as a profinite group and is not determined by its finite quotients. Thus understanding how the groups fit together, as in question (2), is essential to determining the profinite group structure of the fundamental group.

4.4 Translation into field theory. The material in Section 4.3 can be reinterpreted in terms of field extensions as follows. The function field $K(C)$ of C is the same as that of X. A separable closure $K(C)^{sep}$ is an infinite Galois extension of $K(C)$ whose Galois group $\mathrm{Gal}_{K(C)}$ is called the absolute Galois group of C.

Given a G-Galois extension $L/K(C)$, consider an open cover of C by affine opens $U_i = \mathrm{Spec}(R_i)$ and let S_i be the integral closure of R_i in L and let $V_i = \mathrm{Spec}(R_i)$. The affine opens V_i cover a curve V and there is a cover $V \to C$. There is a G-Galois cover $\phi : Y \to X$ where Y is the projective closure of V. However, any point of X could be a branch point of ϕ so this may not correspond to a surjection of $\pi_1(C)$ onto G.

To remedy this, one instead considers the maximal Galois extension $K(C)_{un,B}$ of $K(C)$ unramified outside of the set of places in $K(C)$ for points in B. Then there is a bijection between surjections $\pi : \mathrm{Gal}(K(C)_{un,B}/K(C) \twoheadrightarrow G$ and G-Galois covers $\phi : Y \to X$ with branch locus in B. Furthermore, π factors through a surjection $\pi' : \mathrm{Gal}(K(C)_{un,B}/K(C) \twoheadrightarrow \Gamma$ if and only if the G-Galois cover ϕ can be dominated by a Γ-Galois cover ϕ' with branch locus in B. Thus, $\mathrm{Gal}(K(C)_{un,B}/K(C) = \pi_1(C)$.

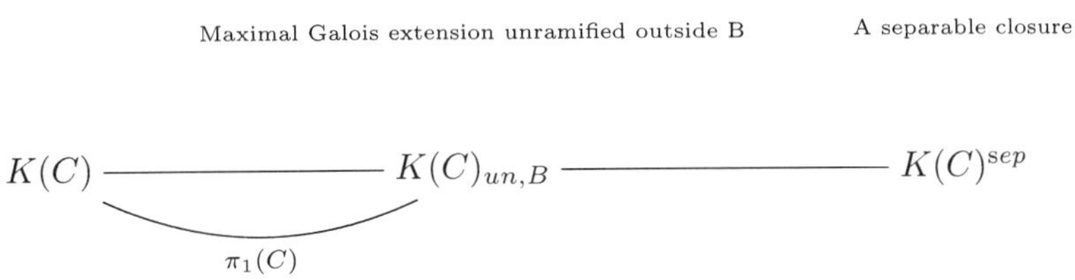

4.5 Higher ramification groups. There is extra ramification information at a wildly ramified point Q, including a filtration of I_Q called the filtration of higher ramification groups, [36, IV]. If $\phi : Y \to X$ is ramified at Q, consider the complete local ring $\hat{\mathcal{O}}_Q$ of functions at Q and the valuation function ν_Q. For any integer $i \geq -1$ the *ith ramification group* at Q is

$$I_i(Q) = \{\sigma \in D_Q | \nu_Q(\sigma(z) - z) \geq i+1, \forall z \in \hat{\mathcal{O}}_Q\}.$$

The decomposition group at Q is $I_{-1}(Q)$ and the inertia group is $I_0(Q)$. The inertia at a wildly ramified point is usually not cyclic though it is always cyclic-by-p, in that it has a normal Sylow p-subgroup $I_1(Q)$ and the quotient $I_Q/I_1(Q)$ is cyclic and prime-to-p.

The genus of Y now depends on the ramification filtration. The Riemann-Hurwitz formula states that $2g_Y - 2 = |G|(2g_X - 2) + \mathrm{Ram}$ where Ram is the sum of the degrees of the different at each ramified point Q. The degree of the different at Q equals $\sum_{i=0}^{\infty}(|I_i(Q)| - 1)$. If $\mathrm{char}(K) = 0$, then $\mathrm{Ram} = \sum_{\mathrm{Q}\in\phi^{-1}(\mathrm{B})}(|\mathrm{I_Q}| - 1)$ which recovers the equation in part (vii) of Section 2.2. If $\mathrm{char}(K) > 0$, then $\mathrm{Ram} \geq \sum_{\mathrm{Q}\in\phi^{-1}(\mathrm{B})}(|\mathrm{I_Q}| - 1)$. From this, one sees that the genus of Y can grow "more quickly" in positive characteristic than in characteristic zero.

For the Artin-Schreier extension $y^p - y = f(x)$ in Equation 2 where $\deg(f(x)) = d$ and $p \nmid d$, if Q is the point above ∞ then $I_i(Q) = \mathbb{Z}/p$ if $0 \leq i \leq d$ and $I_i(Q) = \{0\}$ if $i > d$, [41, III.7.8(c)]. By the Riemann-Hurwitz formula, $g_Y = (p-1)(d-1)/2$.

4.6 The Jacobian and torsion points. Let X be a smooth projective K-curve of genus g. A divisor on X is a formal sum $\sum_{P\in X} n_P P$ where $n_P \in \mathbb{Z}$ and $n_P = 0$ for all but finitely many $P \in X$. The degree of a divisor $D = \sum_{i=1}^{r} n_i P_i$ is $\sum_{i=1}^{r} n_i$ and $\mathrm{Div}^0(X)$ denotes the abelian group of all divisors of X of degree 0. Given a non-zero element f in the function field $K(X)$ of X, there is a divisor $\mathrm{div}(f) = \sum_{P\in X} \mathrm{ord}_P(f)P$. A divisor D is *principal* if $D = \mathrm{div}(f)$ for some function $f \in K(X)$. Every principal divisor has degree zero. Let $\mathrm{Prin}(X)$ be the set of all principal divisors of X. The sets $\mathrm{Div}^0(X)$ and $\mathrm{Prin}(X)$ are abelian groups under addition and $\mathrm{Prin}(X) \subset \mathrm{Div}^0(X)$. The algebraic definition of the Jacobian J_X of X is $J_X := \mathrm{Div}^0(X)/\mathrm{Prin}(X)$. This is an abelian group which is naturally isomorphic to the K-points of an abelian variety of dimension g over K, which we also denote J_X.

For a prime ℓ, consider the multiplication-by-ℓ morphism m_ℓ on J_X. The ℓ-torsion $J_X[\ell]$ of the Jacobian is the kernel of m_ℓ. The K-points of $J_X[\ell]$ can be identified with the set

$$\{[D] \in J_X \mid \text{there exists f such that } \ell\mathrm{D} = \mathrm{div(f)}\}.$$

If $\ell \neq p$, then m_ℓ is separable of degree ℓ^{2g}. Thus $J_X[\ell] \simeq (\mathbb{Z}/\ell)^{2g}$ [23, pg. 64].

For example, suppose $\mathrm{char}(K) \neq 2$ and Y is a hyperelliptic K-curve with equation $y^2 = \prod_{i=1}^{2g+1}(x - b_i)$. Let Q_∞ be the point at infinity of Y and let Q_i be the point $(x, y) = (b_i, 0)$ for $1 \leq i \leq 2g+1$. The divisors $D_i = Q_i - Q_\infty$ are 2-torsion points of J_Y since $2D_i = \mathrm{div}(x - b_i)$. There is a relation $0 = \mathrm{div}(y) = \sum_{i=1}^{2g+1} D_i$ in J_Y. There is no nontrivial linear relation $\sum_{i=1}^{2g} a_i D_i = 0$ with $a_i \in \{0, 1\}$ because $\prod_{i=1}^{2g}(x - b_i)^{a_i}$ is not a square in $k[x, y]/(y^2 - \prod_{i=1}^{2g+1}(x - b_i))$. Thus the set $\{D_1, \ldots, D_{2g}\}$ is linearly independent and hence is a basis for $J_Y[2]$.

In contrast, if $p = \mathrm{char}(K)$, the multiplication-by-p morphism m_p factors as the composition of the Frobenius morphism, which is inseparable of degree p^g and the

Verschiebung morphism which is separable of degree p^g. This implies that $J_X[p]$ is a group scheme of rank p^{2g}. The number of points in $J_X[p](K)$ equals p^s for some integer s such that $0 \leq s \leq g$. Here s is called the *p-rank* of X.

For example, an elliptic curve defined over an algebraically closed field of characteristic p can be either ordinary ($s = 1$) or supersingular ($s = 0$). If $p > 2$ and $f(x) = x(x-1)(x-\lambda)$, the elliptic curve $y^2 = f(x)$ is supersingular if and only if λ is a root of the coefficient of x^{p-1} in $f(x)^{(p-1)/2}$ [40, V, Thm. 4.1]. It can be computationally difficult to determine the p-rank of a curve of higher genus. For Jacobians of hyperelliptic curves, an algorithm to compute the p-rank can be found in [44]. Another situation where the p-rank can be computed is when $\phi : Y \to X$ is a Galois cover whose Galois group G is a p-group. If $|G| = p^a$, then the Deuring-Shararevich formula [7, Cor. 1.8] states that

$$s_Y - 1 = p^a(s_X - 1) + \sum_{Q \in Y} (|I_Q| - 1).$$

4.7 Unramified covers and the Jacobian. In this section, we describe a connection between the ℓ-torsion points of the Jacobian of X and unramified $\mathbb{Z}/\ell$-Galois covers of X.

For some intuition about this connection, consider the example of an elliptic curve E over K. If ℓ is prime-to-p, then the multiplication-by-ℓ morphism $m_\ell : E \to E$ is a separable cover of degree ℓ^2 [40, III, Cor. 5.4]. Suppose $Q \in E$ is an ℓ-torsion point. If $R \in E$, then $m_\ell(R + Q) = m_\ell(R)$. Thus there is an automorphism σ_Q of E of order ℓ defined by $\sigma_Q(R) = R + Q$ and $m_\ell \circ \sigma_Q = m_\ell$. In other words, m_ℓ is a Galois cover, whose Galois group can be identified with $J_E[\ell]$. After choosing a basis for $J_E[\ell]$, then m_ℓ is a $(\mathbb{Z}/\ell)^2$-Galois cover.

Continuing this example, suppose Q' is a point of order ℓ on the Jacobian of E. Then Q' can be canonically identified with a point Q of order ℓ on E, [40, X, Thm. 3.8]. Consider the subgroup $H_Q = \langle \sigma_Q \rangle \subset J_E[\ell]$. Note that $J_E[\ell]/H_Q$ is a cyclic group of order ℓ. Let E_Q be the quotient of E by H_Q. The quotient cover $E_Q \to E$ is a $\mathbb{Z}/\ell$-Galois cover, which is unramified by the Riemann-Hurwitz theorem. To summarize, every ℓ-torsion point on the Jacobian of E yields an unramified $\mathbb{Z}/\ell$-Galois cover of E. Using this, one can prove that

$$\pi_1(E) \simeq \hat{\mathbb{Z}}_p^s \times \prod_{\ell \neq p} (\hat{\mathbb{Z}}_\ell \times \hat{\mathbb{Z}}_\ell)$$

where $s = 1$ if E is ordinary, and $s = 0$ if E is supersingular [17, IV, exercise 4.8]. This gives a counterexample to fact (ii) from Section 2 for a projective curve defined over an algebraically closed field of characteristic $p > 0$.

For a projective K-curve X of higher genus, the bijection between ℓ-torsion points of J_X and unramified $\mathbb{Z}/\ell$-Galois covers of X is harder to construct. As in [20, III, Section 4], one defines $\pi^1(X, \mathbb{Z}/\ell)$ to be the set of isomorphism classes of unramified $\mathbb{Z}/\ell$-Galois covers of X. By [20, Remarks following Section III, Prop 4.11], for ℓ prime to char(K), the group $\pi^1(X, \mathbb{Z}/\ell) \cong H^1(X, \mathbb{Z}/\ell)$ is isomorphic to $J_X[\ell](K)$. Similarly, $\pi^1(X, \mathbb{Z}/p) \cong H^1(X, \mathbb{Z}/p) \cong J_X[p](K)$ for $p = \text{char}(K)$ by [20, Remarks following Section III, Prop 4.13]. Thus the p-rank equals the maximum rank of a p-group which occurs as the Galois group of an unramified cover of X [20, Cor. 4.18].

5 Major results

Let k be an algebraically closed field of characteristic $p > 0$. Let X be a smooth connected projective k-curve of genus g. Let $B \subset X$ be a finite subset of r points.

5.1 The prime-to-p fundamental group. The main point of this section is Grothendieck's result that the prime-to-p groups that occur for Galois covers in characteristic p are exactly the same as those that occur in characteristic 0, namely those generated by $2g + r - 1$ elements. Recall the definitions of prime-to-p, tame, and p-tame covers from Section 4.2. The *prime-to-p fundamental group* $\pi_1^{p'}(X - B)$ is the inverse limit of the Galois groups of finite Galois prime-to-p covers of X with branch locus in B. The *tame fundmental group* $\pi_1^t(X - B)$ is the inverse limit of the Galois groups of finite Galois covers of X with branch locus in B and tamely ramified over B. The result also shows that $\pi^t(X - B)$ and $\pi_1^{p'}(X - B)$ are finitely generated as profinite groups and, as such, are determined by their finite quotients [10, Prop. 15.4].

The basic idea behind Grothendieck's proof is that tame covers in characteristic p lift to p-tame covers in characteristic 0. More precisely, let A be a complete local ring with residue field k. By [1, Exp. V, Cor. 7.4], there exists a smooth projective A-curve $\mathcal{X}_A$ such that the closed fibre is X. In particular, taking A to be a complete discrete valuation ring of mixed characteristic with residue field k, then the generic fibre $\mathcal{X}$ of $\mathcal{X}_A$ is a lift of X to characteristic 0. Let $\mathcal{B}_A$ be a set of horizontal sections specializing to B and let $\mathcal{B}$ be its generic fibre. The subset $\mathcal{B} \subset \mathcal{X}$ is a lift of $B \subset X$ to characteristic 0. Given a tame G-Galois cover $\phi : Y \to X$ with branch locus in B, there exists a G-Galois cover $\varphi_A : \mathcal{Y}_A \to \mathcal{X}_A$ with branch locus in $\mathcal{B}_A$ whose special fibre is isomorphic to ϕ. The generic fibre is a p-tame G-Galois cover $\varphi : \mathcal{Y} \to \mathcal{X}$ with branch locus in $\mathcal{B}$ in characteristic zero.

Let $\pi_1^{p-\mathrm{tame}}(\mathcal{X} - \mathcal{B})$ be the inverse limit of the system of finite groups that occur as Galois groups of p-tame covers of $\mathcal{X}$ with branch locus in $\mathcal{B}$. Also, consider the prime-to-p fundamental group $\pi_1^{p'}(\mathcal{X} - \mathcal{B})$. The previous paragraph summarizes the main ideas of the proof of the following result.

Theorem 5.1 [1, XIII, Cor. 2.12] *With notation as above,*

$$\pi_1^{p-\mathrm{tame}}(\mathcal{X} - \mathcal{B}) \twoheadrightarrow \pi_1^t(X - B)$$

and

$$\pi_1^{p'}(\mathcal{X} - \mathcal{B}) \simeq \pi_1^{p'}(X - B).$$

In particular, taking $B = \emptyset$ in Theorem 5.1 yields a surjection $\pi_1(\mathcal{X}) \twoheadrightarrow \pi_1(X)$ and thus $\pi_1(X)$ is finitely generated as a profinite group. The kernel of this homomorphism is not well-understood.

5.2 The pro-p fundamental group. Unlike the case for prime-to-p fundamental groups, the structure of the pro-p fundamental group of a curve changes significantly in characteristic p, and depends crucially on whether the curve is projective or affine. Let $\pi_1^p(X - B)$ be the inverse limit of the system of finite p-groups that occur as Galois groups of covers of X with branch locus in B.

Theorem 5.2 [38], [7, Thm. 1.9] *If X is a projective k-curve with p-rank s_X, then the pro-p fundamental group $\pi_1^p(X)$ is a free pro-p group on s_X generators.*

Proof (Outline following [7].) If π is a pro-p group then (1) the minimal number of generators of π is equal to $\dim_{\mathbb{F}_p}\mathrm{Hom}(G, \mathbb{Z}/p)$; and (2) π is free if and only if $H^2(G, \mathbb{Z}/p) = 0$ ([39, Thm. 12 and Cor. 2 to Prop. 23]). Thus it suffices to show that $\dim_{\mathbb{F}_p}\mathrm{Hom}(\pi_1(X), \mathbb{Z}/p) = s_X$ and $H^2(\pi_1(X), \mathbb{Z}/p) = 0$. Item (1) is discussed in Section 4.7. Item (2) follows from the fact that the p-cohomological dimension of X is less than two [2, IX, 3.5]. □

Theorem 5.3 [38] *If C is an affine k-curve, then the pro-p fundamental group $\pi_1^p(C)$ is infinitely generated.*

The proof of Theorem 5.3 relies on cohomological arguments and Artin-Schreier theory. Since the pro-p fundamental group is a quotient of the fundamental group, Theorem 5.3 implies that the fundamental group $\pi_1(C)$ is infinitely, not finitely, generated when C is an affine k-curve. Thus the fundamental group of an affine k-curve is not determined by its finite quotients [10, Prop. 15.4]. Moreover, as the p and prime-to-p parts are respectively infinitely and finitely generated as profinite groups, this shows $\pi_1(C)$ is not free and thus fact (i) is false.

5.3 Abhyankar's Conjecture. Remarkably, the finite quotients of the fundamental group of every affine k-curve are known, even if it is not clear how they fit together. The next result shows that a finite group G occurs as a Galois group of a cover of X with branch locus in B if and only if the maximal prime-to-p quotient of G occurs as a Galois group of a cover of an arbitrary genus g curve with r branch points in characteristic 0.

Theorem 5.4 [3, 13, 31] *Let X be a projective k-curve of genus g and let $B \subset X$ be a finite set of cardinality $r > 0$. A finite group G is a quotient of $\pi_1(X-B)$ if and only if every prime-to-p quotient of G can be generated by $2g+r-1$ elements.*

Theorem 5.4 was conjectured by Abhyankar in 1957, based on his experience working with covers of the affine line. The proof was completed in 1993 by Raynaud and Harbater. It is worth noting that the collection of groups which occur as Galois groups of affine k-curves is vast. Not only does every finite p-group occur as a quotient of the fundamental group of every affine k-curve, but every finite simple group of order divisible by p does as well. An immediate consequence of Theorem 5.4 is the following corollary.

Corollary 5.5 *A finite group is a quotient of $\pi_1(\mathbb{A}^1_k)$ if and only if it has no nontrivial prime-to-p quotient.*

A finite group with no nontrivial prime-to-p quotient is called *quasi-p*. Equivalently, a finite group is quasi-p if it is generated by its elements of p-power order, or by its Sylow p-subgroups.

A first step in proving Theorem 5.4 is the following result of Serre. In particular, it shows that every finite p-group occurs as a Galois group of a cover of the affine line.

Theorem 5.6 [37, Thm. 1] *Suppose $\tilde{G}$ is a finite quasi-p group and $N \subset \tilde{G}$ is a normal subgroup which is solvable. Let $G = \tilde{G}/N$. If G is a quotient of $\pi_1(\mathbb{A}^1_k)$ then so is $\tilde{G}$.*

Proof (Outline following [37].) Let π denote $\pi_1(\mathbb{A}^1_k)$. By hypothesis, there exists a surjection $\psi : \pi \twoheadrightarrow G$. Since N is solvable, using representation theory, one can reduce to the case where N is an elementary abelian group and the action of G on N is irreducible. Then $\tilde{G}$ is an extension of G by N, so this yields a cohomology class $e \in H^2(\tilde{G}, N)$. There are two cases: (a) when $e \neq 0$ and (b) when $e = 0$.

In case (a) the theorem follows by using the fact that π has cohomological dimension at most 1. This allows one to lift the surjection ψ to a homomorphism $\tilde{\psi} : \pi \to \tilde{G}$, and thus it suffices to show that $\tilde{\psi}$ is surjective. The group $H = \mathrm{Im}(\tilde{\psi})$ is a subgroup of $\tilde{G}$ such that $N \cdot H = \tilde{G}$. Because $e \neq 0$, $N \cap H$ is a non-trivial sub-$\tilde{G}/H$-module of N. By the irreducibility of N, then $N \cap H = N$ and thus $H = \tilde{G}$ and $\tilde{\psi}$ is surjective.

In case (b) when $e = 0$, then $\tilde{G}$ is a semi-direct product. A surjection $\psi : \pi \to G$ induces a π-module structure N_ψ on N. Then ψ factors through a surjection $\tilde{\psi} : \pi \to \tilde{G}$ if and only if the étale cohomology group $H^1(\pi, N_\psi)$ is strictly larger than the cohomology group $H^1(G, N)$. When N is an elementary abelian p-group, then $H^1(\pi, N_\psi)$ has infinite dimension, which completes the proof. When N is an elementary abelian ℓ-group for a prime $\ell \neq p$, Serre uses the Grothendieck-Ogg-Shafarevich formula to calculate the dimension $H^1(\pi, N_\psi)$ in terms of the filtration of higher ramification groups. It is possible that the dimension is not large enough, in which case it is necessary to change ψ to complete the proof. □

Proof (Outline of proof of Theorem 5.4 following [13, 31].) The forward direction of Abhyankar's Conjecture follows from Theorem 5.1. Here is a sketch of the converse in the case that $X = \mathbb{P}^1_k$ and $B = \{\infty\}$. Suppose G is a finite quasi-p group. The proof proceeds by induction on the order of G. By Theorem 5.6, one can assume that G has no normal p-group subgroup. Let S be a fixed Sylow p-subgroup of G.

Case A: G is generated by two proper quasi-p subgroups G_1 and G_2 satisfying the extra condition that $G_i \cap S$ is a Sylow p-subgroup of G_i. Inductively, one can suppose that G_1 and G_2 are each a quotient of $\pi_1(\mathbb{A}^1_k)$. Then there exists a G_i-Galois cover $\phi_i : Y_i \to \mathbb{P}^1_k$ branched only at ∞ for $i = 1, 2$. Harbater's contribution to the proof was to develop a theory of formal patching and use it to patch the two covers ϕ_1 and ϕ_2 together. In this way, a G-Galois cover $\phi : Y \to \mathbb{P}^1_k$ branched only at ∞ is produced. The basic idea is to build a $k[[t]]$-curve W whose generic fibre is a projective line and whose special fibre is a chain of two projective lines W_1 and W_2 intersecting in exactly one ordinary double point. One can construct a G-Galois cover of the special fibre such that its restriction to W_i is $\mathrm{Ind}^G_{G_i}(\phi_i)$. The condition on the Sylow p-subgroups allows one to do this compatibly near the ordinary double point. One shows that the cover can be deformed over $k[[t]]$ near the ordinary double point. Using formal patching, one can deform the cover of projective curves over $k[[t]]$. (A similar technique with formal patching allows one to reduce the proof of the converse direction of Abhyankar's Conjecture for an arbitrary affine k-curve to the case of the affine line.)

Case B: G is a finite quasi-p group, with no normal p-subgroup, not generated by proper quasi-p subgroups satisfying the above condition on their Sylow p-subgroups. While the conditions seem awkward, this turns out to be *exactly* the case that can be handled by Raynaud's analysis of semi-stable reduction of covers. The idea is to consider a G-Galois cover φ of the projective line over $\mathrm{Frac}(W(k))$ whose inertia groups are p-groups. The cover φ exists by Riemann's Existence Theorem since G is quasi-p and thus can be generated by elements of p-power order. The special fibre ϕ_s of the semi-stable reduction of φ is a G-Galois cover of a tree of projective lines. The cover ϕ_s is inseparable exactly over the interior components of the tree. Over each terminal component of the tree, ϕ_s is ramified only over the node η at which the terminal component intersects the interior of the tree. This yields a cover of a projective line branched only at one point η. The unusual group theoretic conditions insure that there is at least one terminal component over which the restriction of ϕ_s is connected. Thus there exists a G-Galois cover $\phi : Y \to \mathbb{P}^1_k$ branched only at one point. □

5.4 Anabelian results. Theorem 5.4 implies that the fundamental group of an affine k-curve is an infinitely generated profinite group. An interesting consequence of this result is that, for an affine k-curve $X - B$, the structure of $\pi_1(X - B)$ is not determined by its finite quotients. This is different than the situation for projective k-curves or for curves defined over algebraically closed fields of characteristic 0.

Grothendieck's anabelian conjecture predicts that the isomorphism class of a hyperbolic curve (a smooth curve whose geometric fundamental group is non-abelian) defined over a number field is determined by the structure of its arithmetic fundamental group. This conjecture has been settled in large part by Mochizuki [22]. Specifically, let K be a field that can be embedded in a finitely generated field extension of $\mathbb{Q}_p$. Let X be a smooth K-variety and let Y be a hyperbolic K-curve. Then there is a natural bijection between the set of dominant K-morphisms $X \to Y$ and the set of (conjugacy classes of) open homomorphisms $\pi_1(X) \to \pi_1(Y)$ compatible with the action of the absolute Galois group G_K. As an application, he proves a birational version of the anabelian conjecture for function fields of arbitrary dimension over K. The result builds upon the work of others, especially Tamagawa [42], who introduced a characteristic p version of the Grothendieck anabelian conjecture in characteristic p and proved it for affine curves defined over finite fields.

Most anabelian theorems are too technical to include here, but as a very special case, consider the following result.

Theorem 5.7 [43, Cor. 1.8, Thm. 1.9] *Suppose X is a smooth projective k-curve and $B \subset X$ is a finite set of points. The fundamental group $\pi_1(X - B)$ determines the genus of X, the cardinality of B, and the p-rank of X.*

Proof (Outline following [43].) The fundamental group $\pi_1(X - B)$ determines the pro-p and the prime-to-p fundamental groups. By taking the quotient by the commutator, one can determine the abelianization $\pi_1^{ab}(X - B)$ of $\pi_1(X - B)$. If H is an open subgroup of $\pi_1(X - B)$, let $\phi_H : U_H \to X - B$ denote the corresponding cover. The degree of ϕ_H equals the index $[\pi_1(X - B) : H]$ and the Galois group of ϕ_H is abelian if and only if H contains the commutator. Thus these attributes of ϕ_H are determined by the fundamental group.

Let g_X be the genus of X, let $r_X = \#B$, and let s_X be the p-rank of X. By Theorems 5.2 and 5.3, the structure of $\pi_1^p(X - B)$ determines whether $r_X = 0$. If $r_X = 0$, then the pro-p fundamental group determines the p-rank s_X by Theorem 5.2. Consider the rank of the maximal elementary abelian ℓ-group quotient of the abelianization $\pi_1^{ab}(X)$ for a prime $\ell \neq p$. By Section 4.7, this is the rank of $J_X[\ell]$, which equals $2g_X$ by Section 4.6.

Now suppose $r_X > 0$. By Theorem 5.1, the fundamental group determines the quantity $2g_X - 2 + r_X$. Suppose H is an open subgroup of $\pi_1(X - B)$. The corresponding cover $\phi_H : U_H \to X - B$ has degree equal to the index $[\pi_1(X - B) : H]$. Note that H is the fundamental group of $\pi_1(U_H)$ and that U_H is an open subset of a smooth projective k-curve Y_H of genus g_H. Let $r_H = \#(Y_H - U_H)$. It follows that the quantity $2g_H - 2 + r_H$ can be determined for any open subgroup $H \subset \pi_1(X - B)$. Applying the Riemann-Hurwitz formula, one can determine whether ϕ_H is wildly ramified. Thus $\pi_1(X - B)$ determines the tame fundamental group. Now s_X equals the rank of the maximal pro-p quotient of $\pi_1^t(X - B)$. Similarly, one can determine the p-rank s_H of Y_H for any open subgroup $H \subset \pi_1(X - B)$.

Continuing in the case when $r_X > 0$, the Deuring-Shafarevich formula states that $s_H - 1 + r_H = p(s_X - 1 + r_X)$ for any open normal subgroup $H \subset \pi_1(X - B)$ of index p. Thus one can determine the quantity $pr_X - r_H$ which is a multiple of $p - 1$. By the Riemann-Roch Theorem, there is a function $f \in k(X)$ whose set of poles is B and such that $p \nmid \mathrm{ord}_b(f)$ for each $b \in B$. The Artin-Schreier equation $y^p - y = f$ determines a Galois degree p cover $\phi : Y \to X$ which is totally ramified above each $b \in B$. If $H \subset \pi_1(X - B)$ is the corresponding open normal subgroup, then $r_H = r_X$ and $pr_X - r_H$ equals $(p-1)r_X$. Since no smaller value of $pr_X - r_H$ can occur for any open normal subgroup $H \subset \pi_1(X - B)$ of index p, the fundamental group determines the value of r_X and thus of g_X as well. □

Theorem 5.7 finally provides a counterexample to fact (ii) for affine k-curves. An element $b \in k - \{0, 1\}$ is called supersingular if the elliptic curve E with equation $y^2 = x(x - 1)(x - b)$ is supersingular. By a result of Igusa [40, V, Thm. 4.1(c)], there are $(p - 1)/2$ supersingular values of b if p is odd.

Proposition 5.8 *Suppose p is an odd prime. Let $X = \mathbb{P}^1_k$ and $B = \{0, 1, b, \infty\}$. The structure of the fundamental group $\pi_1(X - B)$ depends on whether b is a supersingular value.*

Note that the finite quotients of $\pi_1(X - B)$ in Proposition 5.8 do not depend on whether b is a supersingular value by Theorem 5.4.

Proof The branch locus of a degree 2 cover of $\mathbb{P}^1_k$ has even cardinality, so there are 7 subgroups of $\pi_1(X - B)$ of index two. Of these, exactly one corresponds to a degree two cover $\phi : Y \to \mathbb{P}^1_k$ with branch locus in B such that Y has positive genus, namely the cover with equation $y^2 = x(x - 1)(x - b)$. By Theorem 5.7, one can distinguish the corresponding subgroup $\pi' \subset \pi_1(X - B)$ of index two. Now π' is the fundamental group of $Y - \phi^{-1}(B)$. Applying Theorem 5.7 again, one can determine the p-rank of Y, and in particular determine whether Y is ordinary or supersingular. This determines whether b is a supersingular value. □

For more results along the lines of Proposition 5.8, see [5], [43].

5.5 Freeness results and embedding problems. To understand the structure of the fundamental group of a k-curve, it is crucial to understand how its finite quotients fit together. Harbater and Pop independently proved the following result.

Theorem 5.9 [14, Thm. 3.5] [27, Thm. B] *The absolute Galois group $G_{k(X)}$ of the function field of a projective k-curve X is free of rank* $\mathrm{card}(k)$.

Before describing the proof of Theorem 5.9, it is useful to introduce the terminology of embedding problems. Suppose there exists a G-Galois cover $\phi : Y \to X$ with branch locus in B corresponding to a surjection $\beta : \pi_1(X - B) \twoheadrightarrow G$ and a surjection $\alpha : \Gamma \twoheadrightarrow G$ of finite groups. By Galois theory, a surjection $\lambda : \pi_1(X - B) \twoheadrightarrow \Gamma$

where $\alpha \circ \lambda = \beta$ corresponds to a (connected) Γ-Galois cover $Z \to X$ with branch locus in B that dominates ϕ.

Given a group Π, a pair of surjections $(\beta : \Pi \twoheadrightarrow G, \alpha : \Gamma \twoheadrightarrow G)$ where G and Γ are finite groups is a *finite embedding problem* for Π. In the case where α has a splitting $s : G \to \Gamma$, the pair is a *finite split embedding problem.* A *weak solution* to an embedding problem (β, α) is a group homomorphism $\lambda : \Pi \to \Gamma$ such that $\alpha \circ \lambda = \beta$. A weak solution λ is a *proper solution* if it is a surjection. A group Π is *projective* if every finite embedding problem for Π has a weak solution. Notice that any finite split embedding problem automatically has a weak solution given by $s \circ \beta$.

Solutions to embedding problems are tightly connected to the property of freeness. For example, given a profinite group Π and an infinite cardinal m, by [33, Chapter 8], Π is free of rank m if and only if every finite embedding problem has exactly m distinct proper solutions. Moreover, by [16, Thm. 2.1] this is equivalent to Π being projective and satisfying the property that every non-trival finite split embedding problem for Π has exactly m distinct solutions.

Proof (Outline following [14, 27].) Let $(\beta : G_{k(X)} \twoheadrightarrow G, \alpha : \Gamma \twoheadrightarrow G)$ be a finite embedding problem for $G_{k(X)}$ and let N be the kernel of α. It suffices to show that (β, α) has card(k) distinct proper solutions. The surjection β corresponds to a G-Galois cover $\phi : Y \to X$ whose branch locus is contained in a non-empty set $B \subset X$. Thus there is an induced embedding problem $(\beta : \pi_1(X - B) \twoheadrightarrow G, \alpha : \Gamma \twoheadrightarrow G)$. As in Theorem 5.6, representation theory implies that obstructions for weak solutions lie in $H^2(\pi_1(X - B, N)$. By [37, Prop. 1] the fundamental group of an affine k-curve has cohomological dimension at most 1, and hence it is a projective group [35, I.5.9, Prop. 45]. As a result, there exists a weak solution $\lambda : \pi_1(X - B) \to \Gamma$. This defines a (possibly disconnected) Γ-Galois cover $\psi' : Z' \to X$ with branch locus in B. The cover ψ' can be patched with a branched N-Galois cover $Z'' \to \mathbb{P}^1_k$ in such a way as to produce a proper solution. Specifically, the patched cover produces a (connected) Γ-Galois cover $\psi : Z \to X$ dominating ϕ which is unramified away from a finite set B' containing B. $\square$

In the proof above, the additional branching of the Γ-Galois cover ψ at $B' - B$ is not a problem as ψ still corresponds to a surjection $\lambda : G_{k(X)} \twoheadrightarrow \Gamma$. It is useful to remark here that if the kernel N is a quasi-p group then no additional ramification is required (i.e., $B' = B$) [15, Thm. 4.6] because the wild ramification can be enlarged at one point. In general, the number of additional branch points depends on the number of generators of the maximal prime-to-p quotient of N. Moreover, the location of the additional branching cannot be prescribed. As a result, the freeness result Theorem 5.9 for $G_{k(X)}$ does not translate into a freeness result for $\pi_1(X - B)$ because in the latter case the covers cannot have additional ramification outside of B. It is still true that $\pi_1(X - B)$ has cohomological dimension at most 1, and thus it is projective. However, as we saw at the end of Section 5.2, it is not free.

6 Open questions and results

Let k be an algebraically closed field of characteristic $p > 0$. At this time, the full structure of the fundamental group is not known for any affine k-curve or for any projective curve of genus $g \geq 2$. The fundamental group depends on towers of covers of k-curves and on the geometry of the k-curves in these towers. The goal of understanding fundamental groups provides a strong motivation to answer new questions about these towers. Let X be a smooth connected projective k-curve of genus g. Let $B \subset X$ be a finite set of points.

6.1 Subgroups of fundamental groups of curves. It is interesting to measure the extent to which the fundamental group of a k-curve is not free. In this section, we study this topic in terms of subgroups of the fundamental group and in the next section we will address this topic in terms of quotients of the fundamental group and embedding problems.

Question 6.1 If $X - B$ is an affine k-curve, which closed normal subgroups of $\pi_1(X - B)$ are free?

For example, the commutator subgroup of $\pi_1(X - B)$ is free [18, Thm. 6.12] for every affine k-curve $X - B$. This is a natural subgroup to study for this question since the quotient of $\pi_1(X - B)$ by the commutator subgroup is the maximal abelian quotient of $\pi_1(X - B)$. Additional examples of a similar type can be seen in [26, Thm. 1.1].

Here is an example of an affine k-curve $X - B$ and a closed normal subgroup of $\pi_1(X - B)$ that is not free.

Example 6.2 Consider the affine line $\mathbb{A}^1_k$. Let N be the intersection of all open normal subgroups of $\pi_1(\mathbb{A}^1_k)$ of index p such that, for the corresponding cover $Y \to \mathbb{P}^1_k$ branched only at ∞, the curve Y has genus zero. These subgroups correspond to Artin-Schreier covers $y^p - y = cx$ with $c \in k$. A computation shows that two such covers are linearly disjoint as long as $c_1 - c_2$ is not a $(p-1)^{st}$ root of unity. Thus N has infinite index and it is a closed normal subgroup of $\pi_1(\mathbb{A}^1_k)$.

Let $k(x)_\infty$ be the maximal Galois extension of the function field $k(x)$ which is unramified outside $\{\infty\}$. Then $\mathrm{Gal}(k(x)_\infty/k(x)) = \pi_1(\mathbb{A}^1_k)$. Let F_N be the fixed field of N in $k(x)_\infty$.

Assume that N is free of (possibly infinite) rank r. Then, for any finite group G with at most r generators, there would exist a surjection $\beta : N \twoheadrightarrow G$. Such a surjection would correspond to a G-Galois field extension $\mathcal{L}_\beta/F_N$. Since G is finite, this extension and its Galois action would be defined by a finite set S of polynomials with coefficients in F_N. Thus S would be defined over some field E_β where $k(x) \subset E_\beta \subset F_N$ and where $E_\beta/k(x)$ has finite degree. The Galois group of $k(x)_\infty$ over E_β is an open subgroup N_β of $\pi_1(\mathbb{A}^1_k)$. Since $E_\beta \subset F_N$, one sees that $N \subset N_\beta$. Moreover, there exists a G-Galois extension L_β/E_β such that $L_\beta \subset k(x)_\infty$ and $L_\beta \cdot F_N = \mathcal{L}_\beta$. This process is called "descending" the extension.

Another computation shows that the fiber product of two linearly disjoint covers $y^p - y = c_1 x$ and $y^p - y = c_2 x$ yields a cover $Y \to \mathbb{P}^1_k$ totally ramified over ∞ where Y again has genus 0. Thus, for any open normal subgroup of $\pi_1(\mathbb{A}^1_k)$ containing N, the corresponding cover of $\mathbb{P}^1_k$ has genus zero and is totally ramified over ∞. In particular, consider the (not necessarily Galois) cover $U_\beta \to \mathbb{P}^1_k$ corresponding to the subgroup N_β. Then U_β has genus 0 and the fibre of U_β over ∞ consists of one point P_∞. Since E_β is the function field of U_β, the existence of a G-Galois extension L_β/E_β with $L_\beta \subset k(x)_\infty$ implies that there exists a G-Galois cover $V \to U_\beta$ branched only at P_∞. Choosing G to be prime-to-p and generated by $r \geq 1$ elements, this leads to a contradiction with the fact that $\pi_1^{p'}(\mathbb{A}^1_k)$ is trivial.

6.2 Quotients of fundamental groups of curves. Another approach to understanding the fundamental group is to study how its finite quotients fit together by solving embedding problems. This topic is especially important for the fundamental group of a k-curve X which is projective. The reason is that, when X is projective, then $\pi_1(X) = \pi_1^t(X)$ and so Theorem 5.1 implies that $\pi_1(X)$ is finitely generated as a profinite group. As such, by [10, Prop. 15.4], it is determined by its finite quotients (i.e., it is determined by the answer to question (1) from Section 4.3). Unfortunately, Abhyankar's Conjecture (Theorem 5.4) does not apply to projective curves and for $g \geq 2$, the finite quotients of $\pi_1(X)$ are unknown. However, by Theorems 5.1 and 5.2 the maximal prime-to-p and pro-p quotients of the fundamental group of every projective k-curve are known. Thus the question becomes, how do these prime-to-p and pro-p quotients fit together?

A first step is to determine which finite groups G having a normal p-Sylow subgroup P occur as a quotient of $\pi_1(X)$. Such a quotient corresponds to an unramified G-Galois cover $Z \to X$ which factors as $Z \to Y \to X$ where $\mathrm{Gal}(Z/Y) = P$ and $\mathrm{Gal}(Y/X)$ is the prime-to-p group $H = G/P$. In [25, Thm. 7.5], a necessary and sufficient condition is given for such G-Galois covers of X to occur. The result essentially says that the H-module structure of P must be compatible with the H-module structure of $J_Y[p]$ for some H-Galois cover $Y \to X$. The compatibility is measured in terms of a generalization of the p-rank called the Hasse-Witt invariants [34, Section 2] of Y. Since H is prime-to-p and the prime-to-p quotients of $\pi_1(X)$ are known, this result gives insight into the structure of $\pi_1(X)$. The result was extended by Borne [4, Thm. 1.1] to the case where $|H|$ is not necessarily prime-to-p. The proof in that case uses modular representation theory.

Nevertheless, the structure of $\pi_1(X)$ and its finite quotients are still unknown when $g \geq 2$. A complete analysis of this problem seems beyond reach for now. The results in [4] and [25] give conditions to solve the embedding problems when the kernel is a p-group. Thus, there is a natural question to ask next.

Question 6.3 Given a projective curve X and an embedding problem $(\beta : \pi_1(X) \twoheadrightarrow G, \alpha : \Gamma \twoheadrightarrow G)$ with $|\ker(\alpha)|$ prime-to-p, what conditions on Γ and X will ensure the existence of a proper solution?

6.3 Ramification of covers of curves. Given X, B, and G, only in special cases is it known what ramification data can occur for G-Galois covers $\phi : Y \to X$ with branch locus in B. Answering this question is necessary to determine which values will occur for the genus of Y. This is important for the goal of understanding the fundamental group $\pi_1(X)$, because the finite quotients of $\pi_1(Y)$ will depend on invariants like the genus or the p-rank of Y.

This topic is most interesting for the case of wildly ramified covers of affine k-curves because, in this case, there is the extra structure of the filtration of higher ramification groups to consider. One result is that a cover can always be deformed using formal patching to lengthen the filtration of higher ramification groups at a wildly ramified point. Since the degree of the different depends on the ramification filtration, this leads to the following result.

Theorem 6.4 [30, Cor. 3.4] *Suppose $X - B$ is an affine k-curve and G is a finite quotient of $\pi_1(X - B)$ such that p divides $|G|$. Let $N \in \mathbb{N}$. Then there exists a G-Galois cover $\phi : Y \to X$ with branch locus in B such that the genus of Y is greater than N.*

An open problem is to determine the smallest genus that can occur for a G-Galois cover of X with branch locus in B. Because of results like Proposition 5.8, the smallest genus will often depend on the subset B, not just on its cardinality.

Question 6.5 Given an affine k-curve $X-B$ and a finite quotient G of $\pi_1(X-B)$, what is the smallest positive integer $g = g(X,B,G)$ which occurs as the genus of Y for a G-Galois cover $\phi : Y \to X$ with branch locus in B?

A crucial case is to understand Galois covers of the affine line. By Abhyankar's Conjecture, there exists a G-Galois cover of the affine line if and only if G is quasi-p, which means that G is generated by p-groups. For the affine line, if G is an abelian p-group, then the answer to Question 6.5 can be determined by class field theory. There are many other quasi-p groups, including all simple groups with order divisible by p. When G is the projective special linear group $\mathrm{PSL}_2(\mathbb{F}_p)$, then the answer to Question 6.5 is $(p-1)^2/4$, [6]. Under certain group theoretic conditions, an upper bound for the minimal genus can be found in [28, Thm. 3.5].

One example of a quasi-p group is a non-abelian semi-direct product G of the form $(\mathbb{Z}/\ell)^a \rtimes \mathbb{Z}/p$ where ℓ and p are distinct primes and a is the order of ℓ modulo p. In a group project supervised by the authors at the WIN conference in Banff, November 2008, the group calculated the minimal genus that can occur for a Galois cover of the affine line with this group G. Specifically, in [12, Thm. 4.1], the group proved that there is a $(\mathbb{Z}/\ell\mathbb{Z})^a \rtimes \mathbb{Z}/p\mathbb{Z}$-Galois cover $Z \to \mathbb{P}^1_k$ branched only at ∞ with genus $g_Z = 1 + \ell^a(p-3)/2$ if p is odd. In addition, the group proved that this is the minimal genus and that there are only finitely many curves of this minimal genus which are Galois covers of the affine line with this Galois group. For the proof, the group determined the action of an automorphism of order p on $J_Y[\ell]$ where Y is the Artin-Schreier curve $y^p - y = x^d$. This gave insight into the unramified elementary abelian ℓ-group covers of Y that are Galois over $\mathbb{P}^1_k$. We now extend this result to a more general class of quasi-p groups.

For a finite group G, let $\Phi(G)$ denote the Frattini subgroup of G (the intersection of all proper maximal subgroups of G). This is the set of "non-generators" of G. If ℓ is a prime and L is an ℓ-group then $\Phi(L) = L^\ell[L,L]$ and $\mathcal{L} = L/\Phi(L)$ is an elementary abelian ℓ-group. We will need the following lemma.

Lemma 6.6 *Let ℓ and p be distinct primes and let a be the order of ℓ modulo p. There is a unique non-abelian semi-direct product of the form $(\mathbb{Z}/\ell)^a \rtimes \mathbb{Z}/p$ up to isomorphism.*

Proof Let G be a non-abelian semi-direct product of the form $(\mathbb{Z}/\ell)^a \rtimes \mathbb{Z}/p$. Then G is determined by a non-trivial homomorphism $\gamma : \mathbb{Z}/p \to \mathrm{Aut}((\mathbb{Z}/\ell)^a)$. The isomorphism type of G depends only on $\mathrm{Im}(\gamma)$ because of the flexibility of choice of a generator for $\mathbb{Z}/p$. Furthermore, it depends only on the conjugacy class of $\mathrm{Im}(\gamma)$ because of the choice of basis for $(\mathbb{Z}/\ell)^a$. Thus, to show that G is unique up to isomorphism, it suffices to show that all subgroups of order p in $\mathrm{Aut}((\mathbb{Z}/\ell)^a) \simeq \mathrm{GL}_a(\mathbb{Z}/\ell)$ are conjugate. Let $H \subset \mathrm{GL}_a(\mathbb{Z}/\ell)$ be a subgroup of order p and let $h \in H$ be a generator. Up to conjugacy, h can be chosen in rational canonical form. Since a is the order of ℓ modulo p, the vector space $(\mathbb{Z}/\ell)^a$ is indecomposable under the semi-direct product action. The matrix h consists of one block since the action is indecomposable. Thus h is determined by its characteristic polynomial $f_h(x)$. Then $f_h(x)$ is an irreducible (degree a) factor of the cyclotomic polynomial $\Phi_p(x)$. After possibly changing the generator $h \in H$, then $f_h(x)$ is the minimal polynomial for a fixed pth root of unity ζ_p. Thus the conjugacy class of H is uniquely determined. □

Here is the answer to Question 6.5 for groups of the form $L \rtimes \mathbb{Z}/p$ where L is an ℓ-group whose maximal elementary abelian quotient is $(\mathbb{Z}/\ell)^a$.

Proposition 6.7 *Let ℓ and p be distinct primes with p odd. Suppose L is an ℓ-group such that the quotient $L/\Phi(L)$ is elementary abelian of rank $a = ord_p(\ell)$. Suppose Γ is a quasi-p group which is a semi-direct product of the form $L \rtimes \mathbb{Z}/p$. Then there exists a Γ-Galois cover $W \to \mathbb{P}^1_k$ branched only at ∞ such that the genus of W is $g_W = 1+|L|(p-3)/2$. This is the minimal genus that occurs for a Γ-Galois cover of $\mathbb{P}^1_k$ branched only at ∞.*

Before proving Proposition 6.7, we need some information about Frattini covers. A surjective group homomorphism $\phi : G \twoheadrightarrow H$ is a *Frattini cover* if $\ker(\phi) \subset \Phi(G)$. For each finite (even profinite) group H, there exists a cover $\tilde{\phi} : \mathcal{H} \to H$, unique up to isomorphism, such that $\tilde{\phi}$ is the largest Frattini cover of H. The group $\mathcal{H}$ is the *universal Frattini cover* of H (see [10, Chapter 20, sections 6 and 7] or [9, 22.11 and 22.12] for definitions and details). A group N is a normal subgroup of $\mathcal{H}$ if and only if it is a Frattini cover of H.

The universal Frattini cover of H is in fact the smallest cover of H that is projective. In other words, every embedding problem $(\tilde{\phi} : \mathcal{H} \twoheadrightarrow H, \alpha : G \twoheadrightarrow H)$ has a *weak* solution λ. When α is a Frattini cover, then λ is automatically a proper solution (i.e., surjective).

Proof The group Γ has a quotient H which is a semi-direct product $(\mathbb{Z}/\ell)^a \rtimes \mathbb{Z}/p$. Since Γ is quasi-p, the group H is non-abelian. By Lemma 6.6, the structure of H is uniquely determined up to isomorphism.

Let $\mathcal{L}$ be the universal Frattini cover of $(\mathbb{Z}/\ell\mathbb{Z})^a$. This is a free pro-$\ell$ group of rank a. Because $\mathcal{L}/\Phi(\mathcal{L}) = (\mathbb{Z}/\ell\mathbb{Z})^a$ and $\Phi(\mathcal{L})$ is the set of non-generators of $\mathcal{L}$, the infinite group $\mathcal{L}$ can be generated by a elements.

The semi-direct product H is determined by an action of $\mathbb{Z}/p$ on $(\mathbb{Z}/\ell\mathbb{Z})^a$. This induces an action of $\mathbb{Z}/p$ on $\mathcal{L}$ [10, Prop. 22.12.2]. Let $\mathcal{L} \rtimes \mathbb{Z}/p$ be the resulting semi-direct product. Then $\mathcal{L} \rtimes \mathbb{Z}/p$ is the universal Frattini ℓ-cover of $(\mathbb{Z}/\ell\mathbb{Z})^a \rtimes \mathbb{Z}/p\mathbb{Z}$ and Γ is a quotient of $\mathcal{L} \rtimes \mathbb{Z}/p$. That is, there exists a normal subgroup N of $\mathcal{L}$ that is $\mathbb{Z}/p$-invariant with $(\mathcal{L}/N) \rtimes \mathbb{Z}/p = \Gamma$.

By [12, Thm. 4.1], there is a $(\mathbb{Z}/\ell\mathbb{Z})^a \rtimes \mathbb{Z}/p\mathbb{Z}$-Galois cover $Z \to \mathbb{P}^1_k$ branched only at ∞. Furthermore, it factors through the Artin-Schreier cover $\phi : Y_2 \to \mathbb{P}^1_k$ with equation $y^p - y = x^2$. Also the $(\mathbb{Z}/\ell\mathbb{Z})^a$-Galois cover $Z \to Y_2$ is unramified. This yields a surjection $\psi_1 : \pi_1(Y_2) \twoheadrightarrow (\mathbb{Z}/\ell\mathbb{Z})^a$. Since ℓ is prime-to-p and $a = \mathrm{ord}_p(\ell) \leq p-1 = 2g(Y_2)$, by [1, Cor. 2.12] there exists a surjection $\psi_2 : \pi_1(Y_2) \twoheadrightarrow \mathcal{L}$ that dominates ψ_1. This induces an infinite unramified $\mathcal{L}$-Galois extension F of the function field $k(Y_2)$ of Y_2.

As $k(Y_2)$ is a $\mathbb{Z}/p$-Galois extension of $k(x) = k(\mathbb{P}^1_k)$ branched only at ∞, the extension $F/k(x)$ is algebraic and branched only at ∞. Let F' be the Galois closure of $F/k(x)$. Then $F'/k(Y_2)$ is a Galois extension with pro-l Galois group that surjects onto $\mathcal{L}$ and thus also onto $(\mathbb{Z}/\ell\mathbb{Z})^a$. But $\mathcal{L}$ is universal for all pro-l groups surjecting onto $(\mathbb{Z}/\ell\mathbb{Z})^a$ [9, Remark 22.11.19] so $F' = F$ and the extension $F/k(x)$ is Galois. By Schur-Zassenhaus the Galois group is $\mathcal{L} \rtimes \mathbb{Z}/p$. Thus there is a surjection $\psi_2' : \pi_1(\mathbb{A}^1_k) \to \mathcal{L} \rtimes \mathbb{Z}/p$.

Taking the composition of ψ_2' with the natural surjection $\mathcal{L} \rtimes \mathbb{Z}/p \to \Gamma$, this yields a surjection $\lambda : \pi_1(\mathbb{A}^1_k) \to L \rtimes \mathbb{Z}/p$. This induces an unramified Γ-Galois cover $W \to \mathbb{P}^1_k$ branched only at ∞ and dominating ϕ. Moreover, the cover $W \to Y_2$ is unramified.

By the Riemann-Hurwitz formula, the genus of W is $1 + |L|(p-3)/2$. The statement that this is the minimal genus follows just as in [12, Thm. 4.1], since the minimal genus will be realized when the L-Galois subcover is unramified and the genus of the $\mathbb{Z}/p$-Galois quotient is the smallest positive number possible. □

6.4 An open question on arithmetic invariants of Galois covers. As discussed in Section 4.7, there is a connection between unramified $\mathbb{Z}/p$-Galois covers of a projective curve and the p-torsion of its Jacobian. As a result (see Theorem 5.7), the fundamental group $\pi_1(X - B)$ will depend on the p-rank s_Y when $\phi : Y \to X$ is a Galois cover with branch locus in B. For this reason, there is good motivation to understand the values that occur for the p-rank associated with covers. Even for the case when G is cyclic and $X = \mathbb{P}^1_k$, there are many papers on this subject, e.g., [5], [44].

There are arithmetic invariants of the Jacobian of a k-curve other than its p-rank, including the Newton polygon and the p-torsion group scheme (see [8] and [24] respectively). As an example, recall that an elliptic k-curve E can be either ordinary or supersingular. The two cases can be distinguished by the number of points in $E[p](k)$, which is either p or 1. If E is ordinary, then its Newton polygon has slopes 0 and 1. The p-torsion group scheme of an ordinary elliptic curve is $E[p] \simeq \mathbb{Z}/p \oplus \mu_p$ where μ_p is the kernel of Frobenius on $\mathbb{G}_m$. If E is supersingular, then its Newton polygon has slopes 1/2. The p-torsion group scheme of a supersingular elliptic curve fits into a (non-split) short exact sequence $1 \to \alpha_p \to E[p] \to \alpha_p \to 1$ where α_p is the kernel of Frobenius on $\mathbb{G}_a$. Let $E_{ss}[p]$ denote the (unique) isomorphism class of the p-torsion group scheme of a supersingular elliptic curve.

While the connection between these other invariants and the fundamental group is not clear, it still raises the following question.

Question 6.8 Given a finite group G which is a quotient of $\pi_1(X - B)$, what are the possibilities for the p-rank, Newton polygon, and p-torsion group scheme of J_Y for G-Galois covers $\phi : Y \to X$ with branch locus in B?

Here is a new result about this question, building upon the group result in [12, Thm. 4.1]. We find a Galois cover $Z \to \mathbb{P}^1_k$ branched only at ∞ with Galois group $(\mathbb{Z}/\ell\mathbb{Z})^a \rtimes \mathbb{Z}/p\mathbb{Z}$ such that Z has small genus and large p-rank.

Proposition 6.9 *Let ℓ and p be distinct primes with p odd and $\ell \geq -1 + (p-1)^2/2$. Let a be the order of ℓ modulo p. Suppose G is the non-abelian semi-direct product $(\mathbb{Z}/\ell)^a \rtimes \mathbb{Z}/p$. Then there exists a Galois cover $Z \to \mathbb{P}^1$ branched only at ∞, with Galois group G, genus $g_Z = 1 + \ell^a(p-3)/2$ and p-rank $s_Z = (\ell^a - 1)(p-3)/2$. Furthermore, $J_Z[p]$ decomposes completely into s_Z copies of $\mathbb{Z}/p \oplus \mu_p$ and $(p-1)/2$ copies of $E_{ss}[p]$, the p-torsion group scheme of a supersingular elliptic curve. In particular, the Newton polygon of J_Z only has slopes 0, 1/2, and 1.*

Proof Consider the cover $\phi : Y \to \mathbb{P}^1_k$ with affine equation $y^p - y = x^2$. Then Y has genus $g_Y = (p-1)/2$ and p-rank 0. By [29, Cor. 3.3], J_Y is superspecial, i.e., $J_Y[p]$ decomposes into g_Y copies of $E_{ss}[p]$. In particular, J_Y is supersingular, i.e., the slopes of its Newton polygon all equal 1/2.

If $Z_1 \to Y$ is an unramified $\mathbb{Z}/\ell$-Galois cover, then Z_1 has genus $g_{Z_1} = 1 + \ell(p-3)/2$ by the Riemann-Hurwitz formula. Suppose $\ell \neq p$ is prime such that $\ell+1 \geq (p-1)^2/2$. By [32, 4.3.1], there exists an unramified $\mathbb{Z}/\ell$-Galois cover $Z_1 \to Y$ such that the new part of J_{Z_1} is ordinary. Thus Z_1 has p-rank $s_{Z_1} = (\ell-1)(p-3)/2$ and $J_{Z_1}[p]$ contains a factor isomorphic to $(\mathbb{Z}/p \oplus \mu_p)^{s_{Z_1}}$.

Also J_Y is isogenous to a factor of J_{Z_1}. Since the cover $Z_1 \to Y$ has degree ℓ, the degree of the isogeny is prime-to-p. As a result, $J_{Z_1}[p]$ contains a factor isomorphic to $J_Y[p]$. Thus $J_{Z_1}[p]$ decomposes into s_{Z_1} copies of $(\mathbb{Z}/p \oplus \mu_p)$ and $(p-1)/2$ copies of $E_{ss}[p]$. In particular, the Newton polygon of J_{Z_1} has slopes 0, 1/2, and 1.

Consider the action of an automorphism σ of Y of order p on the set of unramified cyclic $\mathbb{Z}/\ell$-Galois covers of Y. If Z_2 is in the orbit of Z_1 under the action of σ, then Z_2 and Z_1 are isomorphic, and so every invariant of the curves is the same. Consider the Galois closure $\psi : Z \to Y \to \mathbb{P}^1$ of $Z_1 \to Y \to \mathbb{P}^1$. The Galois group of ψ is isomorphic to G since it is a semi-direct product of the form $(\mathbb{Z}/\ell)^a \rtimes \mathbb{Z}/p$, by Lemma 6.6. The genus is $g_Z = 1 + \ell^a(p-3)/2$ by the Riemann-Hurwitz formula.

Relative to the cover $Z \to Y$, the new part of $J_Z[p]$ is ordinary and the old part of $J_Z[p]$ is isomorphic to $Y[p]$. Thus the curve Z has p-rank $s_Z = (\ell^a - 1)(p-3)/2$ and $J_Z[p]$ decomposes completely into s_Z copies of $\mathbb{Z}/p \oplus \mu_p$ and $(p-1)/2$ copies of $E_{ss}[p]$. In particular, the Newton polygon of J_Z has slopes 0, 1/2, and 1. □

References

[1] *Revêtements étales et groupe fondamental.* Springer-Verlag, Berlin, 1971. Séminaire de Géométrie Algébrique du Bois Marie 1960–1961 (SGA 1), Dirigé par Alexandre Grothendieck. Augmenté de deux exposés de M. Raynaud, Lecture Notes in Mathematics, Vol. 224.

[2] *Théorie des topos et cohomologie étale des schémas. Tome 3.* Lecture Notes in Mathematics, Vol. 305. Springer-Verlag, Berlin, 1973. Séminaire de Géométrie Algébrique du Bois-Marie 1963–1964 (SGA 4), Dirigé par M. Artin, A. Grothendieck et J. L. Verdier. Avec la collaboration de P. Deligne et B. Saint-Donat.

[3] Shreeram Abhyankar. Coverings of algebraic curves. *Amer. J. Math.*, 79:825–856, 1957.

[4] Niels Borne. A relative Shafarevich theorem. *Math. Z.*, 248(2):351–367, 2004.

[5] Irene I. Bouw. The p-rank of ramified covers of curves. *Compositio Math.*, 126(3):295–322, 2001.

[6] Irene I. Bouw and Stefan Wewers. Stable reduction of modular curves. In *Modular curves and abelian varieties*, volume 224 of *Progr. Math.*, pages 1–22. Birkhäuser, Basel, 2004.

[7] Richard M. Crew. Etale p-covers in characteristic p. *Compositio Math.*, 52(1):31–45, 1984.

[8] Michel Demazure. *Lectures on p-divisible groups.* Springer-Verlag, Berlin, 1972. Lecture Notes in Mathematics, Vol. 302.

[9] Michael D. Fried and Moshe Jarden. *Field arithmetic*, volume 11 of *Ergebnisse der Mathematik und ihrer Grenzgebiete. 3. Folge. A Series of Modern Surveys in Mathematics [Results in Mathematics and Related Areas. 3rd Series. A Series of Modern Surveys in Mathematics].* Springer-Verlag, Berlin, second edition, 2005.

[10] Michael D. Fried and Moshe Jarden. *Field arithmetic*, volume 11 of *Ergebnisse der Mathematik und ihrer Grenzgebiete. 3. Folge. A Series of Modern Surveys in Mathematics [Results in Mathematics and Related Areas. 3rd Series. A Series of Modern Surveys in Mathematics].* Springer-Verlag, Berlin, third edition, 2008. Revised by Jarden.

[11] Phillip Griffiths and Joseph Harris. *Principles of algebraic geometry.* Wiley-Interscience [John Wiley & Sons], New York, 1978. Pure and Applied Mathematics.

[12] Linda Gruendken, Laura Hall-Seelig, Bo-Hae Im, Ekin Ozman, Rachel Pries, and Katherine Stevenson. Semi-direct Galois covers of the affine line. This volume.

[13] David Harbater. Abhyankar's conjecture on Galois groups over curves. *Invent. Math.*, 117(1):1–25, 1994.

[14] David Harbater. Fundamental groups and embedding problems in characteristic p. In *Recent developments in the inverse Galois problem (Seattle, WA, 1993)*, volume 186 of *Contemp. Math.*, pages 353–369. Amer. Math. Soc., Providence, RI, 1995.

[15] David Harbater. Abhyankar's conjecture and embedding problems. *J. Reine Angew. Math.*, 559:1–24, 2003.

[16] David Harbater and Katherine F. Stevenson. Local Galois theory in dimension two. *Adv. Math.*, 198(2):623–653, 2005.

[17] Robin Hartshorne. *Algebraic geometry.* Springer-Verlag, New York, 1977. Graduate Texts in Mathematics, No. 52.

[18] Manish Kumar. Fundamental group in positive characteristic. *J. Algebra*, 319(12):5178–5207, 2008.

[19] Ariane Mézard. Fundamental group. In *Courbes semi-stables et groupe fondamental en géométrie algébrique (Luminy, 1998)*, volume 187 of *Progr. Math.*, pages 141–155. Birkhäuser, Basel, 2000.

[20] James S. Milne. *Étale cohomology*, volume 33 of *Princeton Mathematical Series.* Princeton University Press, Princeton, N.J., 1980.

[21] Rick Miranda. *Algebraic curves and Riemann surfaces*, volume 5 of *Graduate Studies in Mathematics.* American Mathematical Society, Providence, RI, 1995.

[22] Shinichi Mochizuki. The local pro-p anabelian geometry of curves. *Invent. Math.*, 138(2):319–423, 1999.

[23] David Mumford. *Abelian varieties.* Tata Institute of Fundamental Research Studies in Mathematics, No. 5. Published for the Tata Institute of Fundamental Research, Bombay, 1970.

[24] F. Oort. A stratification of a moduli space of abelian varieties. In *Moduli of abelian varieties (Texel Island, 1999)*, volume 195 of *Progr. Math.*, pages 345–416. Birkhäuser, Basel, 2001.

[25] Amílcar Pacheco and Katherine F. Stevenson. Finite quotients of the algebraic fundamental group of projective curves in positive characteristic. *Pacific J. Math.*, 192(1):143–158, 2000.

[26] Amílcar Pacheco, Katherine F. Stevenson, and Pavel Zalesskii. Normal subgroups of the algebraic fundamental group of affine curves in positive characteristic. *Math. Ann.*, 343(2):463–486, 2009.

[27] Florian Pop. Étale Galois covers of affine smooth curves. The geometric case of a conjecture of Shafarevich. On Abhyankar's conjecture. *Invent. Math.*, 120(3):555–578, 1995.

[28] Rachel J. Pries. Conductors of wildly ramified covers. I. *C. R. Math. Acad. Sci. Paris*, 335(5):481–484, 2002.

[29] Rachel J. Pries. Jacobians of quotients of Artin-Schreier curves. In *Recent progress in arithmetic and algebraic geometry*, volume 386 of *Contemp. Math.*, pages 145–156. Amer. Math. Soc., Providence, RI, 2005.

[30] Rachel J. Pries. Wildly ramified covers with large genus. *J. Number Theory*, 119(2):194–209, 2006.

[31] M. Raynaud. Revêtements de la droite affine en caractéristique $p > 0$ et conjecture d'Abhyankar. *Invent. Math.*, 116(1-3):425–462, 1994.

[32] Michel Raynaud. Sections des fibrés vectoriels sur une courbe. *Bull. Soc. Math. France*, 110(1):103–125, 1982.

[33] Luis Ribes and Pavel Zalesskii. *Profinite groups*, volume 40 of *Ergebnisse der Mathematik und ihrer Grenzgebiete. 3. Folge. A Series of Modern Surveys in Mathematics [Results in Mathematics and Related Areas. 3rd Series. A Series of Modern Surveys in Mathematics].* Springer-Verlag, Berlin, 2000.

[34] H.-G. Rück. Class groups and L-series of function fields. *J. Number Theory*, 22(2):177–189, 1986.

[35] Jean-Pierre Serre. *Cohomologie galoisienne*, volume 1965 of *With a contribution by Jean-Louis Verdier. Lecture Notes in Mathematics, No. 5. Troisième édition.* Springer-Verlag, Berlin, 1965.

[36] Jean-Pierre Serre. *Local fields*, volume 67 of *Graduate Texts in Mathematics.* Springer-Verlag, New York, 1979. Translated from the French by Marvin Jay Greenberg.

[37] Jean-Pierre Serre. Construction de revêtements étales de la droite affine en caractéristique p. *C. R. Acad. Sci. Paris Sér. I Math.*, 311(6):341–346, 1990.

[38] I. Shafarevitch. On p-extensions. *Rec. Math. [Mat. Sbornik] N.S.*, 20(62):351–363, 1947.

[39] Stephen S. Shatz. *Profinite groups, arithmetic, and geometry.* Princeton University Press, Princeton, N.J., 1972. Annals of Mathematics Studies, No. 67.

[40] Joseph H. Silverman. *The arithmetic of elliptic curves*, volume 106 of *Graduate Texts in Mathematics*. Springer-Verlag, New York, 1992. Corrected reprint of the 1986 original.

[41] Henning Stichtenoth. *Algebraic function fields and codes*, volume 254 of *Graduate Texts in Mathematics*. Springer-Verlag, Berlin, second edition, 2009.

[42] Akio Tamagawa. The Grothendieck conjecture for affine curves. *Compositio Math.*, 109(2):135–194, 1997.

[43] Akio Tamagawa. On the fundamental groups of curves over algebraically closed fields of characteristic > 0. *Internat. Math. Res. Notices*, (16):853–873, 1999.

[44] N. Yui. On the Jacobian varieties of hyperelliptic curves over fields of characteristic $p > 2$. *J. Algebra*, 52(2):378–410, 1978.

Fields Institute Communications
Volume **60**, 2011

Covers of the Affine Line in Positive Characteristic with Prescribed Ramification

Irene I. Bouw
Institute of Pure Mathematics
Ulm University
D-89069 Ulm, Germany
bouw@math.uni-duesseldorf.de

Abstract. This note concerns the existence of Galois covers of curves in positive characteristic with prescribed ramification. We fix a quasi-p group G and ask what the minimal genus of an unramified G-Galois cover of the affine line is. We answer this question in the case that $G = A_d$ is the alternating group on $p + 2 \leq d < 2p$ letters.

Introduction

Let k be an algebraically closed field of characteristic $p > 0$. In this note we consider Galois covers $Y \to \mathbb{P}^1_k$ which are only branched at ∞. We call such covers *unramified covers of the affine line*. It follows from Abhyankar's conjecture for the affine line proved by Raynaud ([7]) that such a cover exists for a given group G if and only if G is a quasi-p group, i.e. can be generated by elements of p-power order. Many groups satisfy this property, for example all simple groups of order divisible by p are quasi-p groups. For a given quasi-p group G there are many unramified covers of the affine line. For example, if $G = \mathbb{Z}/p\mathbb{Z}$ there are infinitely many such covers.

In this note we fix a quasi-p group G and ask what the minimal genus of an unramified G-Galois cover of the affine line is. This question has been motivated by the work of Muskat–Pries ([6]). Muskat and Pries compute the genus of many alternating-group covers which had been found by Abhyankar ([1]). It turns out that the covers they consider have "small" genus in a sense which can be made precise. We show that for $p+2 \leq d < 2p$ the covers of [6] are indeed the unramified A_d-covers of the affine line of minimal genus.

Another motivation for studying the existence of covers with given ramification comes from the theory of stable reduction. One method for showing that covers with given tame ramification exist in positive characteristic is to show that not all covers in characteristic zero have bad reduction to characteristic p. Associated with

2010 *Mathematics Subject Classification.* Primary 14H25; Secondary 11G20.

a cover with bad reduction is a set of tail covers. Essentially these are the restriction to certain irreducible components of the stable reduction of the cover. A technique due to Wewers ([9]) sometimes allows one to count the number of covers with bad reduction in terms of sufficient knowledge of these tail covers. This technique has been applied in [4] to covers of degree p which are tamely branched at 4 points. To generalize the result of [4] to degree $p < d < 2p$ one needs to know the existence of covers of the affine line of degree d of small genus.

The strategy of the proof of our main result is as follows. Rather than considering covers with fixed Galois group, we consider unramified covers $f : X \to \mathbb{P}^1$ of the affine line which are non-Galois. Let d be the degree of f. As an application of the Riemann–Hurwitz formula in positive characteristic, we compute the genus of the Galois closure of f in terms of the ramification of f (Proposition 1.3). We conclude that for $p+2 \leq d < 2p$ the covers considered in [6] indeed have minimal genus. For $d = p+1$ the cover of minimal genus is another well-known cover. This cover is in fact given by the same equation as the covers in degree $d \geq p+2$, but the Galois group of the Galois closure is $\mathrm{PSL}_2(p)$ (Proposition 2.1). For $d = p$, we use the theory of stable reduction to construct covers with minimal genus. In this case, we were not able to specify the Galois group G of the Galois closure.

In this paper, we always suppose that $d < 2p$. This implies that p strictly divides the order of the Galois group. This condition limits the ramification indices; this makes all arguments combinatorially rather simple. This condition is also natural for the applications to stable reduction: it is a necessary condition to apply the results of Raynaud ([8]) and Wewers ([9]).

I thank the referee for pointing out a mistake in the proof of Proposition 2.1 in the original version of this paper.

1 The ramification invariant

Let $p > 2$ be a prime number and k an algebraically closed field of characteristic p. We fix an integer $p \leq d < 2p$. Suppose that $f : X \to \mathbb{P}^1_k$ is a non-Galois cover of degree d, defined over k, only branched at $z = \infty$. We call such covers *unramified degree-d covers of the affine line*, for simplicity.

The situation for unramified Galois covers of the affine line whose degree is less than $2p$ is well understood: all such covers are Artin–Schreier covers, and have degree p. In this case everything is known, therefore we exclude it.

Let $\tilde{f} : Y \to \mathbb{P}^1_k$ be the Galois closure of f, and let $G = \mathrm{Gal}(Y, \mathbb{P}^1)$ be its Galois group. It is easy to see that $\tilde{f}$ is also only branched at $z = \infty$. In particular, G is a quasi-p group. The assumption that $p \leq d < 2p$ implies, moreover, that p strictly divides the order of G.

We choose a ramification point $y \in Y$ of $\tilde{f}$, and let $I = I_y$ be its inertia group. The point y is necessarily wildly ramified. Since p strictly divides the order of G, it follows that I is the extension of a cyclic group C of order prime to p by a cyclic group P of order p.

Let h be the conductor of $\tilde{f}$ and m the order of C. The *ramification invariant* of f (or of $\tilde{f}$) is defined as $\sigma = h/m$. It is the last jump of the higher ramification groups of y in the upper numbering. The following well-known lemma is proved in [8, Lemme 1.1.6].

Lemma 1.1 *Let σ be as above, then $\sigma > 1$.*

The goal of this paper is to determine the minimal ramification invariant for certain classes of covers of the affine line. More concretely, we give partial answers to the following questions.

Question 1.2 (a) Let $p \le d < 2p$. Determine the minimal σ such that there exists a cover $f : X \to \mathbb{P}^1$ of degree d, only branched at ∞, with ramification invariant σ.

(b) Let G be a quasi-p group. Determine the minimal σ such that there exists a G-Galois cover $f : X \to \mathbb{P}^1$ only branched at ∞ with ramification invariant σ.

We start by studying Question 1.2.(a). We write $\sigma_{d,\min}$ for the minimal ramification invariant occurring for an unramified degree-d cover of the affine line. The following proposition yields a lower bound for $\sigma_{d,\min}$. Before stating it, we need some more notation.

Let $f : X \to \mathbb{P}^1$ be an unramified degree-d cover of the affine line with $p \le d < 2p$. We write $\tilde{f} : Y \to \mathbb{P}^1$ for the Galois closure of f and G for the Galois group of $\tilde{f}$. Recall that G may be naturally considered as a transitive subgroup of the symmetric group S_d. We normalize the inclusion $G < S_d$ such that $\mathrm{Gal}(\tilde{X}, X)$ is the stabilizer in G of 1.

Let $I < G$ be the inertia group of a wild ramification point of $\tilde{f}$, and write $|I| = pm$. Choose elements $\theta \in I$ of order m and $\tau \in I$ of order p. Without loss of generality, we may suppose that $\tau = (1\ 2 \cdots p)$. Then θ acting on $\{1, 2, \ldots, p\}$ has one orbit of length 1 and $(p-1)/m_2$ orbits of length m_2, for some integer m_2. We write $n_1, \ldots, n_t$ for the lengths of the orbits of θ acting on $\{p+1, p+2, \ldots, d\}$. Note that $\sum_{i=1}^t n_i = d - p$. We note that $m = \mathrm{lcm}(m_2, n_1, \ldots, n_t)$.

Proposition 1.3 (a) *The inverse image $f^{-1}(\infty) \subset X$ consists of $t+1$ points, with ramification indices $p, n_1, \ldots, n_t$, respectively.*

(b) *The ramification invariant satisfies*

$$\sigma = \frac{2g(X) + t + d - 1}{p - 1}.$$

Proof The ramification indices of $f : X \to \mathbb{P}^1$ are the lengths of the orbits of I acting on $\{1, 2, \ldots, d\}$. The statement of (a) follows therefore immediately from the definition of the n_i and the structure of I.

For (b), we compare the Riemann–Hurwitz formula for $\tilde{f} : Y \to \mathbb{P}^1$ with that for $Y \to X$. Let (h, m) be as above and G the Galois group of $\tilde{f}$. Then

$$\begin{aligned} 2g(Y) - 2 &= -2|G| + \frac{|G|}{pm}(pm - 1 + h(p-1)) \\ &= (2g(X) - 2)\frac{|G|}{d} + \frac{|G|}{pmd}\left(\sum_{i-1}^{t} n_i(p\frac{m}{n_i} - 1 + h(p-1))\right) + \frac{|G|}{pmd}p(m-1). \end{aligned}$$

The statement of (b) follows immediately from this, since $\sigma = h/m$. □

Corollary 1.4 (a) *Suppose that $d = p$. Then $\sigma_{d,\min} \ge (d+1)/(p-1)$.*

(b) *Suppose that $p + 1 \le d < 2p$. Then $\sigma_{d,\min} \ge d/(p-1)$.*

Proof Let $d = p$. Let $f : X \to \mathbb{P}^1$ be a degree-d cover of the affine line with ramification invariant σ. Its Galois group is a subgroup of A_p, hence $t = 0$. We conclude that $\sigma = (2g(X)+d-1)/(p-1)$. If $g(X)=0$ then $\sigma = (d-1)/(p-1) = 1$. But this value is not possible by Lemma 1.1. Part (a) follows.

If $p+1 \leq d < 2p$, we have that $t \geq 1$. As above, we find that $\sigma = (2g(X)+t+d-1)/(p-1) \geq d/(p-1)$. Part (b) follows. $\square$

Remark 1.5 We note that Proposition 1.3 may easily be generalized to covers branched at more than one point. (See for example Lemma 3.2 below.) Another straightforward generalization is to groups G such that p strictly divides the order of the group without the assumption that the degree d of the cover is strictly less than $2p$. In this situation, the permutation representation of the elements τ and θ is different, but the argument extends.

2 Computation of the minimal ramification invariants

In this section, we show that the lower bounds of Corollary 1.4 are indeed sharp for $p+1 \leq d < 2p$. Here we rely on realization results found in the literature. The analogous result for $d=p$ is proved in § 3.

Proposition 2.1 *Let $p+1 \leq d < 2p$. There exists a unique degree-d cover $f: X \to \mathbb{P}^1$ with $\sigma = d/(p-1)$. The Galois group of the Galois closure of f is*

$$\begin{cases} \mathrm{PSL}_2(p) & \text{if } d = p+1, \text{ and } p \geq 5, \\ \mathrm{PSL}_2(8) & \text{if } p = 7 \text{ and } d = p+2, \\ A_d & \text{otherwise.} \end{cases}$$

Proof We first suppose that $d \geq p+2$. The cover $f: X \to \mathbb{P}^1$ is given by the equation

$$x^d - zx^{d-p} + 1, \qquad (x,z) \mapsto z. \tag{2.1}$$

This equation has been discovered by Abhyankar ([1]), who also showed the statement on the Galois group.

The computation of the ramification invariant can be found in [6, Theorem 4.9]. In that paper Muskat–Pries also compute the precise structure of the inertia group of a ramification point of the Galois closure of f. Alternatively, one may observe that $g(X)=0$ and apply Proposition 1.3.(b).

Suppose $p \geq 5$. There exists a $\mathrm{PSL}_2(p)$-Galois cover $\tilde{f}: Y \to \mathbb{P}^1$ with ramification invariant $\sigma = (p+1)/(p-1)$, see for example [3, Cor. 4.2.6]. Let $H < \mathrm{PSL}_2(p)$ be the normalizer of a Sylow p-subgroup of $\mathrm{PSL}_2(p)$. Then H is a subgroup of order $p(p-1)/2$. Therefore $f: X := Y/H \to \mathbb{P}^1$ is a degree-$(p+1)$ cover which is only branched at one point. Alternatively, this may also be seen by considering (2.1) for $d = p+1$.

We now show the uniqueness of the cover. The assumptions $\deg(f) = d$ and $\sigma = d/(p-1)$ imply that $t=1$ and $g(X)=0$, where t is as in Proposition 1.3. In particular, $|f^{-1}(\infty)|$ consists of two points, with ramification indices p and $d-p$, respectively.

Choose a parameter x on X such that $x=\infty$ (resp. $x=0$) is the ramification point of order p (resp. of order $d-p$). We also choose a coordinate z on $\mathbb{P}^1$ such that $z=\infty$ is the branch point. Then f may be given by

$$z = \frac{h(x)}{x^{d-p}},$$

where $h(x)$ is a polynomial of degree d. The assumption that f is unbranched outside $z = \infty$ implies that $xh' - dh$ is a nonzero constant. We conclude that $h(x) = c_0 + c_1 x^{d-p} + c_2 x^d$ with $c_0 c_2 \not\equiv 0 \pmod p$. After suitably renormalizing the coordinates z and x, we may assume that $h(x) = 1 + x^d$.

□

The following corollary is an immediate consequence of Proposition 2.1.

Corollary 2.2 *Let $G = A_d$ with $p + 2 \leq d < 2p$. We exclude $d = p + 2$ in the case that $p = 7$. The minimal ramification invariant of an unramified cover of the affine line with Galois group A_d is $\sigma = d/(p-1)$.*

Proof Let $G = A_d$ with d as in the statement of the corollary. Then d is the minimal degree an unramified G-Galois cover of the affine line can have. Therefore we conclude from Proposition 1.3.(b) that the ramification invariant σ of such a cover satisfies

$$\sigma \geq \frac{d}{p-1}.$$

By Proposition 2.1 this value is attained. □

Remark 2.3 Let p and d be as in the statement of Corollary 2.2. Corollary 2.2 immediately yields an expression for the minimal genus of an A_d-Galois cover $\tilde{f} : Y \to \mathbb{P}^1$ which is only branched at $z = \infty$. For the concrete expression, we refer to [6, Theorem 4.9].

Corollary 2.4 (a) *There does not exist an unramified A_{p+1}-cover of the affine line with ramification invariant $\sigma = (p+1)/(p-1)$ and degree $p + 1$.*
(b) *There does not exist an unramified A_9-cover of the affine line with ramification invariant $\sigma = 9/6 = 3/2$ and degree 9.*

Proof Proposition 2.1 implies that there is a unique unramified cover of the affine line with degree $(p+1)$ and ramification invariant $\sigma = (p+1)/(p-1)$. But this cover has Galois group $\mathrm{PSL}_2(p)$. This proves (a). Part (b) is proved similarly. □

3 Application of stable reduction

In this section, we compute the minimal ramification invariant $\sigma_{p,\min}$ in degree $d = p$, where $p > 3$ is a prime. We exclude the case that $p = 3$, since all degree-3 covers of the affine line in characteristic 3 are Galois.

We use the theory of stable reduction. Rather than giving an overview of the theory, we illustrate the kind of results one may or may not expect to prove. For a quick introduction to stable reduction in the situation we need it here, we refer to [4, § 4]. To understand the statements of this section, it suffices to accept the statement of the description of the so called primitive tail covers as a black box.

The reason for only considering $d = p$ in this case is that this is much easier than the general case. For a thorough introduction to stable reduction of Belyi maps which is needed to understand the case $p < d < 2p$, we refer the reader to [9].

A *Belyi map* is a finite morphism $f : X \to \mathbb{P}^1_{\bar{\mathbb{Q}}}$ defined over $\bar{\mathbb{Q}}$ which is branched exactly at the three ordered points $0, 1, \infty$. Here X is a smooth projective curve. Belyi maps can be described topologically as *dessins d'enfant* or combinatorially in terms of generating systems.

Definition 3.1 Fix an integer $d > 1$. A *generating system* of degree d is a triple $\boldsymbol{\rho} = (\rho_1, \rho_2, \rho_3)$ of permutations $\rho_i \in S_d$ which satisfy

- $\rho_1\rho_2\rho_3 = 1$,
- $G := \langle \rho_1, \rho_2, \rho_3 \rangle \subset S_d$ acts transitively on $\{1, 2, \ldots, d\}$.

The *combinatorial type* of $\boldsymbol{\rho}$ is a tuple $(d; C(\rho_1), C(\rho_2), C(\rho_3))$, where d is the degree and $C(\rho_i)$ is the conjugacy class of ρ_i in S_d.

Two generating systems $\boldsymbol{\rho}$ and $\boldsymbol{\rho}'$ are *equivalent* if there exists a permutation $\tau \in S_d$ such that $\rho_i' = \tau\rho_i\tau^{-1}$ for $i = 1, 2, 3$. A general criterion for the existence of generating systems of given combinatorial type can be found in [2].

Let $\boldsymbol{\rho}$ be a generating system. The conjugacy class $C_i := C(\rho_i)$ of S_d corresponds to a partition $\sum_{i=1}^{r_i} n_i = d$ of d. The *length* $r_i = r(C_i)$ of C_i is the number of cycles of the elements of C_i. Note that we include the 1-cycles. The nonnegative integer $g := (d + 2 - r_1 - r_2 - r_3)/2$ we call the *genus* of the generating system. Note that $g = g(X)$ is the genus of the corresponding Belyi map $X \to \mathbb{P}^1$ and $r_i = |f^{-1}(x_i)|$ is the cardinality of the inverse image of the ith branch point.

Let p be an odd prime number. In this section, we consider Belyi maps $f : X \simeq \mathbb{P}^1 \to \mathbb{P}^1$ of combinatorial type $(p; p, 2^s, C)$, where p is the conjugacy class of a p-cycle, 2^s is the conjugacy class of the product of s disjoint transpositions, and C is any conjugacy class such that $g(X) = 0$. Such generating systems clearly exist for all s. The following generating system satisfies the criteria:

$$\rho_1 = (1\,2\,\cdots\,p), \quad \rho_2 = (1\,2)(3\,4)\cdots(2s-1\;s),$$
$$\rho_3 = (p\;p-1\;\cdots\;2s+1\;2s\;\cdots 4\,2).$$

Alternatively, one may apply the criterion from [2].

Since p divides one of the ramification indices, the cover f has bad reduction to characteristic p ([9, Lemma 1.4.(4)]). We refer to [9, § 1.1] for a detailed definition of bad reduction.

With f we may associate two so-called *primitive tail covers* $\bar{f}_2, \bar{f}_3$ ([4, Lemma 4.2]). The primitive tail cover $\bar{f}_i : \bar{X}_i \simeq \mathbb{P}^1 \to \mathbb{P}^1_k$ is a degree-p cover which is totally branched at $z = \infty$, tamely branched at $z = 0$ and unbranched elsewhere. The tame ramification above $z = 0$ is described by the conjugacy class $C(\rho_i)$, i.e. it is the *same* as the ramification above the ith branch point of the cover f we started with.

Let σ_i be the ramification invariant corresponding to the wild branch point of $\bar{f}_2$.

The following lemma is a straightforward generalization of Proposition 1.3.

Lemma 3.2 *Let $p \leq d < 2p$ and $f : X \to \mathbb{P}^1$ be a degree-d cover in characteristic p which is wildly branched above ∞, tamely branched above 0, and unbranched elsewhere. Let σ be the ramification invariant of ∞ and define t as in § 2. Let $n_1, \ldots, n_r$ with $\sum_{i=1}^{r} n_i = d$ be the ramification indices above 0. Then*

$$\sigma = \frac{2g(X) + r + t - 1}{p - 1}.$$

Lemma 3.2 implies in our situation that

$$\sigma_i = \frac{r(C_i) - 1}{p - 1}.$$

In particular, $\sigma_2 = (p - s - 1)/(p - 1)$.

Proposition 3.3 *Let $p > 3$ be a prime number. There exists a cover $\tilde{f} : \bar{Z} \to \mathbb{P}^1$ of degree p in characteristic p which is only branched at ∞ with ramification invariant $\sigma = (p+1)/(p-1)$.*

In particular, $\sigma_{p,\min} = (p+1)/(p-1)$.

Proof Let $\varphi : \mathbb{P}^1 \to \mathbb{P}^1$, $w \mapsto w^2$. Let $\bar{Z}$ be the normalization of $\bar{X}_2 \times_{\mathbb{P}^1,\varphi} \mathbb{P}^1$. We obtain a commutative square

$$\begin{array}{ccc} \bar{Z} & \longrightarrow & \bar{X}_2 \\ {\scriptstyle \tilde{f}_2}\downarrow & & \downarrow{\scriptstyle \bar{f}_2} \\ \mathbb{P}^1 & \xrightarrow{\varphi} & \mathbb{P}^1. \end{array}$$

The natural map $\tilde{f}_2 : \bar{Z} \to \mathbb{P}^1$ of degree p has ramification invariant $2\sigma_2 = 2(p-s-1)/(p-1)$ and is only branched above ∞. This follows from Abhyankar's Lemma (in the version found in [3, Lemma 2.1.2] or [6, Lemma 4.1]). We apply Lemma 3.2 to $\tilde{f}_2$, using that $\sigma = 2(p-s-1)/(p-1)$, $t = 0$, and $r = p$. Solving for $g(\bar{Z})$ yields that $g(\bar{Z}) = (p-1-2s)/2$.

The first statement follows by taking $s = (p-3)/2$. The second statement follows from the first and Corollary 1.4.(b).

□

In [10] one finds an alternative construction of degree-p maps branched at one point between an elliptic curve and the projective line in characteristic p. Zapponi ([10, Prop. 5]) shows that a given elliptic curve E admits such a map if and only if E is supersingular. Moreover, for given E the map is unique. Therefore a degree-p cover of the affine line with ramification invariant $\sigma = (p+1)/(p-1)$ is unique if and only if $p < 12$. This is very different from what we showed in Proposition 2.1 for $p+1 \leq d < 2p$.

We already constructed a $\mathrm{PSL}_2(p)$-Galois cover $\tilde{f} : Y \to \mathbb{P}^1$ of the affine line with ramification invariant $\sigma = (p+1)/(p-1)$ in Proposition 2.1. The group $\mathrm{PSL}_2(p)$ admits a subgroup of index p if and only if $p \in \{5, 7, 11\}$ ([5]). For these primes there exists therefore a degree-p cover of the affine line with $\sigma = (p+1)/(p-1)$ and Galois group $\mathrm{PSL}_2(p)$. The previous argument implies that it is unique. For large p I expect that there always exists a degree-p cover of the affine line with $\sigma = (p+1)/(p-1)$ whose Galois group is the alternating group A_p.

The results in this note seem to pose more new questions than they solve. Let me just mention a few.

- What is the minimal ramification invariant for $G = A_p$ or $G = A_{p+1}$?
- What happens for other transitive permutation groups which are also quasi-p groups? Natural groups to consider are the noncommutative simple groups, like the Mathieu groups or groups of Lie type.
- What is the minimal ramification invariant for A_9-covers in characteristic 7? (I guess this should be $\sigma = 11/6$.)

References

[1] S.S. Abhyankar, *Galois theory on the line in nonzero characteristic.* Bull. Amer. Math Soc. 27:68–133, 1992.

[2] G. Boccara, *Cycles comme produit de deux permutations de classes données.* Discrete Math. 23:129–142, 1982.

[3] I.I. Bouw and R.J. Pries, *Rigidity, reduction, and ramification.* Math. Ann. 326:803–824, 2003.

[4] I.I. Bouw and B. Osserman, *Some 4-point Hurwitz numbers in positive characteristic.* To appear in Trans. Amer. Math. Soc.

[5] B. Huppert, *Endliche Gruppen 1.* Grundlehren Math. Wiss. 134, Springer-Verlag, Berlin, 1967.

[6] J. Muskat and R. Pries, *Alternating group covers of the affine line.* To appear in Israel J. Math.

[7] M. Raynaud, *Revêtements de la droite affine en caractéristique* $p > 0$ *et conjecture d'Abhyankar.* Invent. Math. 116:425–462, 1994.

[8] M. Raynaud, *Spécialisation des revêtements en caractéristique* $p > 0$. Ann. Scient. Éc. Norm. Sup 36:87–126, 2003.

[9] S. Wewers, *Three point covers with bad reduction.* J. Amer. Math. Soc 16:991–1032, 2003.

[10] L. Zapponi, *On the 1-pointed curves arising as étale covers of the affine line in positive characteristic.* Math. Z. 258:711–727, 2008.

Fields Institute Communications
Volume **60**, 2011

Semi-Direct Galois Covers of the Affine Line

Linda Gruendken
Department of Mathematics, University of Pennsylvania
David Rittenhouse Lab, 209 South 33rd Street, Philadelphia, PA 19104-6395 USA
lindagr@math.upenn.edu

Laura Hall-Seelig
Department of Mathematics and Statistics, University of Massachusetts
Lederle Graduate Research Tower, Amherst, MA 01003-9305 USA
hall@math.umass.edu

Bo-Hae Im
Department of Mathematics, Chung-Ang University
221, Heukseok-dong, Dongjak-gu, Seoul 156-756, South Korea
imbh@cau.ac.kr

Ekin Ozman
Department of Mathematics, University of Wisconsin-Madison
480 Lincoln Drive, Madison, WI 53706 USA
ozman@math.wisc.edu

Rachel Pries
Department of Mathematics, Colorado State University
Fort Collins, CO 80523-1874 USA
pries@math.colostate.edu

Katherine Stevenson
Department of Mathematics, California State University
18111 Nordhoff St, Northridge, CA 91330-8313 USA
katherine.stevenson@csun.edu

Abstract. Let k be an algebraically closed field of characteristic $p > 0$. Let G be a semi-direct product of the form $(\mathbb{Z}/\ell\mathbb{Z})^b \rtimes \mathbb{Z}/p\mathbb{Z}$ where b is a positive integer and ℓ is a prime distinct from p. In this paper, we study Galois covers $\psi : Z \to \mathbb{P}^1_k$ ramified only over ∞ with Galois group G. We find the minimal genus of a curve Z which admits a covering map of this form and we give an explicit formula for this genus in terms of ℓ and p. The minimal genus occurs when b equals the order a of ℓ modulo b and we also prove that the number of curves Z of this minimal genus which admit such a covering map is at most $(p-1)/a$ when p is odd.

2010 *Mathematics Subject Classification.* Primary 14H30; Secondary 14E20, 14F40.

1 Introduction

Let k be an algebraically closed field of characteristic $p > 0$. In sharp contrast with the situation in characteristic 0, there exist Galois covers $\psi : Z \to \mathbb{P}^1_k$ ramified only over infinity. By Abhyankar's Conjecture [2], proved by Raynaud and Harbater [9], [4], a finite group G occurs as the Galois group of such a cover ψ if and only if G is quasi-p, i.e., G is generated by p-groups. This result classifies all the finite quotients of the fundamental group $\pi_1(\mathbb{A}^1_k)$. It does not, however, determine the profinite group structure of $\pi_1(\mathbb{A}^1_k)$ because this fundamental group is an infinitely generated profinite group.

There are many open questions about Galois covers $\psi : Z \to \mathbb{P}^1_k$ ramified only over infinity. For example, given a finite quasi-p group G, what is the smallest integer g for which there exists a cover $\psi : Z \to \mathbb{P}^1_k$ ramified only over infinity with Z of genus g? As another example, suppose G and H are two finite quasi-p groups such that H is a quotient of G. Given an unramified Galois cover ϕ of $\mathbb{A}^1_k$ with group H, under what situations can one dominate ϕ with an unramified Galois cover ψ of $\mathbb{A}^1_k$ with Galois group G? Answering these questions will give progress towards understanding how the finite quotients of $\pi_1(\mathbb{A}^1_k)$ fit together in an inverse system. These questions are more tractible for quasi-p groups that are p-groups since the maximal pro-p quotient $\pi_1^p(\mathbb{A}^1_k)$ is free (of infinite rank) [10].

In this paper, we study Galois covers $\psi : Z \to \mathbb{P}^1_k$ ramified only over ∞ whose Galois group is a semi-direct product of the form $(\mathbb{Z}/\ell\mathbb{Z})^b \rtimes \mathbb{Z}/p\mathbb{Z}$, where ℓ is a prime distinct from p. Such a cover ψ must be a composition $\psi = \phi \circ \omega$ where $\omega : Z \to Y$ is unramified and $\phi : Y \to \mathbb{P}^1_k$ is an Artin-Schreier cover ramified only over ∞. The cover ϕ has an affine equation $y^p - y = f(x)$ for some $f(x) \in k[x]$ with degree s prime-to-p. The ℓ-torsion $\mathrm{Jac}(Y)[\ell]$ of the Jacobian of Y is isomorphic to $(\mathbb{Z}/\ell\mathbb{Z})^{2g_Y}$. When $f(x) = x^s$, we determine how an automorphism τ of Y of order p acts on $\mathrm{Jac}(Y)[\ell]$. This allows us to construct a Galois cover $\psi_a : Z_a \to \mathbb{P}^1_k$ ramified only over ∞ which dominates ϕ, such that the Galois group of ψ_a is $(\mathbb{Z}/\ell\mathbb{Z})^a \rtimes \mathbb{Z}/p\mathbb{Z}$ where a is the order of ℓ modulo p (Section 3). We prove that the genus of Z_a is minimal among all natural numbers that occur as the genus of a curve Z which admits a covering map $\psi : Z \to \mathbb{P}^1_k$ ramified only over ∞ with Galois group of the form $(\mathbb{Z}/\ell\mathbb{Z})^b \rtimes \mathbb{Z}/p\mathbb{Z}$. We also prove that the number of curves Z of this minimal genus which admit such a covering map is at most $(p-1)/a$ when p is odd (Section 4).

2 Quasi-p semi-direct products

We recall which groups occur as Galois groups of covers of $\mathbb{P}^1_k$ ramified only over ∞.

Definition 2.1 A finite group is a quasi p-group if it is generated by all of its Sylow p-subgroups.

It is well-known that there are other equivalent formulations of the quasi-p property, such as the next result.

Lemma 2.2 *A finite group is a quasi p-group if and only if it has no nontrivial quotient group whose order is relatively prime to p.*

The importance of the quasi-p property is that it characterizes which finite groups occur as Galois groups of unramified covers of the affine line.

Theorem 2.3 *A finite group occurs as the Galois group of a Galois cover of the projective line $\mathbb{P}^1_k$ ramified only over infinity if and only if it is a quasi-p group.*

Proof This is a special case of Abhyankar's Conjecture [2] which was jointly proved by Harbater [4] and Raynaud [9]. □

We now restrict our attention to groups G that are semi-direct products of the form $(\mathbb{Z}/\ell\mathbb{Z})^b \rtimes \mathbb{Z}/p\mathbb{Z}$. The semi-direct product action is determined by a homomorphism $\iota : \mathbb{Z}/p\mathbb{Z} \to \mathrm{Aut}\,((\mathbb{Z}/\ell\mathbb{Z})^b)$.

Lemma 2.4 *Suppose a quasi-p group G is a semi-direct product of the form $(\mathbb{Z}/\ell\mathbb{Z})^b \rtimes \mathbb{Z}/p\mathbb{Z}$ for a positive integer b.*

1. *Then G is not a direct product.*
2. *Moreover, $b \geq \mathrm{ord}_p(\ell)$ where $\mathrm{ord}_p(\ell)$ is the order of ℓ modulo p.*

Proof Part (1) is true since $(\mathbb{Z}/\ell\mathbb{Z})^b$ cannot be a quotient of the quasi-p group G. For part (2), the structure of a semi-direct product $(\mathbb{Z}/\ell\mathbb{Z})^b \rtimes \mathbb{Z}/p\mathbb{Z}$ depends on a homomorphism $\iota : \mathbb{Z}/p\mathbb{Z} \to \mathrm{Aut}\,(\mathbb{Z}/\ell\mathbb{Z})^b$. By part (1), ι is an inclusion. Thus $\mathrm{Aut}\,(\mathbb{Z}/\ell\mathbb{Z})^b \simeq \mathrm{GL}_b(\mathbb{Z}/\ell\mathbb{Z})$ has an element of order p. Now

$$|\mathrm{GL}_b(\mathbb{Z}/\ell\mathbb{Z})| = (\ell^b - 1)(\ell^b - \ell)\cdots(\ell^b - \ell^{b-1}).$$

Thus $\ell^\beta \equiv 1 \bmod p$ for some positive integer $\beta \leq b$ which implies $b \geq \mathrm{ord}_p(\ell)$. □

Lemma 2.5 *If $a = \mathrm{ord}_p(\ell)$, then there exists a semi-direct product of the form $(\mathbb{Z}/\ell\mathbb{Z})^a \rtimes \mathbb{Z}/p\mathbb{Z}$ which is quasi-p. It is unique up to isomorphism.*

Proof If $a = \mathrm{ord}_p(\ell)$, then there is an element of order p in $\mathrm{Aut}\,((\mathbb{Z}/\ell\mathbb{Z})^a)$ and so there is an injective homomorphism $\iota : \mathbb{Z}/p\mathbb{Z} \to \mathrm{Aut}\,((\mathbb{Z}/\ell\mathbb{Z})^a)$. Thus there exists a non-abelian semi-direct product G of the form $(\mathbb{Z}/\ell\mathbb{Z})^a \rtimes \mathbb{Z}/p\mathbb{Z}$. To show that G is quasi-p, suppose N is a normal subgroup of G whose index is relatively prime to p. Then N contains an element τ of order p. By [3, 5.4, Thm. 9], since G is not a direct product and $(\mathbb{Z}/\ell\mathbb{Z})^a$ is normal in G, the subgroup $\langle\tau\rangle$ is not normal in G. Thus $\langle\tau\rangle$ is a proper subgroup of N. It follows that ℓ divides $|N|$ and so N contains an element h of order ℓ by Cauchy's theorem. Recall that $\mathrm{Aut}\,((\mathbb{Z}/\ell\mathbb{Z})^\beta)$ contains no element of order p for any positive integer $\beta < a$. Thus the group generated by the conjugates of h under τ has order divisible by ℓ^a. Thus $N = G$ and G has no non-trivial quotient group whose order is relatively prime to p. By Lemma 2.2, G is quasi-p.

The uniqueness follows from [8, Lemma 6.6]. □

3 Explicit construction of $(\mathbb{Z}/\ell\mathbb{Z})^a \rtimes \mathbb{Z}/p\mathbb{Z}$-Galois covers of $\mathbb{A}^1_k$

In this section, we give concrete examples of Galois covers $\psi : Z \to \mathbb{P}^1_k$ ramified only over ∞ with Galois group of the form $(\mathbb{Z}/\ell\mathbb{Z})^a \rtimes \mathbb{Z}/p\mathbb{Z}$. To compute the genus of the covering curve Z, we will need to determine the higher ramification groups of ψ.

Definition 3.1 Let L/K be a Galois extension of function fields of curves with Galois group G and let P, P' be primes of K and L such that $P'|P$. Let $\nu_{P'}$ and $\mathcal{O}_{P'}$ be the corresponding valuation function and valuation ring for P'. For any integer $i \geq -1$, the ith ramification group of $P'|P$ is

$$I_i(P'|P) = \{\sigma \in G \mid \nu_{P'}(\sigma(z) - z) \geq i + 1, \forall z \in \mathcal{O}_{P'}\}.$$

Lemma 3.2 *Suppose $f(x) \in k[x]$ is a polynomial of degree s for a positive integer s prime to p. Let $\phi : Y \to \mathbb{P}^1_k$ be the cover of curves corresponding to the field extension*

$$k(x) \hookrightarrow k(x)[y]/(y^p - y - f(x)).$$

1. *Then $\phi : Y \to \mathbb{P}^1_k$ is a Galois cover with Galois group $\mathbb{Z}/p\mathbb{Z}$ ramified only at the point P_∞ over ∞.*
2. *The ith ramification group at P_∞ satisfies*

$$I_i = \begin{cases} \mathbb{Z}/p\mathbb{Z} & \text{if } i \leq s \\ 0 & \text{if } i > s. \end{cases}$$

3. *The genus g_Y of Y is equal to*

$$g_Y = (p-1)(s-1)/2.$$

Proof For part (1), note that the extension $k(x) \hookrightarrow k(x)[y]/(y^p - y - f(x))$ is cyclic of degree p, with Galois group generated by the automorphism $\tau : y \mapsto y+1$ of order p. Let P be a finite prime of $k(x)$ and let ν_P be the corresponding valuation. Then $\nu_P(f(x)) \geq 0$, hence P is unramified by [12, Prop. III.7.8(b)]. For the infinite prime ∞ with corresponding valuation υ_∞, we have

$$\nu_\infty(f(x) - (z^p - z)) \leq 0$$

for all $z \in k[x]$ thus P_∞ is totally ramified by [12, Prop. III.7.8(c)].

To prove part (2), we note that furthermore

$$\upsilon_{P_\infty}(y^p - y) = \upsilon_{P_\infty}(f(x)) = \upsilon_{P_\infty}(x^s) = -sp,$$

which implies that

$$\upsilon_{P_\infty}(y) = -s.$$

Now let $\widehat{\theta}$ be the completion of the valuation ring of $k(x)[y]/(y^p - y - f(x))$ at P_∞, and let π_∞ be a generator of the unique prime in $\widehat{\theta}$. Then write $y = \pi_\infty^{-s} u$, where u is a unit in $\widehat{\theta} \simeq k[[\pi_\infty]]$. Since k is algebraically closed, $\sqrt[s]{u} \in \widehat{\theta}$, and so $\sqrt[s]{y} \in \widehat{\theta}$. After possibly changing π_∞, we can assume without loss of generality that $\sqrt[s]{y} = \pi_\infty^{-1}$. Recalling that τ acts on y by $\tau(y) = y + 1$, we have

$$\begin{aligned} \tau(\pi_\infty) &= \tau\left(1/y\right)^{1/s} = \left(\pi_\infty^s/(1+\pi_\infty^s)\right)^{1/s} \\ &= \pi_\infty(1 - \pi_\infty^s + \pi_\infty^{2s} - + \ldots)^{\frac{1}{s}} \\ &= \pi_\infty - (1/s)\pi_\infty^{s+1} + a_{2s+1}\pi_\infty^{2s+1} - + \ldots. \end{aligned}$$

Thus $\upsilon_{P_\infty}(\tau(\pi_\infty) - \pi_\infty) = s + 1$, which completes the proof of part (2).

To find the genus g_Y of Y for part (3), we make use of the Riemann-Hurwitz formula

$$2g_Y - 2 = p(-2) + \sum_{i=0}^{\infty} (|I_i| - 1),$$

where I_i denotes the ith ramification group at P_∞, [5, Thms. 7.27 & 11.72]). From part (2), we then obtain that $g_Y = (p-1)(s-1)/2$. □

Recall the following facts about the pth cyclotomic polynomial $\Phi_p(t) := t^{p-1} + \cdots + 1$, which is the minimal polynomial over $\mathbb{Q}$ of a primitive pth root of unity ζ_p. Now $\mathbb{Q}(\zeta_p)$ is a Galois extension of $\mathbb{Q}$, unramified over ℓ since $\ell \neq p$, and all primes over ℓ have the same residue field degree. The irreducible factors of $\Phi_p(t)$ modulo ℓ are in one-to-one correspondence with the primes of $\mathbb{Z}[\zeta_p]$ over ℓ, and each of their degrees is equal to the residue field degree of the corresponding prime over ℓ. The latter equals the order $a = \mathrm{ord}_p(\ell)$ of ℓ modulo p [3, Ch. 12.2, Exercise #20].

We shall soon explicitly construct a cover of $\mathbb{P}^1_k$ ramified only over ∞ with Galois group $(\mathbb{Z}/\ell\mathbb{Z})^a \rtimes \mathbb{Z}/p\mathbb{Z}$. But before we do so, we start with a specific example.

Example 3.3 Let p be an odd prime. Consider the Artin-Schreier cover $\phi : Y_2 \to \mathbb{P}^1_k$ corresponding to the field extension $k(x) \hookrightarrow k(x)[y]/(y^p - y - x^2)$. By Lemma 3.2(3), the genus of Y_2 is $g_Y = (p-1)/2$.

Let $\mathrm{Jac}\,(Y)$ be the Jacobian of Y. The automorphism τ of Y given by $\tau(y) = y + 1$ defines an automorphism of $\mathrm{Jac}\,(Y)$ of order p.

Now we describe the action of τ on the subgroup $\mathrm{Jac}\,(Y)[2]$ of 2-torsion points of $\mathrm{Jac}\,(Y)$ explicitly. Note that since $2g_Y = (p-1)$, then $\mathrm{Jac}\,(Y)[2]$ is isomorphic to $(\mathbb{Z}/2\mathbb{Z})^{p-1}$ by [7, pg. 64]. Thus we can represent τ as an element of $\mathrm{GL}_{p-1}(\mathbb{Z}/2\mathbb{Z})$.

For $0 \le i \le p-1$, let P_i denote the closed point of Y at which the function $y - i$ vanishes. For each i, the divisors P_i and $D_i = P_i - P_\infty$ on Y can be identified with elements of $\mathrm{Jac}\,(Y)$. Let O be the identity element of $\mathrm{Jac}\,(Y)$, i.e., the linear equivalence class of principal divisors. Then the divisor $2D_i$ is equivalent to O since $\mathrm{div}(y-i) = 2D_i$. Moreover since $\mathrm{div}(x) = D_0 + D_1 + \cdots + D_{p-1}$ is equivalent to 0, we have $D_i \in \mathrm{Jac}\,(Y)[2]$ with the only relation $D_{p-1} = -(D_0 + D_1 + \cdots + D_{p-2})$. In particular, $D_0, \ldots, D_{p-2}$ form a basis of $\mathrm{Jac}\,(Y)[2]$. With respect to this basis, the action of τ can be represented by the $(p-1) \times (p-1)$-matrix

$$\begin{pmatrix} 0 & 0 & \ldots & 0 & -1 \\ 1 & 0 & \ldots & 0 & -1 \\ 0 & 1 & \ldots & 0 & -1 \\ \vdots & \vdots & \ddots & 0 & -1 \\ 0 & 0 & \ldots & 1 & -1 \end{pmatrix}.$$

The characteristic polynomial of τ is $\Phi_p(t) = 1 + t + \ldots t^{p-1} \in (\mathbb{Z}/2\mathbb{Z})[t]$, which factors into irreducible polynomials each of degree equaling the order of 2 modulo p. In particular, τ acts irreducibly on $\mathrm{Jac}\,(Y)[2]$ if and only if 2 is a primitive root modulo p, i.e., if and only if p is an Artin prime.

For example, if $p = 3$, then τ acts irreducibly on $\mathrm{Jac}\,(Y)[2]$ with minimal polynomial $\Phi_3(t) = t^2 + t + 1$. If $p = 7$, then 2 has order 3 modulo 7 and the factorization of $\Phi_7(t)$ into irreducible polynomials is $\Phi_7(t) \equiv (x^3+x^2+1)(x^3+x+1)$ modulo 2. Thus the action of τ on $\mathrm{Jac}\,(Y)[2]$ can be represented by the 6×6-matrix

$$\begin{pmatrix} A_1 & 0 \\ 0 & A_2 \end{pmatrix}$$

where A_1 and A_2 are the irreducible 3-dimensional companion matrices of x^3+x^2+1 and $x^3 + x + 1$ respectively.

For the rest of the paper, let $\phi_s : Y_s \to \mathbb{P}^1_k$ be the Artin-Schreier cover corresponding to the field extension

$$k(x) \hookrightarrow k(x)[y]/(y^p - y - x^s).$$

We show that ϕ_s can be dominated by a Galois cover of $\mathbb{P}^1_k$ with Galois group of the form $(\mathbb{Z}/\ell\mathbb{Z})^a \rtimes \mathbb{Z}/p\mathbb{Z}$ for a equal to the order of ℓ modulo p.

Proposition 3.4 *Let s and ℓ be primes distinct from p. Let $\phi_s : Y_s \to \mathbb{P}^1_k$ be the Artin-Schreier cover with affine equation $y^p - y = x^s$. Let $a = \mathrm{ord}_p(\ell)$ be the order of ℓ modulo p. Then there exists an unramified Galois cover $\omega : Z_a \to Y_s$*

with Galois group $(\mathbb{Z}/\ell\mathbb{Z})^a$ *such that* $\psi_a = \phi_s \circ \omega : Z_a \to \mathbb{P}^1_k$ *is a Galois cover of* $\mathbb{P}^1_k$ *ramified only over* ∞ *whose Galois group is a semi-direct product of the form* $(\mathbb{Z}/\ell\mathbb{Z})^a \rtimes \mathbb{Z}/p\mathbb{Z}$.

Proof By Lemma 3.2(1), $\phi_s : Y_s \to \mathbb{P}^1_k$ is a Galois cover with Galois group $\mathbb{Z}/p\mathbb{Z}$ ramified only at the point P_∞ over ∞. The genus g_s of Y_s is $(p-1)(s-1)/2$. Consider two commuting automorphisms of Y_s defined by

$$\tau : \begin{cases} x \mapsto x, \\ y \mapsto y+1, \end{cases} \qquad \sigma : \begin{cases} x \mapsto \zeta_s x, \text{ where } \zeta_s \text{ is a primitive } s\text{th root of unity}, \\ y \mapsto y. \end{cases}$$

Let Jac(Y_s) be the Jacobian of Y_s. Then τ and σ define commuting automorphisms of Jac(Y_s) of orders p and s respectively. Therefore, End(Jac(Y_s)) contains a ring isomorphic to $\mathbb{Z}[\zeta_p, \zeta_s] \cong \mathbb{Z}[\zeta_{ps}]$, which is a $\mathbb{Z}$-module of rank $\phi(ps) = (p-1)(s-1) = 2g_s$. Then $\mathbb{Q}(\zeta_{ps})$ is contained in End(Jac(Y_s)) $\otimes\, \mathbb{Q}$. In other words, Jac(Y_s) has complex multiplication by $\mathbb{Q}(\zeta_{ps})$.

For a prime ℓ distinct from p, the automorphism τ induces an action on the subgroup Jac$(Y_s)[\ell]$ of ℓ-torsion points of Jac(Y_s). Recall that there is a bijection between ℓ-torsion points D of Jac(Y_s) and unramified $(\mathbb{Z}/\ell\mathbb{Z})$-Galois covers $\omega_D : Z_D \to Y_s$ [6, Prop. 4.11]. Also D has order ℓ if and only if Z_D is connected. This induces a bijection between orbits of τ on the set of unramified $(\mathbb{Z}/\ell\mathbb{Z})$-Galois covers $\omega_D : Z_D \to Y_s$ and on the set of ℓ-torsion points of Jac(Y_s). For a point D of order ℓ of Jac(Y_s), consider the compositum $\omega : Z \to Y_s$ of all of the conjugates $\omega_{\tau^j(D)} : Z_{\tau^i(D)} \to Y_s$ for $0 \le j \le p-1$:

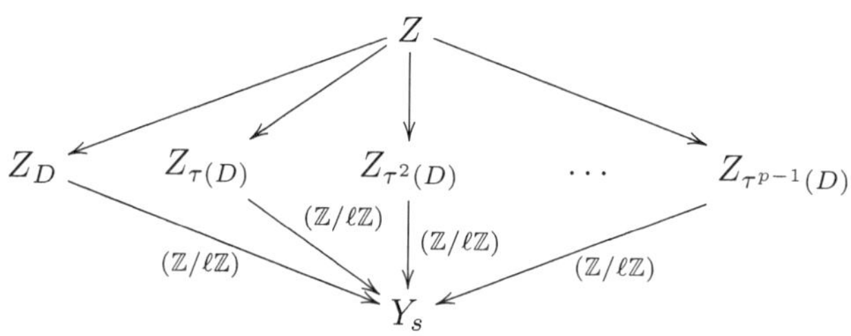

Then Z is invariant under τ and so $\phi_s \circ \omega : Z \to \mathbb{P}^1_k$ is Galois. Moreover, $\phi_s \circ \omega$ is the Galois closure of $\phi_s \circ \omega_D : Z_D \to \mathbb{P}^1_k$.

Suppose there is a non-trivial one-dimensional τ-invariant subspace of Jac$(Y_s)[\ell]$ with eigenvalue 1; i.e. τ acts trivially on this subgroup of order ℓ. This yields a cover $\psi_s \circ \omega_1 : Z_1 \to Y_s \to \mathbb{P}^1_k$. Since the action of τ is trivial, $\psi_s \circ \omega_1$ is Galois, ramified only over ∞, with abelian Galois group $\mathbb{Z}/\ell\mathbb{Z} \times \mathbb{Z}/p\mathbb{Z}$. This contradicts Lemma 2.4.

Since τ has order p, the minimal polynomial $m_\tau(t)$ of τ divides $t^p - 1 = (t-1)(t^{p-1} + \cdots + 1)$ in $(\mathbb{Z}/\ell\mathbb{Z})[t]$. From the preceding paragraph, there is no non-trivial one-dimensional τ-invariant subspace of Jac$(Y_s)[\ell]$ with eigenvalue 1. This implies that $m_\tau(t)$ divides the pth cyclotomic polynomial $\Phi_p(t) = t^{p-1} + \cdots + 1$ in $(\mathbb{Z}/\ell\mathbb{Z})[t]$. The irreducible factors of $\Phi_p(t)$ in $(\mathbb{Z}/\ell\mathbb{Z})[t]$ all have degree a. Thus the degree of $m_\tau(t)$ equals a.

Since $2g_s = (p-1)(s-1)$, we have Jac$(Y_s)[\ell] \cong (\mathbb{Z}/\ell\mathbb{Z})^{(p-1)(s-1)}$, so we can represent τ as an element of $\mathrm{GL}_{(p-1)(s-1)}(\mathbb{Z}/\ell\mathbb{Z})$. We can choose a basis of Jac$(Y_s)[\ell]$ such that τ is represented as an element of $\mathrm{GL}_{(p-1)(s-1)}(\mathbb{Z}/\ell\mathbb{Z})$ in block form. The first irreducible subrepresentation of τ has dimension a. Moreover, since

$\mathbb{Q}(\zeta_{p^s})$ is a Galois extension of $\mathbb{Q}$, the block form of τ consists entirely of irreducible blocks of the same size. In particular, the number of irreducible blocks is $(p-1)(s-1)/a$. In other words, τ can be represented by an element of $\mathrm{GL}_{(s-1)(p-1)}(\mathbb{Z}/\ell\mathbb{Z})$ of the form

$$\begin{pmatrix} A_1 & & & 0 \\ & A_2 & & \\ & & \ddots & \\ 0 & & & A_{(p-1)(s-1)/a} \end{pmatrix},$$

where A_i is an $a \times a$ matrix representing an a-dimensional irreducible subrepresentation of τ on $\mathrm{Jac}\,(Y_s)[\ell]$.

Using the bijection between orbits of $\mathrm{Jac}\,(Y_s)[\ell]$ and orbits of $(\mathbb{Z}/\ell\mathbb{Z})$-covers of Y_s under τ and the above observation for the action of τ on $\mathrm{Jac}\,(Y_s)[\ell]$, there exists an unramified $(\mathbb{Z}/\ell\mathbb{Z})^a$-Galois cover $\omega : Z_a \to Y_s$ such that $\psi_a = \phi_s \circ \omega : Z_a \to \mathbb{P}^1_k$ is a Galois cover of $\mathbb{P}^1_k$ with Galois group of the form $(\mathbb{Z}/\ell\mathbb{Z})^a \rtimes \mathbb{Z}/p\mathbb{Z}$. Also ψ_a is ramified only over infinity since ϕ_s is ramified only over ∞ and since ω is unramified. □

4 Minimal genus of $(\mathbb{Z}/\ell\mathbb{Z})^b \rtimes \mathbb{Z}/p\mathbb{Z}$-Galois covers of $\mathbb{A}^1_k$

In this section, we find the minimal genus of a curve Z that admits a covering map $\psi : Z \to \mathbb{P}^1_k$ ramified only over ∞, with Galois group of the form $(\mathbb{Z}/\ell\mathbb{Z})^b \rtimes \mathbb{Z}/p\mathbb{Z}$. The minimal genus depends only on ℓ and p. We consider the cases p odd and $p = 2$ separately. We also prove that the number of curves Z of this minimal genus which admit such a covering map is at most $(p-1)/a$ when p is odd and at most $\ell + 1$ when $p = 2$. The following lemma will be useful.

Lemma 4.1 *Let G be a semi-direct product of the form $(\mathbb{Z}/\ell\mathbb{Z})^b \rtimes \mathbb{Z}/p\mathbb{Z}$ where ℓ is a prime distinct from p. If $\psi : Z \to \mathbb{P}^1_k$ is a Galois cover ramified only over ∞ with Galois group G, then the subcover $\omega : Z \to Y$ with Galois group $(\mathbb{Z}/\ell\mathbb{Z})^b$ is unramified.*

Proof The quotient of G by the normal subgroup $N = (\mathbb{Z}/\ell\mathbb{Z})^b$ is $\mathbb{Z}/p\mathbb{Z}$. Thus the cover ψ is a composition $\psi = \phi \circ \omega$ where $\phi : Y \to \mathbb{P}^1_k$ has Galois group $\mathbb{Z}/p\mathbb{Z}$ and is totally ramified at the unique point P_∞ over ∞ and where $\omega : Z \to Y$ has Galois group N and is branched only over P_∞. Then ω is a prime-to-p abelian cover of Y. Let g be the genus of Y. Then by [1, XIII, Cor. 2.12], the prime-to-p fundamental group of $Y - \{P_\infty\}$ is isomorphic to the prime-to-p quotient Γ of the free group on generators $\{a_1, b_1, \ldots, a_g, b_g, c\}$ subject to the relation $\prod_{i=1}^g [a_i, b_i] = c^{-1}$. The cover ω corresponds to a surjection of Γ onto N where c maps to the canonical generator of inertia γ of a point of Z over P_∞. Thus N is generated by elements $\{\alpha_1, \beta_1, \ldots, \alpha_g, \beta_g, \gamma\}$ subject to the relation $\prod_{i=1}^g [\alpha_i, \beta_i] = \gamma^{-1}$. Then $\gamma = 1$ since N is abelian and so ω is unramified. □

Theorem 4.2 *Let p be an odd prime. Let ℓ be a prime distinct from p and let a be the order of ℓ modulo p. Then:*

1. *There exists a Galois cover $\psi_a : Z_a \to \mathbb{P}^1_k$ ramified only over ∞ whose Galois group is a semi-direct product of the form $(\mathbb{Z}/\ell\mathbb{Z})^a \rtimes \mathbb{Z}/p\mathbb{Z}$ such that $g_{Z_a} = 1 + \ell^a(p-3)/2$.*

2. *The integer* g_{Z_a} *is the minimal genus of a curve* Z *which admits a covering map* $\psi : Z \to \mathbb{P}^1_k$ *ramified only over* ∞ *with Galois group of the form* $(\mathbb{Z}/\ell\mathbb{Z})^b \rtimes \mathbb{Z}/p\mathbb{Z}$ *for any positive integer* b.
3. *There are at most* $(p-1)/a$ *isomorphism classes of curves* Z *which admit a Galois covering map as in part (1) with minimal genus* g_{Z_a}.

Proof By the construction in Proposition 3.4, there exists a Galois cover $\psi_a : Z_a \to \mathbb{P}^1_k$ ramified only over ∞ whose Galois group is a semi-direct product of the form $(\mathbb{Z}/\ell\mathbb{Z})^a \rtimes \mathbb{Z}/p\mathbb{Z}$. We compute the genus of the curve Z_a. Recall that ψ_a is a composition $\psi = \phi_2 \circ \omega$ where $\omega : Z \to Y_2$ is an unramified $(\mathbb{Z}/\ell\mathbb{Z})^a$-Galois cover and $\phi_2 : Y_2 \to \mathbb{P}^1_k$ has Artin-Schreier equation $y^p - y = x^2$. Then Y_2 has genus $g_{Y_2} = (p-1)/2$ by Lemma 3.2(3). By the Riemann-Hurwitz formula, $2g_{Z_a} - 2 = \ell^a(2g_{Y_2} - 2) = \ell^a(p-3)$, i.e., $g_{Z_a} = 1 + \ell^a(p-3)/2$. This completes part (1).

For part (2), suppose $\psi : Z \to \mathbb{P}^1_k$ is a Galois cover ramified only over ∞ with Galois group of the form $(\mathbb{Z}/\ell\mathbb{Z})^b \rtimes \mathbb{Z}/p\mathbb{Z}$. If g is the genus of Z, we will show that $g \geq g_{Z_a}$. As described in the proof of Lemma 4.1, the cover ψ is a composition $\psi = \phi \circ \omega$ where $\phi : Y \to \mathbb{P}^1_k$ has Galois group $\mathbb{Z}/p\mathbb{Z}$ and is ramified only over ∞ and where ω is unramified with group $(\mathbb{Z}/\ell\mathbb{Z})^b$. By the Riemann-Hurwitz formula, $2g - 2 = \ell^b(2g_Y - 2)$.

By Artin-Schreier theory, ϕ is given by an equation $y^p - y = f(x)$ where $f \in k[x]$ has degree s for some integer s relatively prime to p. Since the genus g_Y of Y is $(p-1)(s-1)/2$ by Lemma 3.2 (3), we should make s as small as possible. The value $s = 1$ is impossible since then Y is a projective line and there do not exist Galois covers of the projective line ramified only over one point with Galois group $\mathbb{Z}/\ell\mathbb{Z}$. Thus $s = 2$ yields the smallest possible value for g_Y, namely $(p-1)/2$. Recall that $b \geq a$ by Lemma 2.4. Thus $g \geq 1 + \ell^a(p-3)/2 = g_{Z_a}$.

For part (3), suppose $\psi : Z \to \mathbb{P}^1_k$ is a Galois cover ramified only over ∞ with Galois group of the form $(\mathbb{Z}/\ell\mathbb{Z})^a \rtimes \mathbb{Z}/p\mathbb{Z}$ and the genus of Z satisfies $g_Z = 1 + \ell^a(p-3)/2$. As in part (2), ψ factors as $\phi \circ \omega$ where $\omega : Z \to Y$ is an unramified $(\mathbb{Z}/\ell\mathbb{Z})^a$-Galois cover, where $\phi : Y \to \mathbb{P}^1_k$ is an Artin-Schreier cover ramified only over ∞, and where Y has genus $(p-1)/2$. By Lemma 3.2(3), Y has an affine equation $y^p - y = a_2x^2 + a_1x + a_0$ for some $a_0, a_1 \in k$ and $a_2 \in k^*$. Since p is odd and k is algebraically closed, it is possible to complete the square and write $a_2x^2 + a_1x + a_0 = x_1^2 + \epsilon$ where $x_1 = \sqrt{a_2}x + a_1/2\sqrt{a_2}$. After modifying by an automorphism of the projective line, specifically by the affine linear transformation $x \mapsto x_1$, the equation for Y can be rewritten as $y^p - y = x_1^2 + \epsilon$. Since k is algebraically closed, there exists $\delta \in k$ such that $\delta^p - \delta = \epsilon$. Let $y_1 = y - \delta$. After the change of variables $y \mapsto y_1$, the curve Y is isomorphic to the curve Y_2 with affine equation $y_1^p - y_1 = x_1^2$. Thus there is a unique possibility for the isomorphism class of the curve Y.

From the proof of Proposition 3.4, there is a bijection between τ-invariant connected unramified $(\mathbb{Z}/\ell\mathbb{Z})^a$-Galois covers of Y_2 and orbits of τ on points D of order ℓ on $\mathrm{Jac}(Y_2)$. The action of τ on $\mathrm{Jac}(Y_2)[\ell]$ decomposes into $(p-1)/a$ irreducible subrepresentations. Each of these is distinct, because the irreducible factors of $\Phi_p(t) \in (\mathbb{Z}/\ell\mathbb{Z})[t]$ are distinct. Thus there are $(p-1)/a$ choices for a τ-invariant unramified $(\mathbb{Z}/\ell\mathbb{Z})^a$-Galois cover of Y_2. Thus there are at most $(p-1)/a$ isomorphism classes of curves Z which admit a Galois covering map as in part (1) with minimal genus g_{Z_a}. □

We note that the set of curves which are unramified $(\mathbb{Z}/\ell\mathbb{Z})^a$-Galois covers of Y_2 may contain fewer than $(p-1)/a$ isomorphism classes of curves.

Theorem 4.3 *Let $p = 2$ and let ℓ be an odd prime. Then:*

1. *There exists a Galois cover $\psi : Z \to \mathbb{P}^1_k$ ramified only over ∞ with Galois group of the form $\mathbb{Z}/\ell\mathbb{Z} \rtimes \mathbb{Z}/2\mathbb{Z}$.*
2. *The minimal genus of a curve Z which admits a covering map as in part (1) is $g_Z = 1$.*
3. *There are at most $\ell + 1$ isomorphism classes of curves Z which admit a Galois covering map as in part (1) with minimal genus $g_Z = 1$.*

Proof Note that the order of ℓ modulo 2 is $a = 1$. For part (1), Lemma 2.5 shows that there exists a semi-direct product of the form $\mathbb{Z}/\ell\mathbb{Z} \rtimes \mathbb{Z}/2\mathbb{Z}$ which is quasi-2. The result is then immediate from Theorem 2.3.

Suppose $\psi : Z \to \mathbb{P}^1_k$ is a Galois cover ramified only over ∞ with Galois group as in part (1). As before, ψ factors as a composition $\phi \circ \omega$. where $\omega : Z \to Y$ has Galois group $\mathbb{Z}/\ell\mathbb{Z}$ and $\phi : Y \to \mathbb{P}^1_k$ is an Artin-Schreier extension with affine equation $y^2 - y = f(x)$ for some $f(x) \in k[x]$ of odd degree s. By Lemma 4.1, ω is unramified. The minimal genus for Z will thus occur when s is as small as possible. As before, $s = 1$ is impossible, and so $s = 3$ is the smallest choice. In this case, by Lemma 3.2(3), $g_Y = 1$, i.e., Y is an elliptic curve. By the Riemann-Hurwitz formula, the minimal genus for Z is $g_Z = 1 + \ell(g_Y - 1) = 1$, which completes part (2).

For part (3), since k is algebraically closed, we can complete the cube of $f(x)$ and make the corresponding change of variables, which is a scaling and translation of x. So we can assume that Y has affine equation $y^2 - y = x^3 + a_1 x + a_0$ for some $a_0, a_1 \in k$. Then it follows from [11, Appendix A, Prop. 1.1c] that the j-invariant of Y is $j(Y) = 0$ and that the discriminant is $\Delta(Y) = (-1)^4 = 1$. Since k is algebraically closed, by [11, Appendix A, Prop. 1.2b], all elliptic curves Y with $j(Y) = 0$ are isomorphic over k. Thus there is a unique choice for Y up to isomorphism. Without loss of generality, we may assume that $Y = Y_3$ has affine equation $y^2 - y = x^3$.

From the proof of Proposition 3.4, the action of τ on $\mathrm{Jac}\,(Y_3)[\ell]$ decomposes into the direct sum of two 1-dimensional subrepresentations. In other words, the action of τ is diagonal with both eigenvalues equal to -1. The number of non-trivial τ-invariant subgroups of $\mathrm{Jac}\,(Y_3)[\ell]$ is the number of subgroups of order ℓ in $(\mathbb{Z}/\ell\mathbb{Z})^2$, which is $\ell+1$. As in Theorem 4.2, this implies that there are at most $\ell+1$ isomorphism classes of curves Z which admit a Galois covering map as in part (1) with minimal genus $g_Z = 1$. □

We note that the set of curves which are unramified $\mathbb{Z}/\ell\mathbb{Z}$-Galois covers of Y_3 may contain fewer than $\ell + 1$ isomorphism classes of curves.

Acknowledgments. This project was initiated at the workshop WIN Women in Numbers in November 2008. The authors would like to thank the Banff International Research Station for hosting the workshop and the National Security Agency, the Fields Institute, the Pacific Institute for the Mathematical Sciences, Microsoft Research, and University of Calgary for their financial support. Author Pries was partially supported by NSF grant 07-01303. Author Im was partially supported by the Korea Science and Engineering Foundation (KOSEF) Grant (No. R01-2007-000-10660-0) funded by the Korea government (MOST). The authors would also like to thank the referee for helpful comments.

References

[1] *Revêtements étales et groupe fondamental.* Springer-Verlag, Berlin, 1971. Séminaire de Géométrie Algébrique du Bois Marie 1960–1961 (SGA 1), Dirigé par Alexandre Grothendieck. Augmenté de deux exposés de M. Raynaud, Lecture Notes in Mathematics, Vol. 224.

[2] Shreeram Abhyankar. Coverings of algebraic curves. *Amer. J. Math.*, 79:825–856, 1957.

[3] David S. Dummit and Richard M. Foote. *Abstract algebra.* John Wiley & Sons Inc., Hoboken, NJ, third edition, 2004.

[4] David Harbater. Abhyankar's conjecture on Galois groups over curves. *Invent. Math.*, 117(1):1–25, 1994.

[5] J. W. P. Hirschfeld, G. Korchmáros, and F. Torres. *Algebraic curves over a finite field.* Princeton Series in Applied Mathematics. Princeton University Press, Princeton, NJ, 2008.

[6] James S. Milne. *Étale cohomology*, volume 33 of *Princeton Mathematical Series.* Princeton University Press, Princeton, N.J., 1980.

[7] David Mumford. *Abelian varieties.* Tata Institute of Fundamental Research Studies in Mathematics, No. 5. Published for the Tata Institute of Fundamental Research, Bombay, 1970.

[8] Rachel Pries and Katherine Stevenson. *A survey of Galois theory of curves in characteristic p.* WIN - Women In Numbers, Fields Communication Volume.

[9] M. Raynaud. Revêtements de la droite affine en caractéristique $p > 0$ et conjecture d'Abhyankar. *Invent. Math.*, 116(1-3):425–462, 1994.

[10] I. Shafarevitch. On p-extensions. *Rec. Math. [Mat. Sbornik] N.S.*, 20(62):351–363, 1947.

[11] Joseph H. Silverman. *The arithmetic of elliptic curves*, volume 106 of *Graduate Texts in Mathematics.* Springer-Verlag, New York, 1992. Corrected reprint of the 1986 original.

[12] Henning Stichtenoth. *Algebraic function fields and codes*, volume 254 of *Graduate Texts in Mathematics.* Springer-Verlag, Berlin, second edition, 2009.

Part IV – Zeta Functions of Graphs

Fields Institute Communications
Volume **60**, 2011

Looking into a Graph Theory Mirror of Number Theoretic Zetas

Audrey Terras
Department of Mathematics
U.C.S.D.
La Jolla, CA 92093-0112 USA
aterras@ucsd.edu

Abstract. We survey the Ihara zeta function for irregular graphs, emphasizing connections with number theory zetas and open problems.

1 Introduction to the Ihara zeta

We always assume that our graphs **X** *are finite, connected, possibly irregular, of rank greater than one with no danglers* (i.e., degree 1 vertices). Here "rank" means the rank of the fundamental group of the graph. Let us recall some of the basic definitions. References include Hashimoto [6], Stark and Terras [25, 26, 27] and my book [30]. For the generalization to zeta functions of complexes, see Kang, Li and Wang [10].

The Ihara zeta function arose in a paper of Ihara from the 1960's where it was viewed as a Selberg zeta function for p-adic groups. However the graph theory interpretation of such a Selberg zeta function requires the graph to be regular of degree $1 + p^e$, where p is a prime. Here we consider general, possibly irregular, graphs.

If X is any connected finite undirected graph with vertex set V and (undirected) edge set E, we *orient its edges arbitrarily and obtain* $2|E|$ *oriented edges labelled*

$$e_1, e_2, \cdots, e_n, e_{n+1} = e_1^{-1}, ..., e_{2n} = e_n^{-1}. \tag{1.1}$$

"*Primes*" $[P]$ in X are equivalence classes of closed backtrackless tailless primitive paths P. A prime is the graph analog of a geodesic in a Riemannian manifold. Write a path $C = a_1 a_2 \cdots a_s$, where a_j is an oriented edge of X. The *length* of C is $\nu(C) = s$. *Backtrackless* means that $a_{i+1} \neq a_i^{-1}$, for all i. Tailless means that $a_s \neq a_1^{-1}$. The *equivalence class* $[C]$ is the set

$$[C] = \{a_1 a_2 \cdots a_s,\ a_2 a_3 \cdots a_s a_1,\ \ldots,\ a_s a_1 \cdots a_{s-1}\}.$$

2010 *Mathematics Subject Classification.* Primary 11N99; Secondary 05C25, 05C50.

P is *primitive* means $P \neq D^m$, for any integer $m \geq 2$ and path D in X. Here r_X will denote the rank of the fundamental group of X. We have $r_X - 1 = |E| - |V|$. Then r_X is the number of edges deleted from X to form a spanning tree.

The *Ihara zeta function* of X is defined at $u \in \mathbb{C}$, for $|u|$ sufficiently small, by

$$\zeta(u, X) = \prod_{[P]} \left(1 - u^{\nu(P)}\right)^{-1}, \tag{1.2}$$

where $[P]$ runs over the primes of X. This zeta function can be thought of as an analog of Selberg's zeta function which is a similar product over primitive geodesics in a Riemannian manifold.

The generalization of *Ihara's formula* (proved by Hashimoto and Bass) says

$$\zeta(u, X)^{-1} = (1 - u^2)^{r_X - 1} \det\left(I - A_X u + Q_X u^2\right), \tag{1.3}$$

where r_X is the rank of the fundamental group of X, A_X is the adjacency matrix of X, Q_X is the diagonal matrix whose jth diagonal entry is ($-1+$ degree of jth vertex).

Definition 1.1 R_X is the *radius of the largest circle of convergence* of the Ihara zeta function.

When X is a $(q+1)$-regular graph, $R_X = 1/q$. As with the Dedekind zeta function, $\zeta(u, X)$ has a meromorphic continuation to the entire complex u-plane, but now $\zeta(u, X)^{-1}$ is entire. In fact, $\zeta(u, X)^{-1}$ is a polynomial of degree $2m$, where $m = |E|$, the number of edges of X. Thus if we want to formulate a version of the RH=Riemann Hypothesis, we must consider the poles of $\zeta(u, X)$, not the zeros. *The Graph theory RH* says $\zeta_X(u)$ is pole free for

$$R_X < |u| < \sqrt{R_X}. \tag{1.4}$$

Experiments show that perhaps one should not expect this to hold exactly but rather approximately for a random graph. See Angel et al [1], Horton et al [8], [9].

When a graph is regular, the Riemann hypothesis is equivalent to the statement that the graph is Ramanujan and thus a good expander. See Lubotzky, Phillips and Sarnak [16] or [30].

Definition 1.2 We define the *0,1 edge matrix* W_1 by orienting the m edges of X and labeling them as in formula (1.1). Then W_1 is the $2m \times 2m$ matrix with ij entry 1 if edge e_i feeds into e_j provided that $e_j \neq e_i^{-1}$, and ij entry 0 otherwise. By *"a feeds into b,"* we mean that the terminal vertex of edge a is the same as the initial vertex of edge b.

One can prove that (see [30])

$$\zeta(u, X)^{-1} = \det\left(I - W_1\, u\right). \tag{1.5}$$

It follows that the poles of zeta are the reciprocals of the eigenvalues of W_1. Kotani and Sunada [13] show that, for a graph X satisfying our hypotheses, if $q+1$ is the maximum degree of X and $p+1$ is the minimum degree of X, then every non-real pole u of $\zeta(u, X)$ satisfies the inequality

$$q^{-1/2} \leq |u| \leq p^{-1/2}. \tag{1.6}$$

Moreover, they show that every pole u of $\zeta_X(u)$ satisfies $R_X \leq |u| \leq 1$ and

$$q^{-1} \leq R_X \leq p^{-1}. \tag{1.7}$$

The Ihara zeta function leads to a proof of the graph prime number theorem by a similar but happily less tortuous path than that one must travel from the properties of Riemann zeta to a proof of the ordinary prime number theorem. In order to state the graph theory prime number theorem, we need some definitions.

Definition 1.3 The *prime counting function* is

$$\pi(n) = \#\{\text{primes } [P] \mid n = \nu(P) = \text{length of } P\}.$$

Definition 1.4 The *greatest common divisor of the prime path lengths* is

$$\Delta_X = g.c.d.\{\nu(P) \mid [P] \text{ prime of } X\}.$$

Theorem 1.5 (The Graph Prime Number Theorem.) *For a graph X satisfying the assumptions stated in the first line of this paper, if Δ_X divides m, then*

$$\pi(m) \sim \frac{\Delta_X}{mR_X^m}, \quad \text{as } m \to \infty.$$

If Δ_X does not divide m, then $\pi(m) = 0$.

The proof can be found in [30].

2 Problems

There are graph theory analogs of many familiar topics in number theory: zeta functions, formulas for special values of derivatives at 0, the Riemann hypothesis, explicit formulas, the prime number theorem, the Chebotarev density theorem, factorization of zeta functions as products of Artin L-functions.

This has been observed by many researchers as have the parallels between differential geometry and number theory. For example, consider the following passage from the biography of Sunada [11]. "The motivation behind this work came up also during his [Sunada's] stay in Bonn. At that time, he wanted to understand the class field theory because it has been his dream to prove something related to this theory ever since he read Takagi's book on algebraic number theory. For this sake, he tried to find a geometric model of the class field theory. Thus the path he took is the reverse of what Hilbert had taken up to speculate a correct formulation of his "absolute" class field theory by looking at the theory of covering surfaces. Sunada observed that closed geodesics under covering maps behave like prime ideals under field extensions, and could soon formulate the class field theory in the Riemannian geometric setting. He once said, "'It is a simple toy model, but I enjoy playing with it very much'." See also Sunada [28, 29].

I have much the same background, since I learned number theory from Sigekatu Kuroda who was using Takagi's book on algebraic number theory. This book should be translated into English. Harold Stark and I were both motivated to explore the number theory analogies. But there are still many things left to do. See problem 6) below. Or see my website for the talks from the I.A.S. women's program in 2006 with lots of project suggestions: www.math.ucsd.edu/˜aterras/zeta stroll.pdf.

My problem list

1) Perform experiments on differences between properties of zetas of weighted or quantum and unweighted graphs. See Horton et al. [8, 9]. In particular, consider the connections with random matrix theory. See also Smilansky [23].

2) a) Investigate poles of zeta and the RH for random graphs. How does the distribution of poles of zeta depend on the probability of an edge or the degree sequence?

b) Can one find a Galois graph covering Y of a base graph X such that the poles of the Ihara zeta of Y behave like those of a random cover of X? See [30] for some experiments with abelian Galois groups. Pole distributions of zetas for abelian covers were seen to be very different from those of random covers. It would be interesting to consider non-abelian Galois groups. For example, one could imitate the work of A. Lubotzky, R. Phillips and P. Sarnak [16] using the group $SL(2, F)$, F=finite field. Angel, Friedman, and Hoory [1] give a method for finding the region one expects to approximately enclose the spectrum of the edge adjacency matrix of a random cover of a base graph X.

3) Can you prove an analog of the theorem of Katz and Sarnak [12] for function field zeta functions at least in the case of regular graphs? This theorem says that for almost all curves over a finite field, as the genus and order of the field go to infinity, the imaginary parts of the zeros of zetas approach GUE level spacing. GUE spacing means the spacings look like those Odlyzko found for the high zeros of Riemann zeta as in Figure 23 in [30]. A reference giving background on this subject is the book by S. J. Miller and R. Takloo-Bighash [18]. One expects pole spacings of regular graph zetas to be related to GOE spacings (i.e. those of eigenvalues of real symmetric matrices) here. See the experiments of D. Newland [21].

4) Connect the zeta polynomials of graphs to other polynomials associated to graphs and knots such as Tutte, Alexander, and Jones polynomials. See Lin et al [15].

5) *Graph Theoretic Analogs of Number Theoretic Results.*

Zeta functions of graph coverings can be viewed as analogs of zeta functions of extensions of algebraic number (or function) fields. Thus one would expect to find analogs of many number-theoretic results. See Lang [14] for Dedekind zeta functions of algebraic number fields and Rosen [22] for zeta functions of function fields over finite fields. Ihara zeta functions are perhaps closer to zeta functions of function fields than to the zetas of number fields, since both are rational functions.

There is an analog of the prime number theorem as we have seen. There is also an analog of the Chebotarev density theorem (see [30]). Finally, there is an analog of Brauer-Siegel theory (see [27]).

a) Find the analogs of the explicit formulas of Weil from analytic number theory. See the next section.

b) The *complexity* κ_X of a graph is defined to be the number of spanning trees in X. One can use the matrix-tree theorem (see Biggs [5]) to prove that

$$\left[\frac{d^r}{du^r}\zeta_X^{-1}(u)\right]\Bigg|_{u=1} = r!(-1)^{r+1}2^r(r-1)\kappa_X. \tag{2.1}$$

This is an analog of the formula for derivatives of the Dedekind zeta function of a number field at 0 (a formula involving the class number and the regulator of the number field). The ideal class group is the Jacobian of a graph and has order equal to the number of spanning trees. See R. Bacher, P. de la Harpe, and Tatiana Nagnibeda [2].

c) Are there analogs of the Stark Conjectures (see [24]) on values of derivatives of Artin L-functions of representations of the Galois group of extensions of number fields?

d) Are there analogs of class field theory for abelian graph coverings? Or more simply is there an analog of the a quadratic reciprocity law?

e) What is a ramified graph covering Y/X? Is $\zeta(u,Y)$ divisible by $\zeta(u,X)$? If not, what can be said? More on this question is found in the last section of this paper and Malmskog and Manes [17] or M. Baker and S. Norine [3].

6) Look at Avi Wigderson's website (http://www.math.ias.edu/˜avi/) and find out what zig-zag products of graphs are. Compute their zeta functions. Are there any divisibility properties? Infinite families of regular expanders of arbitrary constant degree are obtained via the modified zig-zag product by Cristina M. Ballantine and Matthew D. Horton [4].

3 Explicit formula for Ihara zeta

First we need an analog of the von Mangoldt function $\Lambda(n)$. So we make the following definition.

Definition 3.1 $N_m = N_m(X)$ is the *number of closed paths of length m in the graph X without backtracking and tails.*

With N_m as in the preceding definition, we have the generating function:

$$u\frac{d}{du}\log\zeta(u,X) = \sum_{m\geq 1} N_m u^m. \tag{3.1}$$

Thus N_m is our analog of the von Mangoldt function $\Lambda(n)$ from number theory. In proving the prime number theorem, one shows that, assuming Δ_X divides m, we have $N_m \sim \frac{\Delta_X}{R_X^m}$, as $m \to \infty$.

Using formula (1.5), we find that

$$u\frac{d}{du}\log\zeta(u,X) = -u\frac{d}{du}\sum_{\lambda\in Spec(W_1)}\log(1-\lambda u)$$
$$= \sum_{\lambda\in Spec(W_1)}\frac{\lambda u}{1-\lambda u} = -\sum_{\substack{\rho \\ \text{poles } \zeta}}\frac{u}{u-\rho}.$$

Note that the reciprocals of the $\lambda \in Spec(W_1)$ are the poles of $\zeta(u,X)$.

Proposition 3.2 (An Explicit Formula.) *Suppose* $0 < a < R$ *(where R is the radius of convergence of* $\zeta(u,X)$ *from Definition 1.1). Assume* $h(u)$ *is meromorphic in the plane and holomorphic outside the circle of center* 0 *and radius* $a-\varepsilon$, *for small* $\varepsilon > 0$. *Assume also that* $h(u) = O(|u|^p)$ *as* $|u| \to \infty$, *for some* $p < -1$. *Finally assume that its transform* $\widehat{h}_a(n)$ *decays rapidly enough for the right hand side of the formula to converge absolutely. Then if* N_m *is as in Definition 3.1, we have*

$$\sum_\rho \rho h(\rho) = \sum_{n\geq 1} N_n \widehat{h}_a(n), \tag{3.2}$$

where the sum on the left is over the poles of $\zeta(u,X)$ *and*

$$\widehat{h}_a(n) = \frac{1}{2\pi i} \int\limits_{|u|=a} u^n h(u) du.$$

Proof We follow the method of Murty [18, p. 109]. Use Cauchy's residue theorem and the fact that

$$u \frac{d}{du} \log \zeta(u,X) = - \sum_{\substack{\rho \\ \text{poles } \zeta}} \frac{u}{u-\rho}$$

to see that if $0 < a < R < 1 < b$,

$$\frac{1}{2\pi i} \int\limits_{|u|=a} \left(u \frac{d}{du} \log \zeta(u,X) \right) h(u) du$$

$$= \frac{1}{2\pi i} \int\limits_{|u|=b} \left(u \frac{d}{du} \log \zeta(u,X) \right) h(u) du + \sum_{\rho} \rho h(\rho).$$

Let $b \to \infty$ to make the integral over the circle of radius b go to 0. Here we use the fact that zeta is the reciprocal of a polynomial and thus $u \frac{d}{du} \log \zeta(u,X) = O(1)$, as $|u| \to \infty$. Since

$$u \frac{d}{du} \log \zeta(u,X) = \sum_{m \geq 1} N_m u^m,$$

upon interchange of sum and integral, we have

$$\frac{1}{2\pi i} \int\limits_{|u|=a} \left(u \frac{d}{du} \log \zeta(u,X) \right) h(u) du = \sum_{n \geq 1} N_n \widehat{h}_a(n).$$

The result follows. □

The formula in Proposition 3.2 is an analog of Weil's explicit formula for the Riemann zeta function. In Weil's original work he used the result to formulate an equivalent statement to the Riemann hypothesis. See Weil [32] and Lang [14].

For example, if you set $h(u) = u^{-(m+1)}$, for $m = 1, 2, 3, \ldots$, the explicit formula (3.2) says N_m is the sum of the mth powers of the eigenvalues of W_1, which is clear from formulas (1.5) and (3.1).

Such explicit formulas are basic to work on the pair correlation of complex zeros of zeta (see Montgomery [19]). They can also be viewed as an analog of Selberg's trace formula. See [7, 31] for discussions of Selberg's trace formula for a $q+1$ regular graph. In these papers, various kernels such as the heat kernel were plugged in to the trace formula. For example, this leads to a proof of McKay's theorem on the distribution of eigenvalues of the adjacency matrix. It would be interesting to do the same sort of thing for irregular graphs. We might also want to write the analogous result to Proposition 3.2 using the Ihara determinant formula (1.3) instead of formula (1.5). Finally it would be nice to weaken the hypotheses in the preceding Proposition. Perhaps one should consider the result in the light of the theory of distributions.

4 Ramification experiments

A *neighborhood* N of a vertex v in a directed graph X is obtained by taking one–third of each edge at v. The labels and directions are to be included.

Definition 4.1 An undirected finite graph Y is an *unramified covering* of an undirected graph X if, after arbitrarily directing the edges of X, there is an assignment of directions to the edges of Y and an onto *covering map* $\pi : Y \longrightarrow X$ sending neighborhoods of Y 1-1, onto neighborhoods of X preserving directions.

For unramified covers one has the following divisibility property. The analogous fact for number fields is still only a conjecture.

Proposition 4.2 *For an unramified cover* Y *of* X,

$$\zeta(u,X)^{-1} \text{ divides } \zeta(u,Y)^{-1}.$$

Proof Start with the Ihara formula

$$\zeta(u,Y)^{-1} = (1-u^2)^{r_Y-1}\det\left(I_Y - A_Y u + Q_Y u^2\right).$$

Note that $r_Y - 1 = |E_Y| - |V_Y| = d\left(|E_X| - |V_X|\right)$. Thus $(1-u^2)^{r_X-1}$ divides $(1-u^2)^{r_Y-1}$. Now order the vertices of Y in blocks corresponding to the sheets of the cover, so that A_Y consists of blocks $\widetilde{A}_{ij}$, with $1 \le i,j \le d$ such that $\sum_j \widetilde{A}_{ij} = A$. The same ordering puts Q_Y in block diagonal form with n copies of Q_X down the diagonal. Similarly I_Y has block diagonal form consisting of d copies of I_X down the diagonal. Consider $I_Y - A_Y u + Q_Y u^2$. Without changing the determinant, we can add the right $d-1$ block columns to the first block column. The new first column is

$$\begin{pmatrix} I_X - A_X u + Q_X u^2 \\ \vdots \\ I_X - A_X u + Q_X u^2 \end{pmatrix}.$$

Then subtract the first block row from all the rest of the block rows. The first block column becomes:

$$\begin{pmatrix} I_X - A_X u + Q_X u^2 \\ 0 \\ \vdots \\ 0 \end{pmatrix}.$$

Thus $\zeta(u,X)^{-1}$ divides $\zeta(u,Y)^{-1}$. □

Experiments

I tried some experiments; ramifying vertices and edges. See the examples below.

Example 1. The zeta function of a graph L_n consisting of 1 vertex and n loops can be found from the Ihara determinant formula (1.3). Here the adjacency matrix is 1×1: $A = 2n$. The matrix Q, also 1×1, is $Q = 2n-1$. The rank of the fundamental group is $r = n$. The Ihara formula says

$$\begin{aligned}\zeta(u,L_n)^{-1} &= (1-u^2)^{r_X-1}\det\left(I - A_X u + Q_X u^2\right)\\ &= (1-u^2)^{n-1}\left(1-2nu+(2n-1)u^2\right)\\ &= (1-u^2)^{n-1}(1-u)(1-(2n-1)u)\end{aligned}$$

If we view L_n as a ramified covering of L_1, we are happy since $\zeta(u, L_1)^{-1}$ divides $\zeta(u, L_n)^{-1}$. However $\zeta(u, L_2)^{-1}$ does not divide $\zeta(u, L_{2n})^{-1}$. The good thing is that there is only one bad factor and it is linear.

It is easy to turn this example into a bouquet of n triangles T_n covering a triangle T_1. Since each path in L_n is tripled in length, we see that $\zeta(u, T_n) = \zeta(u^3, L_n)$. Fortunately there is still only one bad factor. See Beth Malmskog and Michelle Manes [17] for a more general result.

Example 2. The graph Y of Figure 1 comes from taking 2 copies of K_4 and identifying 2 vertices. Think of it as a covering of K_4. But there is not much good to say about the divisibility of the zeta functions. We list them below.

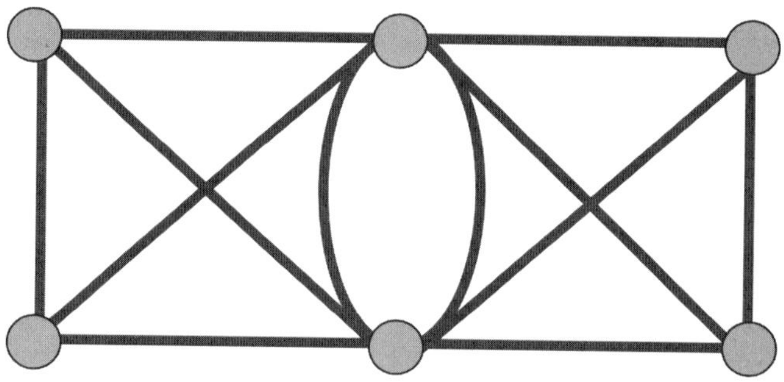

Figure 1 Graph Y obtainined by identifying (fusing) 2 vertices in 2 copies of K_4.

$$\begin{aligned}
\zeta(u, Y)^{-1} &= (u+1)(u-1)(2u+1)^2(2u-1)\left(5u^2-2u-1\right) \\
&\quad \times \left(10u^4+9u^3-13u^2-3u+1\right). \\
\zeta(u, K_4)^{-1} &= \left(1-u^2\right)^2(1-u)(1-2u)(1+u+2u^2)^3.
\end{aligned}$$

Example 3. Next we obtain a graph Z from 2 copies of K_4 and identifying (fusing) an edge. See Figure 2. Here I was looking for an analog of the Riemann surface of $\sqrt{z}$.

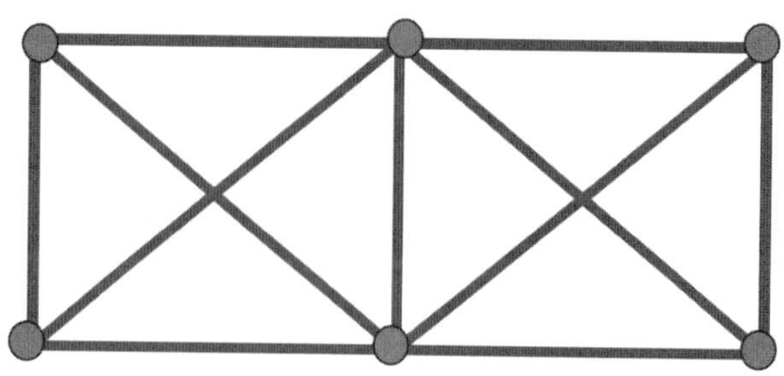

Figure 2 Graph Z obtained by taking 2 copies of K_4 and identifying an edge.

We compute the zeta function to be

$$\begin{aligned}\zeta(u,Z)^{-1} &= (1-u^2)^5(u-1)(4u^2+u+1)\left(2u^2-u+1\right)\left(8u^3+2u^2+u-1\right)\\ &\quad\times\left(2u^2+u+1\right)^2.\end{aligned}$$

The only factor of $\zeta(u,K_4)^{-1}$ that does not divide $\zeta(u,Y)^{-1}$ is $1-2u$. This looks more like the result of Beth Malmskog and Michelle Manes [17].

Example 4. Suppose Z_3 is obtained from 3 copies of K_4 with 1 edge from each copy fused. Then Y is pictured in Figure 3.

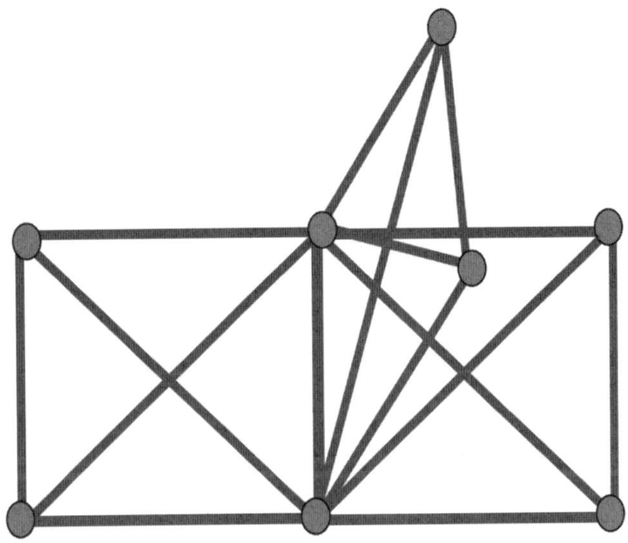

Figure 3 The graph Z_3 is obtained by fusing one edge from each of 3 copies of K_4.

We find that the zeta function is

$$\begin{aligned}\zeta(u,Z_3)^{-1} &= (1-u^2)^8\,(u-1)\left(6u^2+u+1\right)\left(12u^3+4u^2+u-1\right)\\ &\quad\times\left(2u^2-u+1\right)^2\left(2u^2+u+1\right)^3.\end{aligned}$$

Again the only part of $\zeta(u,K_4)^{-1}$ that does not divide $\zeta(u,Z_3)^{-1}$ is $(1-2u)$.

Acknowledgment. I would like to thank the organizers of the WIN conference for creating the first number theory conference of its kind in my experience. It was wonderful. I would also like to express my appreciation to the graph zetas group (Shabnam Akhtari, Habiba Kadiri, Winnie Li, Elizabeth Malmskog, Michelle Manes) for many lively discussions. Finally I am grateful my student Tom Petrillo for many helpful discussions, particularly about Proposition 3.2.

References

[1] O. Angel, J. Friedman, S. Hoory, The Non-backtracking spectrum of the universal cover of a graph, preprint, www.math.ubc.ca/~jf.

[2] R. Bacher, P. de la Harpe, and T. Nagnibeda, The lattice of integral flows and the lattice of integral cuts on a finite graph, *Bull. Math. Soc. France*, 125 (1997), 167-198.

[3] M. Baker and S. Norine, Harmonic morphisms and hyperelliptic graphs, *Int Math Res Notices*, 15 (2009), 2914-2955.

[4] C. M. Ballantine and M. D. Horton, Infinite families of regular expanders of arbitrary constant degree obtained via the modified zig-zag product, preprint.

[5] N. Biggs, *Algebraic Graph Theory*, Cambridge Univ. Press, London, 1974.

[6] K. Hashimoto, Zeta functions of finite graphs and representations of p-adic groups, *Adv. Studies in Pure Math.*, Vol. 15, Academic Press, N.Y., 1989, pp. 211-280.

[7] M. D. Horton, D. B. Newland, and A. Terras, The contest between the kernels in the Selberg trace formula for the (q+1)-regular tree, *Contemporary Math., 398*, A.M.S., Providence, 2006, pages 265-293.

[8] M. D. Horton, H. M. Stark, and A. A. Terras, What are Zeta Functions of Graphs and What are They Good For?, *Contemporary Mathematics*, Volume 415 (2006), Quantum Graphs and Their Applications; Edited by Gregory Berkolaiko, Robert Carlson, Stephen A. Fulling, and Peter Kuchment, pages 173-190.

[9] M. D. Horton, H. M. Stark, and A. A. Terras, Zeta Functions of Weighted Graphs and Covering Graphs, in *Proc. Symp. Pure Math.*, Vol. 77, Analysis on Graphs, Edited by Exner, Keating, Kuchment, Sunada and Teplyaev, A.M.S., 2008.

[10] Kang, M.-H., Li, W.-C. Winnie, and Wang, C.-J., The zeta functions of complexes from PGL(3): A Representation-theoretic Approach, *arXiv*:0809.1401v2 [math.NT] 19 Nov 2008.

[11] A. Katsuda and P. W. Sy, An Overview of Sunada's Work, *Meiji Institute for Advanced Study of Mathematical Sciences Technical Report* No.00013 (200807091) 9 July 2008.

[12] N. Katz and P. Sarnak, *Random Matrices, Frobenius Eigenvalues and Monodromy*, Amer. Math. Soc., Providence, RI, 1999.

[13] M. Kotani and T. Sunada, Zeta functions of finite graphs, *J. Math. Sci. Univ. Tokyo*, 7 (2000), 7-25.

[14] S. Lang, *Algebraic Number Theory*, Addison-Wesley, Reading, Mass., 1968.

[15] X.-S. Lin and Z. Wang, Random walk on knot diagrams, colored Jones polynomial and Ihara-Selberg zeta function. Gilman, Jane (ed.) et al., Knots, braids, and mapping class groups - papers dedicated to Joan S. Birman. Proceedings of a conference in low dimensional topology in honor of Joan S. Birman's 70th birthday, *A.M.S./IP Stud. Adv. Math., Vol.* 24, (2001) Providence, RI: 107-121.

[16] A. Lubotzky, R. Phillips and P. Sarnak, Ramanujan graphs, *Combinatorica*, 8 (1988), 261-277.

[17] B. Malmskog and M. Manes, Almost Divisibility in the Ihara Zeta Functions of Certain Ramified Covers of q+1-regular Graphs, preprint.

[18] S. J. Miller and R. Takloo-Bighash, *An Invitation to Modern Number Theory*, Princeton U. Press, Princeton, 2006.

[19] H. L. Montgomery, The pair correlation of zeros of the zeta function, *Proc. Symp. Pure Math., Vol. 24*, A.M.S., Providence, 1973, pages 181-193.

[20] R. Murty, *Problems in Analytic Number Theory*, Springer-Verlag, N.Y., 2001.

[21] D. Newland, *Kernels in the Selberg Trace Formula on the k-regular tree and Zeros of the Ihara Zeta Function*, Ph.D. Thesis, U.C.S.D., 2005.

[22] M. Rosen, *Number Theory in Function Fields*, Springer-Verlag, N.Y., 2002.

[23] U. Smilansky, Quantum chaos on discrete graphs, *J. Phys. A: Math. Theor.,* 40 (2007), F621-F630.

[24] H. M. Stark, Hilbert's twelfth problem and L series, *Bull. A.M.S.*, 83 (1977), 1072-1074.

[25] H. M. Stark and A. A. Terras, Zeta Functions of Finite Graphs and Coverings, *Advances in Math.*, 121 (1996), 124-165.

[26] H. M. Stark and A. A. Terras, Zeta Functions of Finite Graphs and Coverings, II, *Advances in Math.*, 154 (2000), 132-195.

[27] H. M. Stark and A. A. Terras, Zeta Functions of Finite Graphs and Coverings, III, *Advances in Math.*, 208 (2007), 467-489.

[28] T. Sunada, L-functions in geometry and some applications, Proc. Taniguchi Symp. 1985, "Curvature and Topology of Riemannian Manifolds", *Springer Lect. Notes in Math.* 1201 (1986), 266-284.

[29] T. Sunada, Number theoretic analogues in spectral geometry, Proc. The 6th Symp. "Differential Geometry and Differential Equations", held in Shanghai, 1985, *Springer Lect. Notes in Math.* 1255 (1987), 96-108.

[30] A. Terras, *Zeta Functions of Graphs: A Stroll Through the Garden*, to appear (Cambridge U. Press, in press). A preliminary version can be found at: http://math.ucsd.edu/~aterras/newbook.pdf.

[31] A. Terras and D. Wallace, Selberg's Trace Formula on k-Regular Trees and Applications. *Internatl. Journal of Math. and Math. Sciences*, Vol. 2003, No. 8 (Feb. 2003), 501-526.

[32] A. Weil, Sur les 'formules explicites' de la théorie des nombres premiers, *Comm. Sém. Math. Univ. Lund, 1952, Tome supplémentaire* (1952), 252-265.

Fields Institute Communications
Volume **60**, 2011

Zeta Functions of Group Based Graphs and Complexes

Wen-Ching W. Li
Department of Mathematics
Pennsylvania State University
University Park, PA 16802-6401, USA
and
National Center for Theoretical Sciences
Mathematics Division
Third General Building, National Tsing Hua University
Hsinchu, Taiwan 300, R.O.C.
wli@math.psu.edu

Abstract. In this survey article, we focus on graphs and complexes arising from quotients of the Bruhat-Tits buildings associated to $\mathrm{PGL}_2(F)$ and $\mathrm{PGL}_3(F)$, respectively. As such, the combinatorial objects, like vertices, edges and chambers, are parametrized algebraically by cosets, and the combinatorial adjacency operators can be interpreted as operators supported on suitable double cosets acting on certain L^2-spaces. The algebraic structure provides links to group theory and number theory. We show interesting connections between combinatorics and number theory, mainly through zeta functions.

1 Introduction

Let F be a nonarchimedean local field with q elements in its residue field. In this review article, we focus on regular graphs and complexes arising from quotients by discrete torsion-free cocompact subgroups of $\mathrm{PGL}_2(F)$ and $\mathrm{PGL}_3(F)$, respectively. As such, the combinatorial objects, like vertices, edges and chambers, are parametrized algebraically by cosets, and the combinatorial adjacency operators can be interpreted as operators supported on suitable double cosets acting on certain L^2-spaces. The algebraic structure provides links to group theory and number theory. The emphasis of this paper is to draw interesting connections between combinatorics and number theory, mainly through zeta functions.

The graph zeta function is defined as a function counting tailless geodesic cycles of given lengths in a graph, analogous to the curve zeta function which counts points in finite extensions of the base (finite) field. Both can be expressed as an infinite

2010 *Mathematics Subject Classification.* Primary 22E35; Secondary 11F70.
Research supported in part by the NSF grant DMS-0801096 and the National Center for Theoretical Sciences, Hsinchu, Taiwan.

product, and both are rational functions. However, there is a major difference: the Riemann hypothesis holds for all smooth projective curves, while combinatorially it holds precisely for Ramanujan graphs. These properties have been known for twenty years.

On the other hand, despite several attempts [6, 7, 8], the complex zeta function with similar properties was not defined until the work of Kang-Li [23]. Like the graph zeta function, the complex zeta function counts a certain type of geodesic cycles of given lengths in a complex. It has the following nice features:

(1) It is a rational function with a closed form expression;
(2) It provides topological and spectral information of the complex;
(3) The complex is Ramanujan if and only if its zeta function satisfies the Riemann Hypothesis.

As we shall see in §9, the complex zeta identity bears striking resemblance to graph zeta identity, and it is also analogous to the surface zeta function.

This paper is organized as follows. The zeta function for a variety defined over a finite field is recalled in §2, followed by a brief review of modular curves in §3. In §4 we recall the spectral theory for regular graphs, leading to the definition of Ramanujan graphs as spectrally optimal graphs. The zeta function of a graph is introduced in §5, where we discuss Ihara's and Hashimoto's results (Theorems 5.2.1 and 5.3.1), yielding a graph zeta function identity (5.1) expressed in terms of the vertex and edge adjacency operators. In §6 we see that for regular graphs arising from $\mathrm{PGL}_2(F)$, connections to automorphic forms and modular curves can be drawn. This is showcased by an elegant identity of Hashimoto (Theorem 6.3.2), linking the class numbers of the graph, the quaternion algebra, and the function field of the modular curve. It is natural to ask how graphs and associated zeta functions vary with the base field F. Some open questions of this nature are raised in §7. In §8 we introduce combinatorial objects and operators on the Bruhat-Tits building $\mathcal{B}_3$ associated to $\mathrm{PGL}_3(F)$ and their algebraic parametrizations using cosets. The main results on the zeta function attached to a finite quotient of $\mathcal{B}_3$ are summarized in §9. Both graphs and complexes considered in this paper arise as quotients by discrete torsion-free cocompact subgroups Γ of $\mathrm{PGL}_n(F)$, ($n = 2$ or 3), and their zeta functions are defined combinatorially in terms of cycles. In the final section §10, we discuss and compare connections between the zeta function and the group Γ.

The reader is referred to [28] for some omitted details.

2 Zeta functions of varieties

2.1 Let V be a smooth irreducible projective variety of dimension d defined over a finite field $\mathbb{F}_q$ of q elements. The zeta function of V counts the number $N_n = \#V(\mathbb{F}_{q^n})$ of $\mathbb{F}_{q^n}-$rational points on V. More precisely, it is defined as

$$Z_V(u) = \exp(\sum_{n\geq 1} \frac{N_n}{n} u^n),$$

which can also be expressed as an infinite product

$$\prod_{v \text{ closed pts}} \frac{1}{(1 - u^{\deg v})}.$$

Recall that a closed point is the Galois orbit of a point x on V with coordinates in an algebraic closure of $\mathbb{F}_q$; it consists of points on V which are conjugate to x. The number of points in the closed point v is called the degree of v, denoted by $\deg v$. The celebrated Weil conjectures [37] state that the zeta function $Z_V(u)$ is a rational function with coefficients in $\mathbb{Z}$, and it satisfies the Riemann Hypothesis. The rationality was first proved by Dwork [12]. Motivated by the Weil conjectures, in 1960's Grothendieck laid general geometric foundation in a series of lecture notes [2, 16, 10], in which a more detailed expression of the zeta function as a rational function was obtained:

$$Z_V(u) = \frac{P_1(u)P_3(u)\cdots P_{2d-1}(u)}{P_0(u)P_2(u)\cdots P_{2d}(u)},$$

where $P_i(u) \in \mathbb{Z}[u]$ is the characteristic polynomial of the Frobenius element acting on the ith l-adic étale cohomology of V. The constant term of $P_i(u)$ is always 1. The Riemann Hypothesis, proved by Hasse [19] and Weil [36] for curves and Deligne [9] for varieties in general, asserts that if $P_i(u) \neq 1$, then its roots have absolute value equal to $q^{-i/2}$ for $0 \leq i \leq 2d$. The reader is referred to the book by Freitag and Kiehl [14] for a more comprehensive treatment of the subject.

2.2 When V is a curve of genus g, its zeta function has the form

$$Z_V(u) = \frac{P_V(u)}{(1-u)(1-qu)},$$

where

$$P_V(u) = \prod_{1 \leq j \leq 2g} (1 - \alpha_j u) = \prod_{1 \leq j \leq g} (1 - \lambda_j u + qu^2).$$

Since α_j and $\bar{\alpha_j}$ occur simultaneously, we have combined them to get a quadratic factor as shown; further the α_j's have absolute value $q^{1/2}$, hence the real number $\lambda_j = \alpha_j + \bar{\alpha_j}$ satisfies the absolute value bound $|\lambda_j| \leq 2q^{1/2}$. It is well-known that (see, for example, Theorem 1 in Chapter 5 of [29], p. 75)

$$P_V(1) = |Jac_V(\mathbb{F}_q)|,$$

which is the class number of the function field of V.

3 Modular curves

3.1 The modular group $\mathrm{SL}_2(\mathbb{Z})$ acts on the Poincaré upper half-plane $\mathfrak{H}$ by fractional linear transformations. For a positive integer N, define the congruence subgroup

$$\Gamma_0(N) = \left\{ \begin{pmatrix} a & b \\ c & d \end{pmatrix} \in \mathrm{SL}_2(\mathbb{Z}) : N|c \right\}.$$

The quotient $\Gamma_0(N)\backslash\mathfrak{H}$ compactified by adding cusps is the modular curve $X_0(N)$. It is well-known that $X_0(N)$ is a smooth projective curve defined over $\mathbb{Q}$ of genus $g = g_0(N)$. Further, $X_0(N)$ has good reductions at the primes $p \nmid N$ (cf. [20]), with the same genus (see §10.4, Prop. 11 in [33]).

3.2 The space $S_2(\Gamma_0(N))$ of weight 2 cusp forms for $\Gamma_0(N)$ has dimension $g_0(N)$. By the work of Hecke, this space is simultaneously diagonalized by the Hecke operators T_p for all $p \nmid N$.

3.3 Now choose a prime $p \nmid N$ and let $V = X_0(N) \mod p$ be the reduction. By Eichler-Shimura congruence relations [13], the λ_j's occurring in the numerator $P_V(u)$ of $Z_V(u)$ are the eigenvalues of T_p on $S_2(\Gamma_0(N))$.

4 Spectra of regular graphs and Ramanujan graphs

4.1 To a finite undirected graph X we attach the adjacency matrix A, whose rows and columns are indexed by the vertices of X such that the ij entry records the number of edges from the vertex i to the vertex j. The eigenvalues of A are called the spectrum of the graph. Since X is undirected, the matrix A is symmetric and hence eigenvalues are real. If X is $(q+1)$-regular, its eigenvalues lie in the interval $[-q-1,\ q+1]$. Further, $q+1$ is always an eigenvalue and $-(q+1)$ is an eigenvalue if and only if X is a bipartite graph. These are called trivial eigenvalues of X. In order to study the distribution of the remaining (nontrivial) eigenvalues, denote by $\lambda^+(X)$ the largest eigenvalue $< q+1$, and by $\lambda^-(X)$ the smallest eigenvalue $> -(q+1)$. We are interested in the behavior of $\lambda^\pm$, that is, the spread of nontrivial eigenvalues.

4.2 Let X_j be a family of $(q+1)$-regular graphs with $|X_j| \to \infty$ as $j \to \infty$. For $\lambda^+(X_j)$, the following result is well-known:

Theorem 4.1 (Alon-Boppana, see [1] and Prop. 4.2 of [30])

$$\liminf_{j \to \infty} \lambda^+(X_j) \geq 2\sqrt{q}.$$

The symmetric statement for λ^- does not hold unconditionally. A mild assumption is needed.

Theorem 4.2 (Li, [26]) *Assume the length of the shortest odd cycle in X_j tends to infinity as $j \to \infty$. Then*

$$\limsup_{j \to \infty} \lambda^-(X_j) \leq -2\sqrt{q}.$$

4.3 A connected $(q+1)$-regular graph X is called *Ramanujan* if

$$|\lambda^\pm(X)| \leq 2\sqrt{q}.$$

In view of the above theorems, the Ramanujan graphs are spectrally optimal.

The universal cover of $(q+1)$-regular graphs is the infinite $(q+1)$-regular tree $\mathcal{T}$. To this infinite graph, we associate the adjacency operator A acting on the space of L^2-functions on vertices of $\mathcal{T}$. The operator spectrum of A is $[-2\sqrt{q},\ 2\sqrt{q}]$. Thus a graph is Ramanujan if and only if all of its nontrivial eigenvalues fall in the spectrum of its universal cover.

4.4 Extending this viewpoint to higher dimensional complexes, we define a Ramanujan complex in a similar way in [27], where higher dimensional Ramanujan complexes were explicitly constructed.

5 Zeta functions of graphs

5.1 The zeta function of a finite graph X counts the number of geodesic tailless cycles in X. Recall that a cycle in a graph is oriented and has a starting point. It is called tailless if it remains a geodesic regardless the initial point. It is primitive if it is not obtained by tracing a shorter cycle more than once. Two cycles are equivalent if one is obtained from the other by shifting the starting point. Denote by N_n the number of tailless geodesic cycles in X of length n. The precise definition of the graph zeta function of X, first introduced by Ihara, is

$$Z(X,u) := \exp\left(\sum_{n\geq 1}\frac{N_n}{n}u^n\right) = \prod_{[C]}\frac{1}{1-u^{l(C)}},$$

where $[C]$ runs through all equivalence classes of primitive geodesic tailless cycles C in X, and $l(C)$ is the length of C. The equality of the two expressions results from comparing logarithmic derivatives of both functions.

If we regard each cycle as a "point" and each equivalence class as an "orbit", then the size of the equivalence class of a primitive cycle C is equal to the length of C. With this viewpoint, the zeta functions of graphs and varieties are quite alike. Thus it is not surprising that they share some common properties.

5.2 Similar to zeta functions of curves, zeta functions of graphs are rational functions with coefficients in $\mathbb{Z}$. This was first shown by Ihara for regular graphs.

Theorem 5.1 (Ihara, [21]) *Let X be a finite $(q+1)$-regular graph with vertex set V and edge set E. Then its zeta function $Z(X,u)$ is a rational function of the form*

$$Z(X,u) = \frac{(1-u^2)^{\chi(X)}}{\det(I - Au + qu^2 I)},$$

where $\chi(X) = \#V - \#E = -\frac{q-1}{2}\#V$ is the Euler characteristic of X and A is the (vertex) adjacency matrix of X.

This result extends to arbitrary finite undirected graphs, see [3] and [34] and the references therein.

Note that the numerator of the graph zeta function gives topological information, namely, the Euler characteristic of the graph, while the denominator gives the spectral information of the graph. Note also that the fundamental group $\pi_1(X, pt)$ with a base point of the graph X is free of rank $1-\chi(X)$.

The Ramanujan condition on X can be rephrased as nontrivial poles of $Z(X,u)$ having absolute value $q^{-1/2}$; in this case we say that its zeta function $Z(X,u)$ satisfies the Riemann Hypothesis (RH). This indicates that the graph zeta functions are analogous to curve zeta functions on the one hand. On the other hand, RH always holds for curves, but it is not so for regular graphs. In fact, RH characterizes Ramanujan graphs.

5.3 Endow two opposite orientations on each edge of X. Define the neighbors of the directed edge $u \to v$ to be the edges $v \to w$ with $w \neq u$. The edge adjacency matrix A_e has its rows and columns indexed by the oriented edges $\vec{e}$ of X such that the $\vec{e}\,\vec{e'}$ entry is 1 if $\vec{e'}$ is a neighbor of $\vec{e}$, and 0 otherwise.

Theorem 5.2 (Hashimoto, [17]) *For* $n \geq 1$, $N_n = \mathrm{Tr} A_e^n$ *so that*

$$Z(X,u) = \frac{1}{\det(I - A_e u)}.$$

Combined with Ihara's theorem, we obtain the following identity for a $(q+1)$-regular graph X:

$$\frac{(1-u^2)^{\chi(X)}}{\det(I - Au + qu^2 I)} = \frac{1}{\det(I - A_e u)}. \tag{5.1}$$

6 Connections with number theory

6.1 When $q = p^r$ is a prime power, the $(q+1)$-regular tree can be interpreted as a homogeneous space as follows. Choose a nonarchimedean local field F with q elements in its residue field. Denote by $\mathcal{O}_F$ its ring of integers and π a uniformizer. The $(q+1)$-regular tree $\mathcal{T}$ can be identified with

$$\mathcal{T} = \mathrm{PGL}_2(F)/\mathrm{PGL}_2(\mathcal{O}_F).$$

Under this identification, we can express the combinatorial objects algebraically. The vertices of the tree are parametrized by $\mathrm{PGL}_2(\mathcal{O}_F)$-cosets, and the vertex adjacency operator A is the Hecke operator on the double coset

$$\mathrm{PGL}_2(\mathcal{O}_F) \begin{pmatrix} 1 & 0 \\ 0 & \pi \end{pmatrix} \mathrm{PGL}_2(\mathcal{O}_F).$$

The group $\mathrm{PGL}_2(F)$ acts transitively on the vertices by left translations and preserves adjacency. There is an edge between vertices $v_1 = \mathrm{PGL}_2(\mathcal{O}_F)$ and $v_2 = \begin{pmatrix} 1 & 0 \\ 0 & \pi \end{pmatrix} \mathrm{PGL}_2(\mathcal{O}_F)$. The intersection of the stabilizers of these two vertices is the Iwahori subgroup $\mathcal{I}$ consisting of elements in $\mathrm{PGL}_2(\mathcal{O}_F)$ congruent to the upper triangular matrices modulo π. Denote by $\mathcal{I}$ the directed edge from v_1 to v_2. The action of $\mathrm{PGL}_2(F)$ on vertices maps the directed edge $\mathcal{I}$ to all directed edges of $\mathcal{T}$. In particular, left multiplication by $\begin{pmatrix} 0 & 1 \\ \pi & 0 \end{pmatrix}$ permutes v_1 and v_2, hence $\begin{pmatrix} 0 & 1 \\ \pi & 0 \end{pmatrix} \mathcal{I}$ denotes the edge from v_2 to v_1. Consequently, the directed edges are parametrized by the $\mathcal{I}$-cosets. The edge adjacency operator A_e is the Iwahori-Hecke operator on the double coset $\mathcal{I} \begin{pmatrix} 1 & 0 \\ 0 & \pi \end{pmatrix} \mathcal{I}$. We obtain a finite $(q+1)$-regular graph by taking left quotient by a discrete torsion-free cocompact subgroup Γ of $\mathrm{PGL}_2(F)$:

$$X = X_\Gamma = \Gamma \backslash \mathrm{PGL}_2(F)/\mathrm{PGL}_2(\mathcal{O}_F) = \Gamma \backslash \mathcal{T}.$$

Theorem 6.1 (Ihara [21]) *A discrete torsion-free cocompact subgroup* Γ *of* $PGL_2(F)$ *is free of rank* $1 - \chi(X_\Gamma)$.

To prove this theorem, Ihara showed that Γ is the fundamental group of the graph X_Γ. In fact, the graph zeta function originated from this setting; in this case, the equivalence classes $[C]$ of tailless primitive geodesics are parametrized by the primitive conjugacy classes of Γ. It was Serre ([31]) who observed that the definition can be phrased for general graphs.

6.2 Let K be a global field such that F is the completion of K at a nonarchimedean place v, that is, $F = K_v$. Assume K is totally real if F has characteristic zero. Let D be a quaternion algebra over K unramified at v and ramified at all places at ∞ if K is a number field. The unramified condition of D at v implies that $D^\times(K_v)$ mod center is isomorphic to $\mathrm{PGL}_2(K_v)$. Let Γ be the v-integral (i.e., integral everywhere but v) units in $D^\times(K)$ mod center, or a congruence subgroup of this group. Since $D^\times(K)$ is contained in $D^\times(K_v)$, Γ is a discrete cocompact subgroup of $\mathrm{PGL}_2(K_v)$.

For such Γ, functions on $X_\Gamma = \Gamma\backslash\mathrm{PGL}_2(K_v)/\mathrm{PGL}_2(\mathcal{O}_v)$ can be interpreted as automorphic forms for $D^\times$ mod center by using strong approximation theorem; those perpendicular to the constant functions can be further identified with cuspidal automorphic forms for PGL_2 over K under the Jacquet-Langlands correspondence, as explained in the last chapter of [22]. The Hecke operator T_v acts on these cusp forms and its eigenvalues have absolute values at most $2q^{1/2}$, resulting from the Ramanujan conjecture established by Deligne [9] for $K = \mathbb{Q}$, Brylinski-Labesse [5] and Blasius [4] for K totally real, and Drinfeld for K any function field [11].

Thus X_Γ is a Ramanujan graph. If Γ is torsion free, then the graph has no loops and no multiple edges, while if Γ is not torsion free, then loops and/or multiple edges may occur.

6.3 Now consider the special case where $K = \mathbb{Q}$ and $F = \mathbb{Q}_p$. Denote by D_ℓ the definite quaternion algebra ramified only at ∞ and the prime $\ell \neq p$. Pick the discrete subgroup $\Gamma_\ell = D_\ell^\times(\mathbb{Z}[\frac{1}{p}])$ mod center. We obtain a Ramanujan graph X_{Γ_ℓ} with vertex adjacency operator $A = T_p$, the Hecke operator at p. Hashimoto [17] had shown that when $\ell \equiv 1 \pmod{12}$, Γ_ℓ is torsion-free.

The zeta function of the modular curve $V = X_0(\ell)$ mod p discussed in §2.2 has numerator

$$P_V(u) = \frac{\det(I - Au + pu^2)}{(1-u)(1-pu)}.$$

In the above expression, the numerator is equal to the product of $1 - \lambda u + pu^2$ over eigenvalues λ of A, and the quotient simply means that the factor corresponding to the trivial eigenvalue $p+1$ is removed so that the remaining eigenvalues come from the weight 2 cusp forms for $\Gamma_0(\ell)$. See Theorem 7.11 in Chap. 7 of [32]. On the other hand, the class number of the quaternion algebra D_ℓ is equal to $|X_{\Gamma_\ell}| = 1 + g_0(\ell)$, where $g_0(\ell)$ is the genus of $X_0(\ell)$.

On the combinatorial side, the number of spanning trees in a graph X is called the class number $\kappa(X)$ of the graph. Hashimoto showed that the class number of X is a special value of the denominator of the zeta function of X:

Theorem 6.2 (Hashimoto, [17]) *For a $(q+1)$-regular graph X,*

$$\kappa(X)|X| = \frac{\det(I - Au + qu^2)}{(1-u)(1-qu)}|_{u=1}.$$

When $X = X_{\Gamma_\ell}$, this gives a beautiful relation on three class numbers.

Theorem 6.3 (Hashimoto, [17]) *Suppose $\ell \equiv 1 \pmod{12}$. Then*

$$\kappa(X_{\Gamma_\ell})(1 + g_0(\ell)) = |Jac_{X_0(\ell)}(\mathbb{F}_p)|.$$

7 Further connections and some open questions

7.1 The equivalence classes of primitive geodesic cycles $[C]$ in the definition of the zeta function of a graph X represent the primitive conjugacy classes of the fundamental group $\pi_1(X, pt)$. The zeta function $Z(X, u)$ can be viewed as the zeta function attached to the trivial representation of $\pi_1(X, pt)$. One can consider irreducible representations ρ of $\pi_1(X, pt)$ and consider "Artin" L-function, $L(X, \rho, u)$. This was done by Hashimoto [18] and Stark-Terras [35].

7.2 Stark and Terras introduced the notion of finite Galois unramified covers Y of a graph X, studied how primes of X decompose in Y, and showed that $Z(X, u)$ divides $Z(Y, u)$, similar to the behavior of Artin L-functions for global fields [35].

7.3 A curve V defined over $\mathbb{F}_q$ is also defined over $\mathbb{F}_{q^n}$ for integer $n \geq 1$; denote this curve by V_n. Then

$$Z_{V_n}(u^n) = \prod_{1 \leq m \leq n} Z_V(\zeta_n^m u), \qquad \text{where } \zeta_n = exp(2\pi i/n).$$

This is Theorem 1 in Chap. 6 of [29].

Let D be a quaternion algebra over $K = \mathbb{F}_q(x)$ ramified at ∞ and unramified at 0. Let $F = K_0 = \mathbb{F}_q((x))$. Let $K_n = \mathbb{F}_{q^n}(x)$ and suppose D is also a quaternion algebra over K_n, which we denote by D_n to distinguish the base field. As in §6.2, let Γ (resp. Γ_n) be the 0-integral elements in $D(K)^\times$ mod center (resp. $D_n(K_n)^\times$ mod center) so that we obtain two graphs

$$X_\Gamma = \Gamma \backslash \mathrm{PGL}_2(\mathbb{F}_q((x)))/\mathrm{PGL}_2(\mathbb{F}_q[[x]])$$

and

$$X_{\Gamma_n} = \Gamma_n \backslash \mathrm{PGL}_2(\mathbb{F}_{q^n}((x)))/\mathrm{PGL}_2(\mathbb{F}_{q^n}[[x]]).$$

Is there any relation between the graphs X_Γ and X_{Γ_n}, and their zeta functions?

8 The Bruhat-Tits building attached to $\mathrm{PGL}_3(F)$

8.1 Let F be a nonarchimedean local field as in §6. For convenience, write G for $\mathrm{PGL}_3(F)$ and K for $\mathrm{PGL}_3(\mathcal{O}_F)$. The Bruhat-Tits building is $\mathcal{B}_3 = G/K$. It is a 2-dimensional simply connected simplicial complex consisting of vertices, edges, and chambers. The Iwahori subgroup B consists of elements in K congruent to upper triangular matrices mod π, and the parahoric subgroup E consists of elements in K whose last row mod π is a multiple of the row vector $(0, 0, 1)$. They satisfy the relation

$$K \supset E \supset B.$$

Like the tree $\mathcal{T}$, we can parametrize combinatorial objects on $\mathcal{B}_3$ algebraically in terms of K-, E- and B-cosets. More precisely, the vertices of $\mathcal{B}_3$ are parametrized by the K-cosets. Each vertex gK has a type $\tau(gK) = \mathrm{ord}_\pi \det g \pmod 3$, adjacent vertices have different types, and the three vertices in a chamber have distinct types. The type of an edge $gK \to g'K$ is $\tau(g'K) - \tau(gK) = 1$ or 2. Call $g'K$ a type 1 or 2 neighbor of gK, accordingly. Each undirected edge in $\mathcal{B}_3$ has a unique orientation to make it a type 1 edge. The E-cosets parametrize the type 1 edges, and hence also those with reverse orientation, namely the type 2 edges. A chamber together with a choice of one of its three vertices is called a directed chamber. The directed chambers are parametrized by B-cosets.

8.2 Now we discuss the operators on $\mathcal{B}_3$. Acting on the vertices are the two Hecke operators A_1 and A_2. Algebraically they can be expressed as operators supported on K-double cosets and acting on $L^2(G/K)$. Combinatorially, A_i, $i = 1, 2$, is the adjacency matrix of type i neighbors of vertices.

The parahoric operator L_E is supported on an E-double coset and acting on $L^2(G/E)$. Combinatorially, it is the type one edge adjacency matrix : the neighbors of the type one edge $u \to v$ are the type one edges $v \to w$ such that u, v, w do not form a chamber. Its transpose L_E^t is the type two edge adjacency matrix.

The Iwahori-Hecke operator L_B is supported on a B-double coset and acting on $L^2(G/B)$. Combinatorially, it is the adjacency matrix of directed chambers: The neighbors of a directed chamber (C, u) with three distinct vertices u, v, w and a type 1 edge $v \to w$ are (C', v) where the chambers $C' \neq C$ share the edge $v \to w$. See [23] for detailed definitions of these operators algebraically.

9 Zeta functions of finite quotients of $\mathcal{B}_3$

9.1 Similar to graphs, the finite complexes we consider are

$$X_\Gamma = \Gamma\backslash G/K = \Gamma\backslash\mathcal{B}_3,$$

where Γ is a torsion-free discrete cocompact subgroup of G and $\operatorname{ord}_\pi \det \Gamma \subset 4\mathbb{Z}$.

Our goal is to define a suitable zeta function, which counts geodesic cycles up to homotopy in X_Γ, and which has the following properties:

- It is a rational function with a closed form expression;
- It provides topological and spectral information of the complex;
- The complex is Ramanujan if and only if its zeta function satisfies the Riemann Hypothesis.

Described below is joint work with Ming-Hsuan Kang [23], where zeta functions with these three properties are constructed.

9.2 The type 1 edge zeta function of X_Γ is defined as

$$Z_E(X_\Gamma, u) = \prod_{[C]} \frac{1}{1 - u^{l(C)}},$$

where $[C]$ runs through the equivalence classes of primitive tailless geodesic cycles in X_Γ consisting of type 1 edges. It is not hard to prove

Theorem 9.1 ([23]) *$Z_E(X_\Gamma, u)$ is a rational function given by*

$$Z_E(X_\Gamma, u) = \frac{1}{\det(I - L_E u)}.$$

9.3 A sequence of edge-adjacent chambers is called a gallery. The chamber zeta function of X_Γ is defined as

$$Z_B(X_\Gamma, u) = \prod_{[\mathfrak{G}]} \frac{1}{1 - u^{l(\mathfrak{G})}},$$

where $[\mathfrak{G}]$ runs through the equivalence classes of primitive geodesic type 1 closed galleries in X_Γ.

Theorem 9.2 ([23]) *$Z_B(X_\Gamma, u)$ is a rational function given by*

$$Z_B(X_\Gamma, u) = \frac{1}{\det(I - L_B u)}.$$

9.4 Note that the type 1 cycles traveled in reverse direction are the type 2 cycles, while the length is doubled. The zeta function of X_Γ is defined to be

$$Z(X_\Gamma, u) = \prod_{[C]} \frac{1}{1 - u^{l(C)}} = \frac{1}{\det(I - L_E u)} \frac{1}{\det(I - L_E^t u^2)},$$

where $[C]$ runs through the equivalence classes of primitive tailless geodesic cycles in X_Γ consisting of edges of the same type.

Summarized below are our main results on zeta function arising from finite quotients of $\mathcal{B}_3$.

Theorem 9.3 (Kang-Li, [23]) (a) *$Z(X_\Gamma, u)$ is a rational function with closed form*

$$Z(X_\Gamma, u) = \frac{(1 - u^3)^{\chi(X_\Gamma)}}{\det(I - A_1 u + qA_2 u^2 - q^3 u^3 I) \det(I + L_B u)}, \tag{9.1}$$

where $\chi(X_\Gamma)$ is the Euler characteristic of X_Γ.

(b) *X_Γ is a Ramanujan complex if and only if $Z(X_\Gamma, u)$ satisfies the Riemann Hypothesis.*

Remarks. (1) Ramanujan complexes are originally defined in terms of the eigenvalues of A_1 and A_2 (cf. [27]), which can be rephrased as

(i) the nontrivial zeros of $\det(I - A_1 u + qA_2 u^2 - q^3 u^3 I)$ have absolute value q^{-1}.

Kang-Li-Wang showed in [24] that this has two more equivalent statements in terms of other operators:

(ii) the nontrivial zeros of $\det(I - L_E u)$ have absolute values q^{-1} and $q^{-1/2}$;

(iii) the nontrivial zeros of $\det(I + L_B u)$ have absolute values 1, $q^{-1/2}$ and $q^{-1/4}$.

$Z(X_\Gamma, u)$ is said to satisfy the Riemann Hypothesis if the nontrivial zeros of $\det(I - A_1 u + qA_2 u^2 - q^3 u^3 I)$ and $\det(I + L_B u)$ are as described in (i) and (iii). In view of Table 1 in [24], this is equivalent to $Z(X_\Gamma, u)$ having no poles of absolute value equal to $q^{-3/2}$ or $q^{3/4}$.

(2) The zeta identity can be reformulated in terms of operators:

$$\frac{(1 - u^3)^{\chi(X_\Gamma)}}{\det(I - A_1 u + qA_2 u^2 - q^3 u^3 I)} = \frac{\det(I + L_B u)}{\det(I - L_E u) \det(I - L_E^t u^2)},$$

which bears striking resemblance to the identity (5.1) for graph zeta functions.

(3) The original proof in [23] used the combinatorial interpretation of the operators. By regarding the operators as acting on various L^2 spaces, in [24] Kang-Li-Wang gave a totally different representation-theoretical proof of the theorem.

(4) The zeta identity for 2-dimensional complexes is analogous to the surface zeta functions. Indeed, there is a connection to zeta functions of modular surfaces discussed in [25] similar to the connection we saw in §6.3 between graph zeta functions and zeta functions of modular curves.

10 Connections between zeta function and Γ

Let F be a local field as before and let Γ be a discrete torsion-free cocompact subgroup of $\mathrm{PGL}_n(F)$. The 1- and 2-dimensional complexes considered above are expressed as

$$X_\Gamma = \Gamma\backslash \mathrm{PGL}_n(F)/\mathrm{PGL}_n(\mathcal{O}_F)$$

for $n = 2, 3$, respectively. The zeta function for X_Γ is defined combinatorially. There is a close connection with the group Γ, which, however, differs between graphs and complexes. We compare them below.

10.1 The case $n = 2$.

- Every element in Γ has two eigenvalues in F;
- The primitive equivalence classes $[C]$ of geodesic tailless cycles C correspond bijectively to the primitive conjugacy classes $[\gamma]$ of Γ (cf. [34]);
- If $[C]$ corresponds to $[\gamma]$, then the length $l(C) = n$ is the unique nonnegative integer such that

$$\gamma \in \mathrm{PGL}_2(\mathcal{O}_F) \begin{pmatrix} 1 & 0 \\ 0 & \pi^n \end{pmatrix} \mathrm{PGL}_2(\mathcal{O}_F).$$

10.2 The case $n = 3$. The following facts were shown in [23].

- Every element in Γ has one or three eigenvalues in F;
- Γ contains elements having only one eigenvalue in F;
- A primitive conjugacy class $[\gamma]$ of type 1 or 2 is a packet containing finitely many primitive $[C]$, and the packet of a non-primitive conjugacy class $[\gamma]$ of type 1 or 2 may also contain finitely many primitive $[C]$;
- If $[C]$ is contained in the packet $[\gamma]$, then $l(C) = m + n$, where m and n are the unique nonnegative integers such that

$$\gamma \in \mathrm{PGL}_3(\mathcal{O}_F) \begin{pmatrix} 1 & 0 & 0 \\ 0 & \pi^n & 0 \\ 0 & 0 & \pi^{m+n} \end{pmatrix} \mathrm{PGL}_3(\mathcal{O}_F).$$

References

[1] N. Alon, *Eigenvalues and expanders*, Combinatorica, **6** (1986), 83-96.

[2] M. Artin, A. Grothendieck and J. L. Verdier, *Théorie des topos et cohomologie étale des schemas (1963-64).* SGA4. Lecture Notes in Math. **269, 270, 305**, Springer-Verlag, Berlin-Heidelberg-New York, 1972-73.

[3] H. Bass, *The Ihara-Selberg zeta function of a tree lattice*, Internat. J. Math. **3** (1992), no. 6, 717-797.

[4] D. Blasius, *Hilbert modular forms and the Ramanujan conjecture*, Noncommutative geometry and number theory, 35–56, Aspects Math., E37, Vieweg, Wiesbaden, 2006.

[5] J.-L. Brylinski and J.-P. Labesse, *Cohomologie d'intersection et fonctions L de certaines varie'te's de Shimura*, Ann. Sci. E'cole Norm. Sup. **4** 17 (1984), no. 3, 361–412.

[6] A. Deitmar, *Geometric zeta functions of locally symmetric spaces*, Amer. J. Math. **122** (2000), no. 5, 887-926.

[7] A. Deitmar, *Lefschetz formulae and zeta functions*, J. Ramanujan Math. Soc. **18** (2003), no. 2, 139-152.

[8] A. Deitmar and J. W. Hoffman, *The Ihara-Selberg zeta function for PGL(3) and Hecke operators*, Internat. J. Math. **17**(2006), 143-156.

[9] P. Deligne, *La conjecture de Weil I*, Inst. Hautes Etudes Sci Publ. Math. **43** (1974), 273-307.

[10] ______ (with J. F. Boutot, L. Illusie and J. L. Verdier), *Cohomologie étale.* SGA4$\frac{1}{2}$. Lecture Notes in Math. **569**, Springer-Verlag, Berlin-Heidelberg-New York, 1977.

[11] V. G. Drinfel'd, *The proof of Petersson's conjecture for* $GL(2)$ *over a global field of characteristic* p, Functional Anal. Appl. **22** (1988), 28-43.

[12] B. Dwork, *On the rationality of the zeta function of an algebraic variety*, Amer. J.Math. **82** (1960), 631-648.

[13] M. Eichler, *Eine Verallgemeinerung der Abelschen Integrale*, Math. Z. **67** (1957), 267–298.

[14] E. Freitag and R. Kiehl, *Étale Cohomology and the Weil Conjecture*, Ergebnisse der Mathematik und ihrer Grenzgebiete (3), 13. Springer-Verlag, Berlin, 1988.

[15] A. Grothendieck, *Formule de Lefschetz et rationalité des fonctions L*, Seminaire Bourbaki 1964/65, Exposé 279. W. A. Benjamin, New York, 1966.

[16] ———, *Cohomologie* ℓ*-adique et fonctions L (1965-66)*. SGA5. Lecture Notes in Math. **589**, Springer-Verlag, Berlin-Heidelberg-New York, 1977.

[17] K. Hashimoto, *Zeta functions of finite graphs and representations of p-adic groups*. Automorphic Forms and Geometry of Arithmetic Varieties, 211-280, Adv. Stud. Pure Math., 15, Academic Press, Boston, MA, 1989.

[18] ———, *Artin type L-functions and the density theorem for prime cycles on finite graphs*, Internat. J. Math. **3** (1992), no. 6, 809-826.

[19] H. Hasse, *Beweis des Analogons der Riemannschen Vermutung für die Artinschen und F. K. Schmidtschen Kongruenzzetafunktionen in gewissen elliptischen Fällen*, Ges. d. Wiss. Nachrichten. Math. Phys. Klasse, 1933, Heft 3, 253-262.

[20] J. Igusa, *Kroneckerian model of fields of elliptic modular functions*. Amer. J. Math. **81** (1959), 561-577.

[21] Y. Ihara, *On discrete subgroups of the two by two projective linear group over p-adic fields*, J. Math. Soc. Japan **18** (1966), 219-235.

[22] H. Jacquet and R. P. Langlands, *Automorphic forms on* $GL(2)$, Lecture notes in math. **114**, Springer-Verlag, Berlin-Heidelberg-New York, 1970.

[23] M.-H. Kang and W.-C. W. Li, *Zeta Functions of Complexes Arising from* $PGL(3)$, preprint (2010).

[24] M.-H. Kang, W.-C. W. Li and C.-J. Wang, *The Zeta Functions of Complexes from* $PGL(3)$*: a Representation-theoretic Approach*, Israel J. Math. **177** (2010), 335-347.

[25] G. Laumon, M. Rapoport and U. Stuhler, $\mathcal{D}$*-elliptic sheaves and the Langlands correspondence*, Invent. Math. **113** (1993), 217-238.

[26] W.-C. W. Li, *On negative eigenvalues of regular graphs*, C. R. Acad. Sci. Paris, t. **333**, Série I, 907-912, 2001.

[27] ———, *Ramanujan hypergraphs*, Geom. Funct. Anal. textbf14 (2004), 380-399.

[28] ———, *Zeta Functions in Combinatorics and Number Theory*, Proc. of the 4th International Congress of Chinese Mathematicians, Hangzhou, China, 2007, AMS/IP, to appear.

[29] ———, *Number Theory with Applications*, World Scientific, Singapore, 1996.

[30] A. Lubotzky, R. Phillips, and P. Sarnak, *Ramanujan graphs*, Combinatorica **8** (1988), 261-277.

[31] J-P. Serre, *Trees*. (Translated from the French by John Stillwell). Springer-Verlag, 1980.

[32] G. Shimura, *Introduction to the Arithmetic Theory of Auromorphic Functions*, Iwami Shoten and Princeton University Press, 1971.

[33] G. Shimura and Y. Taniyama, *Complex multiplication of abelian varieties and its applications to number theory*. Publ. Math. Soc. Japan, no. 6, 1961.

[34] H. Stark and A. Terras, *Zeta functions of finite graphs and coverings*, Adv. Math. **121** (1996), 124-165.

[35] ———, *Zeta functions of finite graphs and coverings. II*. Adv. Math. **154** (2000), no. 1, 132–195.

[36] A. Weil, *Sur les Courbes Algébriques et les Variétés Qui s'en Deduisent*, Hermann, Paris (1948).

[37] ———, *Numbers of solutions of equations in finite fields*, Bulletin of Amer. Math. Soc. **55** (1949), 497-508.

Fields Institute Communications
Volume **60**, 2011

Ramified Covers of Graphs and the Ihara Zeta Functions of Certain Ramified Covers

Beth Malmskog
Department of Mathematics
Colorado State University
Fort Collins, CO 80523-1874 USA
malmskog@math.colostate.edu

Michelle Manes
Department of Mathematics
University of Hawaii
Honolulu, HI 96822 USA
mmanes@math.hawaii.edu
http://www.math.hawaii.edu/~mmanes/

Abstract. This paper discuses the Ihara zeta functions of regular graphs and certain coverings of regular graphs. As a warm-up example, we calculate the zeta functions of the platonic solids. We outline our recent results, describing a ramified covering of graphs $Y \to X$, where Y is formed by identifying a single vertex on k copies of a $(q+1)$-regular graph X. We find that the reciprocal of the zeta function for X "almost divides" the reciprocal of the zeta function for Y, in the sense that the reciprocal of the zeta function of X divides the product of the reciprocal of the zeta function of Y and some polynomial of bounded degree (which depends only on the graph X). Two specific examples show that in fact "almost divisibility" is the best that can be hoped for. These coverings give examples of the zeta functions of infinite families of non-regular graphs. Non-regular complete bipartite graphs are also considered, and we provide a new proof of Hashimoto's result on the zeta functions of these graphs.

Introduction

The Ihara zeta function of a graph was defined by Ihara in the 1960s [3]. It was modeled on other zeta functions in its form, an infinite product over primes in the graph, and has some analogous properties, for example convergence to a rational function. Ihara introduced the function in the context of groups, and

2010 *Mathematics Subject Classification.* Primary 05C50, 14G10.

The authors thank the Banff International Research Station for supporting this project in its initial stages.

Serre observed that the function could be interpreted in terms of graphs. This connection was first seen for the finite regular graphs that arose as quotients of regular trees. Sunada [4] studied the zeta functions of these regular graphs in the 1980s. Soon after, Hashimoto [2] further developed the theory of zeta functions, with an emphasis on bipartite graphs. Hashimoto also proved that the complexity, i.e. number of spanning trees, of a regular graph can be expressed in terms of the zeta function of the graph [2]. Northshield extended this result to include irregular graphs [6]. Bass extended earlier results to general, non-regular graphs. Terras' paper in this volume surveys some current research areas and presents some open problems.

One focus of this paper is the analogy between Ihara zeta functions for graphs and Artin zeta functions for curves defined over finite fields, defined by E. Artin in his 1921 thesis. Serre proved that the Artin zeta function of a covered curve divides that of the covering curve as long as the covering is defined over the base field of the zeta function.

Terras and Stark have studied Ihara zeta functions extensively [8] [9] [10] and have outlined a notion of unramified coverings of connected graphs. They found a divisibility relationship among the reciprocals of the Ihara zeta functions of the graphs in unramified coverings (see Terras's paper in this volume for a proof, or [11]). In an attempt to extend the theory of Ihara zeta functions and draw further parallels with Artin's zeta function, we sought a notion of ramified coverings of graphs for which some divisibility relations between Ihara zeta functions could be found.

In 2007, Baker and Norine proved a graph theoretic analogue of the Riemann-Hurwitz formula using a definition of ramified covering maps of graphs adopted from Urakawa [1]. Adapting this notion of ramification, we find an "almost divisibility" relationship between the Ihara zeta functions of some ramified covers of graphs.

In this paper, we begin by considering the zeta functions of regular graphs. We calculate the Ihara zeta functions of the platonic solids as an example. We restate the results proven in [5], forming a ramified cover of a $(q+1)$-regular graph X by a second graph Y, made up of k copies of X identified at a single common vertex. We prove in [5] that the reciprocal of the zeta function for X "almost divides" the reciprocal of the zeta function for Y, in the sense that the reciprocal of the zeta function of X divides the product of the reciprocal of the zeta function of Y and some polynomial of bounded degree. Two specific examples show that in fact true divisibility does not hold. These examples also illustrate the zeta functions of two infinite families of non-regular graphs. We also consider non-regular complete bipartite graphs, and provide a different proof of Hashimoto's result on the zeta functions of these graphs

1 Background and definitions

Let H be a finite connected graph with edge set $E(H)$ and vertex set $V(H)$. We arbitrarily orient the edges so that we can refer to directions of travel along the edges, i.e. if e is an edge with an arbitrary orientation, let e^{-1} denote that edge with opposite orientation. Let $P = a_1a_2...a_n$ be a path in H, where $a_i \in E(H)$ for each i. A path is closed if it starts and ends at the same vertex. A path is said to contain a backtrack if $a_i = a_{i+1}^{-1}$ for some i. A closed path is said to contain a tail if $a_1 = a_n^{-1}$. Paths are considered equivalent if they are the same cycle of edges with a different starting point. If C is a closed path in H, let $[C]$ denote the equivalence class of C in H. A prime path in H is a tailless, backtrackless, closed path C such that $C \neq D^s$ for any positive integer s and any path D in H. In other words, a prime is any closed path without any backtracking that is not simply another path traced several times. A prime in a graph is an equivalence class $[C]$ of prime paths in the graph. The length of a prime $[C]$ is the number of edges in any representative of the class and is denoted $\nu(C)$.

Definition 1.1 Let $u \in \mathbb{C}$ with $|u|$ sufficiently small. The Ihara zeta function of a finite connected graph H is

$$\zeta_H(u) := \prod_{[P] \text{ primes in } H} (1 - u^{\nu(P)})^{-1}.$$

To calculate the zeta function for anything but an extremely simple graph we will use the following, known as Ihara's formula. The adjacency matrix of a graph H with $|V(H)| = m$ is an $m \times m$ matrix where the $(i,j)^{\text{th}}$ entry is 1 if vertex i is adjacent to vertex j and 0 otherwise. Also, the rank of the fundamental group of the finite, connected graph H with no vertices of degree 1 is equal to $|E(H)|-|V(H)|+1$.

Theorem 1.2 [See e.g. Ihara, Bass, Hashimoto] Let A be the adjacency matrix of a finite connected graph H which has no vertices of degree 1. Let I be the $m \times m$ identity matrix. Let Q be the diagonal matrix with the $(j,j)^{\text{th}}$ entry equal to one less than the degree of vertex j. Let r be the rank of the fundamental group of H. Then

$$\zeta_H^{-1}(u) = (1-u^2)^{r-1}\det(I - Au + Qu^2). \tag{1.1}$$

1.1 Example: The Platonic Solids. The platonic solids provide pretty examples of graphs and their zeta functions. We create a graph G from each platonic solid P by letting the vertex set and edge set of G be the vertices and edges of P.

The graph of the tetrahedron is K_4, the complete graph on 4 vertices. It has adjacency matrix

$$A_{\text{tetra}} = \begin{pmatrix} 0 & 1 & 1 & 1 \\ 1 & 0 & 1 & 1 \\ 1 & 1 & 0 & 1 \\ 1 & 1 & 1 & 0 \end{pmatrix}.$$

Every vertex has degree 3, so

$$Q_{\text{tetra}} = \begin{pmatrix} 2 & 0 & 0 & 0 \\ 0 & 2 & 0 & 0 \\ 0 & 0 & 2 & 0 \\ 0 & 0 & 0 & 2 \end{pmatrix}.$$

The rank r of the fundamental group for the tetrahedron is $r = 6 - 4 + 1 = 3$. Therefore

$$\begin{aligned} \zeta_{\text{tetra}}^{-1}(u) &= (1-u^2)^2 \det \begin{pmatrix} 1+2u^2 & -u & -u & -u \\ -u & 1+2u^2 & -u & -u \\ -u & -u & 1+2u^2 & -u \\ -u & -u & -u & 1+2u^2 \end{pmatrix} \\ &= (1-u^2)^2 (u-1)(2u-1)\left(2u^2+u+1\right)^3. \end{aligned}$$

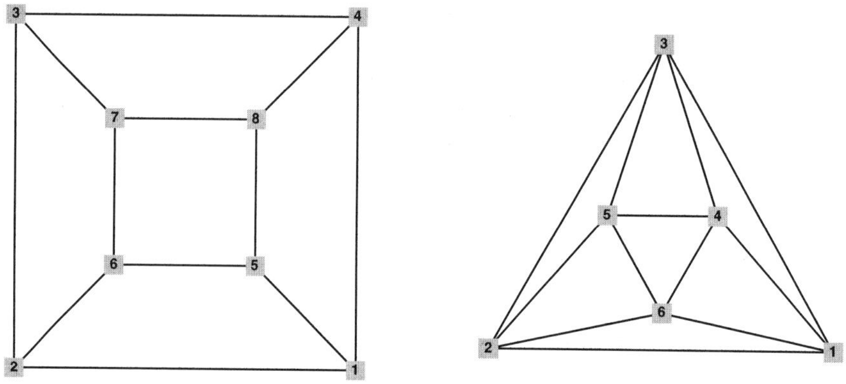

(a) The cube. (b) The octahedron.

The zeta functions for the cube, octahedron, icosahedron, and dodecahedron can be obtained similarly (matrices for the icosahedron and dodecahedron are omitted):

$$\begin{aligned}\zeta_{\text{cube}}^{-1} &= (1-u^2)^4 \\ &\cdot\det\begin{pmatrix} 1+2u^2 & -u & 0 & -u & 0 & -u & 0 & 0 \\ -u & 1+2u^2 & -u & 0 & 0 & 0 & -u & 0 \\ 0 & -u & 1+2u^2 & -u & 0 & 0 & 0 & -u \\ -u & 0 & -u & 1+2u^2 & -u & 0 & 0 & 0 \\ 0 & 0 & 0 & -u & 1+2u^2 & -u & 0 & -u \\ -u & 0 & 0 & 0 & -u & 1+2u^2 & -u & 0 \\ 0 & -u & 0 & 0 & 0 & -u & 1+2u^2 & -u \\ 0 & 0 & -u & 0 & -u & 0 & -u & 1+2u^2 \end{pmatrix} \\ &= (1-u^2)^5(2u+1)(2u-1)(2u^2-u+1)^3(2u^2+u+1)^3.\end{aligned}$$

$$\begin{aligned}\zeta_{\text{octa}}^{-1} &= (1-u^2)^6\det\begin{pmatrix} 1+3u^2 & -u & -u & -u & 0 & -u \\ -u & 1+3u^2 & -u & 0 & -u & -u \\ -u & -u & 1+3u^2 & -u & -u & 0 \\ -u & 0 & -u & 1+3u^2 & -u & -u \\ 0 & -u & -u & -u & 1+3u^2 & -u \\ -u & -u & 0 & -u & -u & 1+3u^2 \end{pmatrix} \\ &= (1-u^2)^6(3u-1)(u-1)(3u^2+2u+1)^2(1+3u^2)^3.\end{aligned}$$

$$\begin{aligned}\zeta_{\text{icos}}^{-1} &= (1-u^2)^{18}(4u-1)(u-1)(1+3u^2+16u^4)^3(4u^2+u+1)^5. \\ \zeta_{\text{dod}}^{-1} &= (1-u^2)^{10}(2u-1)(u-1)(1-u^2+4u^4)^3(1+2u^2)^4 \\ &\quad\cdot(2u^2+2u+1)^4(2u^2-u+1)^5.\end{aligned}$$

1.2 Graph coverings. Terras and Stark considered the notion of an unramified covering of graphs. Intuitively, a covering is a surjective map $f : H \to G$ with $V(H) \mapsto V(G)$ and $E(H) \mapsto E(G)$ which respects the structure of the graph. Let for $a, b \in V(H)$ adjacent, let (a,b) indicate the edge from a to b. A covering map requires that if $(a,b) \in E(H)$, then $(f(a), f(b)) \in E(G)$, and $f(a,b) = (f(a), f(b))$. An unramified covering also has the property that the fiber above each edge and vertex of G has the same number of elements, known as the degree of the covering.

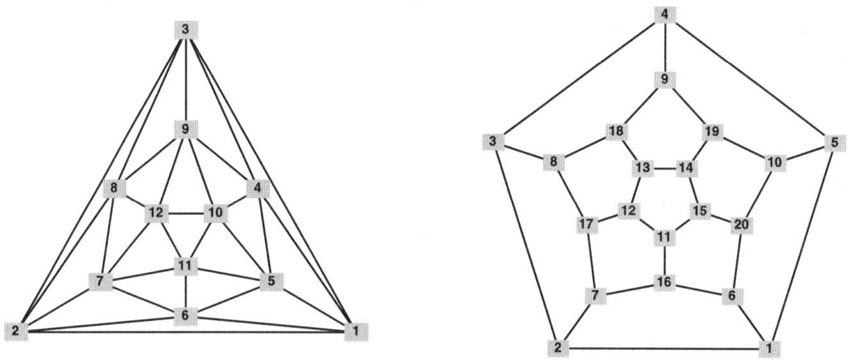

(a) The icosahedron.

(b) The dodecahedron.

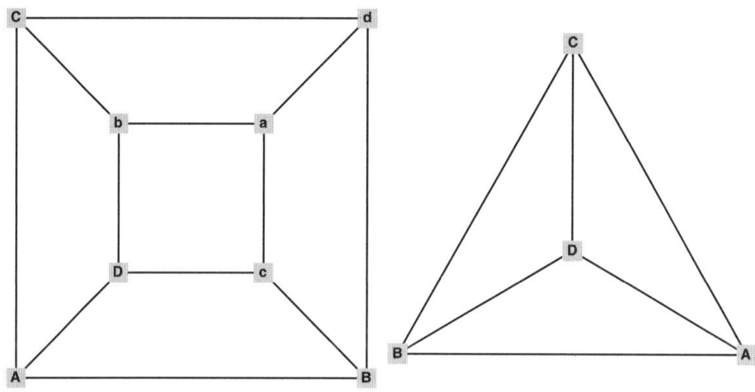

(a) The cube.

(b) The complete graph on 4 vertices, also known as the tetrahedron.

Figure 1: The cube graph is an unramified degree 2 cover of the tetrahedron.

Interesting examples of unramified coverings exist, for instance the map from the cube to the tetrahedron (see Figure 1).

Terras' paper in this volume includes a proof that if such an unramified covering map exists, then $\zeta_G^{-1} \mid \zeta_H^{-1}$. Ihara zeta functions of graphs have some parallels with Artin zeta functions of curves, in which we have zeta function divisibility for ramified covers as well. We wanted to know if the divisibility relation for Ihara zeta functions in an unramified cover also held for ramified coverings of graphs. We used a definition of ramified covering adopted from Urakawa (via Baker and Norine) as outlined below.

Definition 1.3 Let H and G be graphs. A function $\phi : V(H) \cup E(H) \to V(G) \cup E(G)$ is called a morphism if

- $\phi(x) \in V(G)$ for all $x \in V(H)$, and
- for every $x \in V(H)$ and $e \in E(H)$ such that $x \in e$, we have either $\phi(e) \in E(G)$ and $\phi(x) \in \phi(e)$, or $\phi(e) = \phi(x)$.

A morphism $\phi : H \to G$ is said to be *harmonic* if for all $x \in V(H)$ and $e' \in E(G)$, the number of edges e in $E(H)$ such that $x \in e$ with $\phi(e) = e'$ is the same for each e' which contains $\phi(x)$, i.e. ϕ is harmonic if for all $x \in V(H)$, the quantity $|\{e \in E(H) : x \in e, \phi(e) = e'\}|$ is constant for all $e' \in E(G)$ with $\phi(x) \in e'$.

Notice that harmonic morphisms encompass the unramified coverings of graphs studied by Terras and Stark, but allows for the size of fibers to vary (as in ramified covers of curves). Edges in H may also collapse, i.e. be mapped to vertices of G. Urakawa [12] also proved that the following is a well-defined notion of the *degree* of a harmonic morphism.

Definition 1.4 If $\phi : G \to H$ is a harmonic morphism of graphs with $x \in V(G)$, the vertical multiplicity of ϕ at x is given by

$$v_\phi(x) = |e \in E(G) : \phi(e) = \phi(x)|\,.$$

The horizontal multiplicity of ϕ at x is given by

$$m_\phi(x) = |e \in E(G) : x \in e, \phi(e) = e'|$$

for any edge $e' \in E(G)$ with $\phi(x) \in e'$. For ϕ as above, and any vertex $y \in V(H)$, the *degree* of ϕ is given by

$$\deg(\phi) = \sum_{x \in V(G) : \phi(x) = y} m_\phi(x).$$

Based on these definitions of multiplicity and degree, Baker and Norine [1] proved a graph-theoretic analogue of the Riemann-Hurwitz formula, as well as several other results. This indicates that harmonic morphisms could be a good choice for generalizing the idea of covering maps of graphs to include ramification. We consider the simplest version of a ramified covering, namely $k \geq 1$ copies of a fixed graph X, with a single vertex on each graph identified. See Figure 2 for an example. Note that this fits the definition of ramified covering given in [1].

2 "Almost divisibility" of zeta functions

Our main theorem is a general result for the above mentioned type of covering of $(q+1)$-regular graphs. Proofs of all the following results may be found in [5].

Theorem 2.1 *Let X be a finite $(q+1)$-regular graph with vertices labeled v_1, v_2, ..., v_n. Let A_X be the adjacency matrix for X, and suppose that A_X has d distinct eigenvalues. Let Y_k be the ramified cover of X created by identifying vertex v_1 on k copies of X. Then there exists a polynomial $P_{k,q}(u) \in \mathbb{Z}[u]$, with coefficients depending on k and q (and on the original graph X) and with $\deg_u P_{k,q} \leq 2d$, such that for $l \geq k$,*

$$\zeta_{Y_k}^{-1}(u) P_{l,q}(u) \text{ divides } \zeta_{Y_l}^{-1}(u) P_{k,q}(u). \tag{2.1}$$

Remark 2.2 In particular, since $Y_1 = X$, we have that for all $l \geq 1$

$$\zeta_X^{-1}(u) \text{ divides } \zeta_{Y_l}^{-1}(u) P_{1,q}(u).$$

Here, the degree of the extra term $P_{1,q}(u)$ needed for divisibility is independent of both n and k, so the failure of true divisibility is controlled.

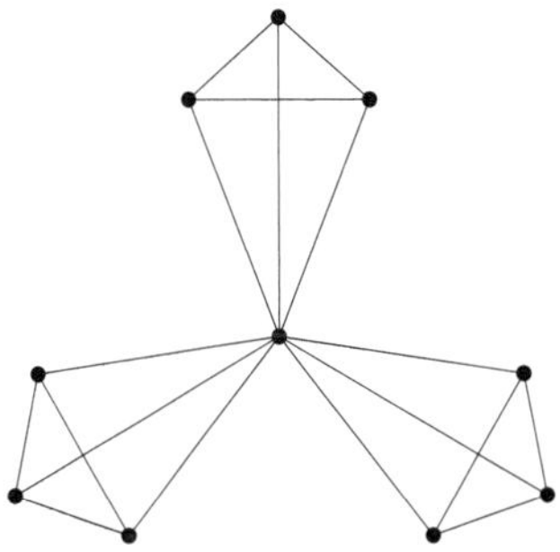

(a) Three copies of K_4.

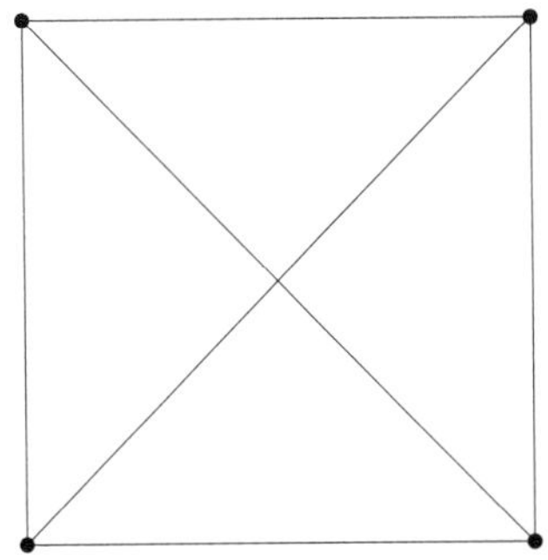

(b) The complete graph on 4 vertices.

Figure 2: K_4 and three copies of the graph around a single identified vertex.

The main tools of the proof come from linear algebra. We use Ihara's formula to change the calculation of the zeta function of a graph into a determinant calculation.

In the case where the base graph X is K_n, the complete graph on n vertices, the adjacency matrices are simple enough that we can say even more. See Figure 2 for the $n = 4$ case.

Proposition 2.3 *The Ihara zeta function for K_n is*

$$\zeta_{K_n}^{-1}(u) = (1-u^2)^{n(n-3)/2}(u-1)\left((n-2)u-1\right)\left((n-2)u^2+u+1\right)^{n-1}. \quad (2.2)$$

Proposition 2.4 *Let $X_{n,k}$ be the graph consisting of k copies of K_n identified at a single identified vertex. Then the Ihara zeta function for $X_{n,k}$ is*

$$\zeta_{X_{n,k}}^{-1}(u) = (1-u^2)^{r-1}(u-1)\left((n-2)(nk-k-1)u^3+(n-3)u-1\right)$$
$$\left((n-2)u^2+u+1\right)^{k(n-2)}\left((n-2)u^2-(n-2)u+1\right)^{k-1}, \quad (2.3)$$

where $r = k(n-1)(n-2)/2$ is the rank of the fundamental group of $X_{n,k}$.

Corollary 2.5 *Let $X_{n,k}$ be the graph consisting of k copies of K_n with a single vertex from each copy identified. If $i \leq j$, then*

$$\zeta_{X_{n,i}}^{-1}(u)\left((n-2)(nj-j-1)u^3+(n-3)u-1\right)$$

divides

$$\zeta_{X_{n,j}}^{-1}(u)\left((n-2)(ni-i-1)u^3+(n-3)u-1\right).$$

In particular, for all $k \geq 1$

$$\zeta_{K_n}^{-1}(u) \text{ divides } \zeta_{X_{n,k}}^{-1}(u)\left((n-2)^2u^3+(n-3)u-1\right).$$

Here we have a specific example of a graph K_n and a ramified cover of the graph $X_{n,k}$, where $\zeta_{K_n}^{-1}(u) \nmid \zeta_{X_{k,n}}^{-1}(u)$. We see that each term of (2.2) divides some term of equation (2.3), except for the term

$$(n-2)^2u^3+(n-3)u-1,$$

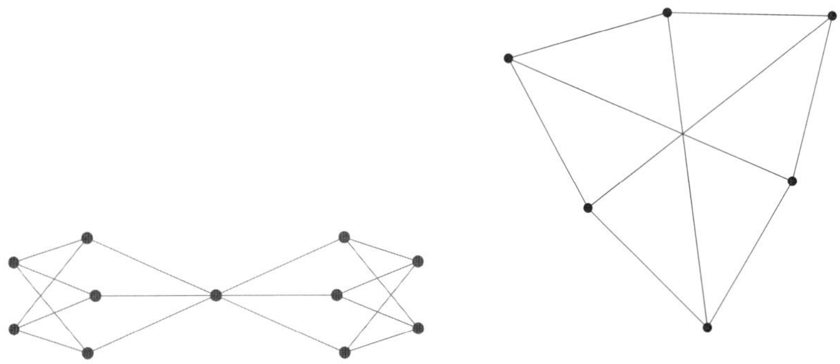

(a) Two copies of $K_{3,3}$. (b) The (3,3)-complete bipartite graph.

Figure 3: $K_{3,3}$ and two copies of the graph around a single identified vertex.

which will never divide

$$(n-2)(nk-k-1)u^3+(n-3)u-1$$

(nor will it divide any other term of $\zeta_{X_{k,n}}^{-1}(u)$ when $n>3$).

In the special case $n=3$ we do actually have $\zeta_{K_3}^{-1}(u) \mid \zeta_{X_{3,k}}^{-1}(u)$ for every $k \geq 1$. The u^3-1 term divides into an "extra" copy of $(u-1)(u^2+u+1)$, where the first term arises from $(1-u^2)^{r-1}$, and the other arises from $(u^2+u+1)^k$.

We turn now to the another fairly simple case: the $(q+1, q+1)$-regular bipartite graph. See Figure 3 for an example.

Proposition 2.6 *The Ihara zeta function for $K_{q+1,q+1}$ is*

$$\zeta_{K_{q+1,q+1}}^{-1}(u) = (1-u^2)^{q^2-1}\left(qu^2-(q+1)u+1\right) \left(qu^2+(q+1)u+1\right)\left(qu^2+1\right)^{2q}. \tag{2.4}$$

Proposition 2.7 *Let $X_{q+1,k}$ be the graph consisting of k copies of $K_{q+1,q+1}$ with a single vertex on each of the k copies identified. Then the Ihara zeta function for $X_{q+1,k}$ satisfies*

$$\zeta_{X_{q+1,k}}^{-1}(u) = (1-u^2)^{kq^2-1}\left(qu^2+1\right)^{k(2q-1)} \left(q^2u^4-q(q-1)u+1\right)^{k-1}(u^2-1)P_{k,q}(u), \text{ where} \tag{2.5}$$

$$P_{k,q}(u) = q^2((q+1)k-1)u^4+q(q-1)u^2-1. \tag{2.6}$$

The results 1.6 and 1.7 again show that "almost divisibility" is the best one may achieve for zeta functions of these special ramified covers of graphs. The polynomial $P_{k,q}(u)$ will not divide any term of $\zeta_{X_{q+1,l}}(u)^{-1}$ when $k<l$.

As true divisibility is false for zeta functions under this definition of ramified cover, it might be useful to consider some other possible definitions of ramification. See Terras' paper in this volume for some experiments in this vein.

3 Complete bipartite graphs

In Proposition 2.6, we gave a formula for the zeta function of the complete balanced bipartite graph. The easiest class of non-regular graphs for which the zeta functions can be easily determined are bipartite graphs. Hashimoto determined the zeta functions for semi-regular bipartite graphs in terms of the eigenvalues of their adjacency matrices [2]. Sato determined the zeta functions and complexities for the line graphs of semi-regular bipartite graphs [7]. Here, we prove a special case of Hashimoto's result to give a sense of why bipartite graphs have accessible zeta functions, and to introduce techniques which can be applied to find the zeta functions of some families of non-regular graphs.

3.0.1 *Example: The unbalanced complete bipartite graph,* $K_{m,n}$. Let $K_{m,n}$ be the bipartite graph with $m+n$ vertices, with vertices partitioned into $P_m = \{a_1, a_2, ...a_m\}$ and $P_n = \{b_1, b_2, ...b_n\}$, and edges $\{(a_i, b_j)$ for all $i, j \in \mathbb{Z}$, $1 \le i \le m$ and $1 \le j \le n\}$.

Proposition 3.1 The zeta function for the complete bipartite graph is given by:

$$\zeta^{-1}_{K_{m,n}}(u) = (1-u^2)^{mn-(m+n)+1}(1+(n-1)u^2)^{m-1} \cdot (1+(m-1)u^2)^{n-1}(1-(m-1)(n-1)u^2).$$

Proof Order the vertices $a_1, a_2, ...a_m, b_1, b_2, ...b_n$ to create the matrices A and Q for $K_{m,n}$. Let $J_{i,j}$ be the $i \times j$ matrix consisting of all ones. Let $\mathbf{0}$ be the appropriately sized matrix of all zeros. We represent A and Q as block matrices:

$$A_{K_{m,n}} = \begin{pmatrix} \mathbf{0} & J_{m,n} \\ J_{n,m} & \mathbf{0} \end{pmatrix}, \qquad Q_{K_{m,n}} = \begin{pmatrix} (n-1)I_m & \mathbf{0} \\ \mathbf{0} & (m-1)I_n \end{pmatrix}.$$

So

$$\zeta^{-1}_H(u) = (1-u^2)^{r-1}\det(I_{|V|} - Au + Qu^2),$$

$$\zeta^{-1}_{B_{m,n}}(u) = (1-u^2)^{mn-(m+n)}\det(I_{m+n} - Au + Qu^2).$$

Let $Z_H := (I - Au + Qu^2)$. Then $Z_{B_{m,n}}$ is the block matrix $\begin{pmatrix} U & V \\ V^T & W \end{pmatrix}$, where

$$U = (1+(n-1)u^2)I_m,$$

$$W = (1+(m-1)u^2)I_n,$$

and V is the $m \times n$ matrix with each entry equal to $-u$.

For a square block matrix $M = \begin{pmatrix} A & B \\ C & D \end{pmatrix}$ where A is invertible, it is known that

$$\det(M) = \det(A)\det(D - CA^{-1}B).$$

In our case, this means that

$$\det(Z_{K_{m,n}}) = \det(U)\det(W - V^TU^{-1}V).$$

Calculation shows that $W - V^TU^{-1}V$ is the $n \times n$ matrix with each diagonal entry equal to

$$t_1 = 1 + (m-1)u^2 - \frac{mu^2}{1+(n-1)u^2},$$

and all other entries equal to

$$t_2 = -\frac{mu^2}{1+(n-1)u^2}.$$

So

$$\begin{aligned}\zeta^{-1}_{K_{m,n}}(u) =&(1-u^2)^{mn-(m+n)}\det(Z_{K_{m,n}})\\ =&(1-u^2)^{mn-(m+n)}(1+(n-1)u^2)^m(\det((1+(m-1)u^2)I_n - t_2J_{n,n}).\end{aligned}$$

Let $A_{K_n} = J_{n,n} - I_n$ be the adjacency matrix for the complete graph on n vertices.

$$\zeta^{-1}_{B_{m,n}}(u) = (1-u^2)^{mn-(m+n)}(1+(n-1)u^2)^m\det((1+(m-1)u^2)I_n - t_2(A_{K_n}+I_n)).$$

It is well known that the eigenvalues of A_{K_n} are $n-1$ with multiplicity 1 and -1 with multiplicity $n-1$. So the eigenvalues of $(1+(m-1)u^2)I_n - t_2(A_{K_n}+I_n)$ are

- $(1+(m-1)u^2) - t_2((n-1)+1)$ with multiplicity 1 and
- $(1+(m-1)u^2) - t_2(-1+1) = (1+(m-1)u^2)$ with multiplicity $n-1$.

Therefore

$$\begin{aligned}\zeta^{-1}_{B_{m,n}}(u) =&(1-u^2)^{mn-(m+n)}(1+(n-1)u^2)^m\\ &((1+(m-1)u^2) - \frac{mu^2}{1+(n-1)u^2})(n)(1+(m-1)u^2)^{n-1}\\ =&(1-u^2)^{mn-(m+n)+1}(1+(n-1)u^2)^{m-1}\\ &(1+(m-1)u^2)^{n-1}(1-(m-1)(n-1)u^2).\end{aligned}$$

□

Acknowledgements. The authors thank the participants of the "Women in Numbers" conference at BIRS for an interesting week, especially to the organizers Kristin Lauter, Rachel Pries, and Renate Scheidler. Thanks also to our working group members, Habiba Kadiri and Shabnam Akhtari. Special thanks to our working group leaders, Audrey Terras and Winnie Li, for asking interesting questions and encouraging us to keep working on them after we left the conference.

References

[1] Matthew Baker and Serguei Norine, *Harmonic morphisms and hyperelliptic graphs*, International Mathematics Research Notices **2009** (2009), no. 15.

[2] Ki-ichiro Hashimoto, *Zeta functions of finite graphs*, Automorphic Forms and Geometry of Arithmetic Varieties (K. Hashimoto and Y. Namikawa, eds.), Academic Press, Harcourt Brace Jovanovich, 1989.

[3] Yasutaka Ihara, *On discrete subgroups of the two by two projective linear group over p-adic fields*, J. Math. Soc. Japan **18** (1966), 219–235.

[4] Motoko Kotani and Toshikazu Sunada, *Zeta functions of finite graphs*, Journal of the Mathematical Society of University of Tokyo **7** (2000).

[5] Beth Malmskog and Michelle Manes. *"Almost divisibility" in the Ihara zeta functions of certain ramified covers of q+1-regular graphs*, Linear Algebra and its Applications, **432** (2010), 2486-2506.

[6] Sam Northshield, *A note on the zeta function of a graph*, Journal of Combinatorial Theory **64** (1998).

[7] Iwao Sato, *Zeta functions and complexities of a semiregular bipartite graph and its line graph*, Discrete Mathematics **307** (2007).

[8] Harold Stark and Audrey Terras, *Zeta functions of finite graphs and coverings*, Adv. Math. **121** (1996), no. 1, 124–165.

[9] ———, *Zeta functions of finite graphs and coverings II*, Adv. Math. **154** (2000), no. 1, 132–195.

[10] ———, *Zeta functions of finite graphs and coverings III*, Adv. Math. **208** (2007), no. 1, 467–489.

[11] Audrey Terras, *A stroll through the garden of graph zeta functions*, Unpublished, available at http://math.ucsd.edu/~aterras/newbook.pdf, 2007.

[12] Hajime Urakawa, *A discrete analogue of the harmonic morphism*, Harmonic Morphisms, Harmonic Maps and Related Topics (Christopher Kum Anand, ed.), CRC Press, 1999.

Part V – Other Topics

Fields Institute Communications
Volume **60**, 2011

Zeros of Classical Eisenstein Series and Recent Developments

Sharon Anne Garthwaite
Department of Mathematics, Bucknell University
Lewisburg, PA 17837 USA
sharon.garthwaite@bucknell.edu

Ling Long
Department of Mathematics, Iowa State University
Ames, IA 50011 USA
linglong@iastate.edu

Holly Swisher
Department of Mathematics, Oregon State University
Corvallis, OR 97301 USA
swisherh@math.oregonstate.edu

Stephanie Treneer
Department of Mathematics, Western Washington University
Bellingham, WA 98225 USA
stephanie.treneer@wwu.edu

Abstract. In this survey, we begin by recalling a beautiful result of F. K. C. Rankin and Swinnerton-Dyer on the location of zeros of the classical Eisenstein series $E_k(z)$ for the full modular group. We then explore more recent studies which have built upon this work to analyze the behavior of zeros of $E_k(z)$, such as work by Nozaki on their separation property. We also review similar results for other classes of modular forms as well as zeros of Eisenstein series for different groups. We conclude with some results of the authors for a family of odd weight Eisenstein series on $\Gamma(2)$, a prototypical genus zero subgroup with a simple fundamental domain.

1 Eisenstein series on $\mathrm{SL}_2(\mathbb{Z})$

We begin this survey by defining modular forms on the full modular group $\Gamma := \mathrm{SL}_2(\mathbb{Z})$. Let $\mathbb{H}$ denote the complex upper half-plane

$$\mathbb{H} = \{z \in \mathbb{C} \mid \mathrm{Im}(z) > 0\}.$$

2010 *Mathematics Subject Classification.* Primary 11F11; Secondary 11F03, 11F37.
The second author was supported by NSA grant H98230-08-1-0076.

The modular group Γ acts on $\mathbb{H}$, or more generally $\widehat{\mathbb{C}} := \mathbb{C} \cup \{\infty\}$, by linear fractional transformations

$$\begin{pmatrix} a & b \\ c & d \end{pmatrix} \cdot z := \frac{az+b}{cz+d}.$$

As the two matrices $\pm \left(\begin{smallmatrix} a & b \\ c & d \end{smallmatrix}\right)$ generate the same action, it is also sometimes appropriate to restrict our view to $\mathrm{PSL}_2(\mathbb{Z}) := \mathrm{SL}_2(\mathbb{Z})/\{\pm I\}$. A fundamental domain for this action is given by

$$\mathcal{F} := \{z \in \mathbb{H} : -1/2 \le \mathrm{Re}(z) \le 1/2, |z| \ge 1\}.$$

We define a modular form f of integer weight k for Γ to be a holomorphic function on $\mathbb{H}$ such that

$$f\left(\frac{az+b}{cz+d}\right) = (cz+d)^k f(z) \tag{1.1}$$

for all $z \in \mathbb{H}$ and $\left(\begin{smallmatrix} a & b \\ c & d \end{smallmatrix}\right) \in \Gamma$, and such that f is holomorphic when extended to ∞.

Note that under the action of Γ the points $\mathbb{Q} \cup \{\infty\}$ form one equivalence class; we often refer to these points collectively as the cusp of Γ. Alternatively, we call an element $\gamma \in \Gamma$ parabolic if γ has only one fixed point on $\widehat{\mathbb{R}} := \mathbb{R} \cup \{\infty\}$, and define the cusps to be the set of equivalence classes of fixed points of the parabolic elements. In Γ, and in fact in $\mathrm{SL}_2(\mathbb{R})$, the parabolic elements are exactly those with trace ± 2.

The above conditions imply that a modular form f has a Fourier expansion of the form

$$\sum_{n \ge 0} a(n) q^n, \tag{1.2}$$

where here and throughout this paper we define $q := e^{2\pi i z}$. Moreover, the set of modular forms of fixed weight k is a finite-dimensional vector space over $\mathbb{C}$, denoted M_k. If $f \in M_{k_1}$ and $g \in M_{k_2}$ then $fg \in M_{k_1+k_2}$. Hence, the collection of all positive integral weight modular forms is a graded algebra. For more basic properties of modular forms, see ([23], [12], [18]).

Of primary importance among modular forms for the full modular group are the Eisenstein series. For even weight $k \ge 4$, the functions

$$G_k(z) := \sum_{\substack{c,d \in \mathbb{Z} \\ (c,d) \ne (0,0)}} (cz+d)^{-k} \tag{1.3}$$

are holomorphic modular forms of weight k. These classical Eisenstein series can be viewed as arising from the Weierstrass $\wp$ function, an elliptic function satisfying the differential equation

$$(\wp'(u))^2 = 4\wp(u)^3 - 60G_4\wp(u) - 140G_6. \tag{1.4}$$

One can solve this differential equation recursively to recover $\wp$ as a generating function for the G_k:

$$\wp(z) = \frac{1}{z^2} + \sum_{k=1}^{\infty} (2k+1) G_{2k+2} z^{2k}.$$

The series $G_k(z)$ are often normalized to have a constant coefficient of 1 by dividing by $2\zeta(k)$. Alternatively, one can write these normalized Eisenstein series, denoted

$E_k(z)$, as

$$E_k(z) := \frac{1}{2} \sum_{\substack{c,d \in \mathbb{Z} \\ (c,d)=1}} (cz+d)^{-k}. \tag{1.5}$$

There are many questions that naturally arise about Eisenstein series and modular forms in general. For example, one might be interested in the number of zeros that these functions have and their location, which is the focus of this survey. A key tool in the study of zeros of modular forms is the valence formula. Define $\nu_\tau(f)$ to be the order of a modular form f at a point $\tau \in \widehat{\mathbb{H}}$, where $\widehat{\mathbb{H}} := \mathbb{H} \cup \{\infty\} \cup \mathbb{Q}$. The valence formula states that if f is a nonzero modular form of weight k, then

$$\nu_\infty(f) + \frac{1}{2}\nu_i(f) + \frac{1}{3}\nu_\omega(f) + \sum_{\substack{\tau \in \Gamma \backslash \mathbb{H} \\ \tau \neq i, \omega}} \nu_\tau(f) = \frac{k}{12}, \tag{1.6}$$

where i and ω are the usual primitive fourth and third roots of unity in $\mathbb{H}$.

We call an element $\gamma \in \Gamma$ elliptic if $\gamma \cdot z = z$ has exactly one solution in $\mathbb{H}$; we refer to these fixed points as elliptic points. The elliptic elements are those with trace strictly less than 2 in absolute value, and, along with $\pm I$, are the elements of Γ with finite order. The corresponding elliptic points are Γ-equivalent to either i or ω, which are of order 2 and 3, respectively.

Notice that equation (1.6) implies that the number of zeros of a modular form is roughly one-twelfth of its weight. In addition, it follows from (1.6) that for $k \geq 4$, the set

$$\{E_4^a E_6^b \mid 4a + 6b = k \text{ and } a, b \geq 0\}$$

forms a basis for M_k. This fact illustrates the fundamental role Eisenstein series play as building blocks in the theory of modular forms.

2 A classical result of Rankin and Swinnerton-Dyer

The study of the location of zeros of Eisenstein series has its origins in work of Wohlfahrt and R. A. Rankin in the 1960s, who showed in [26] and [20] that for even $4 \leq k \leq 38$ (excepting $k = 36$), the zeros of $E_k(z)$ lie on the unit circle. R. A. Rankin also gave partial results towards establishing this property for all $k \geq 4$. Soon after, R. A. Rankin's daughter F. K. C. Rankin, together with Swinnerton-Dyer [22], used an elementary yet elegant argument to prove the following.

Theorem 2.1 *For even $k \geq 4$, all zeros of $E_k(z)$ in the fundamental domain for $\mathrm{SL}_2(\mathbb{Z})$ are located on the arc*

$$A := \{z = e^{i\theta} : \pi/2 \leq \theta \leq 2\pi/3\}.$$

Here we give a brief outline of their argument. Write $k = 12n + s$ with $s = 4$, 6, 8, 10, 0, or 14. Note, by the valence formula (1.6), that the value s determines the minimum number of zeros that $E_k(z)$ must have at the elliptic points i and ω. Thus, it is sufficient to show that $E_k(e^{i\theta})$ has at least n zeros in the open interval $(\pi/2, 2\pi/3)$. For $n > 0$, F. K. C. Rankin and Swinnerton-Dyer consider

$$F_k(\theta) = e^{ik\theta/2} E_k(e^{i\theta}) = 2\cos(k\theta/2) + R, \tag{2.1}$$

where R is the remainder term consisting of all terms in (1.5) with $c^2 + d^2 > 1$. They show that $|R| < 2$ on $[\pi/2, 2\pi/3]$, and hence the number of zeros of $E_k(e^{i\theta})$

in $(\pi/2, 2\pi/3)$ is at least the number of zeros of $2\cos(k\theta/2)$ on the same interval; a short calculation reveals that there are n of them.

Theorem 2.1 can also be framed in terms of the j-function,

$$j(z) := 1728E_4^3(z)/(E_4^3(z) - E_6^2(z)), \tag{2.2}$$

which is an isomorphic map between $\mathcal{F}$ and $\mathbb{C}$. For example, $j(i) = 1728$ and $j(\omega) = 0$. This function is invariant under Γ, with equivalent boundary points identified, and holomorphic on $\mathbb{H}$, with a simple pole at the cusp ∞. In this regard it is a weakly holomorphic modular form; weakly holomorphic modular forms satisfy (1.1) and (1.2), except that finitely many terms with $n < 0$ are allowed in the latter. They form an infinite-dimensional vector space, denoted $\mathcal{M}_k$.

The j-function generates the function field of the genus zero modular curve $X(\Gamma) := \Gamma\backslash\widehat{\mathbb{H}}$. It also parameterizes isomorphism classes of elliptic curves. Theorem 2.1 is equivalent to the statement that the j-values of the zeros of $E_k(z)$ are all real and lie within the interval $[0, 1728]$.

3 Recent results for $\mathbf{SL}_2(\mathbb{Z})$

Following the work of F. K. C. Rankin and Swinnerton-Dyer there have been more recent discoveries about the zeros of the classical Eisenstein series, as well as several generalizations to other classes of modular forms. In this section we investigate some recent results about zeros of modular forms on $\mathrm{SL}_2(\mathbb{Z})$.

In [13] Kohnen derives a closed formula for the precise locations of the zeros of $E_k(z)$ on A, as defined in Theorem 2.1. The exact statement of his result requires additional notation, which we omit for brevity. We do note, however, that the formula is an infinite sum whose terms are related to the Fourier coefficients of $E_k(z)$; in particular, they include Bernoulli numbers and divisor sums.

In [5], Gekeler proves p-adic congruence properties for the following polynomials which encode the j-values of the non-elliptic zeros of $E_k(z)$ for even $k \geq 4$:

$$\varphi_k(x) = \prod_{\substack{j(z) \text{ where } E_k(z)=0 \\ j(z)\neq 0,1728}} (x - j(z)). \tag{3.1}$$

Part of the interest in studying these polynomials is that for primes $p \geq 5$, φ_{p-1} is a canonical lift of essentially the supersingular polynomial in characteristic p, whose zeros are the j-invariants of the supersingular elliptic curves defined over $\overline{\mathbb{F}}_p$ (see e.g. [18]). Gekeler shows that up to a constant multiple, the reduction of φ_k modulo p contains φ_{p-1} as a factor.

Gekeler also cites empirical evidence computed by Kremling [14] that suggests that the $\varphi_k(x)$ are irreducible with full Galois group S_d where d is the degree of $\varphi_k(x)$. Moreover, the zeros of φ_{k+12a} for $a \in \mathbb{Z}_{\geq 0}$ indicate separation properties one would expect from orthogonal polynomials, although they are not orthogonal.

In addition, by studying the coefficients of $\varphi_k(x)$ it is possible to relate the location of the zeros of $E_k(z)$ to special values of the Riemann zeta function $\zeta(s)$. The resulting formula

$$\frac{2}{\zeta(1-k)} = 60k - \sum_{\tau\in\Gamma\backslash\mathbb{H}} e_\tau \mathrm{ord}_\tau(E_k(\tau))j(\tau), \tag{3.2}$$

where $e_i = 1/2, e_\omega = 1/3$, and $e_\tau = 1$ for non-elliptic points τ, is highlighted in [19].

Recently, Nozaki [17] has proved the separation property observed by Gekeler in [5]. His precise statement is as follows.

Theorem 3.1 *Let $k \geq 12$ be an even integer, and let $E_k(z)$ be the Eisenstein series with weight k for $\mathrm{SL}_2(\mathbb{Z})$. Let $\{e^{ia_j} \mid a_1 < a_2 < ... < a_n\}$ be the zeros of $E_k(z)$ on A and let $\{e^{ib_j} \mid b_1 < b_2 < ... < b_{n+1}\}$ be the zeros of $E_{k+12}(z)$ on A. Then $b_j < a_j < b_{j+1}$ for $j = 1, 2, \ldots, n$.*

To prove this result, Nozaki considers $F_k(z)$ as defined in (2.1). In [22], F. K. C. Rankin and Swinnerton-Dyer proved that it was possible to approximate the location of the zeros of F_k and F_{k+12} by the location of the zeros of $2\cos(k\theta/2)$ and $2\cos((k+12)\theta/2)$, respectively. By considering the decreasing size of the main error terms in the argument more closely, Nozaki is able to prove that in fact the zeros of F_k and F_{k+12} do not stray far from the zeros of their associated trigonometric functions. In particular, they stray by less than half the distance of the two closest zeros of $2\cos(k\theta/2)$ and $2\cos((k+12)\theta/2)$.

Shifting focus to another family, Asai, Kaneko and Ninomiya consider the zeros of a family of weakly holomorphic modular forms which are related to the j-function via the Hecke operators. Let T_n be the nth Hecke operator, which acts on a modular form $f(z)$ of weight k by

$$f(z) \mid_k T_n = n^{k-1} \sum_{ad=n} \sum_{b=0}^{d-1} d^{-k} f\left(\frac{az+b}{d}\right)$$

(see, e.g. [23]). Define

$$J_n(z) := n(j(z) - 744) \mid_0 T_n.$$

It is well known that $J_n(z)$ is a monic polynomial in $j(z)$ of degree n. Asai, et al. prove the following in [1], where A is defined as in Theorem 2.1.

Theorem 3.2 *For each $n \geq 1$, all the zeros of $J_n(z)$ are simple and lie on the arc A.*

Equivalently, the zeros of the polynomial $J_n(j)$ lie in the interval $(0, 1728)$. Their method is similar to the F. K. C. Rankin and Swinnerton-Dyer method in that they show that the values of $J_n(z)$ do not stray far from the values of a cosine function, and hence the number of zeros may be easily counted.

Gun similarly extends the results of F. K. C. Rankin and Swinnerton-Dyer to another class of modular forms. In [8], Gun examines zeros for a particular class of cusp forms, which are holomorphic modular forms that vanish at ∞. To state his result, we first define the stabilizer of ∞,

$$\Gamma_\infty := \{\gamma \in \Gamma : \gamma \cdot \infty = \infty\} = \{\pm \left(\begin{smallmatrix} 1 & m \\ 0 & 1 \end{smallmatrix}\right) : m \in \mathbb{Z}\}.$$

In addition, we make the following definition.

Definition 3.3 For a fixed even $k \geq 4$, write $k = 12n + s$, where $s \in \{0, 4, 6, 8, 10, 14\}$, as in the proof of Theorem 2.1. For $1 \leq j \leq n-1$, let $f_{k,j}(z)$ denote any sum

$$f_{k,j}(z) = \sum_{\ell=1}^{j} a(\ell) P_\ell(z)$$

where P_ℓ is the ℓth Poincare series

$$P_\ell(z) = \sum_{\gamma=\left(\begin{smallmatrix} a & b \\ c & d \end{smallmatrix}\right)\in\Gamma_\infty\backslash\Gamma} (cz+d)^{-k} e^{2\pi i\ell(\gamma\cdot z)},$$

and the $a(\ell)$ are real numbers with $a(\ell) \geq 0$ for $1 \leq \ell \leq j-1$ and $a(j) > 0$. The $f_{k,j}$ are cusp forms of weight k.

Gun extends results of R. A. Rankin [21] on the Poincare series $P_\ell(z)$ to prove the following.

Theorem 3.4 *For even $k \geq 26$, $f_{k,j}(z)$ has at least $n-j$ zeros on the interior of A.*

Moreover, following Kohnen [13], he obtains a closed formula for the zeros of $f_{k,1}$ showing that all zeros lie on A except for a simple zero at infinity.

In [3], Duke and Jenkins provide a result on the location of zeros of a class of weakly holomorphic modular forms that unifies some of the above results. The modular forms they investigate are a natural basis of the vector space $\mathcal{M}_k$ of weakly holomorphic modular forms of weight k. In addition, the coefficients of these forms exhibit an intriguing duality property. These basis elements are defined as follows. Let n be defined as in the proof of Theorem 2.1 and Definition 3.3. For each integer $m \geq -n$, there exists a unique $f_{k,m} \in \mathcal{M}_k$ with q-expansion given by

$$f_{k,m}(z) = q^{-m} + O(q^{n+1}).$$

Any weakly holomorphic modular form $f(z) = \sum a(m)q^m \in \mathcal{M}_k$ can be written as

$$f(z) = \sum_{m_0 \leq m \leq n} a(m) f_{k,-m}.$$

This basis $\{f_{k,m} \mid m \geq -n\}$ can be written explicitly in terms of the Eisenstein series $E_s(z)$, $j(z)$, and $\Delta(z) := (E_4^3(z) - E_6^2(z))/1728$, and thus can be shown to have integer coefficients. The following theorem is given in [3].

Theorem 3.5 *Using the notation above, if $m \geq |n| - n$, then all of the zeros of $f_{k,m}$ in $\mathcal{F}$ lie on the unit circle.*

One should note that the restriction that $m \geq 0$ excludes cusp forms from the scope of Theorem 3.5. In particular Duke and Jenkins provide examples to show that the result does not hold for general m. However, the results of F. K. C. Rankin and Swinnerton-Dyer in [22], R. A. Rankin in [21], and Asai, Kaneko, and Ninomiya in [1] all deal with certain subclasses of the $\{f_{k,m}\}$. Duke and Jenkins depart somewhat from the method of proof first employed by F. K. C. Rankin and Swinnerton-Dyer in that they use a circle method argument to bound the error term expressing the distance between their function and an appropriate cosine function. The special case $m = 0$ was previously studied by Getz using an alternative method. See [6, 7].

4 Eisenstein series on other groups

We can expand our definition of modular forms to include functions which transform as in (1.1) for groups other than Γ. A natural generalization is to consider certain subgroups of Γ. In this section we will consider some examples of such variations.

Let $N \geq 1$ be an integer, and define

$$\Gamma(N) := \{M \in \Gamma : M \equiv I \pmod{N}\},$$

where the congruence is taken componentwise. A congruence subgroup of level N is any group Γ' such that $\Gamma(N) < \Gamma' < \Gamma$. A modular form f of weight k for Γ' is a holomorphic function on $\mathbb{H}$ which satisfies (1.1) for all $z \in \mathbb{H}$ and $\left(\begin{smallmatrix} a & b \\ c & d \end{smallmatrix}\right) \in \Gamma'$. In addition, f is holomorphic when extended to the cusps of Γ'. There are finitely many equivalence classes of cusps modulo Γ'. These conditions imply that f has a Fourier expansion at each cusp, of the form

$$\sum_{n \geq 0} \alpha(n) q_N^n, \qquad q_N := e^{2\pi i z/N}.$$

We denote the finite-dimensional complex vector space of these modular forms by $M_k(\Gamma')$.

One can define Eisenstein series associated to each cusp of a congruence subgroup Γ'. For simplicity, we only define the Eisenstein series associated with ∞; for further details about Eisenstein series of level N see [2].

Let $\Gamma'_\infty := \{\gamma \in \Gamma' : \gamma \cdot \infty = \infty\}$. Then

$$E_k(\Gamma'; z) := \sum_{\gamma \in \Gamma'_\infty \backslash \Gamma'} (cz + d)^{-k},$$

where $\gamma = \left(\begin{smallmatrix} a & b \\ c & d \end{smallmatrix}\right)$. Note that this matches (1.5) for $\Gamma' = \Gamma$.

To extend the results of the previous section on the location of zeros of Eisenstein series, it is natural to look for other groups which share some nice properties of Γ. In particular, the Riemann surface associated to Γ has genus zero, and its fundamental domain is quite simple. The principal congruence subgroup $\Gamma(2)$ also has genus zero and a simple fundamental domain. In [4], the authors extend Theorems 2.1 and 3.1 to certain odd weight Eisenstein series for $\Gamma(2)$. We review these results in section 5.

One can also consider modular forms on Fricke groups. For p prime, define

$$\Gamma_0^*(p) := \Gamma_0(p) \cup \Gamma_0(p) W_p, \qquad W_p := \begin{pmatrix} 0 & -1/\sqrt{p} \\ \sqrt{p} & 0 \end{pmatrix}.$$

One can define modular forms for $\Gamma_0^*(p)$ analogously to modular forms for congruence subgroups. Note that $\Gamma_0^*(p)$ is not a subgroup of Γ, but is commensurable with Γ, meaning that the intersection $\Gamma \cap \Gamma_0^*(p)$ has finite index in both groups.

Define

$$E_{k,p}^*(z) := \frac{1}{p^{k/2} + 1} (p^{k/2} E_k(pz) + E_k(z))$$

to be the Eisenstein series associated to $\Gamma_0^*(p)$, where $E_k(z)$ is the normalized Eisenstein series for Γ. The location of zeros of $E_{k,p}^*(z)$ has been investigated for $p = 2$ and 3 by Miezaki, Nozaki and Shigezumi in [16], and for $p = 5$ and 7 by Shigezumi in [24]. In each case, the F. K. C. Rankin and Swinnerton-Dyer method extends by a careful analysis of an error term $F_{k,p}^*(\theta)$, which is an appropriate analog of $F_k(\theta)$ as defined in (2.1).

For $p = 2$ and 3, a fundamental domain for $\Gamma_0^*(p)$ is given by

$$\mathcal{F}^*(p) := \{z \in \mathbb{H} : -1/2 \leq \mathrm{Re}(z) \leq 1/2, |z| \geq 1/\sqrt{p}\}.$$

Miezaki, Nozaki and Shigezumi prove the following in [16].

Theorem 4.1 *Let $p = 2$ or 3, and set*

$$A_p^* := \left\{ z = e^{i\theta}/\sqrt{p} : \pi/2 \leq \theta \leq \frac{\pi(2p-1)}{2p} \right\}.$$

Then for even $k \geq 4$, all the zeros of $E_{k,p}^*(z)$ in $\mathcal{F}^*(p)$ are on the arc A_p^*.

For the cases $p = 5$ and 7, the fundamental domain is a bit more complicated and defined as follows. Let

$$\begin{aligned}\mathcal{F}^*(p) := &\{z \in \mathbb{H} : -1/2 \leq \mathrm{Re}(z) \leq 0, |z| \geq 1/\sqrt{p}, |z + 1/2| \geq 1/(2\sqrt{p})\} \\ &\cup \{z \in \mathbb{H} : 0 \leq \mathrm{Re}(z) \leq 1/2, |z| \geq 1/\sqrt{p}, |z - 1/2| \geq 1/(2\sqrt{p})\}.\end{aligned}$$

Moreover, let α_5 and α_7 be the angles which satisfy $\tan \alpha_5 = 2$ and $\tan \alpha_7 = 5/\sqrt{3}$, respectively, and define

$$A_5^* := \{e^{i\theta}/\sqrt{5} : \pi/2 \leq \theta \leq \pi/2 + \alpha_5\} \cup \{-1/2 + e^{i\theta}/2\sqrt{5} : \alpha_5 < \theta \leq \pi/2\},$$

and

$$A_7^* := \{e^{i\theta}/\sqrt{7} : \pi/2 \leq \theta \leq \pi/2 + \alpha_7\} \cup \{-1/2 + e^{i\theta}/2\sqrt{7} : \alpha_7 - \pi/6 < \theta \leq \pi/2\}.$$

Shigezumi conjectures the following in [24].

Conjecture *For $p = 5$ or 7 and for even $k \geq 4$, all the zeros of $E_{k,p}^*(z)$ in $\mathcal{F}^*(p)$ lie on the arc A_p^*.*

The difficulty in proving this conjecture arises from considering the two distinct arcs that form $\mathcal{A}_p^*$; Shigezumi is able to modify the method of F. K. C. Rankin and Swinnerton-Dyer on each section, but has difficulty bounding the error term of $F_{k,p}^*(\theta)$ at $\theta = \pi/2 + \alpha_p$, the point of intersection of the two arcs. Moreover, whereas F. K. C. Rankin and Swinnerton-Dyer find it sufficient to bound their error term in absolute value, Shigezumi demonstrates that the sum of the absolute values of the first few terms in his errors already exceed the necessary upper bound. He instead considers the arguments of these terms to improve the bound. His approach fails when $k(\pi/2 + \alpha_p)/2 \pmod{\pi}$ lies in a certain narrow range of values. However, Shigezumi is able to achieve the following results, which we summarize as follows.

Proposition 4.2 *Let $p = 5$ or 7. All but at most one of the zeros of $E_{k,p}^*(z)$ in $\mathcal{F}^*(p)$ lie on A_p^*. In addition, if $p = 5$ and $4 \mid k$, or $p = 7$ and $6 \mid k$, then all of the zeros of $E_{k,p}^*(z)$ in $\mathcal{F}^*(p)$ lie on A_p^*.*

We now turn to a general result about the location of the zeros of Eisenstein series. So far we have mainly been interested in subgroups of $\mathrm{SL}_2(\mathbb{Z})$. We can naturally extend our definitions of linear fractional transformations, cusps, and modular forms to subgroups of $\mathrm{SL}_2(\mathbb{R})$, or similarly $\mathrm{PSL}_2(\mathbb{R})$, with the Fricke subgroups as an example.

We define a norm on $\mathrm{SL}_2(\mathbb{R})$ by

$$||\left(\begin{smallmatrix} a & b \\ c & d \end{smallmatrix}\right)|| := a^2 + b^2 + c^2 + d^2,$$

which gives the group a metric topology. We consider a subgroup Γ' of $\mathrm{SL}_2(\mathbb{R})$ to be discrete if for any $\epsilon > 0$ the set $\{\gamma \in \Gamma' : ||\gamma|| < \epsilon\}$ is finite. Alternatively, a subgroup of $\mathrm{SL}_2(\mathbb{R})$ is discrete if and only if it acts discontinuously on $\mathbb{H}$ when viewed as a subgroup of $\mathrm{PSL}_2(\mathbb{R})$. We call such a subgroup a Fuchsian group. Note that as it acts discontinuously on $\mathbb{H}$, the orbits $\Gamma' \cdot z = \{\gamma \cdot z : \gamma \in \Gamma'\}$ for each $z \in \mathbb{H}$ have no limit points in $\mathbb{H}$; however, there may be limit points on the boundary $\widehat{\mathbb{R}}$.

More specifically, we call a discrete subgroup of $\mathrm{SL}_2(\mathbb{R})$ a Fuchsian group of the first kind if every point on $\widehat{\mathbb{R}}$ is in the set of limit points of the orbit $\Gamma' \cdot z$ for some $z \in \mathbb{H}$. For a more complete discussion of modular forms on Fuchsian groups, see [11] or [25].

Hahn [9] provides a general result about the zeros of Eisenstein series for certain genus zero Fuchsian groups of the first kind. She considers, for such a group Γ' with a cusp at ∞, a specific Eisenstein series $E_k(\Gamma') \in M_k(\Gamma')$ (the vector space of holomorphic modular forms on Γ') meeting certain technical conditions. Among these conditions is that $E_k(\Gamma')$ must vanish at all cusps but ∞, have real Fourier coefficients, and have 1 as its constant term. In addition, the corresponding fundamental domains must meet acceptable conditions, including that the Hauptmodul $j_{\Gamma'}$ associated to Γ' is real on the boundary. For details, see [9].

In general, Hahn's results are slightly weaker than the aforementioned results; however her argument, involving divisor polynomials similar to (3.1), applies to a large family of groups.

5 A family of Eisenstein series on $\Gamma(2)$

In this section we explore in greater depth the location and interlacing property of zeros of a certain family of odd weight Eisenstein series on the congruence subgroup $\Gamma(2)$. While we expect that Eisenstein series on other congruence subgroups behave similarly, we choose to focus on $\Gamma(2)$ because in this setting the Eisenstein series have an elegant connection to an elliptic function $cn(u)$ which plays a role analogous to that of $\wp(n)$ for Γ. The following results arose from the collaboration by the authors initiated at the WIN conference.

We consider the congruence subgroup $\Gamma(2)$, where

$$\Gamma(2) := \{M \in \mathrm{SL}_2(\mathbb{Z}) : M \equiv I \pmod 2\}.$$

For $0 \le x, y \le 3$, we define

$$S_k(x,y) = S_k(x,y)(z) := \sum_{\substack{c,d\in\mathbb{Z},\ (c,d)=1 \\ c\equiv x \bmod 4,\ d\equiv y \bmod 4}} \frac{1}{(cz+d)^k}.$$

Then for integers $k \ge 0$ let

$$E_{2k+1,\chi} := \frac{1}{2}(S_{2k+1}(0,1) + S_{2k+1}(2,1) - S_{2k+1}(0,3) - S_{2k+1}(2,3)). \tag{5.1}$$

Each $E_{2k+1,\chi}(z)$ is a weight $2k+1$ modular form on $\Gamma(2)$ with character χ, where

$$\chi\left(\begin{pmatrix} a & b \\ c & d \end{pmatrix}\right) = \chi_{-1}(d) := \left(\frac{-1}{d}\right),$$

and $\left(\frac{\cdot}{\cdot}\right)$ stands for the Jacobi symbol. This means that $E_{2k+1,\chi}(z)$ satisfies

$$E_{2k+1,\chi}\left(\frac{az+b}{cz+d}\right) = \chi(d)(cz+d)^k E_{2k+1,\chi}(z)$$

for each $z \in \mathbb{H}$ and $\left(\begin{smallmatrix} a & b \\ c & d \end{smallmatrix}\right) \in \Gamma(2)$; note that this is a twisted version of (1.1).

Moreover, as in the classical case, $E_{2k+1,\chi}(z)$ has a Fourier expansion expressible in terms of generalized Bernouli numbers and generalized divisor sums. For a Dirichlet character ψ of modulus N, the generalized Bernouli numbers $B_{k,\psi}$ are defined by

$$\sum_{j=1}^{N} \frac{\psi(j)xe^{jx}}{e^{Nx}-1} = \sum_{k=0}^{\infty} B_{k,\psi} \cdot \frac{x^k}{k!}.$$

Similarly, the generalized divisor sums $\sigma_{\psi,k}(n)$ are defined by

$$\sigma_{\psi,k}(n) := \sum_{d|n} \psi(d)d^k.$$

With this notation, the Fourier expansion of $E_{2k+1,\chi}(z)$ at ∞ is

$$1 - \frac{2(2k+1)}{B_{2k+1,\chi}} \sum_{n=1}^{\infty} \sigma_{\chi,2k}(n)q^n.$$

We focus on this particular family for two reasons. First, $\Gamma(2)$ has some nice properties; the corresponding Riemann surface has genus zero, and $\Gamma(2)$ has a simple fundamental domain given by

$$D = \{z \in \mathbb{H} : \ -1 \leq \mathrm{Re}(z) \leq 1, \ |z - 1/2| \geq 1/2, \quad |z + 1/2| \geq 1/2\}.$$

Note that this subgroup has three inequivalent cusps at $0, 1$, and ∞. This is a prototypical group, and so we expect that our results generalize to similar simple genus zero congruence subgroups.

The second reason we focus on this family is that it is possible to study the location of the zeros of these series using the classical Jacobi elliptic function $cn(u)$ [10], which satisfies

$$\left(\frac{\mathrm{d}\, cn(u)}{\mathrm{d}\, u}\right)^2 = \left(1 - cn^2(u)\right)\left(1 - \lambda + \lambda cn^2(u)\right). \tag{5.2}$$

Here λ is a Hauptmodul for $\Gamma(2)$; it plays a similar role to that of $j(z)$ defined in (2.2), parameterizing all isomorphism classes of elliptic curves with full 2-torsion structure, and can be defined in terms of Jacobi theta functions. Note the similarity of (5.2) to the differential equation (1.4) involving $\wp(n)$.

From results of the second author and Yang in [15] we obtain a recursive formula for the λ-values of the zeros of the expansion of $E_{2k+1,\chi}$ at the cusp 1 given by

$$\begin{aligned} G_{2k+1}(z) :&= E_{2k+1,\chi} \mid_{\gamma_0} (z) \\ &= \frac{1}{2}\left(S_{2k+1}(1,0) + S_{2k+1}(3,2) - S_{2k+1}(3,0) - S_{2k+1}(1,2)\right), \end{aligned}$$

where $\gamma_0 = \left(\begin{smallmatrix} 1 & -1 \\ 1 & 0 \end{smallmatrix}\right)$. The recursion is given in Proposition 2.3 of [4].

Data from the first few dozen k values indicates that all λ-values of the zeros of $G_{2k+1}(z)$ are real and are within the interval $(-\infty, 0]$, or equivalently, the zeros of $G_{2k+1}(z)$ lie on $\mathrm{Re}(z) = 1$. Moreover, this data indicates that the zeros of $G_{2k-1}(z)$ interlace with the zeros of $G_{2k+1}(z)$. Thus, there is numerical evidence that these Eisenstein series do in general satisfy the same properties proved by Rankin and Swinnerton-Dyer [22] and Nozaki [17] for the classical case.

In order to procure a general result, we can again permute the boundaries of the fundamental domain D, sending $\mathrm{Re}(z) = 1$ to the circular arc

$$A_2 := \left\{z = \frac{1}{2}e^{i\theta} - \frac{1}{2} : 0 \leq \theta \leq \pi\right\},$$

by considering

$$G_{2k+1}\mid_{\gamma_1}(z) = \frac{1}{2}\left(S_{2k+1}(0,3) + S_{2k+1}(2,1) - S_{2k+1}(0,1) - S_{2k+1}(2,3)\right),$$

where $\gamma_1 = \left(\begin{smallmatrix} 0 & -1 \\ 1 & 0 \end{smallmatrix}\right)$. This allows us to apply the argument of Rankin and Swinnerton-Dyer to verify the location of zeros. A direct application proves that at least one third of the zeros of $G_{2k+1}\mid_{\gamma_1}(z)$ do indeed lie on this arc; however, we can modify the proof to do better. The main obstacle in this case is in splitting each Eisenstein series into a main term and an error term sufficiently small, since these error terms are difficult to bound near the cusps, i.e., near $\theta = 0, \pi$.

With a more careful consideration of the error terms, it is possible to obtain the following result.

Theorem 5.1 ([4] Theorem 1.1) *For $k \geq 1$, at least 90% of the zeros of $G_{2k+1}(z)$ in D lie on A_2.*

Theorem 5.1 is equivalent to the statement that at least 90% of the zeros have real λ-values in the range $(-\infty, 0]$. In addition, by restricting our domain slightly we obtain the following result about the separation property of the zeros of $G_{2k-1}(z)$ and $G_{2k+1}(z)$ following an approach of Nozaki [17]. We state this result in terms of the related function of θ

$$F_{2k+1}(z_\theta) := (e^{\frac{-i\theta}{2}})^{-2k-1} G_{2k+1}\mid_{\gamma_1}(z_\theta),$$

where $z_\theta := \frac{1}{2}e^{i\theta} - \frac{1}{2}$, and in terms of the following intervals for integers $k > 15$ defined by

$$I_{j,2k-1} = \left(\alpha_{j,k} - \frac{2\pi}{(2k+1)(2k-1)}, \alpha_{j,k} + \frac{2\pi}{(2k+1)(2k-1)}\right)$$

for each $j = 1, \ldots, k-1$, where $\alpha_{j,k} := 2\pi j/(2k-1)$. Let

$$I_{2k-1} := \bigcup_j I_{j,2k-1}.$$

We have the following result.

Theorem 5.2 ([4] Theorem 1.2) *Let $k > 15$ be an integer, and $I_{j,2k-1}$, I_{2k-1} as defined above. Then the zeros of $F_{2k-1}(z_\theta)$ and $F_{2k+1}(z_\theta)$ in $[\pi/10, 9\pi/10]$ are restricted to I_{2k-1} and I_{2k+1} respectively (in fact each $I_{j,2k-1}$ and $I_{j,2k+1}$ in $[\pi/10, 9\pi/10]$ contains an odd number of zeros). Moreover, the intervals are pairwise disjoint, and the $I_{j,2k-1}$ are separated by the $I_{j,2k+1}$.*

In addition to results about the location and interlacing properties of zeros, we also include an application to L-series which echoes (3.2).

Theorem 5.3 ([4] Theorem 4.1) *Let $k \geq 0$ be an integer and $E_{2k+1,\chi}(z)$ as in (5.1). Then*

$$\frac{2(-1)^k}{(2k)!L(2k+1,\chi)}\left(\frac{\pi}{2}\right)^{2k+1} = 4(2k+1) - 16 \sum_{\tau \in \mathbb{H}\backslash\Gamma(2)} \mathrm{ord}_\tau(E_{2k+1,\chi})\frac{1}{\lambda(\tau)}.$$

Similarly, we also have the following.

Corollary 5.4 ([4] Corollary 4.2) *Let $k \geq 0$ be an integer. Then*

$$\frac{2(-1)^k}{(2k)!L(2k+1,\chi)}\left(\frac{\pi}{2}\right)^{2k+1} = 4(2k+1) - 16 \sum_{\tau \in \mathbb{H}\backslash\Gamma(2)} \mathrm{ord}_\tau(G_{2k+1,\chi})\frac{1}{\lambda(\gamma_0\tau)}.$$

Similar results to Theorem 5.3 and Corollary 5.4 for even weight Eisenstein series on $\Gamma(2)$ can be found in [4].

It is clear that F. K. C. Rankin and Swinnerton-Dyer's beautiful result on the location of zeros of classical Eisenstein series has implications for many other classes of modular forms, including certain cusp forms and weakly holomorphic modular forms on $SL_2(\mathbb{Z})$, as well as Eisenstein series for $\Gamma(2)$, certain Fricke groups, and Fuchsian groups of genus zero in general. However, the literature thus far considers only modular forms on groups of genus zero, and we see in the work of Shigezumi [24] that even in genus zero, a more complex fundamental domain complicates the argument. In contrast, if p is prime, the genus of $\Gamma_0(p)$ grows with p, but the dimension of the Eisenstein space for $\Gamma_0(p)$ is only two, with $E_k(z)$ and $E_k(pz)$ forming a basis. The zero locations are known for these two series from the classical case. It would be very interesting to see if a general result exists for any genus.

References

[1] Asai,T., Kaneko, M., Ninomiya, H. *Zeros of certain modular functions and an application*, Comment. Math. Univ. St. Paul. **46** (1997), no. 1, 93–101.

[2] Diamond, F., Shurman, J. *A first course in modular forms*, Graduate Texts in Mathematics, 228. Springer-Verlag, New York, 2005.

[3] Duke, W., Jenkins, P. *On the zeros and coefficients of certain weakly holomorphic modular forms*, Pure Appl. Math. Q. **4** (2008), no. 4, part 1, 1327–1340.

[4] Garthwaite, S., Long, L., Swisher, H., Treneer, S. *Zeros of some level 2 Eisenstein series*, Proc. Amer. Math. Soc., **138** (2010), 467-480.

[5] Gekeler, E. U. *Some observations on the arithmetic of Eisenstein series for the modular group* $SL(2,\mathbb{Z})$, Arch. Math. (Basel) **77** (2001), no. 1, 5–21.

[6] Getz, J. *A generalization of a theorem of Rankin and Swinnerton-Dyer on zeros of modular forms*, Proc. Amer. Math. Soc. **132** (2004), no. 8, 2221–2231 (electronic).

[7] Getz, J. *Corrigendum to "A generalization of a theorem of Rankin and Swinnerton-Dyer on zeros of modular forms"*, Proc. Amer. Math. Soc. **138** (2010), no. 3, 1159 (electronic).

[8] Gun, S. *On the zeros of certain cusp forms*, Math. Proc. Cambridge Philos. Soc. **141** (2006), no. 2, 191–195.

[9] Hahn, H. *On zeros of Eisenstein series for genus zero Fuchsian groups*, Proc. Amer. Math. Soc. **135** (2007), no. 8, 2391–2401 (electronic).

[10] Hancock, H. *Lectures on the theory of elliptic functions: Analysis*, Dover Publications Inc., New York, 1958.

[11] Iwaniec, H. *Topics in Classical Automorphic Forms*, Graduate Studies in Mathematics, 17. American Mathematical Society, Providence, RI, 1997.

[12] Koblitz, N. *Introduction to Elliptic Curves and Modular Forms*,Graduate Texts in Mathematics, No. 97, Springer-Verlag, New York, 1993.

[13] Kohnen, W. *Zeros of Eisenstein series*, Kyushu J. Math. **58** (2004), no. 2, 251–256.

[14] Kremling, B. *Diskriminanten und Resultanten einer arithmetisch definierten Folge von Polynomen über* $\mathbb{F}_q(T)$ (German), Diplomarbeit Saarbrücken 1999.

[15] Long, L., Yang, Y. *A short proof of Milne's formulas for sums of integer squares*, Int. J. Number Theory **1** (2005), no. 4, 533–551.

[16] Miezaki, T., Nozaki, H., Shigezumi, J. *On the zeros of Eisenstein series for* $\Gamma_0^*(2)$ *and* $\Gamma_0^*(3)$, J. Math. Soc. Japan **59** (2007), no. 3, 693–706.

[17] Nozaki, H. *A separation property of the zeros of Eisenstein series for* $SL(2,\mathbb{Z})$, Bull. Lond. Math. Soc. **40** (2008), no. 1, 26–36.

[18] Ono, K. *The web of modularity: arithmetic of the coefficients of modular forms and q-series*, CBMS Regional Conference Series in Mathematics, vol. 102, Published for the Conference Board of the Mathematical Sciences, Washington, DC, 2004.

[19] Ono, K., Papanikolas, M. A. *p-adic properties of values of the modular j-function*, Galois theory and modular forms, Dev. Math., vol. 11, Kluwer Acad. Publ., Boston, MA, 2004, pp. 357–365.

[20] Rankin, R. A. *The zeros of Eisenstein series*, Publ. Ramanujan Inst. No. **1** (1968/1969), 137–144.
[21] Rankin, R. A. *The zeros of certain Poincaré series*, Compositio Math. **46** (1982), 255–272.
[22] Rankin, F. K. C., Swinnerton-Dyer, H. P. F. *On the zeros of Eisenstein series*, Bull. London Math. Soc. **2** (1970), 169–170.
[23] Serre, J.-P. A Course in Arithmetic, Graduate Texts in Mathematics, No. 7, Springer-Verlag, New York, 1973.
[24] Shigezumi, J. *On the zeros of the Eisenstein series for* $\Gamma_0^*(5)$ *and* $\Gamma_0^*(7)$, Kyushu J. Math. **61** (2007), no. 2, 527–549.
[25] Shimura, G. *Introduction to the arithmetic theory of automorphic functions*, Kan Memorial Lectures, No. 1. Publications of the Mathematical Society of Japan, No. 11. Iwanami Shoten, Publishers, Tokyo; Princeton University Press, Princeton, N.J., 1971.
[26] Wohlfahrt, K. *Über die Nullstellen einiger Eisensteinreihen.* (German), Math. Nachr. **26** (1963/1964), 381-383.

Fields Institute Communications
Volume **60**, 2011

On the μ-Invariant in Iwasawa Theory

Sujatha Ramdorai
School of Mathematics
Tata Institute of Fundamental Research
Homi Bhabha Road
Mumbai 400 005, India
sujatha@math.tifr.res.in

Abstract. The μ-invariant is a classical, yet subtle invariant attached to Iwasawa modules in classical, commutative Iwasawa theory. It has been generalised to more general Iwasawa modules that occur in non-commutative Iwasawa theory. We review this invariant in this article.

Introduction

The aim of this expository article is to discuss the μ-invariant associated to finitely generated modules over Iwasawa algebras. This is an important invariant which was first discovered by Iwasawa in the 1960's and occurs in his seminal work on cyclotomic fields [10], [11]. In fact, Iwasawa conjectured that his μ-invariant was always zero for the p-primary subgroup of the ideal class group of the field obtained by adjoining all p-power roots of unity to a given finite extension F of $\mathbb{Q}$, p being a fixed prime number. This was subsequently proven by Ferrero and Washington when F is an abelian extension of $\mathbb{Q}$, but remains open in general. Mazur found the first simple examples of a positive μ-invariant in the case of the Iwasawa theory of elliptic curves. For example, let A denote the 5-primary subgroup of the Tate-Shafarevich group of the elliptic curve

$$X_0(11) : y^2 + y = x^3 - x^2 - 10x - 20$$

over the field obtained by adjoining all 5-power roots of unity to $\mathbb{Q}$. Then Mazur showed that the μ-invariant of A is equal to 2; in fact, in this case, $5.A = 0$, but A is not a finitely generated abelian group since it has positive μ-invariant.

In recent years, the progress made in non-commutative Iwasawa theory has led to a revival of interest in this subtle invariant, and the open problems associated with it. As we shall see below, even though the definition of the μ-invariant is algebraic, all known deep results about its calculation are by purely analytic means! The article is organised as follows: Section 1 is preliminary in nature, where we also fix notation. Section 2 introduces the μ-invariant and we recall the important

2010 *Mathematics Subject Classification.* Primary 11R23; Secondary 19A31.

classical results in this set up. In Section 3, we consider the μ-invariant for elliptic curves over number fields. Section 4 discusses the relatively recent definition of the μ-invariant for modules over Iwasawa algebras of p-adic Lie groups which are not necessarily pro-p. Finally in Section 5, we mention one of the main open problems related to this invariant, which seems quite difficult to attack. This problem is especially interesting in the context of the non-commutative Iwasawa theory and the formulation of the main conjecture.

The author is grateful to the editors of the 'Women in Numbers' volume for the invitation to contribute an article, and also for their patience with the final submission. She would also like to thank J. Coates for helpful discussions and valuable comments.

1 Notation and preliminaries

Throughout, p will denote an odd prime and $\mathbb{Z}_p$ (respectively $\mathbb{Q}_p$), will denote the ring of p-adic integers (resp. the field of p-adic numbers). Given any number field F, we shall denote by F^{cyc} the cyclotomic $\mathbb{Z}_p$-extension of F, which we recall, is the unique subfield of $F(\mu_{p^\infty})$ with Galois group over F isomorphic to $\mathbb{Z}_p$. Recall that an extension $\mathcal{L}$ of F is said to be a p-adic Lie extension if the Galois group $\mathrm{Gal}(\mathcal{L}/\mathbb{Q})$ is a p-adic Lie group [12]. The cyclotomic $\mathbb{Z}_p$-extension F^{cyc} is a basic example of a p-adic Lie extension of F, and has been studied extensively beginning with Iwasawa's seminal works [10], [11]. For any finite set S of primes of F, F_S will denote the maximal extension of F unramified outside S. We shall always assume that S contains the set S_p of primes in F that lie above p, and the archimedean primes. Given a profinite group G, the p-cohomological dimension of G will be denoted by $\mathrm{cd}_p(G)$, which we recall is the least integer n such that $H^{n+i}(G, A) = 0$ for all $i \geq 1$, and any discrete p-primary torsion module A. For a compact p-adic Lie group with no elements of order p, it is a result of Serre and Lazard that $\mathrm{cd}_p(G) = d$, where d is the dimension of G as a Lie group. If G is a profinite group, the Iwasawa algebra of G is denoted by $\Lambda(G)$, and is defined as

$$\Lambda(G) = \varprojlim_{U} \mathbb{Z}_p[G/U],$$

where U runs over all open normal subgroups of G, with the inverse limit being taken with respect to the natural surjections. For any number field F, we shall denote the Galois group $\mathrm{Gal}(F^{\mathrm{cyc}}/F)$ by Γ. Given a discrete module M over the Iwasawa algebra $\Lambda(G)$, we denote its compact Pontryagin dual by $M^\vee = \mathrm{Hom}\,(M, \mathbb{Q}_p/\mathbb{Z}_p)$. For any module M over the Iwasawa algebra $\Lambda(G)$, $M(p)$ denotes the p-primary torsion submodule of M.

Let us briefly discuss how the μ-invariant was discovered by Iwasawa in his study of class numbers. Let F be a number field and suppose that

$$F = F_0 \subset F_1 \subset \cdots \subset F_n \cdots \subset F_\infty = \bigcup F_n,$$

is a sequence of number fields with $\mathrm{Gal}(F_n/F_0) \simeq \mathbb{Z}/p^n\mathbb{Z}$. Then the Galois group

$$\mathrm{Gal}(F_\infty/F_0) = \varprojlim\,(\mathbb{Z}/p^n\mathbb{Z}) = \mathbb{Z}_p,$$

so that the extension F_∞/F_0 is a $\mathbb{Z}_p$-extension. The most natural example of such an extension is of course the cyclotomic $\mathbb{Z}_p$-extension F^{cyc}. Let p^{e_n} be the exact order of p that divides the order of the class number of F_n. Iwasawa proved that there exist non-negative integers λ, μ, ν such that

$$e_n = \lambda n + \mu p^n + \nu \tag{1.1}$$

for all sufficiently large n. The invariants λ and μ are respectively known as the λ and μ-invariants, which we denote by $\lambda(F_\infty/F)$ and $\mu(F_\infty/F)$ respectively. If F_∞ is the cyclotomic $\mathbb{Z}_p$-extension F^{cyc} of F, and the base field F is clear from the context, we shall simply denote them by λ and μ respectively. In fact, for $F = \mathbb{Q}$ and $F_\infty = \mathbb{Q}^{\text{cyc}}$, it is easy to prove (as Iwasawa did) that $\lambda = \mu = \nu = 0$. The argument breaks down as soon as one replaces $\mathbb{Q}$ by a finite extension, even a quadratic extension, if p splits in the extension. Iwasawa's celebrated "$\mu = 0$ conjecture" asserts that the invariant μ in (1.1) is zero for any number field F and its cyclotomic $\mathbb{Z}_p$-extension F^{cyc}. A deep theorem of Ferrero and Washington proves this conjecture for abelian extensions F of $\mathbb{Q}$. Iwasawa himself constructed examples of non-cyclotomic $\mathbb{Z}_p$-extensions with $\mu > 0$, (see Washington's book [21] for more details). In the next sections, we shall see a more algebraic definition of the μ-invariant.

2 The μ-invariant in Iwasawa theory

In this section, we give two other definitions of the μ-invariant in the more general context of finitely generated modules over Iwasawa algebras. Let therefore G be a pro-p p-adic Lie group of dimension d, which has no elements of order p. For example, if G is isomorphic to $\mathbb{Z}_p^d$, then it is well-known that there is an isomorphism

$$\Lambda(G) \simeq \mathbb{Z}_p[[T_1, \ldots, T_d]]$$

which sends a (fixed) independent set of topological generators $e_1, \ldots, e_d$ of G to $(1+T_1), \ldots, (1+T_d)$ respectively. Thus $\Lambda(G)$ is a commutative, regular local ring of Krull dimension $d+1$, in this case. For a general non-commutative p-adic Lie group G of dimension d as above, the ring $\Lambda(G)$ is a non-commutative, left and right Noetherian local domain. It has additional nice properties, for instance it is Auslander regular (see [20]) of global dimension $d+1$, and hence affords a good dimension theory for finitely generated modules. Recall that a torsion module over $\Lambda(G)$ is said to be *pseudo-null* if it has dimension less than or equal to $d-1$. A homomorphism $f : M \to N$ between two finitely generated $\Lambda(G)$-modules is a *pseudo-isomorphism* if both its kernel and cokernel are pseudo-null. Let M be a finitely generated torsion module over $\Lambda(G)$ and $M(p)$ be its p-primary torsion submodule. The structure theorem for finitely generated torsion $\Lambda(G)$- modules, was first proved by Iwasawa and then generalised by Serre, (cf. [3, Chap. VII, §4]) in the commutative case, and extended to general Iwasawa algebras $\Lambda(G)$, and finitely generated modules over $\Lambda(G)$ that are killed by a power of p by Susan Howson ([8]) and Venjakob [20] independently:

Theorem 2.1 *Let M be a finitely generated torsion module over $\Lambda(G)$, where G is a pro-p p-adic Lie group with no elements of order p. Then there is a pseudo-isomorphism*

$$f : M(p) \to \bigoplus_{i=1}^{k} \Lambda(G)/p_i^{n_i}.$$

The set of integers n_i is unique up to ordering.

The importance of this theorem lies in the fact that it enables us to define a key invariant for finitely generated $\Lambda(G)$-modules. Let M be a finitely generated $\Lambda(G)$-module. Then the *μ-invariant* of M is defined as

$$\mu_G(M) = \sum_j n_j,$$

where the integers n_j are as in the structure theorem for $M(p)$. If $G \simeq \mathbb{Z}_p$, then by the general structure theorem of Serre-Iwasawa for finitely generated torsion $\Lambda(G)$-modules M, it follows that $M/M(p)$ is a finitely generated $\mathbb{Z}_p$-module. In this case, the *λ-invariant* of M, denoted $\lambda(M)$, is defined to be the $\mathbb{Z}_p$-rank of $M/M(p)$.

Let F be a number field and F^{cyc} be the cyclotomic $\mathbb{Z}_p$-extension of F, with $\Gamma = \mathrm{Gal}(F^{\mathrm{cyc}}/F) \simeq \mathbb{Z}_p$. We denote by $M(F^{\mathrm{cyc}})$ the maximal abelian, p-extension of F^{cyc} which is unramified everywhere. In other words, $M(F^{\mathrm{cyc}})$ is the p-Hilbert class field of F^{cyc}. Let $X^{\mathrm{cyc}} = X^{\mathrm{cyc}}(F)$ denote the Galois group $\mathrm{Gal}(M(F^{\mathrm{cyc}})/F^{\mathrm{cyc}})$. As X^{cyc} is abelian, it has a natural structure of a Γ-module. Indeed, given an element γ in Γ and an element x in X^{cyc}, the action is defined by

$$\gamma.x = \tilde{\gamma}x\tilde{\gamma}^{-1},$$

where $\tilde{\gamma}$ is any lift of γ to $\mathrm{Gal}(M(F^{\mathrm{cyc}})/F)$. It is easily checked that the action is independent of the lift since X^{cyc} is abelian. Further, as any compact $\mathbb{Z}_p$-module with a continuous Γ-action has a natural structure of a $\mathbb{Z}_p[[\Gamma]]$-module, it is plain that X^{cyc} is a compact module over the Iwasawa algebra $\Lambda(\Gamma)$. The following theorem is due to Iwasawa [10]:

Theorem 2.2 *(Iwasawa) X^{cyc} is a finitely generated torsion $\Lambda(\Gamma)$-module.*

Iwasawa proved that the invariants λ and μ occurring in (1.1) exist by combining this structure theorem with an arithmetic argument, and indeed the μ and λ invariants appearing in (1.1) are precisely the μ-invariant and λ- invariant that arise from the structure theorem for the finitely generated torsion module X^{cyc}.

There is another equivalent definition for the μ-invariant. Let G be as above and $\Lambda(G)$ be the Iwasawa algebra. Suppose that M is a finitely generated module over $\Lambda(G)$. Then the Euler characteristic $\chi(G, M(p))$ is defined as

$$\chi(G, M(p)) = \prod_i \#\left(H_i(G, M(p))\right)^{(-1)^i}, \tag{2.1}$$

where $H_i(G, M(p))$ are the homology groups $\mathrm{Tor}_i^{\Lambda(G)}(\mathbb{Z}_p, M(p))$. Note that as $\Lambda(G)$ is Auslander regular, the homology groups vanish for $i > (d+1)$ and hence the Euler characteristic is well-defined. We then have (see [9], [20])

$$\mu_G(M) = \mathrm{ord}_p(\chi(G, M(p))). \tag{2.2}$$

Thus the μ-invariant can be interpreted also as an Euler characteristic. It is obvious from the above discussion that a finitely generated p-primary torsion module over $\Lambda(G)$ is pseudo-null if and only if it has μ-invariant zero. Further, the μ-invariant is additive along exact sequences of torsion $\Lambda(G)$-modules.

Again, suppose G is a pro-p, p-adic Lie group of dimension d with no elements of order p. Let $\Omega(G)$ denote the Iwasawa algebra over $\mathbb{F}_p$, i.e.

$$\Omega(G) := \varprojlim_U \mathbb{F}_p[G/U], \tag{2.3}$$

where the inverse limit is taken over open, normal subgroups as before, with respect to the natural maps. The ring $\Omega(G)$ is again a left and right Noetherian local domain which is Auslander regular (cf. [20]) with finite global dimension equal to d. The rank of a finitely generated module over $\Omega(G)$ is well-defined since $\Omega(G)$ is a domain. For any finitely generated $\Lambda(G)$-module M, there exists an integer $r(M)$ such that the p-primary torsion submodule $M(p)$ is annnihilated by $p^{r(M)}$. Let $\mathrm{gr}_p\, M(p)$ denote the p-graded module of $M(p)$, i.e.

$$\mathrm{gr}_p\, M(p) := \bigoplus_{k=1}^{r(M)} M_{p^k}/M_{p^{k-1}}, \tag{2.4}$$

where M_{p^t} denotes the submodule of p^t-torsion elements of M. Then each graded piece is an $\Omega(G)$-module, and $\mu(M)$ is easily seen to be the sum of the ranks of the graded pieces.

Let S be a finite set of primes of F as in §1 and F_S be the maximal extension of F unramified outside S. It is plain that F^{cyc} is contained in F_S. Let $F_S(p)$ denote the subfield of F_S containing F such that the Galois group $G_S(p,F) := \mathrm{Gal}(F_S(p)/F)$ is the maximal pro-p quotient of $\mathrm{Gal}(F_S/F)$. It was proven by Iwasawa that for any number field F, we have

$$H_2(\mathrm{Gal}(F_S/F^{\mathrm{cyc}}), \mathbb{Z}_p) = 0. \tag{2.5}$$

Recall that a pro-p group G is *free* if $\mathrm{cd}_p(G) \leq 1$. Let $G_S(p, F^{\mathrm{cyc}}) = \mathrm{Gal}(F_S(p)/F^{\mathrm{cyc}})$. Using (2.5), it can be shown that $G_S(p, F^{\mathrm{cyc}})$ is free if and only if $\mu(F^{\mathrm{cyc}}/F) = 0$ (see [14, Chap. XI, §3] and [19] for more details). This provides an interesting connection between the classical Iwasawa μ-invariant for a number field F and the freeness of the pro-p Galois group $G_S(p, F^{\mathrm{cyc}})$.

3 The μ-invariant for elliptic curves

The discussion in the previous section may be viewed as the theory of the μ-invariant for the trivial Tate motive $\mathbb{Z}_p$. We shall consider the motive associated to an elliptic curve and describe the μ-invariants that arise in this context. Let E be an elliptic curve defined over a number field F such that E has good ordinary reduction at all primes of F that lie above p. We first describe the arithmetic module that corresponds to the module X^{cyc} considered in the previous section. The set S is now any finite set of primes of F that contains the primes above p and all the primes of bad reduction for E. Put $G_F = \mathrm{Gal}(\bar{F}/F)$ for the absolute Galois group of F. Let

$$E_{p^\infty} := \bigcup_{n\geq 0} E_{p^n}$$

where $E_{p^n} := E_{p^n}(\bar{F})$ is the discrete G_F-module of p^n-torsion points on $E(\bar{F})$. The Tate module of E, denoted $T_p(E)$, is defined by

$$T_p(E) = \varprojlim E_{p^n},$$

and $V_p(E) = T_p(E) \otimes \mathbb{Q}_p$ is the corresponding two dimensional $\mathbb{Q}_p$-vector space with a continuous action of G_F. For the purposes of this article, we consider the G_F-module $V_p(E)$ as the motive of the elliptic curve E.

Let F^{cyc} be the cyclotomic $\mathbb{Z}_p$-extension of F. Classical descent theory for elliptic curves suggests that the $\Lambda(\Gamma)$-module to be considered in this context is the Pontryagin dual of the *Selmer group* over the cyclotomic extension. Recall that for a finite Galois extension L of F, the Selmer group $S(E/L)$ of E over L is a discrete $\mathrm{Gal}(L/F)$-module and is defined by

$$S(E/L) = \mathrm{Ker}\left(H^1(\mathrm{Gal}(F_S/L), E_{p^\infty})) \to \underset{v\in S}{\oplus} J_v(E/L)\right), \tag{3.1}$$

where

$$J_v(E/L) = \underset{w|v}{\oplus} H^1(\mathrm{Gal}(\bar{L}_w/L_w), E)(p).$$

The Selmer group of E over F^{cyc} is defined as the direct limit of $S(E/L)$ as L varies over finite extensions of F in F^{cyc}. Its Pontryagin dual is denoted by $\mathfrak{X}(E/F^{\mathrm{cyc}})$ and is easily seen to be a finitely generated module over $\Lambda(\Gamma)$. The Selmer group and its dual are defined analogously for any infinite extension L_∞ of F, and denoted by $S(E/L_\infty)$ and $\mathfrak{X}(E/L_\infty)$ respectively. It is a deep conjecture due to Mazur that since E has good ordinary reduction at the primes above p, the module $\mathfrak{X}(E/F^{\mathrm{cyc}})$ is $\Lambda(\Gamma)$-torsion. There are plenty of numerical examples known where this conjecture holds. There is also the important theoretical result due to Kato who proves Mazur's conjecture for $\mathfrak{X}(E/F^{\mathrm{cyc}})$, for number fields F that are abelian over $\mathbb{Q}$, provided E is defined over $\mathbb{Q}$. One is thus led naturally to wonder if $\mathfrak{X}(E/F^{\mathrm{cyc}})$ is a finitely generated $\mathbb{Z}_p$-module, which would be the analogue of Iwasawa's $\mu = 0$ conjecture. However, Mazur already gave examples where this is not true, and the fact that the μ-invariant of $\mathfrak{X}(E/F^{\mathrm{cyc}})$ is not always zero causes endless technical difficulties in the Iwasawa theory of elliptic curves! An explicit example is the elliptic curve $E/\mathbb{Q}$ of conductor 11, defined by

$$E : y^2 + y = x^3 - x^2 - 10x - 20.$$

For the prime $p = 5$, Mazur showed that $\mathfrak{X}(E/\mathbb{Q}^{\mathrm{cyc}})$ is not a finitely generated $\mathbb{Z}_5$-module, although it is $\Lambda(\Gamma)$-torsion.

Another interesting case is that of elliptic curves with complex multiplication. Let E be an elliptic curve with complex multiplication by the ring of integers $\mathcal{O}_K$ of an imaginary quadratic field K of class number one. Assume that E is defined over K and that the odd prime p splits in K, say $p\mathcal{O}_K = \mathfrak{p}\mathfrak{p}^*$. Assume also that E has good reduction at $\mathfrak{p}$ and $\mathfrak{p}^*$. Let $K(E_{\mathfrak{p}^n})$ denote the field extension of K obtained by adjoining the co-ordinates of the $\mathfrak{p}^n$-division points of E and consider the infinite Galois extension

$$K_\infty := K(E_{\mathfrak{p}^\infty}) = \bigcup_{n\geq 0} K(E_{\mathfrak{p}^n}).$$

Then the Galois group $G = \mathrm{Gal}(K(E_{\mathfrak{p}^\infty})/K)$ is isomorphic to $\mathbb{Z}_p^\times$ and K_∞ is a $\mathbb{Z}_p$-extension of $K(E_\mathfrak{p})$, and we write $\Lambda(G)$ for the associated Iwasawa algebra. Let X_∞ be the maximal abelian p-extension of K_∞ that is unramified outside of the set of primes above $\mathfrak{p}$. It is known, thanks to results of Coates and Wiles [6], [15] that X_∞ is a finitely generated torsion $\Lambda(G)$-module. Let $S^\mathfrak{p}(E/K_\infty)$ denote the $\mathfrak{p}$-Selmer group of E over K_∞. To be precise, this is defined by taking $E_{\mathfrak{p}^\infty}$ instead of E_{p^∞} in (3.1). Let $\mathfrak{X}^\mathfrak{p}(E/K_\infty)$ be the compact dual of the Selmer group. We then have [15]

$$S^\mathfrak{p}(E/K_\infty) = \mathrm{Hom}\,(X_\infty, E_{\mathfrak{p}^\infty}),$$

and hence it follows that $\mathfrak{X}^\mathfrak{p}(E/K_\infty)$ is a finitely generated $\Lambda(G)$-torsion module. Put $\Gamma = \mathrm{Gal}(K_\infty/K(E_p))$, which is isomorphic to $\mathbb{Z}_p$. The following theorem was proven by Gillard and Schneps independently ([7], [17]) using analytic methods:

Theorem 3.1 *Viewing* $\mathfrak{X}^{\mathfrak{p}}(E/K_\infty)$ *as a* $\Lambda(\Gamma)$*-module, we have* $\mu(\mathfrak{X}^{\mathfrak{p}}(E/K_\infty)) = 0$.

Let K'_∞ be the Galois extension of K obtained by adjoining the co-ordinates of the p-power division points of E, i.e.

$$K'_\infty := K(E_{p^\infty}) = \bigcup_{n\geq 0} K(E_{p^n}).$$

Then K'_∞ is a Galois extension of $K(E_p)$, whose Galois group G' is isomorphic to $\mathbb{Z}_p \times \mathbb{Z}_p$. Let $\Lambda(G')$ be its Iwasawa algebra, so that we have

$$\Lambda(G') \simeq \mathbb{Z}_p[[T_1, T_2]].$$

The Pontryagin dual $\mathfrak{X}^{\mathfrak{p}}(E/K'_\infty)$ of the discrete Selmer group $S^{\mathfrak{p}}(E/K'_\infty)$ is a compact, finitely generated torsion $\Lambda(G')$-module, and as the extension K'_∞/K_∞ is pro-p, the following result is an easy consequence of a generalisation of Theorem 3.1, where $\mu_{G'}$ denotes the μ-invariant for $\Lambda(G')$-modules.

Proposition 3.2 *We have* $\mu_{G'}(\mathfrak{X}^{\mathfrak{p}}(E/K'_\infty)) = 0$.

For a general elliptic curve E over a number field F, and a p-adic Lie extension L_∞/F with pro-p Galois group G having no element of order p, the μ-invariant $\mu_G(\mathfrak{X}(E/L_\infty))$ is a big mystery and we discuss some aspects of this in §5.

4 The μ-invariant for p-adic Lie groups

In the previous sections, the μ-invariant was defined for modules over Iwasawa algebras of pro-p, p-adic Lie groups G with the additional hypothesis that G has no elements of order p. It is clear from the discussion in §2, that the μ-invariant is well-defined on Grothendieck groups. Let $\mathcal{C}$ be a small abelian category. Recall that its Grothendieck group $K_0(\mathcal{C})$, is an abelian group with generators $[N]$ where N runs over all the objects of $\mathcal{C}$, and relations $[N] = [N'] + [N'']$ for any short exact sequence

$$0 \to N' \to N \to N'' \to 0$$

of objects in $\mathcal{C}$. Now consider the abelian categories $\mathcal{M}(G)$ and $\mathcal{P}(G)$ which are respectively the categories of finitely generated (resp. finitely generated, projective) modules over $\Omega(G)$. The additional hypothesis above on G ensures that the Grothendieck groups $K_0(\mathcal{M}(G))$ and $K_0(\mathcal{P}(G))$ are naturally isomorphic. Further, it is plain that the μ-invariant gives a homomorphism from $K_0(\mathcal{C})$ into $\mathbb{Z}$. where $\mathcal{C}$ denotes either of the two categories above. As noted before, $\mu([N])$ is zero if and only if N is pseudo-null.

The K-theoretic context is important from an other point of view as well. The centrepiece of Iwasawa theory is the formulation of (and attempts to prove!) the *Main Conjecture*, which we shall not discuss in detail here. However, we do mention that the formulation of the main conjecture in non-commutative Iwasawa theory involves the characteristic element of a certain class of $\Lambda(G)$-modules, which is an element of a localisation of the Iwasawa alegbra $\Lambda(G)$ (see [4] for details). From this perspective, it is then natural to wonder if the theory, and in particular, the μ-invariant can be extended to more general compact p-adic Lie groups. Let G be a p-adic Lie group with no element of order p and suppose N is a finitely generated $\Omega(G)$-module. We can also view N as a module over $\Lambda(G)$ that is killed by p. The characteristic element of N, denoted ξ_N, is an element of $K_1(\Lambda(G)_T)$ where T is the *central Ore set* in $\Lambda(G)$ defined by $T = \{1, p, p^2, \cdots\}$, and $\Lambda(G)_T$ denotes the localisation of $\Lambda(G)$ at T. For a ring R, let $R^\times$ denote the multiplicative group of units of R. Recall that the group $K_1(R)$ is isomorphic to $\mathrm{GL}(R)/E(R)$, where $\mathrm{GL}(R) = \underset{n}{\cup}\,\mathrm{GL}_n(R)$ and $\mathrm{E}(R) = \underset{n}{\cup}\,E(R)$ consists of the stable elementary matrices (see [2]). The natural map from $R^\times$ to $\mathrm{GL}_1(R)$ induces a homomorphism

$$R^\times \to K_1(R).$$

In particular, we have a natural map

$$\theta : (\Lambda(G))_T^\times \to K_1(\Lambda(G)_T). \tag{4.1}$$

We have the following result (see [8], [20]).

Theorem 4.1 *Let G be a pro-p, p-adic Lie group with no elements of order p, and suppose that N is a finitely generated $\Omega(G)$- module. Then*

$$\xi_N = \theta\left(p^{\mu(N)}\right).$$

Moreover, the following assertions are equivalent:
(i) N is pseudo-null (ii) $\chi(G,N) = 1$ (iii) $\xi_N = 1$.

Ardakov and Wadsley [1] study the characteristic elements for modules over $\Omega(G)$ where they allow G to be an arbitrary compact p-adic Lie group which has no elements of order p. The Iwasawa algebra $\Lambda(G)$ has finitely many simple modules up to isomorphism; if G is pro-p with no elements of order p, then of course, there is a unique simple module, namely $\mathbb{F}_p$. Let $V_1, \cdots, V_s$ be the finitely many simple modules, each V_i being a finite dimensional $\mathbb{F}_p$-vector space. Assume that G has no elements of order p (but is not necessarily pro-p). Then the Euler characteristic definition of the μ-invariant, given in §2 is valid and allows us to define the μ-invariant $\mu(N)$ for any finitely generated $\Lambda(G)$-module N killed by a power of p. Ardakov and Wadsley also introduce *twisted μ-invariants*, denoted $\mu_i(N)$, for $1 \le i \le s$, defined by the formula

$$\mu_i(N) = \frac{\log_p \chi(G, \mathrm{gr}_p N \otimes_{\mathbb{F}_p} V_i^*)}{\dim_{\mathbb{F}_p} \mathrm{End}_{\Omega(G)}(V_i)}.$$

Here V_i^* is the dual vector space V_i and $\mathrm{gr}_p(N)$ is the graded module defined by (2.4).

Theorem 4.2 *(Ardakov-Wadsley) Assume that G is a compact p-adic analytic group with no elements of order p, and let N be a finitely generated p-primary $\Lambda(G)$-module. Then the twisted μ-invariants $\mu_i(N)$ $(1 \le i \le s)$ are integers and*

$$\xi_N = \theta\left(\prod_{i=1}^{s} f_i^{\mu_i(N)}\right),$$

where $f_i = 1 + (p-1)e_i$, and e_i is an idempotent in $\Lambda(G)$ such that V_i is the unique simple quotient module of $e_i\Lambda(G)$.

For a general p-adic Lie group G which has no element of order p, but is not necessarily pro-p, even the integrality of the Euler characteristic $\chi(G,N)$ is unclear for a finitely generated p-primary $\Lambda(G)$-module. In [1, 9.6], there is an explicit example of a dimension one p-adic analytic group G and a module N over $\Omega(G)$ for which the Euler characteristic $\chi(G,N)$ is not integral. It is also known that in this generality, it is no longer true that N is pseudo-null if and only $\xi_N = 1$. In [1], there is a characterisation of those groups for which this assertion holds.

Theorem 4.3 *(Ardakov-Wadsley) Let G be a compact p-adic analytic group with no elements of order p. Then the following assertions are equivalent: (i) $\xi_N = 1$ for all finitely generated p-primary pseudo-null $\Lambda(G)$-modules N (ii) $\chi(G, N) \in \mathbb{Z}$ for all finitely generated p-primary $\Lambda(G)$-modules N (iii) G is a semidirect product of a finite group of order prime to p with a pro-p group.*

5 Open problems

In this final section, we describe some open problems related to the μ-invariant. As these arise naturally in the formulation of the main conjecture in non-commutative Iwasawa theory (see [4]), we begin by discussing the background. For brevity, we shall restrict ourselves to the case of elliptic curves, though it is obvious that the problems can be posed in the wider context of more general motives. Let therefore E be an eliptic curve over a number field F without complex multiplication, such that E has good ordinary reduction at the primes of F that lie above p. Recall that this means that the endomorphism ring $\mathrm{End}(E)$ of the elliptic curve is isomorphic to $\mathbb{Z}$, and by a deep result of Serre, [18], the image of the Galois representation

$$\rho_E : \mathrm{Gal}(\bar{\mathbb{Q}}/F) \to GL_2(V),$$

where $V = V_p(E)$ is as in §3, is an open subgroup of $GL_2(\mathbb{Z}_p)$. Let G denote this image. Then G is the Galois group F of the extension

$$F_\infty := F(E_{p^\infty})$$

obtained by attaching the coordinates of all the p-power division points of E to F. Henceforth assume that $p \geq 5$, as this guarantees that G has no elements of order p. We shall further assume that F contains the p-division points of E, i.e. $F = F(E_p)$, so that $G = \mathrm{Gal}(F_\infty/F)$ is a pro-p, p-adic Lie group. Let $\Lambda(G)$ be the $\mathbb{Z}_p$-Iwasawa algebra of G. As before, we consider the compact Pontryagin dual of the Selmer group $\mathfrak{X}(E/F_\infty)$. It is not difficult to prove that $\mathfrak{X}(E/F_\infty)$ is a finitely generated module over the Iwasawa algebra $\Lambda(G)$. By the Weil pairing, the extension F_∞ contains the cyclotomic $\mathbb{Z}_p$-extension F^{cyc} and we denote by H the Galois group $\mathrm{Gal}(F_\infty/F^{\mathrm{cyc}})$, and by Γ the quotient $G/H = \mathrm{Gal}(F^{\mathrm{cyc}}/F)$. We now make the following definition.

Definition 5.1 The category $\mathfrak{M}_H(G)$ consists of all finitely generated compact modules X over $\Lambda(G)$ such that $X/X(p)$ is finitely generated over $\Lambda(H)$.

Put

$$\mathfrak{Y}(E/F_\infty) = \mathfrak{X}(E/F_\infty)/\mathfrak{X}(E/F_\infty)(p). \tag{5.1}$$

A key hypothesis that is essential for the formulation of the non-commutative main conjecture is the following [4]:

Conjecture 5.2 *($\mathfrak{M}_H(G)$ Conjecture) The $\Lambda(G)$-module $\mathfrak{X}(E/F_\infty)$ is in $\mathfrak{M}_H(G)$, i.e. the module $\mathfrak{Y}(E/F_\infty)$ is a finitely generated $\Lambda(H)$-module.*

This conjecture is essential in order to define the characteristic element ξ_E of $\mathfrak{X}(E/F_\infty)$. In fact, it is shown in [4] that there is a surjection

$$\delta : K_1(\Lambda(G)_{S^*}) \to K_0(\mathfrak{M}_H(G)),$$

where S^* is a certain canonical Ore set in $\Lambda(G)$ defined in [4]. The kernel of δ is the natural image of $K_1(\Lambda(G))$ in $K_1(\Lambda(G)_{S^*})$. For any element $[M]$ in $K_0(\mathfrak{M}_H(G))$, the characteristic element ξ_M is defined to be the class of a lift of M under the map δ above. It is well-defined modulo the natural image of $K_1(\Lambda(G))$ in $K_1(\Lambda(G)_{S^*})$.

Let us now explain the relevance of the μ-invariant for this conjecture. As before, $\mathfrak{X}(E/F^{\text{cyc}})$ denotes the dual Selmer group of E over the cyclotomic extension F^{cyc} of F. An easy application of Nakayama's lemma shows that if $\mathfrak{X}(E/F^{\text{cyc}})$ is a finitely generated $\mathbb{Z}_p$-module, i.e. $\mathfrak{X}(E/F^{\text{cyc}})$ has μ-invariant zero, then Conjecture 5.2 holds for E. Indeed, one proves that $\mathfrak{X}(E/F_\infty)$ is itself a finitely generated $\Lambda(H)$-module. More generally, if the elliptic curve E is isogenous over any subfield of F_∞, to a curve whose dual Selmer group over the cyclotomic $\mathbb{Z}_p$-extension of this field has μ-invariant zero, then Conjecture 5.2 is true for E. Thus there are plenty of numerical examples known where the conjecture is valid. For example, consider the three isogenous elliptic curves over $\mathbb{Q}$ of conductor 11, given by the following equations:

$$\begin{aligned} y^2+y &= x^3-x^2, \\ y^2+y &= x^3-x^2-10x-20, \\ y^2+y &= x^3-x^2-7820x-263580. \end{aligned}$$

The field F_∞ obtained by attaching the co-ordinates of the p-power division points for any prime p, is the same for the three curves as they are isogenous. Take $p=5$, and $L=\mathbb{Q}(\mu_5)$. Then $G_L=\text{Gal}(F_\infty/F)$ is pro-5 and $\mathfrak{X}(E/F_\infty)$ is in $\mathfrak{M}_H(G)$ for these three curves, because $\mathfrak{X}(E/L^{\text{cyc}})=0$ for the first curve $E:=y^2+y=x^3-x^2$, see [5]. However it is unknown whether $\mathfrak{X}(E/F_\infty)$ is in $\mathfrak{M}_H(G)$ for any prime of good ordinary reduction greater than 5.

For any finite extension L of F contained in F_∞, let

$$G_L=\text{Gal}(F_\infty/L),\ \Gamma_L=\text{Gal}(F_\infty/L^{\text{cyc}}).$$

Let $\mu_{G_L}(\mathfrak{X}(E/F_\infty))$ (resp. $\mu_{\Gamma_L}(\mathfrak{X}(E/F^{\text{cyc}}))$) denote the μ-invariants of the corresponding dual Selmer groups over the Iwasawa algebras $\Lambda(G_L)$, (resp. Λ_{Γ_L}). Let $H_L=\text{Gal}(F_\infty/L^{\text{cyc}})$. The following result is proved in [4, §5]:

Proposition 5.3 *Let E/F be an elliptic curve as above and assume Mazur's conjecture holds for $\mathfrak{X}(E/L^{\text{cyc}})$ for all finite extensions L of F in F_∞. Suppose in addition, that there exists a finite extension L_1 of F in F_∞ such that $\mu_{\Gamma_{L_1}}(\mathfrak{X}(E/L_1^{\text{cyc}}))=\mu_{G_{L_1}}(\mathfrak{X}(E/F_\infty))$, and that $H_1(H_{L_1},\mathfrak{Y}(E/F_\infty))$ is finite. Then $\mathfrak{X}(E/F_\infty)$ is in $\mathfrak{M}_H(G)$. Conversely, if $\mathfrak{X}(E/F_\infty)$ is in $\mathfrak{M}_H(G)$, then $\mu_{G_L}(\mathfrak{X}(E/F_\infty))$ $=\mu_{\Gamma_L}(\mathfrak{X}(E/L^{\text{cyc}}))$ and $H_1(H_L,\mathfrak{Y}(E/F_\infty))$ is finite for each finite extension L of F in F_∞.*

It is clear that Conjecture 5.2 can be suitably modified and posed in a wider context of *admissible p-adic Lie extensions*, i.e. infinite extensions of F with a pro-p, p-adic Galois group G that does not have elements of order p, and which contain F^{cyc}. Further, one could also consider more general motives and their associated Galois representations. We do not go further into the details and refer the interested reader to [13] for a striking result on these non-commutative main conjectures.

We end this paper with a discussion of the special case of elliptic curves with complex multiplication. As in §3, let K be an imaginary quadratic field of class number one, and E an elliptic curve defined over K with complex multiplication by the ring of integers $\mathcal{O}_K$ of K. Again, let p be a rational prime which splits in K, $p\mathcal{O}_K=\mathfrak{p}\mathfrak{p}^*$, and assume that E has good reduction at $\mathfrak{p}$ and $\mathfrak{p}^*$. Let

$$K_\infty=K(E_{\mathfrak{p}^\infty}),\ K'_\infty=K(E_{p^\infty}),$$

and $G' = \mathrm{Gal}(K'_\infty/K)$. As noted before, the $\mathfrak{p}$-dual Selmer group $\mathfrak{X}^{\mathfrak{p}}(E/K_\infty)$ is a finitely generated $\mathbb{Z}_p$-module. This in turn implies that $\mu_{G'}(\mathfrak{X}(E/K'_\infty)) = 0$. In fact one can show that $\mathfrak{X}(E/K'_\infty)(p) = 0$. Thus the analogue of Conjecture 5.2 in this case would be the assertion that $\mathfrak{X}(E/K'_\infty)$ is a finitely generated $\Lambda(H)$-module, where

$$H = \mathrm{Gal}(K'_\infty/F)$$

with $F = K(E_p)$. It is then easy to see that this latter assertion holds if and only if

$$\mathfrak{X}(E/F^{\mathrm{cyc}}) \text{ is a finitely generated } \mathbb{Z}_p - \text{module.} \tag{5.2}$$

Unfortunately, (5.2) remains unproven today. Note also that if E is defined over $\mathbb{Q}$, assertion (5.2) implies that $\mathfrak{X}(E/\mathbb{Q}^{\mathrm{cyc}})$ is always a finitely generated $\mathbb{Z}_p$-module. In other words, Mazur's examples of positive μ-invariant over $\mathbb{Q}^{\mathrm{cyc}}$ should never occur for elliptic curves with complex multiplication. But we stress that even this special consequence of (5.2) remains unknown, although all the numerical data supports it. Of course, unlike the case of curves without complex multiplication, we here can formulate the main conjecture in Iwasawa theory for E/K_∞ without knowing the assertion (5.2), because in this case the Galois group G' is abelian. Moreover, this main conjecture has been proven by the work of Rubin and Yager [16], [22].

References

[1] K. Ardakov, S. Wadsley, *Characteristic elements for p-torsion Iwasawa modules*, J. Alg. Geom. **15** (2006), 339–377.

[2] H. Bass, Algebraic K-theory, W. A. Benjamin, Inc., New York-Amsterdam (1968).

[3] N. Bourbaki, Elements of Mathematics, Commutative Algebra, Chapters 1–7, Springer (1989).

[4] J. Coates, T. Fukaya, K. Kato, R. Sujatha, O. Venjakob, *The GL_2 main conjecture for elliptic curves without complex multiplication*, Publ. Math. IHES **101** (2005), 163–208.

[5] J. Coates, R. Sujatha Galois cohomology of elliptic curves, Tata Institute of Fundamental Research Lectures on Mathematics **88** Published by Narosa Publishing House, New Delhi; for the Tata Institute of Fundamental Research, Mumbai, (2000).

[6] J.Coates, *Infinite descent on elliptic curves with complex multiplication*, in "Arithmetic Geometry I ", Progress in Mathematics **35** (1983), 107–137, Birkhäuser.

[7] Gillard, *Transformation de Mellin-Leopoldt des fonctions elliptiques*, J. Number Theory **25** (1987), 379–393.

[8] S. Howson, *Structure of central torsion Iwasawa modules*, Bull. Soc. Math. France **130** (2002), 507–535.

[9] S. Howson, *Euler characteristics as invariants of Iwasawa modules*, Proc. London Math. Soc. **85** (2002), 634–658.

[10] K. Iwasawa, *On $\mathbb{Z}_l$-extensions of algebraic number fields*, Ann. of Math. **98** (1973), 246–326.

[11] K. Iwasawa, *On the theory of cyclotomic fields, J. Math. Soc. Japan* **20** (1964), 42–82.

[12] M. Lazard, *Groupes analytiques p-adiques*, Publ. IHES **26** (1965), 389–603.

[13] M. Kakde, *Proof of the Main Conjecture Noncommutative Iwasawa theory for totally real number fields*, Jour. Alg. Geom. (To appear).

[14] J. Neukirch, A. Schmidt, K. Wingberg, Cohomology of number fields, Second Ed., Grundlehren Series **323**, Springer (2008).

[15] B. Perrin-Riou, *Arithmétique des courbes elliptiques et thorie d'Iwasawa*, Mém. Soc. Math. France (N.S.) **17** (1984).

[16] K. Rubin, *The main conjectures of Iwasawa theory for imaginary quadratic fields*, Invent. Math. **103** (1991), 25–68.

[17] L. Schneps, *On the -invariant of p-adic L-functions attached to elliptic curves with complex multiplication*, J. Number Theory **25** (1987), 20–33.

[18] J.-P. Serre, *Propriétés galoisiennes des points d'ordre fini des courbes elliptiques*, Invent. Math. **15** (1972), 259–331.

[19] R. Sujatha, *Elliptic curves and Iwasawa's $\mu = 0$ conjecture*, Quadratic forms, linear algebraic groups, and cohomology edited by J.-L. Colliot-Thélène, Skip Garibaldi, R. Sujatha, and V. Suresh Developments in Mathematics **18**, Springer (2010).
[20] O. Venjakob, *On the structure theory of the Iwasawa algebra of a p-adic Lie group*, Jour. Eur. Math. Soc. (JEMS) **4** (2002), 271–311.
[21] L. Washington, Introduction to Cyclotomic Fields, Second Ed. GTM **83**, Springer (1996).
[22] R. Yager, *On two variable p-adic L-functions*, Ann. Math. **115** (1982), 411–449.

Fields Institute Communications
Volume **60**, 2011

Galois Representations and the Tame Inverse Galois Problem

Sara Arias-de-Reyna
Dept. d'Àlgebra i Geometria
Universitat de Barcelona
Gran Via de les Corts Catalanes, 585, 08007 Barcelona, Spain
ariasdereyna@ub.edu

Núria Vila
Dept. d'Àlgebra i Geometria
Universitat de Barcelona
Gran Via de les Corts Catalanes, 585, 08007 Barcelona, Spain
nuriavila@ub.edu

1 Introduction

In this paper we will focus on a variant of the Inverse Galois Problem over the rationals, emphasizing the progress made through the analysis of the Galois representations arising from arithmetic-geometric objects. The study of the Inverse Galois Problem explores the finite quotients of the absolute Galois group $G_{\mathbb{Q}} = \mathrm{Gal}(\overline{\mathbb{Q}}/\mathbb{Q})$, and sheds light on its structure.

The Inverse Galois Problem over $\mathbb{Q}$, first considered by D. Hilbert around 1890, asks whether, given a finite group G, there exists a finite Galois extension $K/\mathbb{Q}$ with Galois group G. This problem, which remains unsolved today, has given rise to significant mathematical advances, and several different techniques have been developed to address it. A strategy to deal with it is the constructive Galois theory (or rigidity method), which roughly consists in realizing the group G as the Galois group of a polynomial with coefficients in the complex numbers (which is possible thanks to the Riemann existence theorem for compact surfaces) and imposing some conditions that guarantee that, in fact, the polynomial is defined over $\mathbb{Q}(T)$, where T is a variable. This method has been very successful, and thanks to it many simple groups are now known to be realizable as Galois groups over $\mathbb{Q}$. For instance, all sporadic simple groups save the Mathieu group M_{23} have been realized through this method (cf. [20]).

Given a finite group G, when this method is applied one usually obtains a realization of G as the Galois group of a wildly ramified extension. Noting this

---

2010 *Mathematics Subject Classification.* Primary 11F80, 12F12, 11S15; Secondary 11F11.

Research partially supported by MEC grant MTM2009-07024. The first author is also supported by a FPU predoctoral grant AP-20040601 of the MEC.

fact, B. Birch posed the following question around 1994 (see Section 2 of [5], cf. [16]).

Problem 1.1 Given a finite group G, is there a tamely ramified Galois extension $K/\mathbb{Q}$ with Galois group G?

This problem is one of the possible variants of the Inverse Galois Problem, involving a ramification condition, which have been studied. For instance, let us fix a finite set of primes, say S, and let G be a finite group. Can we realize the group G as the Galois group of an extension $K/\mathbb{Q}$ which is unramified in S? Note that this is a stronger question than Problem 1.1, since if it can be solved for a finite group G and any prefixed set S, we can, in particular, choose a set S containing all the primes dividing the order of G. The extension $K/\mathbb{Q}$ obtained will be tamely ramified. In this way it can be proved that Problem 1.1 has an affirmative answer for finite abelian groups, symmetric groups (and all groups such that the Noether Problem has an affirmative answer [31]), finite solvable groups (see [22], [18]) and alternating groups A_n [24].

On the other hand, the Mathieu groups M_{11} and M_{12}, the group of automorphisms of M_{22} (cf. [25]), and the finite central extensions of symmetric groups, alternating groups and the Mathieu groups M_{11} and M_{12} (cf. [23]) can be realized as the Galois group of a finite tamely ramified extension of $\mathbb{Q}$.

A different approach to the Inverse Galois Problem is the study of Galois representations of the absolute Galois group $G_{\mathbb{Q}}$, obtained through an action upon certain geometric objects with arithmetic properties. Let

$$\rho : G_{\mathbb{Q}} \to \mathrm{GL}(V)$$

be a continuous Galois representation, where V denotes a finite-dimensional vector space over a finite field $\mathbb{F}$ of positive characteristic ℓ, endowed with the discrete topology, and where the topology on $G_{\mathbb{Q}}$ is the Krull topology. It holds that

$$\mathrm{Im}\rho \simeq G_{\mathbb{Q}}/\ker \rho \simeq \mathrm{Gal}(K/\mathbb{Q}),$$

where $K/\mathbb{Q}$ is a finite Galois extension. Thus we obtain a realization of $\mathrm{Im}(\rho)$ as a Galois group over $\mathbb{Q}$. This strategy has already been used to address the Inverse Galois Problem for families of linear groups (cf. [35], [10], [11], [36], [12], [17]). In order to make use of techniques of this kind, it is essential to have explicit criteria to determine precisely the image of such a representation.

This strategy can be adapted to address Problem 1.1. Although the realizations obtained are mostly wildly ramified, this ramification usually occurs only at the prime ℓ, or perhaps also at a finite set of primes which is related to the properties of the arithmetic-geometric object. The key point is to choose this object carefully in order to obtain stronger control over the ramification that may occur, and simultaneously ensuring that the image of the Galois representation is large. One needs explicit criteria that ensure that the ramification at every prime is at most tame. It will usually be far easier to tackle the primes p different from ℓ than to ensure tame ramification at the prime ℓ. The construction of Galois representations with prefixed local behaviour is a very active research field today.

To fix some notation, we will denote by I_p the inertia group at the prime p, and by $I_{p,\mathrm{w}}$ the wild inertia group at p (see section 3). Note that the Galois extension $K/\mathbb{Q}$ will be unramified (respectively tamely ramified) at p if $\rho(I_p) = \{\mathrm{Id}\}$ (respectively $\rho(I_{p,\mathrm{w}}) = \{\mathrm{Id}\}$).

First we will consider the Galois representations attached to the ℓ-torsion points of elliptic curves, and we will give a sketch of [1], attaining tame Galois realizations of $\mathrm{GL}_2(\mathbb{F}_\ell)$ for each prime number ℓ. The next section is devoted to present some progress in the tame Galois realization of linear groups in the families $\mathrm{PGL}_2(\mathbb{F}_{\ell^r})$ and $\mathrm{PSL}_2(\mathbb{F}_{\ell^r})$ by means of the Galois representations attached to modular forms. Finally, the last section gives an overview of the results in [3] and [4] (see also [2]), where tame Galois realizations of the groups in the family $\mathrm{GSp}_4(\mathbb{F}_\ell)$ are obtained for every prime $\ell \geq 5$ through the study of the Galois representations attached to the ℓ-torsion points of Jacobians of genus 2 curves. We also present some general criteria concerning the Galois representations attached to the ℓ-torsion points of an abelian variety in any dimension.

2 Elliptic curves

Let $E/\mathbb{Q}$ be an elliptic curve defined over the rationals. Let us consider the group $E[\ell]$ of ℓ-torsion points of E. There is a natural action of the absolute Galois group $G_\mathbb{Q}$ on $E[\ell]$, which gives rise to a continuous Galois representation

$$\varphi_\ell : G_\mathbb{Q} \to \mathrm{Aut}(E[\ell]) \simeq \mathrm{GL}_2(\mathbb{F}_\ell).$$

This representation provides a realization of the image of φ_ℓ as a Galois group over $\mathbb{Q}$, since

$$\mathrm{Im}\ \varphi_\ell \simeq G_\mathbb{Q}/\ker \varphi_\ell \simeq \mathrm{Gal}(\mathbb{Q}(E[\ell])/\mathbb{Q}),$$

where $\mathbb{Q}(E[\ell])/\mathbb{Q}$ denotes the finite Galois extension obtained by adjoining to $\mathbb{Q}$ the coordinates of the ℓ-torsion points of E. Therefore, if φ_ℓ is surjective, we obtain a realization of $\mathrm{GL}_2(\mathbb{F}_\ell)$ as a Galois group over $\mathbb{Q}$. It is a well-known result of J-P. Serre that if E does not have complex multiplication, then φ_ℓ is surjective for all but finitely many primes ℓ (see [32], Theorem 2 of § 4.2). This result suggests that one can use these representations to realize groups of the family $\mathrm{GL}_2(\mathbb{F}_\ell)$ as Galois groups over $\mathbb{Q}$. In fact, Serre establishes in Example 5.5.6 of [32] that the Galois representation attached to the elliptic curve E of conductor $N = 37$ provides a Galois realization of $\mathrm{GL}_2(\mathbb{F}_\ell)$ for all primes ℓ (cf. [28] for an explicit determination of the images of Galois representations associated to elliptic curves). In order to obtain results of this kind, one needs an explicit control of the exceptional primes, i.e., the primes ℓ such that φ_ℓ is not surjective. For instance, if the elliptic curve $E/\mathbb{Q}$ is semistable, then the Galois representation attached to the ℓ-torsion points of E is surjective if $\ell \geq 11$ (see Theorem 4 of [21]).

In order to produce tame Galois realizations, we need explicit conditions upon the elliptic curve E that ensure that the extension $\mathbb{Q}(E[\ell])/\mathbb{Q}$ is tamely ramified. If $p \neq \ell$ is a prime number and E has semistable reduction at p, then the image of the wild inertia group at p by φ_ℓ is trivial. On the other hand, things become more complicated at the prime ℓ. In [32], Serre proves that if the kind of reduction of E at ℓ is good and supersingular, then the image of the wild inertia group at ℓ by φ_ℓ is trivial. Therefore we have the following result.

Theorem 2.1 *Let $\ell \geq 11$ be a prime number, and consider a semistable elliptic curve $E/\mathbb{Q}$ such that ℓ has good supersingular reduction. Then the extension $\mathbb{Q}(E[\ell])/\mathbb{Q}$ is tamely ramified with Galois group* $\mathrm{GL}_2(\mathbb{F}_\ell)$.

Given a prime number $\ell \geq 11$, we want to construct an elliptic curve E satisfying the conditions in Theorem 2.1. The most interesting point in this construction is to produce a supersingular elliptic curve $\widetilde{E}/\mathbb{F}_\ell$. We give a construction in [1], § 3, based on some results of [6] concerning the factorization of the Deuring polynomial. Another construction has appeared in [7].

In fact, it turns out that one can replace the conditions of Theorem 2.1 with others which are more restrictive but also very simple. Namely, it suffices to consider the elliptic curve defined by the equation

$$y^2 + a_1xy + a_3y = x^3 + a_2x^2 + a_4x + a_6,$$

where $a_1, a_2, a_3, a_4, a_6 \in \mathbb{Z}$ satisfy certain congruences modulo powers of different prime numbers. One can find a precise statement in Theorem 4.4 of [1]. The primes 2, 3, 5 and 7 can be handled similarly, applying specific criteria that ensure that the Galois representation φ_ℓ is surjective (namely, see Proposition 2.1 in [28], Theorem 2.3-(iii) in [28], § 5.3 and Proposition 19 in [32]). As a consequence we have the following result (see [1], Theorem 4.6).

Theorem 2.2 *For each prime number ℓ, there exist infinitely many elliptic curves $E/\mathbb{Q}$ such that the Galois extension $\mathbb{Q}(E[\ell])/\mathbb{Q}$ is tamely ramified with Galois group $\mathrm{GL}_2(\mathbb{F}_\ell)$.*

Moreover, the construction is explicit: given ℓ, we can produce such elliptic curves.

3 Modular forms

Let f be a cuspidal form of weight 2 and level N, which is an eigenform for all Hecke operators. Assume that f is normalized, and call $\mathbb{Q}_f$ the field of coefficients of f, that is to say, the number field generated over $\mathbb{Q}$ by adjoining all the coefficients of the Fourier expansion of f. A classical construction of Shimura attaches to f a continuous Galois representation

$$\overline{\rho}_{f,\ell} : G_\mathbb{Q} \to \mathrm{GL}_2(\overline{\mathbb{F}}_\ell).$$

As we saw in the introduction, such a representation provides us with a realization of $\mathrm{Im}(\overline{\rho}_{f,\ell})$ as Galois group over $\mathbb{Q}$. In [29], K. Ribet studies the image of this representation. Under certain hypothesis, he concludes that the image of $\overline{\rho}_{f,\ell}$ coincides with $\{x \in \mathrm{GL}_2(\mathbb{F}_\lambda) : \det(x) \in \mathbb{F}_\ell^*\}$, where $\mathbb{F}_\lambda$ is a certain finite extension of $\mathbb{F}_\ell$. In order to get rid of the condition on the determinant above, we will consider the projection of $\mathrm{Im}(\overline{\rho}_{f,\ell})$ inside the projective space $\mathrm{PGL}_2(\mathbb{F}_\lambda)$. The precise statement is the following (cf. [26]).

Theorem 3.1 *Let $f \in S_2(N)$ be a newform with no complex multiplication and no non-trivial inner twists. For a rational prime ℓ, let $\mathbb{F}_\lambda$ be the corresponding residue field of the coefficient field $\mathbb{Q}_f$, and call $r = [\mathbb{F}_\lambda : \mathbb{F}_\ell]$. For all but finitely many primes ℓ, the projective image of $\overline{\rho}_{\ell,f}$ coincides with $\mathrm{PSL}_2(\mathbb{F}_{\ell^r})$ if r is even and $\mathrm{PGL}_2(\mathbb{F}_{\ell^r})$ if r is odd.*

Therefore, in this section we try to obtain tame Galois realizations of groups of the families $\mathrm{PSL}_2(\mathbb{F}_{\ell^r})$ and $\mathrm{PGL}_2(\mathbb{F}_{\ell^r})$.

Given a continuous, odd, irreducible Galois representation

$$\overline{\rho}_\ell : G_\mathbb{Q} \to \mathrm{GL}_2(\overline{\mathbb{F}}_\ell),$$

there is a way to measure the ramification away from ℓ. In [33], J-P. Serre attaches to each Galois representation as above a *conductor* $N(\overline{\rho}_\ell)$, a *weight* $k(\overline{\rho}_\ell)$ and a *character* $\varepsilon(\overline{\rho}_\ell)$, and he conjectures that $\overline{\rho}_\ell$ is isomorphic to the Galois representation $\overline{\rho}_{g,\ell}$ attached to a modular form g of level $N(\overline{\rho}_\ell)$, weight $k(\overline{\rho}_\ell)$ and character

$\varepsilon(\overline{\rho}_\ell)$. The conductor $N(\overline{\rho}_\ell)$ constitutes a precise measure of the ramification of $\overline{\rho}_\ell$ at the primes $p \neq \ell$.

Let us briefly sketch its definition. Fix a prime number p. Then one can define an integer number $n(p)$ in the following way (see [33], § 1). Let us fix an extension w to $\overline{\mathbb{Q}}$ of the p-adic valuation in $\mathbb{Q}$. This is equivalent to fixing an embedding $\overline{\mathbb{Q}} \hookrightarrow \overline{\mathbb{Q}}_p$.

On the other hand, since $\overline{\rho}_\ell$ is continuous, there exists a finite Galois extension $K/\mathbb{Q}$ such that $\ker \overline{\rho}_\ell \simeq \mathrm{Gal}(\overline{\mathbb{Q}}/K)$. The embedding above restricts to an embedding $K \hookrightarrow K_w$, where K_w denotes the completion of K with respect to the restriction to K of the valuation w. Let $\mathcal{O}_w$ be the ring of integers of K_w with respect to the valuation w, and normalize w so that $w(\mathcal{O}_w) = \mathbb{Z}$. For each $i \in \mathbb{N}$, we can define the *i-th higher ramification group* as

$$G_i = \{\sigma \in \mathrm{Gal}(K_w/\mathbb{Q}_p) \text{ such that } w(\sigma(x) - x) \geq i + 1 \text{ for all } x \in \mathcal{O}_w\}.$$

Since $\mathrm{Gal}(K_w/\mathbb{Q}_p)$ is isomorphic to the decomposition group of w in $\mathrm{Gal}(K/\mathbb{Q})$, we can view G_i inside $\mathrm{Gal}(K/\mathbb{Q})$.

In this way we obtain a decreasing sequence of subgroups of the inertia group at p:

$$I_p(K/\mathbb{Q}) = G_0 \supset I_{p,\mathrm{w}}(K/\mathbb{Q}) = G_1 \supset G_2 \supset \cdots \supset G_i \supset \cdots . \tag{3.1}$$

Let V be a 2-dimensional vector space over $\overline{\mathbb{F}}_\ell$, and consider the action of $G_{\mathbb{Q}}$ given by $\overline{\rho}_\ell$ upon V. For each $i \in \mathbb{N}$, define

$$V_i := \{v \in V : \overline{\rho}_\ell(\sigma)(v) = v \text{ for all } \sigma \in G_i\}$$

A way to measure the size of G_i is to consider the dimension of V_i as $\overline{\mathbb{F}}_\ell$-vector space. More precisely, one can consider the quantity

$$n(p) = \sum_{i=0}^{\infty} \frac{\dim(V/V_i)}{(G_0 : G_i)}, \tag{3.2}$$

where $(G_0 : G_i)$ denotes the index of G_i in G_0. It can be seen that $n(p)$ is an integer number, which is greater than or equal to zero. Moreover, $n(p) = 0$ if and only if $\overline{\rho}_\ell$ is unramified at p. Since $\overline{\rho}_\ell$ can only ramify in the primes which are ramified in the extension $K/\mathbb{Q}$, $n(p) = 0$ for all but finitely many primes p.

Definition 3.2 The *Artin conductor* is defined as

$$N(\overline{\rho}_\ell) = \prod_{\substack{p \neq \ell \\ p \text{ prime}}} p^{n(p)},$$

where $n(p)$ is given by Equation (3.2).

If $n(p) = 1$, it must hold that $\dim(V/V_0) = 1$ and all the remaining terms are zero. In particular, $V_1 = V$, that is to say, the wild inertia group at p acts trivially upon V. Therefore, a sufficient condition for $\overline{\rho}_\ell$ to be tamely ramified at a prime $p \neq \ell$ is that the exponent $n(p)$ of p in the Artin conductor $N(\overline{\rho}_\ell)$ is either 0 or 1. In particular, if $N(\overline{\rho}_\ell)$ is squarefree, then $\overline{\rho}_\ell$ is tamely ramified at all primes $p \neq \ell$.

Let us consider the Galois representation $\overline{\rho}_{f,\ell}$ attached to a newform $f \in S_2(N)$. According to Serre's conjecture, $\overline{\rho}_{f,\ell}$ comes from a modular form g of level $N(\overline{\rho}_{f,\ell})$ (and weight $k(\overline{\rho}_{f,\ell})$, character $\varepsilon(\overline{\rho}_{f,\ell})$). But it may well be that $N \neq N(\overline{\rho}_{f,\ell})$. Luckily, there is a strong relationship between the two values: thanks to the work of Carayol and Livné (cf. [8], remark following Theorem 1 and [19], Proposition

0.1), we know that $N(\overline{\rho}_{f,\ell})$ divides N (when $\overline{\rho}_{f,\ell}$ is irreducible). Therefore, if N is squarefree, then $N(\overline{\rho}_{f,\ell})$ has no choice but to be squarefree too.

Proposition 3.3 *Let $f \in S_2(N)$ be a newform, and let $\overline{\rho}_{f,\ell}$ be the Galois representation attached to f. Assume that $\overline{\rho}_{f,\ell}$ is irreducible. If N is squarefree, then $\overline{\rho}_{f,\ell}$ is tamely ramified at all primes $p \neq \ell$.*

Now that we have dealt with all primes $p \neq \ell$, we address the problem of obtaining some control over the ramification at the prime ℓ. In the case of elliptic curves, this was achieved by requiring the elliptic curve to have good supersingular reduction at ℓ. Is there a similar condition in this general context?

As a matter of fact, the answer is affirmative. There exist the concepts of *ordinary* and *supersingular* (see [30], Chapter 2, § 2.1).

Definition 3.4 Let $f \in S_k(N, \varepsilon)$ be a newform, and call a_p the eigenvalue corresponding to the Hecke operator T_p. Let ℓ be a prime number. We will say that the newform f is *supersingular* at ℓ if there exists a homomorphism φ from the ring of integers of $\mathbb{Q}_f$ to $\overline{\mathbb{F}}_\ell$ such that $\varphi(a_\ell) = 0$

Fontaine has studied the image by $\overline{\rho}_{f,\ell}$ of the inertia group at ℓ when f is supersingular. The following theorem, which is Theorem 2.6 of [13], seems to have appeared in a letter from Fontaine to Serre.

Theorem 3.5 (Fontaine) *Let f be a newform in $S_k(N, \varepsilon)$ with $2 \leq k \leq \ell + 1$. Assume that $\overline{\rho}_{f,\ell}$ is supersingular at ℓ. Then $\overline{\rho}_{f,\ell}$ is irreducible and*

$$\overline{\rho}_{f,\ell}|_{I_\ell} = \begin{pmatrix} \psi^{k-1} & 0 \\ 0 & \psi'^{k-1} \end{pmatrix},$$

where ψ and ψ' are the two fundamental characters of level 2.

As a corollary, we obtain:

Corollary 3.6 *Let $f \in S_2(N)$ be a newform, and let $\overline{\rho}_{f,\ell}$ be the Galois representation attached to f. If $\overline{\rho}_{f,\ell}$ is supersingular, then it is tamely ramified at ℓ.*

Let us now consider the image of $\overline{\rho}_{f,\ell}$. We know that for all but finitely many primes ℓ the projective image of $\overline{\rho}_{f,\ell}$ is $\mathrm{PSL}_2(\mathbb{F}_{\ell^r})$ or $\mathrm{PGL}_2(\mathbb{F}_{\ell^r})$ for a certain exponent r, provided that some good conditions hold (no complex multiplication and no non-trivial inner twists). This result is similar to Serre's result about the image of the Galois representations attached to the ℓ-torsion points of elliptic curves. But recall that this was not enough for us, and we had to resort to a theorem of Mazur, which allowed us to replace the condition "for all but finitely many primes ℓ" by "for all primes $\ell \geq 11$", provided we also required the elliptic curve to be semistable.

A result of this kind is not available in this context. Therefore we will put to use such means as we have. A result of L. Dieulefait and N. Vila (see [10]) provides us with an algorithm which takes as input a modular form f and produces a finite set of primes S such that the representation has large image at all primes outside this set. In this way, they manage to realize the groups $\mathrm{PSL}_2(\mathbb{F}_{\ell^2})$, $\mathrm{PGL}_2(\mathbb{F}_{\ell^3})$, $\mathrm{PSL}_2(\mathbb{F}_{\ell^4})$ for many primes ℓ.

Their method relies on some results of Ribet (see [29]). More precisely, they make use of the following theorem (cf. § 3 of [29]).

Theorem 3.7 *Let $f \in S_2(N)$ be a newform with coefficient field $\mathbb{Q}_f$ and ring of integers $\mathcal{O}$, ℓ a prime number and λ a prime of $\mathcal{O}$ above ℓ. Call $\mathcal{O}_\ell = \mathcal{O} \otimes_{\mathbb{Z}} \mathbb{Z}_\ell$. Let $\rho_{f,\lambda}$ the λ-adic Galois representation attached to f. Then the image of this representation equals $\{x \in \mathrm{GL}_2(\mathcal{O}_\ell) : \det(x) \in \mathbb{Z}_\ell^*\}$ if the following conditions are satisfied:*

(0) *ℓ does not ramify in $\mathbb{Q}_f/\mathbb{Q}$.*
(1) *The determinant map $\det \circ \rho_{f,\lambda}$ is surjective.*
(2) *$\ell \geq 5$.*
(3) *The image of $\rho_{f,\lambda}$ contains an element x_ℓ such that $(\mathrm{trace}\ x_\ell)^2$ generates $\mathcal{O}_\ell$ as a $\mathbb{Z}_\ell$-algebra.*
(4) *For each $\lambda|\ell$, the image of the composition of $\rho_{f,\lambda}$ with the reduction modulo λ is an irreducible subgroup of $\mathrm{GL}_2(\mathbb{F}_\lambda)$ whose order is divisible by ℓ.*

The algorithm of Dieulefait and Vila consists of several steps, which are designed to guarantee that some of the conditions above hold. As a general remark, we can say that each step consists of finding an auxiliary prime p, satisfying perhaps some conditions regarding the level N and the prime ℓ, and such that the coefficient a_p enjoys some suitable property involving λ. If the image of the representation $\rho_{f,\lambda}$ is large, such a prime p will exist. A complete description of the algorithm can be found in section 2.2 of [10].

Our wish is to combine this algorithm with the tameness results presented above in order to obtain tame Galois realizations of linear groups of the form $\mathrm{PSL}_2(\mathbb{F}_{\ell^r})$ or $\mathrm{PGL}_2(\mathbb{F}_{\ell^r})$. Unfortunately, we have not been able to obtain a procedure that, given a prime number ℓ, constructs a suitable modular form yielding a tame Galois realization of the corresponding group. This problem remains a challenging issue for further investigation. Nevertheless, the results we have explained allow us to present a few concrete examples.

In [10], section 3.1, the authors apply their algorithm to some weight 2 newforms whose field of coefficients is a quadratic extension of $\mathbb{Q}$. More precisely, they consider newforms of level 23, 29, 410, 414, 496, 418, 546. For each newform, they provide a finite set of primes such that the Galois representation attached to it has large image (in the sense of Theorem 3.1) whenever ℓ does not belong to this set. We present their results in the following table:

Table 1 Computations of big image

Modular form	Level	Coefficient Field	Set of primes
f_1	23	$x^2 - 5$	5, 11
f_2	29	$x^2 - 2$	7
f_3	410	$x^2 - 3$	5, 41
f_4	410	$x^2 - 17$	5, 41
f_5	414	$x^2 - 7$	7, 23
f_6	496	$x^2 - 33$	11, 31
f_7	418	$x^2 - 13$	11, 13, 19
f_8	546	$x^2 - 57$	7, 13, 19

Our purpose is to take advantage of their computations and try to produce tamely ramified Galois realizations of groups of the form $\mathrm{PSL}_2(\mathbb{F}_{\ell^2})$ from the modular forms $f_1, \ldots, f_8$ presented above. First of all, note that the numbers 23, 29, 410, 418, 546 are squarefree. Therefore, we know that the Galois representations $\overline{\rho}_{f,\ell}$ corresponding to the modular forms $f = f_1, f_2, f_3, f_4, f_7, f_8$ can be wildly ramified only at the prime ℓ. In order to ensure that they are also tamely ramified at ℓ, we will make use of Corollary 3.6.

We have taken each of the modular forms $f_1, f_2, f_3, f_4, f_7, f_8$ and have looked at their coefficients a_ℓ, for all primes $\ell < 5000$. Since we are looking for Galois realizations of groups of the form $\mathrm{PSL}_2(\mathbb{F}_{\ell^2})$, we are only interested in the primes which are inert in the extension $\mathbb{Q}_{f_i}/\mathbb{Q}$. For each modular form f_i we can find the coefficient field $\mathbb{Q}_{f_i}$ in the table above.

For each $i = 1, 2, 3, 4, 7$ and 8, we list below the primes $\ell \leq 5000$ such that ℓ is inert in the extension $\mathbb{Q}_{f_i}/\mathbb{Q}$ and $a_\ell \in (\ell)$:

Table 2 Inert supersingular primes

Modular Form	Primes ℓ
f_1	43, 1033
f_2	
f_3	173
f_4	23, 31, 4391
f_7	59, 149, 1709
f_8	47, 4799

As a consequence of Proposition 3.3, Corollary 3.6 and Tables 1 and 2, we obtain the following result.

Proposition 3.8 *The following groups occur as Galois groups of a tamely ramified extension of* $\mathbb{Q}$*:*

$$\mathrm{PSL}_2(\mathbb{F}_{23^2}), \mathrm{PSL}_2(\mathbb{F}_{31^2}), \mathrm{PSL}_2(\mathbb{F}_{43^2}), \mathrm{PSL}_2(\mathbb{F}_{47^2}), \mathrm{PSL}_2(\mathbb{F}_{59^2}), \mathrm{PSL}_2(\mathbb{F}_{149^2}),$$
$$\mathrm{PSL}_2(\mathbb{F}_{173^2}), \mathrm{PSL}_2(\mathbb{F}_{1033^2}), \mathrm{PSL}_2(\mathbb{F}_{1709^2}), \mathrm{PSL}_2(\mathbb{F}_{4391^2}), \mathrm{PSL}_2(\mathbb{F}_{4799^2}).$$

Remark 3.9 We are developing these techniques in order to obtain tamely ramified extensions of an infinite set of groups in the family $\mathrm{PSL}_2(\mathbb{F}_{\ell^2})$.

4 Abelian varieties

Let $A/\mathbb{Q}$ be an abelian variety of dimension n, and let ℓ be a prime number. As in the case of elliptic curves, there is an action of the absolute Galois group $G_\mathbb{Q}$ upon the group $A[\ell]$ of ℓ-torsion points of A. This action gives rise to a continuous Galois representation

$$\rho_\ell : G_\mathbb{Q} \to \mathrm{Aut}(A[\ell]) \simeq \mathrm{GL}_{2n}(\mathbb{F}_\ell).$$

Let us assume that A is principally polarized. One first variation with respect to the case of dimension 1 is that, if $n > 1$, then ρ_ℓ cannot be surjective. This is due to the existence of the Weil pairing, which is a non-degenerate, Galois compatible symplectic form. This compels the image of ρ_ℓ to be contained in the general symplectic group $\mathrm{GSp}_{2n}(\mathbb{F}_\ell)$. As a matter of fact, if the endomorphism ring of the abelian variety is equal to $\mathbb{Z}$ and the dimension n is either odd or equal to 2 or 6, then $\text{Im}\rho_\ell \simeq \text{GSp}_{2n}(\mathbb{F}_\ell)$ for all but finitely many primes ℓ (see [34], Theorem 3 of 137). Therefore, one can try to obtain tame realizations of this general symplectic group as the Galois group of the extension $\mathbb{Q}(A[\ell])/\mathbb{Q}$, obtained by adjoining to $\mathbb{Q}$ the coordinates of the points of $A[\ell]$. This problem has been addressed by the authors in [3] and [4] (cf. [2]).

Concerning the ramification, let us note that since the wild inertia group at a prime p is a pro-p-group, the only primes we need to take care of are the primes dividing the order of $\text{GSp}_{2n}(\mathbb{F}_\ell)$, that is to say, a finite quantity of primes. If $p \neq \ell$ is a prime number where A has semistable reduction, then it follows from a result of Grothendieck (see [15], Proposition 3.5 of Exposé IX, *Modeles de Néron et monodromie*) that the wild inertia group at p acts trivially on $A[\ell]$. On the other hand, the control of the action of the wild inertia group at the prime ℓ is more subtle. In [4], a condition upon the formal group law of A is found that, combined with supersingularity, allows one to control the wild ramification at ℓ (see Theorem 3.4 in [3], cf. Theorem 4.9 in [2]).

Theorem 4.1 *Let ℓ be a prime number, and let $A/\mathbb{Q}$ be an abelian variety of dimension n with good supersingular reduction at ℓ. Call $\mathbf{F}$ the formal group law attached to A at ℓ, v_ℓ the ℓ-adic valuation on $\overline{\mathbb{Q}}_\ell$ and V the group of ℓ-torsion points attached to $\mathbf{F}$. Assume that there exists a positive $\alpha \in \mathbb{Q}$ such that, for all non-zero $(x_1, \ldots, x_n) \in V$, it holds that $\min\{v_\ell(x_i) : 1 \leq i \leq n\} = \alpha$.*

Then the image of the wild inertia group by the Galois representation attached to the ℓ-torsion points of A is trivial.

If we restrict ourselves to the case of dimension $n = 2$, then we can provide very explicit conditions that ensure that the wild inertia group (at the prime ℓ as well as at all other primes $p \neq \ell$) acts trivially upon the ℓ-torsion points of the Jacobian of a genus 2 curve (see Theorem 2.7 in [4], cf. Theorem 9.2 in [2]).

Theorem 4.2 *Let C be a genus 2 curve defined by a hyperelliptic equation*

$$y^2 = f(x), \tag{4.1}$$

where $f(x) = f_0x^6 + f_1x^5 + f_2x^4 + f_3x^3 + f_2x^2 + f_1x + f_0 \in \mathbb{Z}[x]$ is a polynomial of degree 6 without multiple factors. Let $\ell > 2$ be a prime number, and let $\mathcal{P}$ be the set of prime numbers that divide the order of $\text{GSp}_4(\mathbb{F}_\ell)$. Assume that the following conditions hold:

- *For all $p \in \mathcal{P}$ different from ℓ, C has stable reduction at p.*
- *The reduction of $f(x)$ modulo ℓ is of the form $x^6 + bx^4 + bx^2 + 1$, and the elliptic curve E defined over $\mathbb{F}_\ell$ by $y^2 = x^3 + bx^2 + bx + 1$ is supersingular.*

Then the Galois extension $\mathbb{Q}(A[\ell])/\mathbb{Q}$ is tamely ramified.

Several tools are involved in the proof of this theorem. First of all, the multiplication by ℓ-map in the formal group law attached to the Jacobians of curves defined by an equation (4.1) satisfying the hypothesis of Theorem 4.2 has a very special shape, namely, it is essentially defined by one single equation. The proof of this fact is based upon the explicit computation of the formal group law attached to the Jacobian of a genus 2 curve carried out by V. Flynn in [14] (cf. [9]). This property of the multiplication by ℓ map allows one to compute the value of $\min\{v_\ell(x_i) : (x_1, x_2) \in V\}$, which in turn allows us to apply Theorem 4.1. Note that one needs a genus 2 curve with good supersingular reduction at ℓ of a particular shape, which is constructed as a very specific bielliptic curve.

It turns out that the curves in the family given in Theorem 4.2 have reducible Jacobians. We need to "deform" them a bit in order to sort out this difficulty. This requires a careful analysis of the proximity (with respect to the ℓ-adic valuation) of the solutions of two systems of equations in formal power series that are near coefficientwise.

In the case of Jacobians of genus 2 curves, one can give explicit results to control the image of the Galois representation ρ_ℓ. More specifically, we can make use of the following result (Theorem 3.7 of [4], cf. Theorem 11.9 of [2]):

Theorem 4.3 *Let G be a subgroup of $\mathrm{Sp}_4(\mathbb{F}_\ell)$, and assume that G contains a transvection. Furthermore, assume that it contains two elements whose characteristic polynomials, $P_i(X) = X^4 + a_iX^3 + b_iX^2 + a_iX + 1$ $(i = 1, 2)$, satisfy the following: denoting by α_i, $1/\alpha_i$, β_i, $1/\beta_i$ the four roots of $P_i(X)$,*

- $\alpha_1 + 1/\alpha_1, \beta_1 + 1/\beta_1 \notin \mathbb{F}_\ell$ *and* $\alpha_1 + 1/\alpha_1 + \beta_1 + 1/\beta_1 \neq 0$.
- $\alpha_2 + 1/\alpha_2, \beta_2 + 1/\beta_2 \in \mathbb{F}_\ell$, $a_2^2 - 4b_2 + 8 \neq 0$ *and* $\alpha_2 \notin \mathbb{F}_\ell$.

Then G equals $\mathrm{Sp}_4(\mathbb{F}_\ell)$.

We can combine the conditions on the ramification of ρ_ℓ and the conditions on the image in order to obtain tame Galois realizations of $\mathrm{GSp}_4(\mathbb{F}_\ell)$. The task of constructing genus 2 curves satisfying all the conditions still remains. Actually, as in the case of elliptic curves, one can replace each of the conditions by a congruence modulo a certain power of a prime. That is to say, it is possible to write a statement asserting that, if $\ell \geq 5$ is a prime number and $f_0, f_1, \ldots, f_6 \in \mathbb{Z}$ satisfy certain congruences, then the Jacobian of the genus 2 curve defined by $y^2 = f_6x^6 + f_5x^5 + f_4x^4 + f_3x^3 + f_2x^2 + f_1x + f_0$ provides a tame Galois realization of $\mathrm{GSp}_4(\mathbb{F}_\ell)$. One such statement is Theorem 5.1 of [4] (cf. Theorem 12.25 of [2]), which we have not included here because of its cumbersome length. In order to obtain this theorem, one needs to exploit certain tools. For instance, one needs some results of Q. Liu relating the type of reduction of a genus 2 curve and its Igusa invariants in order to construct a transvection inside the image of ρ_ℓ. The reader can find the details in [4].

To sum up, the result we obtain is the following:

Theorem 4.4 *Let $\ell \geq 5$ be a prime number. There exist infinitely many genus 2 curves C such that the Galois representation attached to the ℓ-torsion points of $J(C)$ provides a tame Galois realization of $\mathrm{GSp}_4(\mathbb{F}_\ell)$.*

References

[1] Arias-de-Reyna, S., Vila, N. *Tame Galois realizations of* $\mathrm{GL}_2(\mathbb{F}_\ell)$ *over* $\mathbb{Q}$. Journal of Number Theory, Volume **129**, Issue 5, May (2009), pages 1056-1065.

[2] Arias-de-Reyna, S. *Galois representations and tame Galois realizations*, Ph.D. Thesis, Barcelona, June 2009. Available at
`http://www.tesisenxarxa.net/TDX-0612109-101019/`

[3] Arias-de-Reyna, S. *Formal groups, supersingular abelian varieties and tame ramification.* Preprint, 2009. arXiv:0910.1212v1 [math.NT]

[4] Arias-de-Reyna, S., Vila, N. *Tame Galois realizations of* $\mathrm{GSp}_4(\mathbb{F}_\ell)$ *over* $\mathbb{Q}$. Preprint, 2009. arXiv:0910.1445v1 [math.NT]

[5] Birch, B. *Noncongruence subgroups, Covers and Drawings*, pages 25–46 in *The Grothendieck theory of dessins d'enfants*, Leila Schneps, editor. Cambridge Univ. Press (1994).

[6] Brillhart, J., Morton, P. *Class numbers of quadratic fields, Hasse invariants of elliptic curves, and the supersingular polynomial*, J. Number Theory **106**, no. 1, pages 79–111 (2004).

[7] Bröker, R. *Constructing supersingular elliptic curves*, Journal of Combinatorics and Number Theory, Volume **1**, Issue 3 (2009), pages 269–273.

[8] Carayol, H. *Sur les représentations galoisiennes modulo l attachées aux formes modulaires.* Duke Math. J. **59** (1989), no. 3, pages 785–801.

[9] Cassels, J. W. S., Flynn, E. V. *Prolegomena to a Middlebrow Arithmetic of Curves of Genus* 2, London Matematical Society Lecture Note Series 230, Cambridge University Press (1996).

[10] Dieulefait, L., Vila, N. *Projective linear groups as Galois groups over Q via modular representations.* Algorithmic methods in Galois theory. J. Symbolic Comput. **30** (2000), no. 6, pages 799–810.

[11] Dieulefait, L. *Explicit determination of the images of the Galois representations attached to abelian surfaces with* $\mathrm{End}(A) = \mathbb{Z}$. Experiment. Math. **11** (2002), no. 4, pages 503–512 (2003).

[12] Dieulefait, L.; Wiese, G. *On Modular Forms and the Inverse Galois Problem.* Preprint, 2009. arXiv:0905.1288v1 [math.NT]

[13] Edixhoven, B. *The weight in Serre's conjectures on modular forms.* Invent. Math. **109** (1992), no. 3, pages 563–594.

[14] Flynn, E. V., *The Jacobian and Formal Group of a curve of genus* 2 *over an arbitrary ground field*, Math. Proc. Cambridge Philos. Soc. **107** (1990), no. 3, pages 425–441.

[15] Grothendieck, A. *Groupes de monodromie en géométrie algébrique I.* Séminaire de Géométrie Algébrique du Bois-Marie 1967–1969 (SGA 7 I). Dirigé par A. Grothendieck. Avec la collaboration de M. Raynaud et D. S. Rim. Lecture Notes in Mathematics, Vol. **288**. Springer-Verlag, Berlin-New York, (1972).

[16] Harbater, D. *Galois groups with prescribed ramification.* Arithmetic geometry (Tempe, AZ, 1993), pages 35–60, Contemp. Math., **174**, Amer. Math. Soc., Providence, RI, (1994).

[17] Khare, C.; Larsen, M.; Savin, G. *Functoriality and the inverse Galois problem.* Compos. Math. **144** (2008), no. 3, 541–564.

[18] Klüners, J., Malle, G. *Counting nilpotent Galois extensions.* J. Reine Angew. Math. **572**, pages 1–26 (2004).

[19] Livné, R. *On the conductors of mod l Galois representations coming from modular forms.* J. Number Theory **31** (1989), no. 2, pages 133–141.

[20] Malle, G., Matzat, B. H. *Inverse Galois theory*, Springer Monographs in Mathematics Springer-Verlag, (1999).

[21] Mazur, B. *Rational Isogenies of Prime Degree*, Inventiones math. **44**, pages 129–162 (1978).

[22] Neukirch, J., Schmidt, A., Wingberg, K. *Cohomology of number Fields*, Grundlehren der mathematischen Wissenschaften **323**, Springer, (2000).

[23] Plans, B., *Central embedding problems, the arithmetic lifting property, and tame extensions of* $\mathbb{Q}$., Int. Math. Res. Not. **23**, pages 1249–1267 (2003).

[24] Plans, B., Vila, N. *Tame* A_n*-extensions of* $\mathbb{Q}$, J. Algebra **266**, no. 1, pages 27–33 (2003).

[25] Plans, B., Vila, N. *Galois covers of* $\mathbb{P}^1$ *over* $\mathbb{Q}$ *with prescribed local or global behavior by specialization.* J. Théor. Nombres Bordeaux **17**, no. 1, pages 271–282. (2005).

[26] Reverter, A., Vila, N. *Some projective linear groups over finite fields as Galois groups over* $\mathbb{Q}$ in *Recent developments in the inverse Galois problem* (Seattle, WA, 1993), 51–63, Contemp. Math., **186**, Amer. Math. Soc., Providence, RI, (1995).

[27] Reverter, A., Vila, N. *Galois representations attached to the product of two elliptic curves.* Rocky Mountain J. Math. **30**, no. 3, pages 1121–1127 (2000).

[28] Reverter, A., Vila, N. *Images of mod p Galois representations associated to elliptic curves.* Canad. Math. Bull. **44**, no. 3, pages 313–322. (2001).

[29] Ribet, K. A. *On l-adic representations attached to modular forms. II.* Glasgow Math. J. **27** (1985), pages 185–194.

[30] Ribet, K. A., Stein, W. A. *Lectures on Serre's conjectures.* Arithmetic algebraic geometry (Park City, UT, 1999), pages 143–232, IAS/Park City Math. Ser., 9, Amer. Math. Soc., Providence, RI, (2001).

[31] Saltman, D. J. *Generic Galois extensions and problems in field theory.* Adv. in Math. **43** (1982), no. 3, pages 250–283.

[32] Serre, J-P. *Proprietes galoisiénnés des points d'ordre fini des courbes elliptiques*, Inventiones math. **15**, pages 259–331 (1972).

[33] Serre, J-P. *Sur les représentations modulaires de degré* 2 *de* Gal($\overline{Q}/Q$). Duke Math. J. **54** (1987), no. 1, pages 179–230.

[34] Serre, J-P. *Œuvres 4*, Springer-Verlag (2000), pages 1–55.

[35] Vila, N. *Arithmetical-geometrical Galois representations and the inverse Galois problem* in *Algebra, arithmetic and geometry with applications* (West Lafayette, IN, 2000), 775–782, Springer, Berlin, (2004).

[36] Wiese, G. *On projective linear groups over finite fields as Galois groups over the rational numbers.* In: *Modular Forms on Schiermonnikoog* edited by Bas Edixhoven, Gerard van der Geer and Ben Moonen. Cambridge University Press, 2008, 343-350.